Nanotechnology in the Life Sciences

Series Editor

Ram Prasad
Department of Botany
Mahatma Gandhi Central University
Motihari, Bihar, India

Nano and biotechnology are two of the 21st century's most promising technologies. Nanotechnology is demarcated as the design, development, and application of materials and devices whose least functional make up is on a nanometer scale (1 to 100 nm). Meanwhile, biotechnology deals with metabolic and other physiological developments of biological subjects including microorganisms. These microbial processes have opened up new opportunities to explore novel applications, for example, the biosynthesis of metal nanomaterials, with the implication that these two technologies (i.e., thus nanobiotechnology) can play a vital role in developing and executing many valuable tools in the study of life. Nanotechnology is very diverse, ranging from extensions of conventional device physics to completely new approaches based upon molecular self-assembly, from developing new materials with dimensions on the nanoscale, to investigating whether we can directly control matters on/in the atomic scale level. This idea entails its application to diverse fields of science such as plant biology, organic chemistry, agriculture, the food industry, and more.

Nanobiotechnology offers a wide range of uses in medicine, agriculture, and the environment. Many diseases that do not have cures today may be cured by nanotechnology in the future. Use of nanotechnology in medical therapeutics needs adequate evaluation of its risk and safety factors. Scientists who are against the use of nanotechnology also agree that advancement in nanotechnology should continue because this field promises great benefits, but testing should be carried out to ensure its safety in people. It is possible that nanomedicine in the future will play a crucial role in the treatment of human and plant diseases, and also in the enhancement of normal human physiology and plant systems, respectively. If everything proceeds as expected, nanobiotechnology will, one day, become an inevitable part of our everyday life and will help save many lives.

More information about this series at http://www.springer.com/series/15921

Inamuddin • Abdullah M. Asiri
Editors

Nanotechnology-Based Industrial Applications of Ionic Liquids

Springer

Editors
Inamuddin
Chemistry Department
King Abdulaziz University
Jeddah, Saudi Arabia

Department of Applied Chemistry
Aligarh Muslim University
Aligarh, India

Abdullah M. Asiri
Chemistry Department
King Abdulaziz University
Jeddah, Saudi Arabia

ISSN 2523-8027 ISSN 2523-8035 (electronic)
Nanotechnology in the Life Sciences
ISBN 978-3-030-44994-0 ISBN 978-3-030-44995-7 (eBook)
https://doi.org/10.1007/978-3-030-44995-7

This Springer imprint is published by the registered company Springer Nature Switzerland AG
The registered company address is: Gewerbestrasse 11, 6330 Cham, Switzerland

Preface

Ionic liquids are often referred to as "Green Solvents" and linked to "Green Chemistry." Ionic liquids have been identified as a beneficial alternative to volatile organic solvents due to the poisonous and unsafe properties of conventional solvents. Over the past decades, the application of ionic liquids in chemical processes as an alternative and attractive solvent has been flourished. The need for environmentally sustainable materials and their processes of preparation have contributed greatly towards the development of ionic liquids as a green solvent.

Nanotechnology-Based Industrial Applications of Ionic Liquids book is intended to discuss the applications of ionic liquids in allied fields such as biomass pretreatment, catalysis, enzyme synthesis, extraction, electrochemical and chemical syntheses, surfactants, water purification, corrosion inhibition, biodiesel production, food, and beverage industries. This book is beneficial for the graduate and postgraduate students, research and development scientists, environmentalists, and industries intended to use green alternative solvents. Based on thematic topics, the book edition contains the following 16 chapters.

Chapter 1 discusses the toxicological aspects of ionic liquids, mainly their impacts on human health, based on scientific data recently reported.

Chapter 2 discusses the history and application of ionic liquids as a green solvent. Lignocellulosic biomass pretreatment by ionic liquids is discussed in detail. The focus is given on lignin removal and reduction in the crystallinity of cellulose in plant biomass. Additionally, the effect of ionic liquid pretreatment on sugarcane bagasse is discussed.

Chapter 3 discusses applications of ionic liquids in biodiesel synthesis. Specifically, it covers the use of ionic liquids as single catalysts for the transesterification and esterification reactions in biodiesel production, also combined with other chemical materials such as metallic catalysts, and as reaction media for enzymatic biodiesel production.

Chapter 4 reviews the potential of ionic liquids as alternative solvents for the enzymatic synthesis of sugar fatty acid esters (SFAEs). The selection of the suitable ionic liquids to produce SFAESs though esterification and transesterification are discussed. Additionally, the tunability of ionic liquids and the diversity of SFAEs are highlighted.

Chapter 5 describes the analytical potential of non-volatile and thermally stable ionic liquids for the ubiquitous class of bioactive compounds, *i.e.,* phenolics. This chapter will categorically explain various benefits of replacing conventional organic solvents with ionic liquids during solvent, microwave, and ultrasound-assisted extraction techniques.

Chapter 6 discusses the ionic liquids for the sustainable development of chemistry. The applications of ionic liquids and ionic liquids to carry out isomerization and dimerization reactions are summarized in detail throughout the chapter.

Chapter 7 presents some aspects of the usefulness of ionic liquids for enhanced enzymatic saccharification of different type cellulosic substrates. Cellulose saccharification under enzyme action in ionic liquids represents only one of the most interesting ways to modify its structure. The changes in the crystalline structure of the cellulose after such treatments are presented and discussed.

Chapter 8 depicts the current overview of biological applications of ionic liquids (ILs)-based surfactant for drug delivery, biomolecular extraction, compound separations, and enzyme catalysis. Self-assembly features of ILs such as micellar formation, IL-microemulsion, and vesicle/gel IL cooperative systems help us to understand the ILs behavior in the aqueous phase.

Chapter 9 explains how exceptional thermodynamics stability, non-volatile character, and plenty of structural arrangements or designs for ionic liquid can be followed for the purification of the basic necessity of life/water. Moreover, this chapter will critically analyze the attempts undertaken to purify water using ionic liquids, ionic liquids-based polymers, catalysts, or membranes.

Chapter 10 discusses the progress and perspective in experimental and theoretical studies of the structure, dynamics, and property of a double layer of the ionic liquid–electrode interface in ionic liquids. Additionally, a brief introduction of ionic liquids and their properties, the basic theory of the double layer, and the further directions of research are also presented.

Chapter 11 deals with the fundamentals of hydrate and corrosion formation, occurrence, and latest gas hydrate and corrosion inhibition (GHCI) chemical methods along with literature review on the application of ionic liquids in this field. The challenge of the application of GHCl and understanding the chemistry of ionic liquids is also discussed.

Chapter 12 details the use of ionic liquids as solvents in the reaction and purification steps of the biodiesel production process. Ionic liquids application as catalysts, co-solvent, or extracting solvents is presented in detail. Additionally, the role of deep eutectic solvents in the potential replacement of ionic liquids is discussed.

Chapter 13 describes the synthesis of different nanoparticles using plant extracts and their characterization using various analytical techniques. It also provides information about the influences of synthesis parameters like pH, temperature, reaction time, type of extracts, including the antimicrobial activity and the mechanism of nanoparticles formation using bio-extracts.

Chapter 14 discusses recent development in the recovery of precious metals and REEs using ionic liquids (IL) and newer division of ILs known deep eutectic solvents (DES). It briefs introduction about ILs and DES as well as their application in solvent extraction/recovery of precious metals and REEs process.

Chapter 15 describes the role of ionic liquids in the synthesis of various heterocyclic systems both as medium and catalyst. The focus is given to discuss the wider spectrum of heterocyclic scaffolds reported during the last decade.

Chapter 16 presents the future applications of ionic liquids (ILs) in food and biowaste industries. As well as, the classification of ILs, biodiesel production, development of suitable and sustainable methods for solving the problems about the production of food waste management are discussed.

Jeddah, Saudi Arabia Inamuddin
Aligarh, India
Jeddah, Saudi Arabia Abdullah M. Asiri

Contents

About the Editors

Inamuddin is currently working as Assistant Professor in the Chemistry Department, Faculty of Science, King Abdulaziz University, Jeddah, Saudi Arabia. He is a permanent faculty member (Assistant Professor) at the Department of Applied Chemistry, Aligarh Muslim University, Aligarh, India. He obtained Master of Science degree in Organic Chemistry from Chaudhary Charan Singh (CCS) University, Meerut, India, in 2002. He received his Master of Philosophy and Doctor of Philosophy degrees in Applied Chemistry from Aligarh Muslim University (AMU), India, in 2004 and 2007, respectively. He has extensive research experience in multidisciplinary fields of Analytical Chemistry, Materials Chemistry, and Electrochemistry and, more specifically, Renewable Energy and Environment. He has worked on different research projects as project fellow and senior research fellow funded by University Grants Commission (UGC), Government of India, and Council of Scientific and Industrial Research (CSIR), Government of India. He has received Fast Track Young Scientist Award from the Department of Science and Technology, India, to work in the area of bending actuators and artificial muscles. He has completed four major research projects sanctioned by University Grant Commission, Department of Science and Technology, Council of Scientific and Industrial Research, and Council of Science and Technology, India. He has published 173 research articles in international journals of repute and 18 book chapters in knowledge-based book editions published by renowned international publishers. He has published 105 edited books with Springer (UK), Elsevier, Nova Science Publishers, Inc. (USA), CRC Press Taylor & Francis in Asia Pacific, Trans Tech Publications Ltd. (Switzerland), IntechOpen Limited (UK), Wiley-Scrivener (USA), and Materials Research Forum LLC (US.A). He is a member of various journals' editorial boards. He is also serving as Associate Editor for journals (Environmental Chemistry Letter, Applied Water Science and Euro-Mediterranean Journal for Environmental Integration, Springer-Nature), Frontiers Section Editor (Current Analytical Chemistry, Bentham Science Publishers), Editorial Board Member (Scientific Reports-Nature), Editor (Eurasian Journal of Analytical Chemistry), and Review Editor (Frontiers in Chemistry, Frontiers, UK). He is also guest editing various special thematic special issues to the journals of Elsevier, Bentham Science

Publishers, and John Wiley & Sons, Inc. He has attended as well as chaired sessions in various international and national conferences. He has worked as a Postdoctoral Fellow, leading a research team at the Creative Research Initiative Center for Bio-Artificial Muscle, Hanyang University, South Korea, in the field of renewable energy, especially biofuel cells. He has also worked as a Postdoctoral Fellow at the Center of Research Excellence in Renewable Energy, King Fahd University of Petroleum and Minerals, Saudi Arabia, in the field of polymer electrolyte membrane fuel cells and computational fluid dynamics of polymer electrolyte membrane fuel cells. He is a life member of the Journal of the Indian Chemical Society. His research interest includes ion exchange materials, a sensor for heavy metal ions, biofuel cells, supercapacitors, and bending actuators.

Abdullah M. Asiri is the Head of the Chemistry Department at King Abdulaziz University since October 2009, and he is the founder and the Director of the Center of Excellence for Advanced Materials Research (CEAMR) since 2010 till date. He is the Professor of Organic Photochemistry. He graduated from King Abdulaziz University (KAU) with B.Sc. in Chemistry in 1990 and a Ph.D. from the University of Wales, College of Cardiff, UK, in 1995. His research interest covers color chemistry, synthesis of novel photochromic and thermochromic systems, synthesis of novel coloring matters and dyeing of textiles, materials chemistry, nanochemistry and nanotechnology, polymers and plastics. Prof. Asiri is the principal supervisor of more than 20 M.Sc. and six Ph.D. theses. He is the main author of ten books of different chemistry disciplines. Prof. Asiri is the Editor-in-Chief of King Abdulaziz University Journal of Science. A major achievement of Prof. Asiri is the research of tribochromic compounds, a new class of compounds which change from slightly or colorless to deep colored when subjected to small pressure or when grind. This discovery was introduced to the scientific community as a new terminology published by International Union of Pure and Applied Chemistry (IUPAC) in 2000. This discovery was awarded a patent from European Patent office and from UK patent. Prof. Asiri involved in many committees at the KAU level and in the national level. He took a major role in the advanced materials committee working for King Abdulaziz City for Science and Technology (KACST) to identify the national plan for science and technology in 2007. Prof. Asiri played a major role in advancing the chemistry education and research in KAU. He has been awarded the best researchers from KAU for the past five years. He was also awarded the Young Scientist Award from the Saudi Chemical Society in 2009 and also got the first prize for the distinction in science from the Saudi Chemical Society in 2012. He also received a recognition certificate from the American Chemical Society (Gulf region Chapter) for the advancement of chemical science in the Kingdome. He received a Scopus certificate for the most publishing scientist in Saudi Arabia in chemistry in 2008. He is also a member of the editorial board of various journals of international repute. He is the Vice-President of Saudi Chemical Society (Western Province Branch). He holds four USA patents, more than one thousand publications in international journals, several book chapters and edited books.

Contributors

Hilal Acıdereli Sen Research Group, Department of Biochemistry, Dumlupınar University, Kütahya, Turkey

Ahmad Adnan Department of Chemistry, Government College University Lahore, Lahore, Pakistan

Sumia Akram Division of Science and Technology, University of Education Lahore, Lahore, Pakistan

P. Andreo-Martínez Department of Chemical Engineering, University of Murcia, Espinardo, Murcia, Spain

Department of Agricultural Chemistry, University of Murcia, Espinardo, Murcia, Spain

Ayşenur Aygün Sen Research Group, Department of Biochemistry, Dumlupınar University, Kütahya, Turkey

John Bartlett Department of Environmental Science, School of Science, Institute of Technology Sligo, Sligo, Ireland

N. Raphael Biata Department of Chemical Sciences, University of Johannesburg, Johannesburg, South Africa

Kemal Cellat Sen Research Group, Department of Biochemistry, Dumlupınar University, Kütahya, Turkey

Sankar Chakma Department of Chemical Engineering, Indian Institute of Science Education and Research, Bhopal, M.P., India

Bruna Galdorfini Chiari-Andréo Departamento de Ciências Biológicas e da Saúde, Universidade de Araraquara, Araraquara, São Paulo, Brazil

Regina Maria Barretto Cicarelli Departamento de Ciências Biológicas, Faculdade de Ciências Farmacêuticas, Universidade Estadual Paulista, Araraquara, São Paulo, Brazil

Sara Isabel da Cruz Silvério CEB-Centre of Biological Engineering, Universidade do Minho, Braga, Portugal

Miguel Iglesias Departamento de Engenharia Química, Escola Politécnica, Universidade Federal da Bahia, Salvador, Bahia, Brazil

Ehiaze Ehimen Department of Environmental Science, School of Science, Institute of Technology Sligo, Sligo, Ireland

Uroosa Ejaz Department of Microbiology, University of Karachi, Karachi, Sindh, Pakistan

Ardhendu Sekhar Giri Department of Chemical Engineering, Indian Institute of Science Education and Research, Bhopal, M.P., India

Haydar Göksu Kaynasli Vocational College, Duzce University, Duzce, Turkey

Syeda Mariam Hasany Kinnaird College for Women University, Lahore, Pakistan

Mokhtar Che Ismail Department of Mechanical Engineering, Centre for Corrosion Research, Universiti Teknologi PETRONAS, Seri Iskandar, Perak, Malaysia

Kasibhatta Siva Kumar Department of Chemistry, S.V. Arts College, TTD's, Tirupati, India

Bhajan Lal Chemical Engineering Department, Universiti Teknologi PETRONAS, Seri Iskandar, Perak, Malaysia

CO2 Research Centre (CO2RES), Universiti Teknologi PETRONAS, Seri Iskandar, Perak, Malaysia

Nomvano Mketo Department of Chemistry, College of Science and Engineering and Technology, Florida Science Campus, University of South Africa, Johannesburg, South Africa

Anele Mpupa Department of Chemical Sciences, University of Johannesburg, Johannesburg, South Africa

Muhammad Mushtaq Department of Chemistry, Government College University Lahore, Lahore, Pakistan

Mărioara Nechifor Polyaddition and Photochemistry Department, "Petru Poni" Institute of Macromolecular Chemistry, Iasi, Romania

Philiswa N. Nomngongo Department of Chemical Sciences, University of Johannesburg, Johannesburg, South Africa

DST/Mintek Nanotechnology Innovation Centre, University of Johannesburg, Johannesburg, South Africa

DST/NRF SARChI Chair: Nanotechnology for Water, University of Johannesburg, Johannesburg, South Africa

Seán O'Connor Department of Environmental Science, School of Science, Institute of Technology Sligo, Sligo, Ireland

V. M. Ortiz-Martínez Department of Chemical and Environmental Engineering, Technical University of Cartagena, Cartagena, Murcia, Spain

Department of Chemical Engineering, University of Murcia, Espinardo, Murcia, Spain

Suresh C. Pillai Nanotechnology and Bio-Engineering Research Group, Department of Environmental Science, School of Science, Institute of Technology Sligo, Sligo, Ireland

Centre for Precision Engineering, Materials and Manufacturing Research (PEM), Institute of Technology Sligo, Sligo, Ireland

Ali Qasim Chemical Engineering Department, Universiti Teknologi PETRONAS, Seri Iskandar, Perak, Malaysia

CO2 Research Centre (CO2RES), Universiti Teknologi PETRONAS, Seri Iskandar, Perak, Malaysia

J. Quesada-Medina Department of Chemical Engineering, University of Murcia, Espinardo, Murcia, Spain

Lígia Raquel Marona Rodrigues CEB-Centre of Biological Engineering, Universidade do Minho, Braga, Portugal

Fatih Şen Sen Research Group, Department of Biochemistry, Dumlupınar University, Kütahya, Turkey

Azmi Mohammad Shariff Chemical Engineering Department, Universiti Teknologi PETRONAS, Perak, Malaysia

CO2 Research Centre (CO2RES), Universiti Teknologi PETRONAS, Perak, Malaysia

Masixole Sihlahla Department of Chemical Sciences, University of Johannesburg, Johannesburg, South Africa

Rebecca S. Andrade Centro de Ciência e Tecnologia em Energia e Sustentabilidade, Universidade Federal do Recôncavo da Bahia, Feira de Santana, Bahia, Brazil

Magaret Sivapragasam Centre of Research on Ionic Liquid, Universiti Teknologi PETRONAS, Bandar Seri Iskandar, Perak, Malaysia

School of Biological Science, Faculty of Science and Technology, QUEST International University Perak, (QIUP), Ipoh, Perak, Malaysia

Muhammad Sohail Department of Microbiology, University of Karachi, Karachi, Sindh, Pakistan

Cristina-Magdalena Stanciu Natural Polymers, Bioactive and Biocompatible Materials Department, "Petru Poni" Institute of Macromolecular Chemistry, Iasi, Romania

Fulga Tanasă Polyaddition and Photochemistry Department, "Petru Poni" Institute of Macromolecular Chemistry, Iasi, Romania

Carmen-Alice Teacă Advanced Research Center for Bionanoconjugates and Biopolymers, "Petru Poni" Institute of Macromolecular Chemistry, Iasi, Romania

Guocai Tian State Key Laboratory of Complex Nonferrous Metal Resources Clean Utilization, Faculty of Metallurgical and Energy Engineering, Kunming University of Science and Technolgy, Kunming, Yunnan Province, China

Jayathirtha Rao Vaidya Hetero Research Foundation, Hyderabad, India

Gabriela Brasil Romão Veloso Departamento de Engenharia Química, Escola Politécnica, Universidade Federal da Bahia, Salvador, Bahia, Brazil

Cecilia Devi Wilfred Centre of Research on Ionic Liquid, Universiti Teknologi PETRONAS, Bandar Seri Iskandar, Perak, Malaysia

Faculty of Fundamental and Applied Science, University Teknologi PETRONAS, Bandar Seri Iskandar, Perak, Malaysia

Venkata Durga Nageswar Yadavalli Former Chief Scientist, Medicinal Chemistry and Pharmacology Division, CSIR - Indian Institute of Chemical Technology, Hyderabad, India

Nursefa Zengin Kaynasli Vocational College, Duzce University, Duzce, Turkey

Chapter 1
Ionic Liquids as "Green Solvents": Are they Safe?

Gabriela Brasil Romão Veloso, Rebecca S. Andrade, Regina Maria Barretto Cicarelli, Miguel Iglesias, and Bruna Galdorfini Chiari-Andréo

Contents

Abbreviations

[BMPY] [TFSI]	1-butyl-1-methylpyrrolidinium bis(trifluoromethanesulfonyl) imide
[C_{16}mim] Cl	1-hexadecyl-3-methylimidazolium chloride
[C_8mim] Br	1-octyl-3-methylimidazolium bromide
[Chol]	Choline-based ionic liquids

G. B. R. Veloso · M. Iglesias
Departamento de Engenharia Química, Escola Politécnica, Universidade Federal da Bahia, Salvador, Bahia, Brazil
e-mail: miguel.iglesias@ufba.br

R. S. Andrade
Centro de Ciência e Tecnologia em Energia e Sustentabilidade, Universidade Federal do Recôncavo da Bahia, Feira de Santana, Bahia, Brazil
e-mail: rebecca.andrade@ufrb.edu.br

R. M. B. Cicarelli
Departamento de Ciências Biológicas, Faculdade de Ciências Farmacêuticas, Universidade Estadual Paulista, Araraquara, São Paulo, Brazil
e-mail: regina.cicarelli@unesp.br

B. G. Chiari-Andréo (✉)
Departamento de Ciências Biológicas e da Saúde, Universidade de Araraquara, Araraquara, São Paulo, Brazil
e-mail: bgchiari-andreo@uniara.edu.br

Inamuddin, A. M. Asiri (eds.), *Nanotechnology-Based Industrial Applications of Ionic Liquids*, Nanotechnology in the Life Sciences,
https://doi.org/10.1007/978-3-030-44995-7_1

[DMEtA] [Oct]	N,N-dimethylethanolammonium octanoate
[emim (Ms)]	1-ethyl-3-methylimidazolium methanesulfonate
[EMIM] [BF_4]	1-ethyl-3-methylimidazolium tetrafluoroborate
[EMIM] [DCA]	1-ethyl-3-methylimidazolium dicyanamide
[EMIM] [PF_6]	1-ethyl-3-methylimidazolium hexafluorophosphate
[EMIM] [TFSI]	1-ethyl-3-methylimidazolium bis(trifluoromethanesulfonyl) imide
[Mim] Ac	1-methylimidazolium acetate
[TBA] [TFSI]	Tributylmethylammonium bis(trifluoromethanesulfonyl)imide
[tbph (Ms)]	Tetrabutylphosphonium methanesulfonate
[Zn] $[TFSI]_2$	Zinc di[bis(trifluoromethanesulfonyl)imide
2-HDEAA	2-hydroxydiethanolamine acetate
2-HDEAAd	2-hydroxydiethanolamine adipate
2-HDEABe	2-hydroxydiethanolamine benzoate
2-HDEACi	2-hydroxydiethanolamine citrate
2-HDEAL	2-hydroxydiethanolamine lactate
2-HDEAMa	2-hydroxydiethanolamine maleate
2-HDEAPr	2-hydroxydiethanolamine propionate
2-HDEASa	2-hydroxydiethanolamine salicylate
2-HEAA	2-hydroxyethanolamine acetate
2-HEAAd	2-hydroxyethanolamine adipate
2-HEACi	2-hydroxyethanolamine citrate
2-HEAF	2-hydroxyethanolamine formate
2-HEAL	2-hydroxyethanolamine lactate
2-HEAPr	2-hydroxyethanolamine propionate
2-HTEAPe	2-hydroxytriethanolamine pentanoate
8-OhDG	8-hydroxy-2′-deoxyguanosine
AILs	Aprotic ionic liquids
BmPy-Cl	1-butyl-1-methylpyrrolidinium chloride
BOD	Biochemical oxygen demand
CaCo-2	Human colon carcinoma
COD	Chemical oxygen demand
DCA	Dicyanamide anion
DNA	Deoxyribonucleic acid
EAN	Ethylammonium nitrate
EC_{50}	Half maximal effective concentration
HaCat	Keratinocytes
HEK	Normal human embryonic kidney cells
HeLa	Human cervical carcinoma epithelial cells
HepG2	Liver hepatocellular cells
Hs68	Fibroblast cell lines
IC_{50}	Half maximal inhibitory concentration
ILs	Ionic liquids
IPC-81	Rat leukemic cells
m-2-HEAA	N-methyl-2-hydroxyethylammonium acetate

m-2-HEAB	N-methyl-2-hydroxyethylammonium butyrate
m-2-HEAP	N-methyl-2-hydroxyethylammonium pentanoate
m-2-HEAPr	N-methyl-2-hydroxyethylammonium propionate
MCF7	Human breast cancer cells
MTT	3-[4,5-dimethylthiazol-2-yl]-2,5-diphenyltetrazolium bromide
NTf2	Bis(trifluoromethanesulfonyl)amide anion
PILs	Protic ionic liquids
QSAR	Quantitative structure–activity relationship
REACH	Registration, Evaluation, Authorisation and Restriction of Chemicals
rMSC	Rat mesenchymal stem cells
SIA	Sequential injection analysis
T98G	Human brain glioblastoma cell line
TA100	*Salmonella typhimurium* strain
TA98	*Salmonella typhimurium* strain
WST-1	2-(4-iodophenyl)-3-(4-nitrophenyl)-5-(2,4-disulfophenyl)-2H-tetrazolium monosodium salt

1.1 Introduction

The concept of sustainability relates technology and industrial processes to policies aimed at reducing the environmental impact caused by economic activities (Collin and Collin 2010). Interest in this field has increased in recent years, and guidelines have also been set to design and improve chemical products and processes that reduce or eliminate the use of substances that are potentially toxic to the environment, such as organic solvents.

Most organic solvents are fat soluble, volatile, and flammable compounds that put the environment at risk. But, in addition, it is important to emphasize that the action of these solvents in the human body is also deleterious. It is reported that their effects are similar to anesthetics; that is, they are capable of inhibiting brain and spinal cord activity, decreasing the functional capacity of the central nervous system, making it less sensitive to stimuli. Because they are generally lipophilic substances, they easily overcome biological barriers and accumulate in adipose tissue and organs of the body, interacting with molecular and cellular targets. Once deposited, the solvents alter the excitability of cells, suppressing normal conduction of nerve impulses, making the nervous system sensitive to their toxicity (Thriel 2014; Schwarzenbach et al. 2016).

As a result, initiatives aimed at avoiding this situation have been developed, such as the search for alternative solvents that fit the principles of Green Chemistry and are safe for the population (Lenardão et al. 2003).

Several studies have shown the diversity of industrial applications of ionic liquids (ILs), such as use in separation and extraction techniques (Li et al. 2014;

Larriba et al. 2016; Shang et al. 2017; Berthod et al. 2018), catalysis (Muzalevskiy et al. 2016; Vekariya 2017; Li et al. 2018), optical and electronic devices (Borra et al. 2007; Hagiwara and Lee 2007; Li et al. 2011; Liu et al. 2012; Zhang et al. 2016), nanotechnology and bioprocesses (He and Alexandridi 2017; Grewal and Khare 2018), energy (Chang et al. 2014; Forsyth et al. 2016; Zhang et al. 2018), analytical techniques and applications in fine chemistry (Riduan and Zhang 2013; Clark et al. 2016; Nawata et al. 2018), and textile medium dyeing (Yuan et al. 2010a, 2010b; Rouette 2001; Cheng 2011; Dong 2011; Kantouch et al. 2011; Qingdao University 2012; Zhuang et al. 2014; Bianchini et al. 2015; Andrade et al. 2017), among others. It is noteworthy that the recovery of ILs is a great advantage because they can be effectively isolated by distillation or extraction and reused, significantly reducing the cost of their applications (Bubalo et al. 2017; Rykowska et al. 2018).

In recent years, researches about the biological activities of ILs regarding their possible applications in the biotechnology and pharmaceutical industries have begun (Le Bideau et al. 2011; Balk et al. 2015; Egorova et al. 2017; Egorova and Ananikov 2018b), and knowledge about their interactions with biological systems began with the determination of various toxic effects of ILs (Ranke et al. 2007; Petkovic et al. 2011; Egorova and Ananikov 2014). Their use as efficient solvents or co-solvents in bio-catalysis processes and drug delivery has received interest from researchers (Adawiyah et al. 2016; Egorova et al. 2017; Kunov-Kruse et al. 2017; Claus et al. 2018; Egorova and Ananikov 2018b; Elgharbawy et al. 2018).

Basically, ILs are organic salts that are liquid under normalized conditions. They are composed of ions and are held together mainly by electrostatic or coulombic attraction. They are obtained by combining organic cations containing a positively charged nitrogen, sulphur, or phosphorus atom with a wide variety of inorganic and organic anions. Other denominations, such as molten salts or organic liquid salts, are used to describe this group of compounds (Welton 1999; Chiappe and Pieraccini 2005; Handy 2011; Kokorin 2011; Bubalo et al. 2017). The first synthesized IL was ethylammonium nitrate (EAN), which has a melting point of 12 °C (Walden 1914).

The cations commonly used for the synthesis of ILs come from ammonium, sulfonium, imidazolium, triazolium, pyridinium, phosphonium, pyrazolium, and guanidinium ions, with different substituents (Bhattacharjee et al. 2014; Elsheikh 2014; Kordala-Markiewicz et al. 2014; Zheng et al. 2014; Fall et al. 2015; Neale et al. 2016; Tankov et al. 2017; Tian et al. 2017; Calza et al. 2018; Chen et al. 2018; Kishimura et al. 2018; Nehra et al. 2018; Rogalsky et al. 2018; Xiong et al. 2018), and the most commonly used anions are $AlCl_4^-$, Al_2Cl_7, BF_4^-, Br^-, Cl^-, $ZnCl_4^{2-}$, PF_6^-, $CF_3CO_2^-$, $CF_3SO_3^-$, $CH_3CO_2^-$, HSO_2^-, RSO_4^-, RSO_3^-, $H_2PO_3^-$, and $R_2PO_4^-$ (Gilbert et al. 2007; Ochedzan-Siodkak et al. 2008; Bertoti and Netto-Ferreira 2009; Mayoral et al. 2009; Sen et al. 2016; Sharma and Ghorai 2016; Deyab et al. 2017; Fazlali et al. 2017; Lei et al. 2017; Lopes et al. 2017; Basu et al. 2018; Decaen et al. 2018; Kakaei et al. 2018; Zec et al. 2018; Zhou et al. 2018). Figure 1.1 shows some cations and anions commonly used in the synthesis of ILs.

Theoretically, the different combinations of cations and anions can result in the synthesis of approximately 10^{18} new ILs. Because of these numerous combinations, ILs can be synthesized with the desired physical and chemical properties, such as

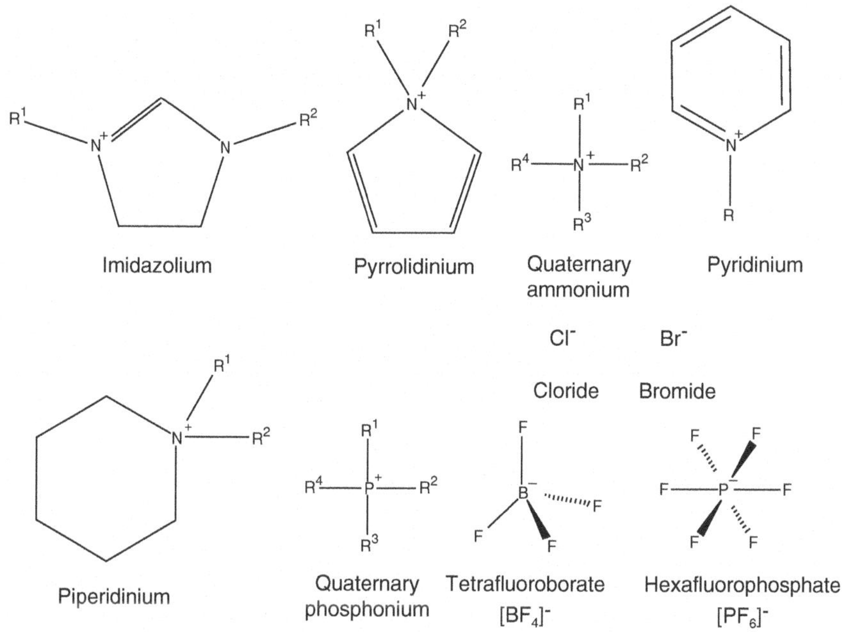

Fig. 1.1 Cations and anions commonly used for the synthesis of ionic liquids (ILs). Source: Authors

melting point, viscosity, density, and solubility. Thus, they may be developed, for example, to be soluble or not soluble in water or certain types of organic solvents in various composition ranges of the mixtures (Niedermeyer et al. 2012).

Most ILs are non-flammable and, given their non-volatility, are often presented as "green solvents" or "green alternatives" to replace the organic solvents. However, their toxicity, biodegradability, and environmental impacts have not been sufficiently investigated (Ranke et al. 2004; Kumar et al. 2006).

Historically, research has been directed at aprotic ionic liquids (AILs), and only in recent years has there been a greater interest in studying protic ionic liquids (PILs) and their industrial applications (Álvarez et al. 2010a, 2010b, 2011). There are references in the literature of their use in electrochemical devices (Greaves et al. 2006; Markusson et al. 2007; Tang et al. 2019), dyes of acrylic fibres (Opwis et al. 2017), textile medium dyeing (Andrade et al. 2017), proton-conducting membranes (Tigelaar et al. 2006; Martinelli et al. 2007; Oliveira et al. 2011), proton-conducting electrolyte fuel cells (Noda et al. 2003), biodiesel separation (Wu et al. 2007), catalysts (Du and Tian 2006; Cota et al. 2014; Serra et al. 2016), renewable and biodegradable solvents (Zhao et al. 2007), corrosion on carbon steel (Dos Santos et al. 2014), cellulose dissolution and regeneration (Meenatchi et al. 2017), carbon dioxide capture (Vijayaraghavan et al. 2018), sugarcane bagasse pre-treatment (Pin et al. 2019), extraction of phycobiliproteins (Rodrigues et al. 2019), and other sustainable processes.

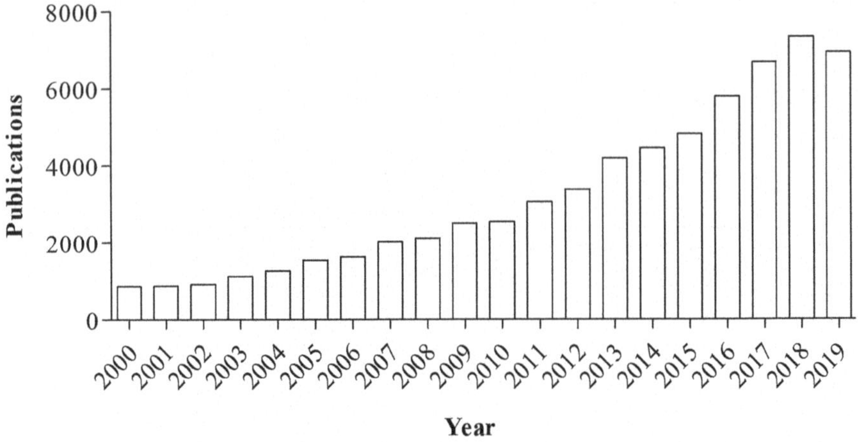

Fig. 1.2 Number of publications regarding ionic liquids toxicity in the last 19 years. Source: Science Direct (accessed on July 2019)

Structurally, PILs are formed by the proton transfer of a Brönsted acid (AH) to a Brönsted base (B) to produce [BH+] [A–] species. AILs, on the other hand, contain substituents other than the proton, generally an alkyl group, and, therefore, tend to be lipophilic, presenting less mobility in the environment, which facilitates their accumulation in organisms of aquatic and terrestrial systems, as well as their biomagnification through food chains if ingested inappropriately. Contrary to this, PILs are extremely hydrophilic, which aids in their dispersion in a medium. However, there are no reasons for ILs to be exempt from causing harm to health or environmental problems, and there is a need for supplementary information in this area (Docherty and Kulpa 2005; Peric et al. 2013).

Considering the use of PILs in industrial processes for import and export, it is necessary to follow rules of regulatory agencies, such as the Registration, Evaluation, Authorisation and Restriction of Chemicals (REACH), which has been operating in the European Union since 2010, and requires chemical manufacturers and users to record physicochemical characterization information for new materials, as well as ecotoxicity analyses and possible impacts on human health. Without this registration, the importation, exportation, manufacture, circulation, and industrial use of any product is not authorized. The main goal is to have a higher level of information about chemicals so that the risks can be controlled and the use of toxic chemicals can be minimized (Reach 2006).

Information on the potential toxicity of ILs, information required to comply with regulatory requirements, and the assessment of the safety and hygiene aspects derived from the handling, use, and transport of ILs in the industrial sector are still limited. The end product and effluents generated in industrial processes may contain residual ILs that must be assessed for environmental and human impacts. Figure 1.2 shows how only in recent years has there been greater interest in studying the toxic-

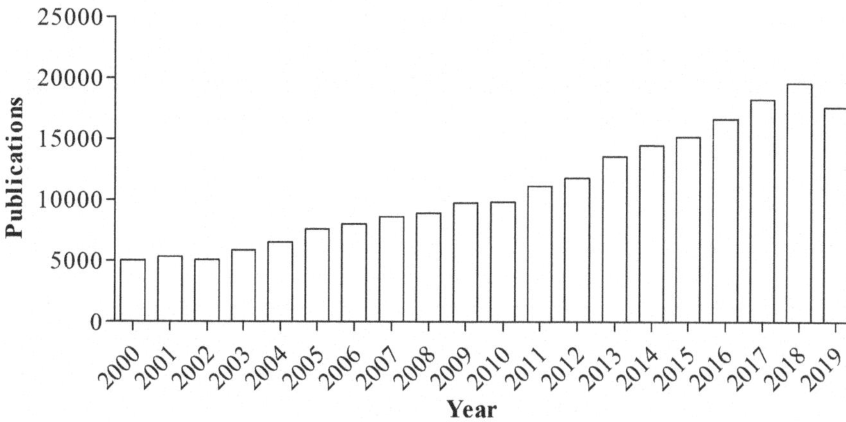

Fig. 1.3 Number of publications regarding ionic liquids in the last 19 years. Source: Science Direct (accessed on July 2019)

ity of ILs, and there are also increasing numbers of publications dealing with ILs in general (Fig. 1.3).

There are few studies on the ecotoxicity of ILs, mostly directed at AILs, derived from cationic groups, such as pyridinium and imidazolium. From the findings in the literature, contradictory conclusions are often drawn to some extent as a consequence of the use of compounds with distinct molecular structures or ecotoxicological assays of different typologies, resulting in unreliable comparisons (Docherty and Kulpa 2005; Zhao et al. 2007; Peric et al. 2013; Matsumoto et al. 2004a, 2004b; Bernot et al. 2005a, 2005b; Latala et al. 2005; Couling et al. 2006; Matsuo and Lamberti 2006; Docherty et al. 2007; Salminen et al. 2007; Stolte et al. 2007; Peric et al. 2011, 2014, 2015; Reid et al. 2018).

According to Docherty and Kulpa (2005), the toxicity of AILs is directly related to the number and size of the cationic substituents, and the variation of the anionic group does not significantly change their toxicity. Luis et al. (2007) evaluated the toxicity of some AILs by quantitative structure–activity relationship (QSAR) studies with different cationic groups, specifically, pyrrolidinium, imidazolium, and pyridinium groups, which conferred about 3%, 20%, and 33% toxicity to the compound, respectively. Furthermore, it was observed that each carbon atom added to the chain in the radicals R_1 and R_2 increased the toxicity by 11%. As reported by some researchers, long chain ILs are even more toxic than classic solvents (benzene, toluene, ethylbenzene, xylene) (Bernot et al. 2005b; Cho et al. 2008), while other authors have indicated that they are less toxic than the cited substances (Couling et al. 2006). However, this depends on the combination of cations and anions and their structures since each IL has a specific characteristic.

Most studies of toxicity-related AILs affirm that the cationic group is primarily responsible for the increase in toxicity; however, Matzke et al. (2007) observed that among the anions, including Cl^-, BF_4^-, $(CF_3SO_2)_2N^-$, $(CF_3)_2N^-$, octylsulfate, and bis(1,2-benzenediolato)borate with the imidazolium cationic group, the toxic effect of $(CF_3)_2N^-$ was significantly larger than that of the cationic group. Egorova et al. (2017) indicated that the behaviour and impact of the anionic group of the IL depend on its interaction with water; for example, hydrophilic anions such as Cl^-, remains in solution, while hydrophobic anions, such as PF_6^-, constitute a film between the lipid/water barrier.

According to work done by Pretti et al. (2009), PILs with quaternary ammonium cations have lower toxicity in aquatic organisms than those with aromatic rings (pyridinium and imidazolium), and, as reported by Ferraz et al. (2014), some ILs, also with quaternary ammonium, showed antibacterial activity.

Regarding the commented results, there are clear contradictions that, in part, can be attributed to the use of distinct ILs and ecotoxicological assays. There are a wide variety of assays that have been used to assess the environmental toxicity of these compounds and most of them make use of a single species of aquatic or terrestrial organisms to determine ecotoxicity although the sensitivity of the different organisms is very different from contaminants.

Assessing the toxicity of ILs on microorganisms may be useful for understanding wastewater treatment possibilities through secondary purification systems, which could be ineffective due to the toxicity of the ILs (Docherty et al. 2007; Azimova et al. 2009; Peric et al. 2013, 2014).

Given the scarcity of information regarding the toxicity of ILs, mainly related to human health, it is necessary to investigate this area in more detail, a matter to be discussed in this chapter, which is extremely relevant as these promising green solvents are being increasingly explored and used in various branches of science and industry.

1.2 (Eco)toxicity of Ionic Liquids

ILs are often described as versatile and less environmentally harmful solvents; however, knowledge of the toxic potential for distinct organisms and the effects in the trophic chain is restricted, particularly when referring to PILs. Usually, they are cited as "green solvents"; however, caution is needed since studies regarding their toxicity, biodegradability, and environmental mobility are still scarce (Oliveira et al. 2016; Meksi and Moussa 2017).

In recent years, the notion that ILs are green and environmentally friendly has changed. These claims were justified by their negligible vapour pressure, which would result in a reduction in atmospheric emission as they are not flammable and not even explosive. Indeed, they have these advantages, but without solid knowledge of their ecotoxicological behaviour, no justification for this classification can be determined. From the studies in the literature, it is now known that there low- or high-risk ILs, and this depends predominantly on their structure (Matzke et al. 2010).

Existing studies on the ecotoxicity of ILs are based on tests with different organisms, such as algae (*O. submarine, P. subcapitata, C. meneghiniana*, and *U. lactuca*), bacteria (*V. fischeri, E. coli, S. aureus, P. phosphoreum*, and *B. subtilis*), yeast

(*S. cerevisiae)*, and aquatic (*L. minor*) and terrestrial plants (*L. sativum*, *T. aestivum* L., and *R. sativus*), among others (Bubalo et al. 2017).

In studies by Peric et al. (2011), assays were performed to evaluate the toxicity of different short-chain PILs, namely 2-hydroxyethanolamine formate (2-HEAF), 2-hydroxydiethanolamine propionate (2-HDEAPr), and 2-hydroxytriethanolamine pentanoate (2-HTEAPe), considering their effect on seedling emergence and seedling growth of three terrestrial plants, *Allium cepa* (onion), *Lolium perenne* (grass), and *Raphanus sativus* (radish), the carbon and nitrogen transformation test; and biodegradable profile. Of the PILs analysed, only 2-HTEAPe showed toxicity to *R. sativus* (EC_{50} = 826 mg/kg), while the other PILs showed no toxicity to any plant, with EC_{50} values above 1000 mg/kg. The three PILs were non-toxic in terms of chronic toxicity to plants in the carbon and nitrogen cycles, and they were also considered biodegradable in soil by respirometric assays and soil pollutant quantification. Thus, the authors concluded that, in general, compounds with more complex structures have a greater tendency to cause inhibition in the organisms tested than those with a simpler and smaller structure.

As there are numerous possible combinations between cations and anions for the synthesis of ILs, Peric et al. (2015) used QSAR studies to predict the ecotoxicity of possible and existing ILs considering the inhibition of *V. fischeri* luminescence assay, *P. subcapitata* and *L. minor* growth inhibition tests, acetylcholinesterase inhibition assay and cytotoxicity using rat leukemic cells IPC-81. Fifty-five ILs, including protic and aprotic ILs, were analysed, and it was observed that the aquatic plant *L. minor* was more sensitive to PILs than AILs due to the hydrophilicity of PILs. The QSAR study was efficient in predicting the toxicity of the ILs, confirming that the cationic group of the AILs has a greater influence because of the longer alkyl chain. For the PILs, both cationic and anionic groups influenced the ecotoxicity of the compounds, and it was with the increase in the carbonic acid chain length in the anionic group that the greatest influence was observed in three of the five ecotoxicity assays.

Ghanem et al. (2017) also used QSAR studies using the multiple linear regression model to evaluate the toxicity of 110 ILs against the bioluminescent bacteria *V. fischeri*, which is the most studied organism to evaluate the toxicity of ILs in an aquatic environment. It was found that the length of the alkyl chain influences the increased toxicity. In addition, the model was able to distinguish between the smallest and the largest effect of hydrophilic and hydrophobic anions, respectively.

There are some studies in the literature that have shown that most PILs that are categorized as biodegradable are synthesized using choline analogues, which is an organic cation, an essential nutrient that is part of the B-complex vitamins. Therefore, their low toxicity and appropriate biodegradability are related to a limited number of chemical compounds, but, in general, combining their properties and advantages is a beneficial environmental alternative considering this class of ILs (Peric et al. 2013; Jordan and Gathergood 2015; Oliveira et al. 2016).

Oliveira et al. (2016) evaluated four PILs, N-methyl-2-hydroxyethylammonium acetate (m-2-HEAA), N-methyl-2-hydroxyethylammonium propionate (m-2-HEAPr), N-methyl-2-hydroxyethylammonium butyrate (m-2-HEAB), and

N-methyl-2-hydroxyethylammonium pentanoate (m-2-HEAP), in relation to their activity against fungi and bacteria (*E. coli*, *S. aureus*, *Fusarium* sp., and *C. albicans*), their luminescence inhibition of *V. fischeri* bacteria, their phytotoxicity of lettuce seeds (*L. sativa*), and their biodegradability by chemical oxygen demand (COD) and biochemical oxygen demand (BOD). Overall, the lengthening of the alkyl chain of PILs has been found to promotes the increase in the negative effect of these compounds on the various microorganisms tested. The low toxicity of the PILs tested in the marine bacteria (*V. fischeri*) was also verified, and this same impact was not confirmed for all microorganisms studied, in particular, for yeast and fungus (*C. albicans* and *Fusarium* sp.). The effect caused by the antibiotic (positive control) was less significant than the effect of some PILs. From the results, it was possible to conclude that m-2-HEAPr and m-2-HEAP represented the least toxic PILs and, as to their biodegradability, the four ILs tested presented low biodegradability.

The studies presented here have provided useful information on the ecotoxicity of some ILs, but due to their large number of distinct combinations and properties, it is necessary to carry out appropriate toxicological tests of ILs in order to determine their toxicity in aquatic and terrestrial mediums, environmental and human levels, and to be able to draw comparisons between conventional and newly developed ILs in order to find those that are capable of having the least environmental and human impact.

1.3 Ionic Liquids and Safety to Humans

As previously mentioned, some studies have been developed related to the toxicity of ILs in the environment, but the potential toxic effects on human health have been hardly investigated, containing only a few references in the literature. Given the great diversity of ILs, it can be expected that they exhibit distinct environmental and toxicological behaviours, depending on their nature, making it necessary to conduct specific studies. The biological activity of a substance can be altered depending on its characteristics, such as the size of the alkyl chain (Peric et al. 2011; Egorova and Ananikov 2014).

An important mechanism for assessing the toxicity of a compound is by enzymatic inhibition, and there are some studies in the literature on acetylcholinesterase inhibition (Stock et al. 2004; Jastorff et al. 2005; Zhang and Malhotra 2005; Matzke et al. 2007; Ranke et al. 2007; Arning et al. 2008; Stasiewicz et al. 2008; Torrecilla et al. 2009), adenosine monophosphate deaminase (Skladanowski et al. 2005), and the antioxidant enzyme system of mouse liver (Yu et al. 2008) by ILs.

Acetylcholinesterase acts on function and nerve responses and catalyses the hydrolysis of choline esters with more specificity for acetylcholine, a common neurotransmitter in many mammalian nervous system synapses (Massoulié et al. 1993; Fulton and Key 2001). Therefore, inhibition of this enzyme causes several adverse effects on neural processes and even heart disease or myasthenia in humans (Chemnitius et al. 1999; Pope et al. 2005).

Ranke et al. (2007) evaluated acetylcholinesterase inhibition of a wide variety of ILs (292 compounds). It was found that the cationic group was responsible for enzymatic inhibition, mainly of those ILs that contained pyridinium cations, and it also caused a greater inhibitory effect than the ILs formed by imidazolium cations. ILs containing phosphonium cations had less inhibitory effects. The anionic group did not cause enzyme inhibition because the linking with the active site of acetylcholinesterase is limited.

Adenosine monophosphate deaminase is an enzyme involved in purine metabolism to produce uric acid. In humans, this enzyme is found at high levels, especially in lymphoid organs such as the spleen, thymus, and lymph nodes. As it is essential for the maintenance of healthy lymphocytes, its inhibition causes the immune system to malfunction (Silva et al. 2016). Skladanowski et al. (2005) evaluated the inhibition of this enzyme from some ILs containing the imidazolium cation. The results showed that imidazolium cations associated with PF_6^-, BF_4^-, p-tosylate, and Cl^- anions caused inhibition of enzymatic activity in a dose-dependent manner, and fluoride-containing ILs had a higher inhibitory effect.

Yu et al. (2008) evaluated the inhibitory effect of the catalase, superoxide dismutase, glutathione S-transferase, and glutathione peroxidase enzymes of an imidazolium cation and bromide anion (1-octyl-3-methylimidazolium bromide—[C_8mim] Br). These enzymes are important in maintaining the body's antioxidant system. The results showed that the administration of 35.7 mg/kg of [C_8mim] Br caused damage and modified the enzymatic activity in the mouse liver.

In a study by Cunha et al. (2013), an automated assay was developed to evaluate the carboxylesterase activity being helpful to predict the toxicity of ILs to human health. For this, a sequential injection analysis (SIA) system based on the hydrolysis of 4-methylumbelliferyl acetate by the carboxylesterase enzyme was implemented to produce the fluorescent compound 4-methylumbelliferone, a method in which the inhibition of enzymatic activity was indicated by decreased fluorescence. Seven commercial ionic liquids were analysed and, among them, the most toxic was tetrabutylphosphonium methanesulfonate—[tbph (Ms)] and the least toxic was 1-ethyl-3-methylimidazolium methanesulfonate—[emim (Ms)].

Costa et al. (2016) evaluated the reduction of cytochrome-c oxidase (or IV complex) enzyme activity in the presence of 15 ILs with different alkyl chains, cationic groups, and anions. This enzyme participates in the respiratory chain and is found in bacterial cytoplasmic membrane and mammalian mitochondria, catalysing the transfer of electrons from cytochrome-c to the oxygen molecule, promoting energy to the cell. It was found that there was considerable inhibition of the enzyme by the BF_4^- anion and by the ILs incorporated by pyrrolidinium and tetrabutylphosphonium non-aromatic cation groups. Choline and acetate groups had a low negative effect on enzymatic activity, demonstrating that the structure of the ILs influences the toxicity, specifically, cationic groups, alkyl side chains, and anions.

ILs may not be damaging to some digestive enzymes, as noted by Bisht and Venkatesu (2017). In their study, some choline-based ILs ([Chol]) were suggested for use as stabilizers of the chymotrypsin enzyme since they protect their enzymatic structure against thermal denaturation. In addition, [Chol] [OAc], [Chol] [Cl], and [Chol] [H_2PO_4] ILs promoted both enzyme stabilization and maintenance of its activity. In another study by Fan et al. (2018), ammonium- and imidazolium-based ILs inhibited the trypsin enzyme, but this inhibition was reversible after removal of the ILs and the enzyme could regain the activity.

Another way to evaluate the toxicity of ILs is by in vitro studies of cytotoxicity, which may provide the first evidence of their impact on the organism. These assays have advantages of shorter assay time and easy handling compared to in vivo studies on multicellular organisms (Egorova et al. 2017). However, the effects of ILs on different cell cultures is still an area to be further investigated (Egorova and Ananikov 2018a).

Stepnowski et al. (2004) evaluated the viability of human cervical carcinoma epithelial cells (HeLa) subjected to treatment with imidazolium cation ILs by the 2-(4-iodophenyl)-3-(4-nitrophenyl)-5-(2,4-disulfophenyl)-2H-tetrazolium monosodium salt (WST-1). The study showed that the toxicities of compounds containing 1-n-butyl-3- methylimidazolium depend directly on the association with the anionic group and, in this case, the EC_{50} values of the compound containing BF_4^- were lower, i.e. it was more toxic to cells at lower concentrations. The increase in the carbonic chain of the methyl, ethyl, or n-hexyl chains did not influence the reduction of EC_{50} values; only the IL with the n-decyl chain, the largest carbonic chain, had a lower EC_{50} value than the one with the n-butyl carbon chain, and it also had higher hydrophobicity. In addition, the effective concentrations of the samples analysed in HeLa cells were lower than the values obtained for conventional solvents, such as dichloromethane, toluene, and xylene.

Regarding HeLa cells, Xia et al. (2018) analysed the toxicity, cell viability, genotoxicity, oxidative stress, and apoptosis of this cell line exposed to 1-hexadecyl-3-methylimidazolium chloride ([C_{16}mim] Cl). It was verified that the sample inhibited or decreased cell growth, as well as induced apoptosis and caused DNA damage, inhibited superoxide dismutase enzyme activity, decreased glutathione content, which is an important antioxidant of the body's antioxidant defence system, and increased the cellular malondialdehyde level, which is one of the products of lipid peroxidation, in HeLa cells. The results showed that [C_{16}mim] Cl cannot be classified as a green solvent since it induced oxidative stress, genotoxicity, and apoptosis in this cell line. Wan et al. (2018) also evaluated this same IL against HepG2 cells, which are metabolizing cells. The results indicated the same behaviour since this IL promoted genotoxicity, oxidative stress, and apoptosis in this cell line.

Zanoni et al. (2019) evaluated the cytotoxic effect on HepG2 and HaCat cells (keratinocytes found on the skin) of 13 PILs obtained from the reaction between a functional amine (2-hydroxy ethylammonium or 2-hydroxy diethylammonium) and a carboxylic acid (acetic acid, 2-HEAA and 2-HDEAA; lactic acid, 2-HEAL and 2-HDEAL; adipic acid, 2-HEAAd and 2-HDEAAd; benzoic acid, 2-HDEABe; citric acid, 2-HEACi and 2-HDEACi; formic acid, 2-HEAF; propanoic acid, 2-HEAPr; salicylic acid, 2-HDEASa; and maleic acid, 2-HDEAMa) by the 3-[4,5-dimethylthi azol-2-yl]-2,5-diphenyltetrazolium bromide (MTT) assay, whose principle is to determine the ability of viable cells to reduce this yellow dye by forming insoluble violet-coloured formazan crystals to determine the viability of the treated cells. Morphological changes in HaCat cells exposed to PILs for 8 h were also evaluated. These changes may include loss of membrane integrity, chromosomal condensation, and apoptotic body formation. Overall, the results showed that the evaluated PILs had a low toxicity profile. The cytotoxic effect was higher in HepG2 cells than in HaCat cells because, possibly, PILs were metabolized by HepG2 cells, which could generate more toxic metabolites. According to the IC_{50} values, the most cytotoxic PILs were 2-HDEASa, 2-HEACi, and 2-HDEACi. PILs with longer carbon chains in the anion group, such as 2-HEAAd and 2-HDEAAd, showed higher cytotoxicity than those with a short chain, such as 2-HEAA and 2-HEAL. Regarding the morphological alterations of the cells, the study was not conclusive in elucidating cytotoxicity since the PILs probably caused osmotic transport due to the highly hypertonic medium; however, cell morphological changes were observed for most PILs.

Frade et al. (2009) evaluated the cytotoxicity using CaCo-2 cells (human colon carcinoma) treated with over 80 ILs with different cation classes composed of imidazolium, guanidinium, ammonium, phosphonium, pyridinium, and pyrrolidinium groups and different anions, such as Cl^-, I^-, BF_4^-, and $FeCl_4^-$, among others. The study showed that, in general, cation toxicity is more expressive as the alkyl chain length increased. Benzyl groups were demonstrated to be toxic, but when carboxyl groups were introduced, there was a significant decrease in toxicity. It was also found that the introduction of ether functionality in the dimethyl-guanidinium decreased the toxicity of this cation. Anion type can also directly affect IL toxicity, and some have had a greater impact than others with bis(trifluoromethanesulfonyl) amide (NTf2) and dicyanamide (DCA) anions.

Another study was conducted by Kumar et al. (2009) regarding the toxicity of ILs with application in metal extraction in a human breast cancer cell (MCF7). The ILs were based on the combination of different cations (imidazolium, piperidinium, pyrrolidinium, orpyridinium, with different sizes of alkyl chains) and anions, such as bromide and bis(trifluoromethanesulfone)imide, among others. It was observed that toxicity significantly depends on the cations and anions of the IL structure, especially for the long chain alkyl cations. The task-specific ILs evaluated in this study were less toxic than classic ones.

Knudsen et al. (2009) characterized the effects of dose and route of administration on the disposition of 1-butyl-1-methylpyrrolidinium chloride (BmPy-Cl) in male rats. In summary, it was found that BmPy-Cl was moderately absorbed, excreted by the kidneys, and eliminated in the urine as a source compound independent of the dose, number, or route of administration, indicating that it may function as a substrate and/or inhibitor of the human organic cation transporter.

Kaushik et al. (2012) synthesized ILs containing ammonium and imidazolium cations and evaluated them in normal cells and brain cancer cells. For this, they carried out the MTT assay. The results demonstrated a potent inhibitory effect of tumour cells (T98G), but low toxicity to normal human embryonic kidney cells (HEK).

Reid et al. (2015) evaluated the mutagenicity of 16 PILs with secondary and tertiary ammonium cations, as well as chloride and carboxylate anions, by the Ames test, which detects mutations in *Salmonella typhimurium* strains, such as TA98 and TA100. Despite being an in vitro test, the correlation of the results is not absolute enough to assert that a sample is mutagenic and later carcinogenic in mammals; several compounds that are mutagenic accordingly to the Ames assay are also carcinogenic in mammals. Of the 16 PILs tested, 15 were negative for the test and were not mutagenic or carcinogenic. The only PIL that could not assess mutagenic potential was N,N-dimethylethanolammonium octanoate ([DMEtA] [Oct]) as it was toxic to the microorganisms tested.

Larangeira et al. (2016) evaluated the cytotoxicity, mutagenicity, and genotoxicity of carotenoids obtained from tomatoes by t-1-butyl-3-methylimidazolium chloride using an in vivo experimental model. The following four groups were evaluated: the control group (untreated), the group treated with 10 mg of carotenoid extracted by an IL, another group treated with 500 mg of these carotenoids, 10 times the recommended dose for humans, and the group treated only with the IL. The livers of the rats treated with IL showed moderate histopathological changes. DNA damage was verified in liver and blood cells of groups that received 500 mg of carotenoid and with the IL. An increase in micronucleated cells and 8-hydroxy-2′-deoxyguanosine (8-OhDG) immunopositive cells was identified in rats treated with 500 mg of carotenoids, indicating levels of oxidative DNA damage. In summary, the results demonstrated that the recommended human dose of carotenoids extracted by the IL (10 mg) was not able to promote cytotoxicity, genotoxicity, or mutagenicity in some rat organs.

Iqbal et al. (2017) used 1-methylimidazolium acetate ([Mim] Ac) for the preparation of collagen and alginate hydrogels for wound application. Antibacterial activity assays were performed by the disc diffusion method against *S. mutans* bacteria and also by cytotoxicity tests using rat mesenchymal stem cells (rMSC) by the MTT method. The study showed that there were zones of inhibition of *S. mutans* with drug loaded samples. In the cell viability assay, no change in the proliferation capacity of rMSC cells was observed, thus indicating that the hydrogels prepared with the IL were non-toxic.

In another study by Hwang et al. (2018), seven ILs (1-ethyl-3-methylimidazolium bis(trifluoromethanesulfonyl)imide—[EMIM] [TFSI]; 1-ethyl-3-methylimidazolium hexafluorophosphate—[EMIM] [PF_6]; 1-ethyl-3-methylimidazolium tetrafluoroborate—[EMIM] [BF_4]; 1-ethyl-3-methylimidazolium dicyanamide—[EMIM] [DCA]; 1-butyl-1-methylpyrrolidinium bis(trifluoromethanesulfonyl) imide—[BMPY] [TFSI]; tributylmethylammonium bis(trifluoromethanesulfonyl) imide—[TBA] [TFSI]; and zinc di[bis(trifluoromethanesulfonyl)imide—[Zn] [TFSI]$_2$) were evaluated regarding their cytotoxicity in keratinocyte and fibroblast cell lines (HaCat and Hs68, respectively) by the MTT assay. Overall, ILs composed of [TFSI] showed higher toxicity than the rest, and significant cytotoxicity was found in [EMIM] [TFSI] and [BMPY] [TFSI], similar to the xylene solvent toxicity used to determine a comparative effect.

Some recent data from the literature, as already noted in this paper, have shown that ILs may have toxicity. Alternatively, some biocomposites from natural sources (organic acids, choline and amino acids) are being studied considering the possibility of use them to the ILs synthesis, as well as substances with recognized properties beneficial to human health, such as vitamin C (Niemczak et al. 2019) and ibuprofen (Wu et al. 2019). This shows that there is evidence that low or non-toxic ILs can be synthesized and even used in food industries (Hijo et al. 2016). Therefore, some research groups are already dedicated to the development of low-toxic ILs, aimed at biological and pharmaceutical applications (Sardar et al. 2018), and, for this purpose, recognizing the factors that interfere with this effect is fundamental.

Literature analysis has shown that there is a small number of studies on ILs targeting human toxicity. In the few research studies found, it was observed that the toxicity of ILs is directly dependent on the nature of the cation or anion of the chemical structure. The toxicology of ILs is a considerably new area, and such versatile solvents that have a wide variety of properties and applications should be further studied for the elucidation of risks to human health as they have numerous advantages. This need for toxicological-based studies of ILs increases with advances in research for their biological applications, one example of which is their use as anticancer agents. This application was pioneered by Kumer et al. (2019) using computational chemistry and by Bansode et al. (2019), who detected effects on breast cancer cells (MCF-7), similar to doxorubicin.

1.4 Conclusion

Studies on the toxicity of ILs are relatively new and have wide applications in industrial processes in several areas since research indicates that they could be alternatives to conventional and innovative solvents and may be part of the development of new greener and more sustainable processes, reducing impacts on the environment.

An essential aspect of ILs that has not been extensively explored is their toxicological profile, both at the environmental and human levels. The studies presented in this chapter have shown a greater amount of research related to AILs, which generally have high toxicity compared to PILs due to their structural aspects and lipophilic character, which favours their accumulation in the environment and in organisms. PILs, on the other hand, because they have hydrophilic character, have greater ease of dispersion in the environment and, consequently, lower environmental impacts and less toxicity. However, due to the limited number of studies, this statement cannot be conclusive.

This chapter has discussed the toxicological aspects of ILs, and it has been found that much remains to be explored on this topic, especially when it comes to impacts on human health.

Acknowledgements The authors would like to acknowledge FAPESB—Fundação de Amparo à Pesquisa do Estado da Bahia (G. B. R. Veloso - Processo 6653/2016) and the National Council for Scientific and Technological Development (M. Iglesias - Chamada CNPq N° 29/2019, Processo 313601/2019-4 and R. S. Andrade - Chamada MCTIC/CNPq N° 28/2018, Processo 438376/2018-8) for its support in developing this research.

References

Adawiyah N, Moniruzzaman M, Hawatulaila S, Goto M (2016) Ionic liquids as a potential tool for drug delivery systems. Med Chem Commun 7:1881–1897. https://doi.org/10.1039/C6MD00358C

Álvarez VH, Iglesias M, Dosil N, Gonzalez-Cabaleiro R, Martin-Pastor M, Mattedi S, Navaza JM (2010a) Brønsted ionic liquids for sustainable processes: synthesis and physical properties. J Chem Eng Data 55:625–632. https://doi.org/10.1021/je900550v

Álvarez VH, Mattedi S, Martin-Pastor M, Aznar M, Iglesias M (2010b) Synthesis and thermophysical properties of two new protic long-chain ionic liquids with the oleate anion. Fluid Phase Equilibr 299:42–50. https://doi.org/10.1016/j.fluid.2010.08.022

Álvarez VH, Mattedi S, Aznar M, Iglesias M (2011) Thermophysical properties of binary mixtures of {ionic liquid 2-hydroxyethylammonium acetate + (water, methanol, or ethanol)}. J Chem Thermodyn 43:997–1010. https://doi.org/10.1590/0104-6632.20180352s20160280

Andrade RS, Torres D, Ribeiro FR, Chiari-Andréo BG, Oshiro JA Jr, Iglesias M (2017) Sustainable cotton dyeing in nonaqueous medium applying protic ionic liquids. ACS Sustain Chem Eng 5:8756–8765. https://doi.org/10.1021/acssuschemeng.7b01555

Arning J, Stolte S, Böschen A, Stock F, Pitner WR, Welz-Biermann U, Jastorff B, Ranke J (2008) Qualitative and quantitative structure activity relationships for the inhibitory effects of cationic head groups, functionalized side chains and anions of ionic liquids on acetylcholinesterase. Green Chem 10:47–58. https://doi.org/10.1039/b712109a

Azimova MA, Morton Iii SA, Frymier PD (2009) Comparison of three bacterial toxicity assays for imidazolium-derived ionic liquids. J Environ Eng 135:1388–1392. https://doi.org/10.1061/(asce)ee.1943-7870.0000092

Balk A, Holzgrabe U, Meinel L (2015) 'Pro et contra' ionic liquid drugs - Challenges and opportunities for pharmaceutical translation. Eur J Pharm Biopharm 94:291–304. https://doi.org/10.1016/j.ejpb.2015.05.027

Bansode P, Patil P, Choudhari P, Bhatia M, Birajdar A, Somasundaram I, Rashinkar G (2019) Anticancer activity and molecular docking studies of ferrocene tethered ionic liquids. J Mol Liq 290:111182–111193. https://doi.org/10.1016/j.molliq.2019.111182

Basu A, Bhattacharya CS, Kumar GS (2018) Influence of the ionic liquid 1-butyl-3-methylimidazolium bromide on amyloid fibrillogenesis in lysozyme: Evidence from photophysical and imaging studies. Int J Biol Macromol 107(Part B):2643–2649. https://doi.org/10.1016/j.ijbiomac.2017.10.152

Bernot RJ, Brueseke MA, Evans-White MA, Lamberti GA (2005a) Acute and chronic toxicity of imidazolium-based ionic liquids on Daphnia magna. Environ Toxicol Chem 24:87–92. https://doi.org/10.1897/03-635.1

Bernot RJ, Kennedy EE, Lamberti GH (2005b) Effects of lionic liquids on the survival, movement, and feeding behaviour of the freshwater snail Physa acuta. Environ Toxicol Chem 24:1759–1765. https://doi.org/10.1897/04-614r.1

Berthod A, Ruiz-Ángel MJ, Broch S (2018) Recent advances on ionic liquid uses in separation techniques. J Chromatogr A 1559:2–16. https://doi.org/10.1016/j.chroma.2017.09.044

Bertoti AR, Netto-Ferreira JC (2009) Ionic liquid [bmim.PF_6]: a convenient solvent for laser flash photolysis studies. Quim Nova 32:1934–1938. https://doi.org/10.1590/s0100-40422009000700040

Bhattacharjee A, Luís A, Santos JH, Lopes-Da-Silva JA, Coutinho JAP (2014) Thermophysical properties of sulfonium- and ammonium-based ionic liquids. Fluid Phase Equilibr 381:36–45. https://doi.org/10.1016/j.fluid.2014.08.005

Bianchini R, Cevasco G, Chiappe C, Pomelli CS, Douton MJR (2015) Ionic liquids can significantly improve textile dyeing: An innovative application assuring economic and environmental benefits. ACS Sustain Chem Eng 3:2303–2308. https://doi.org/10.1021/acssuschemeng.5b00578

Bisht M, Venkatesu P (2017) Influence of cholinium-based ionic liquids on the structural stability and activity of α-chymotrypsin. New J Chem 41:13902–13911. https://doi.org/10.1039/c7nj03023a

Borra EF, Seddiki O, Angel R, Eisenstein D, Hickson P, Seddon KR, Worden SP (2007) Deposition of metal films on an ionic liquid as a basis for a lunar telescope. Nature 447:979–981. https://doi.org/10.1038/nature05909

Bubalo MC, Radošević K, Redovniković IR, Slivac I, Srček VG (2017) Toxicity mechanisms of ionic liquids. Arh Hig Rada Toksiko 68:171–179. https://doi.org/10.1515/aiht-2017-68-2979

Calza P, Fabbri D, Noè G, Santoro V, Medana C (2018) Assessment of the photocatalytic transformation of pyridinium-based ionic liquids in water. J Hazard Mater 341:55–65. https://doi.org/10.1016/j.jhazmat.2017.07.037

Chang LY, Lee CP, Li CT, Ho KC, Lin JJ (2014) Synthesis of a novel amphiphilic polymeric ionic liquid and its application in quasi-solid-state dye-sensitized solar cells. J Mater Chem 2:20814–20822. https://doi.org/10.1039/c4ta05436a

Chemnitius JM, Sadowski R, Winkel H, Zech R (1999) Organophosphate inhibition of human heart muscle cholinesterase isoenzymes. Chem Biol Interact 120:183–192. https://doi.org/10.1016/s0009-2797(99)00027-7

Chen C, Ma Y, Zheng D, Zhang J, Zhang J (2018) Influence of different substitution in pyrazolium ionic liquids on catalytic activity for the fixation of CO_2 under solvent- and metal-free conditions. Tetrahedron 74:1776–1784. https://doi.org/10.1016/j.tet.2018.02.027

Cheng DH (2011) Application of ionic liquid in silk dyeing. Adv Mater Res 331:253–256. https://doi.org/10.4028/www.scientific.net/amr.331.253

Chiappe C, Pieraccini D (2005) Ionic liquids: solvent properties and organic reactivity. J Phys Org Chem 18:275–297. https://doi.org/10.1002/poc.863

Cho CW, Jeon YC, Pham TP, Vijayaraghavan K, Yun YS (2008) The ecotoxicity of ionic liquids and traditional organic solvents on microalga *Selenastrum capricornutum*. Ecotoxicol Environ Saf 71:166–171. https://doi.org/10.1016/j.ecoenv.2007.07.001

Clark KD, Nacham O, Purslow JA, Pierson SA, Anderson JL (2016) Magnetic ionic liquids in analytical chemistry: A review. Anal Chim Acta 934:9–21. https://doi.org/10.1016/j.aca.2016.06.011

Claus J, Sommer FO, Kragl U (2018) Ionic liquids in biotechnology and beyond. Solid State Ionics 314:119–128. https://doi.org/10.1016/j.ssi.2017.11.012

Collin RM, Collin RW (2010) Encyclopedia of Sustainability. ABC-CLIO, LLC, Santa Barbara, California

Costa SPF, Martins BSF, Pinto PCAG, Saraiva LMFS (2016) Automated cytochrome c oxidase bioassay developed for ionic liquids' toxicity assessment. J Hazard Mater 309:165–172. https://doi.org/10.1016/j.jhazmat.2016.02.005

Cota I, Medina F, Gonzalez-Olmos R, Iglesias M (2014) Alanine-supported protic ionic liquids as efficient catalysts for aldol condensation reactions. C R Chim 17(1):18–22. https://doi.org/10.1016/j.crci.2013.06.004

Couling DJ, Bernot RJ, Docherty KM, Dixon JK, Maginn EJ (2006) Assessing the factors responsible for ionic liquid toxicity to aquatic organisms via quantitative structure-property relationship modelling. Green Chem 8:82–90. https://doi.org/10.1039/b511333d

Cunha E, Pinto PCAG, Carbalho JPS, Saraiva MLMFS (2013) Automated carboxylesterase assay for the evaluation of ionic liquids' humantoxicity. J Hazard Mater 244-245:563–569. https://doi.org/10.1016/j.jhazmat.2012.10.045

Decaen P, Rolland-Sabaté A, Guilois S, Jury V, Leroy E (2018) Choline chloride vs choline ionic liquids for starch thermoplasticization. Carbohyd Polym 177:424–432. https://doi.org/10.1016/j.carbpol.2017.09.012

Deyab MA, Zaky MT, Nessim MI (2017) Inhibition of acid corrosion of carbon steel using four imidazolium tetrafluoroborates ionic liquids. J Mol Liq 229:396–404. https://doi.org/10.1016/j.molliq.2016.12.092

Docherty KM, Kulpa CF (2005) Toxicity and antimicrobial activity of imidazolium and pyridinium ionic liquids. Green Chem 7:185–189. https://doi.org/10.1039/b419172b

Docherty KM, Dixon JK, Kulpa CF Jr (2007) Biodegradability of imidazolium and pyridinium ionic liquids by an activated sludge microbial community. Biodegradation 18:481–493. https://doi.org/10.1007/s10532-006-9081-7

Dong ZQ (2011) Influence of ionic liquids on the sorption of acid dyes by nylon fibers. Adv Mater Res 175-176:602–607. https://doi.org/10.4028/www.scientific.net/amr.175-176.602

Dos Santos RF, Falcade T, Antonini LM, Ortega V, María R, Mattedi S, Iglesias M, Malfatti CF (2014) Ionic liquids (IL) corrosion on A285 carbon steel. DYNA 81:122. https://doi.org/10.15446/dyna.v81n184.39512

Du YY, Tian FL (2006) Protic acidic ionic liquids promoted formation of 1,5-benzodiazepines: remarkable effects of cations and anions on their performances. J Chem Res 8:486–489. https://doi.org/10.3184/030823406778256423

Egorova KS, Ananikov VP (2014) Toxicity of ionic liquids: Eco(cyto)activity as complicated, but unavoidable parameter for task-specific optimization. Chem Sus Chem 7:336–360. https://doi.org/10.1002/cssc.201300459

Egorova KS, Ananikov VP (2018a) Fundamental importance of ionic interactions in the liquid phase: A review of recent studies of ionic liquids in biomedical and pharmaceutical applications. J Mol Liq 272:271–300. https://doi.org/10.1016/j.molliq.2018.09.025

Egorova KS, Ananikov VP (2018b) Ionic liquids in whole-cell biocatalysis: a compromise between toxicity and efficiency. Biophys Rev 10:881–900. https://doi.org/10.1007/s12551-017-0389-9

Egorova KS, Gordeev EG, Ananikov VP (2017) Biological activity of ionic liquids and their application in pharmaceutics and medicine. Chem Rev 117:7132–7189. https://doi.org/10.1021/acs.chemrev.6b00562

Elgharbawy AA, Riyadi FA, Alam MZ, Moniruzzaman M (2018) Ionic liquids as a potential solvent for lipase-catalysed reactions: A review. J Mol Liq 251:150–166. https://doi.org/10.1016/j.molliq.2017.12.050

Elsheikh YA (2014) Optimization of novel pyrazolium ionic liquid catalysts for transesterification of bitter apple oil. Process Saf Environ 92:828–834. https://doi.org/10.1016/j.psep.2013.05.001

Fall A, Seck I, Diouf O, Gaye M, Fall Y (2015) Synthesis of new triazolium-based ionic liquids and their use in the Morita–Baylis–Hillman reaction. Tetrahedron Lett 56(36):5128–5131. https://doi.org/10.1016/j.tetlet.2015.07.026

Fan Y, Wang X, Li J, Zhang L, Yang L, Gao P, Zhou Z (2018) Kinetic study of the inhibition of ionic liquids on the trypsin activity. J Mol Liq 252:392–398. https://doi.org/10.1016/j.molliq.2018.01.014

Fazlali A, Shahebrahimi Y, Aliyari N, Mohammadi AH (2017) Oil products desulfurization by 1-butyl-3-methylimidazolium tetrachloroaluminate ionic liquid: Experimental study and thermodynamic modelling. J Mol Liq 237:437–446. https://doi.org/10.1016/j.molliq.2017.04.101

Ferraz R, Teixeira V, Rodrigues D, Fernandes R, Prudencio C, Noronha JP, Petrovski Z, Branco LC (2014) Antibacterial activity of ionic liquids based on ampicillin against resistant bacteria. T Roy Soc Chem 4:4301–4307. https://doi.org/10.1039/c3ra44286a

Forsyth M, Girard GMA, Basile A, Hilder M, Macfarlane DR, Chen F, Howlett PC (2016) Inorganic-Organic Ionic Liquid Electrolytes Enabling High Energy-Density Metal Electrodes for Energy Storage. Electrochim Acta 220:609–617. https://doi.org/10.1016/j.electacta.2016.10.134

Frade RFM, Rosatella AA, Marques CS, Branco LC, Kulkarni PS, Mateus NMM, Afonso CAM, Duarte CMM (2009) Toxicological evaluation on human colon carcinoma cell line (CaCo-2) of ionic liquids based on imidazolium, guanidinium, ammonium, phosphonium, pyridinium and pyrrolidinium cations. Green Chem 11:1660–1665. https://doi.org/10.1039/b914284n

Fulton MH, Key PB (2001) Acetylcholinesterase inhibition in estuarine fish and invertebrates as an indicator of organophosphorus insecticide exposure and effects. Environ Toxicol Chem 20:37–45. https://doi.org/10.1002/etc.5620200104

Ghanem OB, Mutalib MIA, Lévêque JM, El-Harbawi M (2017) Development of QSAR model to predict the ecotoxicity of *Vibrio fischeri* using COSMO-RS descriptors. Chemosphere 170:242–250. https://doi.org/10.1016/j.chemosphere.2016.12.003

Gilbert B, Olivier-Bourbigou H, Favre F (2007) Chloroaluminate Ionic Liquids: from their Structural Properties to their Applications in Process Intensification. Oil Gas Sci Technol 62:745–759. https://doi.org/10.2516/ogst:2007068

Greaves TL, Weerawardena A, Fong C, Krodkiewska I, Drummond CJ (2006) Protic ionic liquids: solvents with tunable phase behavior and physicochemical properties. J Phys Chem B 110:22479–22487. https://doi.org/10.1021/jp0634048

Grewal J, Khare SK (2018) One-pot bioprocess for lactic acid production from lignocellulosic agro-wastes by using ionic liquid stable *Lactobacillus brevis*. Bioresour Technol 251:268–273. https://doi.org/10.1016/j.biortech.2017.12.056

Hagiwara R, Lee JS (2007) Ionic liquids for electrochemical devices. Electrochemistry 75:23–34. https://doi.org/10.5796/electrochemistry.75.23

Handy S (2011) Applications of Ionic Liquids in Science and Technology. InTech, Rijeka. https://doi.org/10.5772/1769

He Z, Alexandridi P (2017) Ionic liquid and nanoparticle hybrid systems: Emerging applications. Adv Colloid Interfac 244:54–70. https://doi.org/10.1016/j.cis.2016.08.004

Hijo AACT, Maximo GJ, Costa MC, Batista EAC, Meirelles AJA (2016) Applications of ionic liquids in the food and bioproducts industries. ACS Sustain Chem Eng 4:5347–5369. https://doi.org/10.1021/acssuschemeng.6b00560

Hwang J, Park H, Choi DW, Nam KT, Lim KM (2018) Investigation of dermal toxicity of ionic liquids in monolayer-cultured skin cells and 3D reconstructed human skin models. Toxicol In Vitro 46:194–202. https://doi.org/10.1016/j.tiv.2017.09.025

Iqbal B, Muhammad N, Jamal A, Ahmad P, Haq Khan ZU, Rahim A, Khan AS, Gonfa G, Iqbal J, Rehman IU (2017) An application of ionic liquid for preparation of homogeneous collagen and alginate hydrogels for skin dressing. J Mol Liq 243:720–725. https://doi.org/10.1016/j.molliq.2017.08.101

Jastorff B, Mölter K, Behrend P, Bottin-Weber U, Filser J, Heimers A, Ondruschka B, Ranke J, Schaefer M, Schröder H, Stark A, Stepnowski P, Stock F, Störmann R, Stolte S, Welz-Biermann U, Ziegert S, Thöming J (2005) Progress in evaluation of risk potential of ionic liquids-basis for an eco-design of sustainable products. Green Chem 7:362–372. https://doi.org/10.1039/b418518h

Jordan A, Gathergood N (2015) Biodegradation of ionic liquids e a critical review. Chem Soc Rev 44:8200–8237. https://doi.org/10.1039/c5cs00444f

Kakaei K, Alidoust E, Ghadimi G (2018) Synthesis of N- doped graphene nanosheets and its composite with urea choline chloride ionic liquid as a novel electrode for supercapacitor. J Alloy Compd 735:1799–1806. https://doi.org/10.1016/j.jallcom.2017.11.310

Kantouch A, Khalil EM, El-Sayed H, Mowafi S (2011) A novel application of ionic liquid in improvement of the felting resistance of wool. Egypt J Chem 54:481–493. https://doi.org/10.21608/ejchem.2011.1406

Kaushik NK, Attri P, Kaushik N, Choi EH (2012) Synthesis and antiproliferative activity of ammonium and imidazolium ionic liquids against T98G brain cancer cells. Molecules 17:13727–13739. https://doi.org/10.3390/molecules171213727

Kishimura H, Kohki E, Nakada A, Tamatani K, Abe H (2018) Ether bond effects in quaternary ammonium and phosphonium ionic liquid-propanol solutions. Chem Phys 502:87–95. https://doi.org/10.1016/j.chemphys.2018.01.015

Knudsen GA, Cheng Y, Kuester RK, Hooth MJ, Sipes IG (2009) Effects of dose and route on the disposition and kinetics of 1-butyl-1-methylpyrrolidinium chloride in male F-344 rats. Drug Metab Dispos 37:2171–2177. https://doi.org/10.1124/dmd.109.029082

Kokorin A (2011) Ionic liquids: applications and perspectives. InTech, Rijeka. https://doi.org/10.5772/1782

Kordala-Markiewicz R, Rodak H, Markiewicz B, Walkiewicz F, Sznajdrowska A, Materna K, Marcinkowska K, Praczyk T, Pernak J (2014) Phenoxy herbicidal ammonium ionic liquids. Tetrahedron 70:4784–4789. https://doi.org/10.1016/j.tet.2014.05.041

Kumar S, Ruth W, Sprenger B, Kragl U (2006) On the biodegradation of ionic liquid 1- Butyl-3-methylimidazolium tetrafluoroborate. Chim Oggi 24:24–26. ISSN: 0392-839x

Kumar RA, Papaïconomou N, Lee JM, Salminen J, Clark DS, Prausnitz JM (2009) *In vitro* cytotoxicities of ionic liquids: effect of cation rings, functional groups, and anions. Environ Toxicol 24:388–395. https://doi.org/10.1002/tox.20443

Kumer A, Sarker M, Paul S (2019) The simulating study of HOMO, LUMO, thermo physical and quantitative structure of activity relationship (QSAR) of some anticancer active ionic liquids. Eur J Environ Res 3:1–10. ISSN: 2602-2990

Kunov-Kruse AJ, Weber CC, Rogers RD, Myerson AS (2017) The a priori design and selection of ionic liquids as solvents for active pharmaceutical ingredients. Chem Eur J 23:5498–5508. https://doi.org/10.1002/chem.201605704

Larangeira PM, Rosso VV, Silva VHP, Moura CFG, Ribeiro DA (2016) Genotoxicity, mutagenicity and cytotoxicity of carotenoids extracted from ionic liquid in multiples organs of Wistar rats. Exp Toxicol Pathol 68:571–578. https://doi.org/10.1016/j.etp.2016.09.003

Larriba M, Navarro P, Delgado-Mellado N, Stanisci V, García J, Rodríguez F (2016) Separation of aromatics from *n*-alkanes using tricyanomethanide-based ionic liquids: Liquid-liquid extraction, vapor-liquid separation, and thermophysical characterization. J Mol Liq 223:880–889. https://doi.org/10.1016/j.molliq.2016.09.017

Latala A, Stepnowski P, Nedzi M, Mrozik W (2005) Marine toxicity assessment of imidazolium ionic liquids: acute effects on the Baltic algae *Oocystis submarina* and *Cyclotella meneghiniana*. Aquat Toxicol 73:91–98. https://doi.org/10.1016/j.aquatox.2005.03.008

Le Bideau J, Viau L, Vioux A (2011) Ionogels, ionic liquid based hybrid materials. Chem Soc Rev 40:907–925. https://doi.org/10.1039/c0cs00059k

Lei H, Tu J, Yu Z, Jiao S (2017) Exfoliation mechanism of graphite cathode in ionic liquids. ACS Appl Mater Interfaces 9:36702–36707. https://doi.org/10.1021/acsami.7b03306

Lenardão EJ, Freitag RA, Dabdoub MJ, Batista ACF, Silveira CC (2003) "Green chemistry" – Os 12 princípios da química verde e sua inserção nas atividades de ensino e pesquisa. Quim Nova 26:123–129. https://doi.org/10.1590/s0100-40422003000100020

Li L, Bu C, Zhang Y, Du J, Lu X, Liu X (2011) Composite system based on biomolecules-functionalized multiwalled carbon nanotube and ionic liquid: electrochemistry and electrocatalysis of tryptophane. Electrochim Acta 58:105–111. https://doi.org/10.1016/j.electacta.2011.08.097

Li W, Sun D, Zhang T, Dai S, Pan F, Zhang Z (2014) Separation of acetone and methanol azeotropic system using ionic liquid as entrainer. Fluid Phase Equilibr 383:182–187. https://doi.org/10.1016/j.fluid.2014.10.011

Li K, Choudhary H, Rogers RD (2018) Ionic liquids for sustainable processes: Liquid metal catalysis. Curr Opin Green Sustain Chem 11:15–21. https://doi.org/10.1016/j.cogsc.2018.02.011

Liu L, Yang P, Li P, Cui MY (2012) Application of bis(trifluoromethanesulfonyl)imide lithium-N-methyl-Nbutylpiperidinium-bis(trifluoromethanesulfonyl)imide-poly(vinylidene difluoride-co-hexafluoropropylene) ionic liquid gel polymer electrolytes in Li/$LiFePO_4$ batteries at different temperatures. Electrochim Acta 85:49–56. https://doi.org/10.1016/j.electacta.2012.08.066

Lopes JM, Bermejo MD, Martín Á, Cocero MJ (2017) Ionic liquid as reaction media for the production of cellulose-derived polymers from cellulosic biomas. Chem Eng 1:1–28. https://doi.org/10.3390/chemengineering1020010

Luis P, Ortiz I, Aldaco R, Irabien A (2007) A novel group contribution method in the development of a QSAR for predicting the toxicity (*Vibrio fischeri* EC_{50}) of ionic liquids. Ecotoxicol Environ Saf 67:423–429. https://doi.org/10.1016/j.ecoenv.2006.06.010

Markusson H, Belieres JP, Johansson P, Angell CA, Jacobsson P (2007) Prediction of macroscopic properties of protic ionic liquids by ab initio calculations. J Phys Chem A 111:8717–8723. https://doi.org/10.1021/jp072036k

Martinelli A, Matic A, Jacobsson P, Borjesson L, Fernicola A, Panero S, Scrosati B, Ohno H (2007) Physical properties of proton conducting membranes based on a protic ionic liquid. J Phys Chem B 111:12462–12467. https://doi.org/10.1021/jp0735029

Massoulié J, Pezzementi L, Bon S, Krejci E, Vallette FM (1993) Molecular and cellular biology of cholinesterases. Prog Neurobiol 41:31–91. https://doi.org/10.1016/0301-0082(93)90040-Y

Matsumoto M, Mochiduki K, Fukunishi K, Kondo K (2004a) Extraction of organic acids using imidazolium-based ionic liquids and their toxicity to *Lactobacillus rhamnosus*. Sep Purif Technol 40:97–101. https://doi.org/10.1016/j.seppur.2004.01.009

Matsumoto M, Mochiduki K, Kondo K (2004b) Toxicity of ionic liquids and organic solvents to lactic acid-producing bacteria. J Biosci Bioeng 98:344–347. https://doi.org/10.1016/s1389-1723(04)00293-2

Matsuo AYO, Lamberti GA (2006) Understanding the mechanism of ionic liquid toxicity to fish and *Daphnia*. Abstr Pap Am Chem S 231:176

Matzke M, Stolte S, Thiele K, Juffernholz T, Arning J, Ranke J, Welz-Biermann U, Jastorff B (2007) The influence of anion species on the toxicity of 1-alkyl-3- methylimidazolium ionic liquids observed in an (eco)toxicological test battery. Green Chem 9:1198–1207. https://doi.org/10.1039/b705795d

Matzke M, Arning J, Johannes R, Jastorff B, Stolte S (2010) Design of inherently safer ionic liquids: toxicology and biodegradation. In: Wasserscheid P, Stark A (eds) Handbook of green chemistry, Ionic liquids, vol 6. Wiley-VCH Verlag GmbH & Co., Weinheim, pp 235–298

Mayoral MJ, Ovejero P, Campo JA, Heras JV, Cano M (2009) Ionic liquid crystals from β-diketonyl containing pyridinium cations and tetrachlorozincate anions. Inorg Chem Commun 12:214–218. https://doi.org/10.1016/j.inoche.2008.12.016

Meenatchi B, Renuga V, Manikandan A (2017) Cellulose dissolution and regeneration using various imidazolium based protic ionic liquids. J Mol Liq 238:582–588. https://doi.org/10.1016/j.molliq.2016.05.008

Meksi N, Moussa A (2017) A review of progress in the ecological application of ionic liquids in textile processes. J Clean Prod 161:105–126. https://doi.org/10.1016/j.jclepro.2017.05.066

Muzalevskiy VM, Shastin AV, Shikhaliev NG, Magerramov AM, Teymurova AN, Nenajdenko VG (2016) Ionic liquids as a reusable media for copper catalysis. Green access to alkenes using catalytic olefination reaction. Tetrahedron 72:7159–7163. https://doi.org/10.1016/j.tet.2016.09.050

Nawata J, Dawidziuk B, Dziedzic D, Gordon D, Popiel S (2018) Applications of ionic liquids in analytical chemistry with a particular emphasis on their use in solid-phase microextraction. Trac-Trend Anal Chem 105:18–36. https://doi.org/10.1016/j.trac.2018.04.010

Neale AR, Murphy S, Goodrich P, Schütter C, Hardacre C, Passerini S, Balducci A, Jacquemin J (2016) An ether-functionalised cyclic sulfonium based ionic liquid as an electrolyte for electrochemical double layer capacitors. *J.* Power Sour 326:549–559. https://doi.org/10.1016/j.jpowsour.2016.06.085

Nehra P, Khungar B, Singh RP, Sivasubramanian SC, Jha PN, Saini V (2018) Synthesis, characterization and applications of imidazolium ionic liquid-tagged zinc(II) complex. Inorg Chim Acta 478:260–267. https://doi.org/10.1016/j.ica.2018.04.001

Niedermeyer H, Hallett JP, Villar-Garcia I, Hunt PA, Welton T (2012) Mixtures of ionic liquids. Chem Soc Rev 41:7780–7802. https://doi.org/10.1039/c2cs35177c

Niemczak M, Kaczmarek DK, Klejdysz T, Gwiazdowska D, Marchwińska K, Pernak J (2019) Ionic liquids derived from vitamin C as multifunctional active ingredients for sustainable stored-product management. ACS Sustain Chem Eng 7:1072–1084. https://doi.org/10.1021/acssuschemeng.8b04696

Noda A, Susan AB, Kudo K, Mitsushima S, Hayamizu K, Watanabe M (2003) Bronsted acid-base ionic liquids as proton-conducting nonaqueous electrolytes. J Phys Chem B 107:4024–4033. https://doi.org/10.1021/jp022347p

Ochedzan-Siodkak W, Dziubek K, Siodkak D (2008) Biphasic ethylene polymerisation using 1-n-alkyl-3-methylimidazolium tetrachloroaluminate ionic liquid as a medium of the Cp_2TiCl_2 titanocene catalyst. Eur Polym J 44:3608–3614. https://doi.org/10.1016/j.eurpolymj.2008.08.037

Oliveira L, Jose NM, Boaventura J, Iglesias M, Mattedi S (2011) Proton conducting polymer membrane using the ionic liquid 2-hydroxyethylammonium lactate for ethanol fuel cells. AIP Conf Proc 1400:149–153. https://doi.org/10.1063/1.3663103

Oliveira MVS, Vidal BT, Melo CM, Miranda RCM, Soares CMF, Coutinho JAP, Ventura SPM, Mattedi S, Lima AS (2016) (Eco)toxicity and biodegradability of protic ionic liquids. Chemophere 147:460–466. https://doi.org/10.1016/j.chemosphere.2015.11.016

Opwis K, Benken R, Knittel D, Gutmann JS (2017) Dyeing of PET fibers in ionic liquids. Int J New Technol Res 3:101–108. ISSN: 2454-4116

Peric B, Martí E, Sierra J, Cruañas R, Iglesias M, Garau MA (2011) Terrestrial ecotoxicity of short aliphatic protic ionic liquids. Environ Toxicol Chem 30:2802–2809. https://doi.org/10.1002/etc.683

Peric B, Sierra J, Marti E, Cruañas R, Garau MA, Arning J, Bottin-Weber U, Stolte S (2013) (Eco)toxicity and biodegradability of selected protic and aproticionic liquids. J Hazard Mater 261:99–105. https://doi.org/10.1016/j.jhazmat.2013.06.070

Peric B, Sierra J, Martí E, Cruañas R, Garau MA (2014) A comparative study of the terrestrial ecotoxicity of selected protic and aprotic ionic liquids. Chemosphere 108:418–425. https://doi.org/10.1016/j.chemosphere.2014.02.043

Peric B, Sierra J, Martí E, Cruañas R, Garau MA (2015) Quantitative structure–activity relationship (QSAR) prediction of (eco)toxicity of short aliphatic protic ionic liquids. Ecotox Environ Safe 115:257–262. https://doi.org/10.1016/j.ecoenv.2015.02.027

Petkovic M, Seddon KR, Rebelo LP, Silva Pereira C (2011) Ionic liquids: a pathway to environmental acceptability. Chem Soc Rev 40:1383–1403. https://doi.org/10.1039/c004968a

Pin TC, Nakasu PYS, Mattedi S, Rabelo SC, Costa AC (2019) Screening of protic ionic liquids for sugarcane bagasse pretreatment. Fuel 235:1506–1514. https://doi.org/10.1016/j.fuel.2018.08.122

Pope C, Karanth S, Liu J (2005) Pharmacology and toxicology of cholinesterase inhibitors: uses and misuses of a common mechanism of action. Environ Toxicol Pharmacol 19:433–446. https://doi.org/10.1016/j.etap.2004.12.048

Pretti C, Chiappe C, Baldetti I, Brunini S, Monni G, Intorre L (2009) Acute toxicity of ionic liquids for three freshwater organisms: *Pseudokirchneriella subcapitata*, *Daphnia magna* and *Danio rerio*. Ecotox Environ Saf 72:1170–1176. https://doi.org/10.1016/j.ecoenv.2008.09.010

Qingdao University (2012) Dyeing method of reactive dye containing ionic liquid. CN Patent 102493222 A

Ranke J, Mölter K, Stock F, Bottin-Weber U, Poczobutt J, Hoffmann J, Ondruschka B, Filser J, Jastorff B (2004) Biological effects of imidazolium ionic liquids with varying chain lengths in acute *Vibrio Fischeri* and WST-1 cell viability assays. Ecotoxicol Environ Saf 58:396–404. https://doi.org/10.1016/s0147-6513(03)00105-2

Ranke J, Stolte S, Stormann R, Arning J, Jastorff B (2007) Design of sustainable chemical products – the example of ionic liquids. Chem Rev 107:2183–2206. https://doi.org/10.1021/cr050942s

Reach R (2006) (Ce) N° 1907/2006 Del Parlamento Europeo Y Del Consejo de 18 de diciembre de 2006 relativo al registro, la evaluación, la autorización y la restricción de las sustancias y preparados químicos (REACH), Diario Oficial de la Unión Europea L 396/1

Reid JESJ, Sullivan N, Swift L, Hernbury GA, Shimizu S, Walker AJ (2015) Assessing the mutagenicity of protic ionic liquids using the mini Ames test. Sustain Chem Process 3:1–11. https://doi.org/10.1186/s40508-015-0044-3

Reid JESJ, Prydderch H, Spulak M, Shimizu S, Gathergood N (2018) Green profiling of aprotic versus protic ionic liquids: Synthesis and microbial toxicity of analogous structures. Sustain Chem Pharm 7:17–26. https://doi.org/10.1016/j.scp.2017.11.001

Riduan SN, Zhang Y (2013) Imidazolium salts and their polymeric materials for biological applications. Chem Soc Rev 42:9055–9070. https://doi.org/10.1039/c3cs60169b

Rodrigues RDP, De Lima PF, De Santiago-Aguiar RS, Rocha MVP (2019) Evaluation of protic ionic liquids as potential solvents for the heating extraction of phycobiliproteins from *Spirulina (Arthrospira) platensis*. Algal Res 38:101391. https://doi.org/10.1016/j.algal.2018.101391

Rogalsky S, Bardeau JF, Makhno S, Babkina N, Brovko O (2018) New proton conducting membrane based on bacterial cellulose/polyaniline nanocomposite film impregnated with guanidinium-based ionic liquid. Polymer 142:183–195. https://doi.org/10.1016/j.polymer.2018.03.032

Rouette HK (2001) Encyclopedia of Textile Finishing. Springer, Berlin

Rykowska I, Ziemblińska J, Nowak I (2018) Modern approaches in dispersive liquid-liquid micro-extraction (DLLME) based on ionic liquids: A review. J Mol Liq 259:319–339. https://doi.org/10.1016/j.molliq.2018.03.043

Salminen J, Papaiconomou N, Kumara RA, Lee JM, Kerr J, Newman J, Prausnitz JM (2007) Physicochemical properties and toxicities of hydrophobic piperidinium and pyrrolidinium ionic liquids. Fluid Phase Equilibri 261:421–426. https://doi.org/10.1016/j.fluid.2007.06.031

Sardar S, Wilfred CD, Mumtaz A, Leveque J, Khan AS, Krishnan S (2018) Physicochemical properties, Brönsted acidity and ecotoxicity of imidazolium-based organic salts: Non-toxic variants of protic ionic liquids. J Mol Liq 269:178–186. https://doi.org/10.1016/j.molliq.2018.08.017

Schwarzenbach RP, Gschwend PM, Imboden DM (2016) The amazing world of anthropogenic organic chemicals. In: Environmental Organic Chemistry, vol 3, 3rd edn. Wiley-Interscience, cap, New Jersey, pp 52–45-79

Sen S, Losey BP, Gordon EE, Argyropoulos DS, Martin JD (2016) Ionic liquid character of zinc chloride hydrates define solvent characteristics that afford the solubility of cellulose. J Phys Chem B 120(6):1134–1141. https://doi.org/10.1021/acs.jpcb.5b11400

Serra JS, Rosario TC, Costa KS, Andrade R, Iglesias M (2016) Synthesis of triacetin by protic ionic liquids as catalysts. World Wide J Multidis Res Develop 2:7–13

Shang D, Liu X, Bai L, Zeng S, Xu Q, Gao H, Zhang X (2017) Ionic liquids in gas separation processing. Curr Opin Green Sustain Chem 5:74–81. https://doi.org/10.1016/j.cogsc.2017.03.015

Sharma A, Ghorai PK (2016) Effect of water on structure and dynamics of [BMIM][PF_6] ionic liquid: An all-atom molecular dynamics simulation investigation. J Chem Phys 144:114505. https://doi.org/10.1063/1.4944083

Silva PS, Junior CTS, Araújo EG, Kanaan S, Xavier AR (2016) Adenosina desaminase: uma enzima extraordinária e onipresente. Pulmão RJ 25(1):11–16

Skladanowski AC, Stepnowski P, Kleszczynski K, Dmochowska B (2005) AMP deaminase in vitro inhibition by xenobiotics. A potential molecular method for risk assessment of synthetic nitro- and polycyclic musks, imidazolium ionic liquids and N-glucopyranosyl ammonium salts. Environ Toxicol Phar 19:291–296. https://doi.org/10.1016/j.etap.2004.08.005

Stasiewicz M, Mulkiewicz E, Tomczak-Wandzel R, Kurmirska J, Siedlecka EM, Golebiowski M, Gajdus J, Czerwicka M, Stepnowski P (2008) Assessing toxicity and biodegradation of novel, environmentally benign ionic liquids (1-alkoxymethyl-3-hydroxypyridinium chloride, saccharine and acesulfamates) on cellular and molecular level. Ecotox Environ Saf 71:157–165. https://doi.org/10.1016/j.ecoenv.2007.08.011

Stepnowski P, Skladanowski AC, Ludwiczak A, Laczynska E (2004) Evaluating the cytocoxicity of ionic liquids using human cell line HeLa. Hum Exp Toxicol 23:513–517. https://doi.org/10.1191/0960327104ht480oa

Stock F, Hoffmann J, Ranke J, Stormann R, Ondruschka B, Jastorff B (2004) Effects of ionic liquids on the acetylcholinesterase–a structure-activity relationship. Green Chem 6:286–290. https://doi.org/10.1039/B402348J

Stolte S, Matzke M, Arning J, Boschen A, Pitner WR, Welz-Biermann U, Jastorff B, Ranke J (2007) Effects of different head groups and functionalised side chains on the aquatic toxicity of ionic liquids. Green Chem 9:1170–1179. https://doi.org/10.1039/b711119c

Tang B, Gondosiswanto R, Hibbert DB, Zhao C (2019) Critical assessment of superbase-derived protic ionic liquids as electrolytes for electrochemical applications. Electrochim Acta 298:413–420. https://doi.org/10.1016/j.electacta.2018.12.100

Tankov I, Yankova R, Genieva S, Mitkova M, Stratiev D (2017) Density functional theory study on the ionic liquid pyridinium hydrogen sulfate. J Mol Struct 1139:400–406. https://doi.org/10.1016/j.molstruc.2017.03.040

Thriel CV (2014) Toxicology of Solvents (Including Alcohol). Ref Mod Biomed Sci. https://doi.org/10.1016/B978-0-12-801238-3.00210-5

Tian T, Hu X, Guan P, Wang S, Ding X (2017) Density and thermodynamic performance of energetic ionic liquids based on 1-alkyl/esteryl-4-amino-1,2,4-triazolium. J Mol Liq 248:70–80. https://doi.org/10.1016/j.molliq.2017.09.024

Tigelaar DA, Waldecker JR, Peplowski KM, Kinder JD (2006) Study of the incorporation of protic ionic liquids into hydrophilic and hydrophobic rigid-rod elastomeric polymers. Polymer 47:4269–4275. https://doi.org/10.1016/j.polymer.2006.04.014

Torrecilla JS, García J, Rojo E, Rodríguez R (2009) Estimation of toxicity of ionic liquids in leukemia rat cell line and acetylcholinesterase enzyme by principal component analysis, neural networks and multiple linear regressions. J Hazard Mater 164:182–194. https://doi.org/10.1016/j.jhazmat.2008.08.022

Vekariya RL (2017) A review of ionic liquids: Applications towards catalytic organic transformations. J Mol Liq 227:44–60. https://doi.org/10.1016/j.molliq.2016.11.123

Vijayaraghavan R, Oncsik T, Mitschke B, Macfarlane DR (2018) Base-rich diamino protic ionic liquid mixtures for enhanced CO_2 capture. Sep Purif Technol 196:27–31. https://doi.org/10.1016/j.seppur.2017.06.057

Walden P (1914) Molecular weights and electrical conductivity of several fused salts. Bull Acad Imper Sci 8:405–422

Wan R, Xia X, Wang P, Huo W, Dong H, Chang Z (2018) Toxicity of imidazoles ionic liquid [C_{16}mim]Cl to HepG2 cells. Toxicol In Vitro 52:1–7. https://doi.org/10.1016/j.tiv.2018.05.013

Welton T (1999) Room-temperature ionic liquids. Solvents for synthesis and catalysis. Chem Rev 99:2071–2084. https://doi.org/10.1021/cr980032t

Wu Q, Chen H, Han MH, Wang DZ, Wang JF (2007) Transesterification of cottonseed oil catalyzed by bronsted acidic ionic liquids. Ind Eng Chem Res 46:7955–7960. https://doi.org/10.1021/ie070678o

Wu H, Deng Z, Zhou B, Qi M, Hong M, Ren G (2019) Improved transdermal permeability of ibuprofen by ionic liquid technology: Correlation between counterion structure and the physicochemical and biological properties. J Mol Liq 283:399–409. https://doi.org/10.1016/j.molliq.2019.03.046

Xia X, Wan R, Wang P, Huo W, Dong H, Du Q (2018) Toxicity of imidazoles ionic liquid [C_{16}mim]Cl to Hela cells. Ecotox Environ Safe 162:408–414. https://doi.org/10.1016/j.ecoenv.2018.07.022

Xiong C, Li Y, Wang S, Zhou Y (2018) Functionalization of nanosilica via guanidinium ionic liquid for the recovery of gold ions from aqueous solutions. J Mol Liq 256:183–190. https://doi.org/10.1016/j.molliq.2018.02.036

Yu M, Li SM, Li XY, Zhang BJ, Wang JJ (2008) Acute effects of 1-octyl-3-methylimidazolium bromide ionic liquid on the antioxidant enzyme system of mouse liver. Ecotox Environ Saf 71:903–908. https://doi.org/10.1016/j.ecoenv.2008.02.022

Yuan J, Wang Q, Fan X, Wang P (2010a) Enhancing dye adsorption of wool fibers with 1-butyl-3-methylimidazolium chloride ionic liquid processing. Text Res J 80:1898–1904. https://doi.org/10.1177/0040517510371865

Yuan J, Wang Q, Fan X (2010b) Dyeing behaviors of ionic liquid treated wool. J Appl Polym Sci 117:2278–2283. https://doi.org/10.1002/app.32020

Zanoni BV, Romão GB, Andrade RS, Cicarelli RMBC, Trovatti E, Chiari-Andrèo BG, Iglesias M (2019) Cytotoxic effect of protic ionic liquids in HepG2 and HaCat human cells: *In vitro* and *In silico* studies. Toxicol Res 8:447–458. https://doi.org/10.1039/c8tx00338f

Zec N, Vranes M, Bester-Rogac M, Trtić-Petrović T, Gadzurić S (2018) Influence of the alkyl chain length on densities and volumetric properties of 1,3-dialkylimidazolium bromide ionic liquids and their aqueous solutions. J Chem Thermodyn 121:72–78. https://doi.org/10.1016/j.jct.2018.02.001

Zhang C, Malhotra SV (2005) Increased paraoxon detection by acetylcholinesterase inactivation with ionic liquid additives. Talanta 67:560–563. https://doi.org/10.1016/j.talanta.2005.03.037

Zhang S, Zhang Q, Zhang Y, Chen Z, Watanabe M, Deng Y (2016) Beyond solvents and electrolytes: Ionic liquids-based advanced functional materials. Prog Mater Sci 77:80–124. https://doi.org/10.1016/j.pmatsci.2015.10.001

Zhang X, Pan L, Wang L, Zou J (2018) Review on synthesis and properties of high-energy-density liquid fuels: Hydrocarbons, nanofluids and energetic ionic liquids. Chem Eng Sci 180:95–125. https://doi.org/10.1016/j.ces.2017.11.044

Zhao DB, Liao YC, Zhang ZD (2007) Toxicity of ionic liquids. Clean Soil Air Water 35:42–48. https://doi.org/10.1002/clen.200600015

Zheng D, Dong L, Huang W, Wu X, Nie N (2014) A review of imidazolium ionic liquids research and development towards working pair of absorption cycle. Renew Sust Energ Rev 37:47–68. https://doi.org/10.1016/j.rser.2014.04.046

Zhou S, Zhu G, Kang X, Li Q, Xu X (2018) Molecular dynamics simulation of the ionic liquid N-octylpyridinium tetrafluoroborate and acetonitrile: Thermodynamic and structural properties. Chem Phys Lett 701:1–6. https://doi.org/10.1016/j.cplett.2018.04.028

Zhuang L, Zheng C, Sun J, Yuan A, Wang G (2014) Performances of ramie fiber pretreated with dicationic imidazolium ionic liquid. Fiber Polym 15:226–233. https://doi.org/10.1007/s12221-014-0226-8

Chapter 2
Ionic Liquids: Green Solvent for Biomass Pretreatment

Uroosa Ejaz and Muhammad Sohail

Contents

Abbreviations

ILs Ionic liquids
LC Lignocellulosic
SB Sugarcane bagasse

U. Ejaz · M. Sohail (✉)
Department of Microbiology, University of Karachi, Karachi, Sindh, Pakistan
e-mail: msohail@uok.edu.pk

Inamuddin, A. M. Asiri (eds.), *Nanotechnology-Based Industrial Applications of Ionic Liquids*, Nanotechnology in the Life Sciences,
https://doi.org/10.1007/978-3-030-44995-7_2

2.1 Introduction

With an increasing environmental pollution caused by energy and chemical industries, there is a social expectation that researchers will provide a sustainable solution that will aim to meet global demands without affecting the ecosystem. Although biomass has been considered as everlasting source of chemicals and energy, however, its valorization without the extensive use of toxic acidic or alkaline solutions and/ or high energy demanding processes is yet to be achieved. In this context, ionic liquids (ILs) have been recognized as a promising solvent with recyclability, non-inflammability, high chemical, and thermal stability (Muhammad et al. 2011). ILs are normally defined as compounds which have low melting point ions (Reddy 2015). Based on cation segments, ILs are usually categorized into four types: (1) phosphonium, (2) dialkylimidazolium, (3) N-alkylpyridinium, and (4) alkyl ammonium-based ionic liquids (Ghandi 2014).

2.2 History

Presumptively, the first ionic liquid (IL) was discovered (with 12 °C melting point) as a molten salt (Welton 2018). Later, Hirao et al. (2000) rediscovered this protic IL as an important subclass of ILs. However, the potential of ILs remained unnoticed and nearly 40 years later, Hurley and Wier (1951) stated the advantages of using lower melting points molten salts. In their work, to make metals electroplated from solution, 1-alkylpyridinium halides were mixed with metal. In this context, Robinson and Osteryoung (1979) used 1-butylpyridinium chloride-aluminum chloride to study the solute-electrochemistry.

Lately, the interest in ILs was renewed and researchers in various fields attempted to explore multidimensional applications of ILs. The discovery of water stable 1-ethyl-3-methylimidazolium-based ILs become a great development in this regard (Cooper and Angell 1983; Wilkes and Zaworotko 1992).

2.3 Application of Ionic Liquids

2.3.1 Material Synthesis

For material synthesis, the use of ionic liquids (ILs) is getting much interest (Torimoto et al. 2010). A biphasic system for separation was made by combining ILs with carbon dioxide (Jutz et al. 2011). This research led the researchers to work on the gas solubility in ILs and eventually its use for carbon capture purpose which could be used for catalysis and synthesis of value-added compounds from carbon dioxide (Zeng et al. 2017).

2.3.2 Lubricant

The physical nature of ionic liquids (ILs) is its main feature to be used in lubricants. Few of the ILs were reportedly found very promising lubricant that can replace oil (Ye et al. 2001). Non-volatile ILs can be used under vacuum while conventional lubricant oils are volatile (Liu et al. 2002). ILs could be used as an additive for conventional lubricants as well (Phillips and Zabinski 2004). For being used as an additive, solubility is not prerequisite and therefore it can be used along with non-polar lubricants. It can also be used in miniaturized devices to make a thin film (Yu et al. 2006).

2.3.3 Separation and Extraction of Materials

Ionic liquids (ILs) have also been used in extractions and separations for compounds (Ventura et al. 2017). In gas-liquid chromatography, it has been used as a stationary phase (Pacholec et al. 1982). In solid-phase microextraction, IL is getting much interest for its commercial use (Ho et al. 2011). The main advantage of using ILs in separation technique is its non-volatility. Hence, ILs have been used for separation of volatile compounds (Salar-Garcia et al. 2017).

2.3.4 Electrolyte

Ionic liquids (ILs) also find application in production and storage of energy (Watanabe et al. 2017). ILs are non-flammable so these can serve as stable electrode material and has good cycle and discharge ability. However, some issues are associated with it, such as disappointing mass and charge transport and high viscosity that still need to be resolved.

2.3.5 Use as a Solvent

One of innovative applications of ionic liquids (ILs) is to use them as solvent for enzyme-based reactions, for instance, lipase-driven reactions can be carried out in ILs (Lau et al. 2000). However, not all ILs can be used for this purpose as some of them provide a hostile environment (Weingaertner et al. 2012). One of the exploited ILs, ethylammonium nitrate, has a network of nitrogen bonds; at lower concentrations, this IL found to have a positive effect on alkaline phosphatase activity (Evans et al. 1981,) but its 80% concentration can inactivate the same enzyme (Magnuson et al. 1984).

2.4 Lignocellulosic Biomass Processing

Lignocellulosic (LC) or plant biomass, which is the most abundant chemical feedstock and renewable resource of energy, mainly consists of cellulose, hemicellulose, and lignin (Brandt et al. 2013). Various challenges hinder in getting chemicals and energy from this biomass due to structural recalcitrance of lignin and cellulose. The biological and chemical degradation becomes difficult because of lignin and highly crystalline structure of cellulose (Yinghuai et al. 2011). Hence, a method is required which can remove or reduce the amount of lignin along with the disruption of crystalline structure (Olivier et al. 2010). The lignin removal or reduction methods have been summarized in Table 2.1. Various researchers attempted to address these issues, consequently different chemicals were studied; however, recent advancements in the field of ionic liquids (ILs) opened the new horizons in biomass prospecting (Behera et al. 2014). A brief summary of ILs used in this regard is given in Table 2.2. Woody biomasses, sugarcane bagasse, wheat straw, and rice straw remained a popular choice of feedstock (Torr et al. 2012). Indeed, ionic liquid-mediated pretreatment reportedly yielded 90% of saccharification efficiency with an enhanced rate of enzymatic digestion (Li et al. 2011). The effectiveness of IL pretreatment is attributed to their ability to disassemble the biomass structure, resulting in decreased and change in cellulose crystalline structures.

2.4.1 Solubility of Cellulose in Ionic Liquid

Cellulose composed of linear chains of glucose joined by1,4 beta glycosidic bonds. These linkages are stacked together by van der Waal forces and hydrogen bonds between the cellulose fibrils. The cellulose crystallinity and degree of polymerization of cellulose are affected by solvent. Cellulose is only soluble in concentrated

Table 2.1 Various pretreatment methods employed in biomass processing

Pretreatment type	Pretreatment method	Reference
Physical	Pyrolysis	Mosier et al. (2005)
	Mechanical disruption	
	Mechanical communication	
Physiochemical	Ammonia Fiber explosion	Mes-Hartree et al. (1988)
	Steam explosion	
Chemical	Acid hydrolysis	Gierer and Norén (1982); Chum et al. (1988)
	Alkaline hydrolysis	
	Oxidative delignification	
	Organic pretreatment at high temperature	
Biological	Lignin degradation by fungi	Hatakka (1983)

Table 2.2 Various ionic liquids and their mode of action which are employed in biomass processing

Ionic liquid	Biomass	Mode of action	References
Methyl sulfate imidazolium-based ILs	Softwood	Dissolve lignin	Pu et al. (2007)
Ethyl-methylimidazolium alkyl benzene sulfonate	Bagasse	Extract lignin	Tan et al. (2009)
1-Ethyl-3-methylimidazolium acetate	Wood	Extract lignin and reduce cellulose crystallinity	Lee et al. (2009)
1-Methyl-3-methylimidazolium methyl sulfate	Wood	Solubilize the lignin	Lee et al. (2009)

Table 2.3 Solvents which are used with ionic liquid in cellulose dissolution

Solvent and ionic liquid	Acronym
Dinitrogen tetraoxide/dimethyl formamide	N2O4/DMF
Lithium chloride/dimethyl acetamide	LiCl/DMAc
Dimethyl sulfoxide/tetrabutyl ammonium fluoride	DMSO/TBAF
Lithium chloride/dimethylimidazolidinone	LiCl/DMI
Dimethyl sulfoxide/paraformaldehyde	DMSO/CH2O
Lithium chloride/N-methyl pyrrolidine	LiCl/NMP

phosphoric acid (Yinghuai et al. 2011). Alternative benign solvents are needed to be discovered because of association of environment-related problems by using concentrated phosphoric acid. Some of the solvents which are commonly used for the cellulose dissolution can be found in Table 2.3 (Heinze and Liebert 2001). By dissolving cellulose, its crystalline structure is disrupted and hence amorphous regions are emerged making it more susceptible to transformation into simple sugars. Cellulose dissolution in ionic liquids (ILs) is dependent on its physical properties (degree of polymerization and crystallinity), the operating conditions (reaction duration, temperature, ratio of cellulose, and IL), and impurities present in water.

Dissolution of cellulose is relied on cations and anions of ILs. The cation acts as an electron acceptor center thus prevents the cross linking of the cellulose. The anion, on the other hand, forms an electron donor electron acceptor complex by interacting in a stoichiometric manner with the OH group of cellulose (Olivier-Bourbigou et al. 2010). The formation of the electron donor electron acceptor complex results in cellulose dissolution by disturbing the hydrogen bonds in glycosidic monomers. Non-dissolution solvent (acetone, water, and ethanol) can precipitate the miscible cellulose in ILs and can be separated either by centrifugation or filtration. Non-dissolution solvent can be separated by distillation of the mixture.

2.4.2 *Association of Lignin Extraction to Cellulose Crystallinity*

There is a promising correlation between extraction of lignin and cellulose crystallinity. The cellulose crystallinity dropped to 45 when lignin extraction was carried out by 1-ethyl-3-methylimidazolium acetate, which ultimately resulted in 90% hydrolysis (Lee et al. 2009). For shorter hydrolysis time, combination of de-crystallization and delignification showed benefits, whereas for longer time duration, delignification is sufficient (Zhu et al. 2008).

2.5 Pretreatment of Sugarcane Bagasse

The extraction of juice from sugarcane results in fibrous residue named as sugarcane bagasse (SB). For bio-based economy, it is considered as a strong candidate of feedstock because of its abundance and high carbohydrate content. Approximately, nearly 100 million tons of SB are produced yearly (Vallejos et al. 2012) that can be utilized for the manufacturing of many value-added products (Al Arni 2018; Achinivu et al. 2014). Currently, major proportion of SB is burned to produce energy, which results in environmental problems and also wastes this valuable bio-resource.

In terms of fiber composition and particle size, SB presents a great morphological heterogeneity (Pandey et al. 2000). Fibers of SB can be used for the production of paper and pulp (Bajpai et al. 2004). While, pith is a non-fibrous residue and has poor morphology and higher ash content. Its low density and hygroscopic property causes drainage and filtration difficulties in paper manufacturing and increases chemical consumption resulting in slowing down the pulping process (Pandey et al. 2000; Sanjuan et al. 2001; Rainey 2009; Aguilar-Rivera 2016). Nonetheless, very limited research has been done to examine the effect of depithing on SB prior to pretreatment and enzymatic hydrolysis (Aghcheh et al. 2016).

Pretreatment technologies for SB need to manage its inhomogeneous fiber morphology, high moisture content (Pandey et al. 2000), and moderate ash content (Sanjuan et al. 2001). Several methods have been reported for the pretreatment of SB, but few showed promising applications for large-scale commercial application. Recently, Rocha et al. (2017) used 10% SB in ionic liquid for 3.5 h and achieved 68% glucose yield. Chen et al. (2014) have designed mono hydrogen sulfate-based protic ionic liquid. As reported in previous studies, mono hydrogen sulfate-anion-based ionic liquids could be successfully recycled by distillation for up to four times and are more stable than 1-ethyl-3-methylimidazolium acetate (Brandt-Talbot et al. 2017; George et al. 2015).

2.6 Problems Associated with Ionic Liquid Pretreatment

Despite encouraging results of ionic liquids for biomass dissolution (Swatloski et al. 2002), its esoteric nature renders a big challenge in its useful applications. The wide-scale uptake of ionic liquids in commercial level hinders by high operating temperature, toxicity, high costs, low solid loading, high viscosities, poor water tolerance, and complex recycling (Tan et al. 2009). More research is required to overcome these hurdles.

2.7 Concluding Remarks

As an industrial feedstock, the use of sugarcane bagasse can displace fossil feedstock and can be utilized for the production of many value-added materials, chemicals, energy, and transportation fuels. High production cost and technological constraints cause limitation of utilizing lignocellulose biomass. Effective pretreatment is required to get high enzymatic saccharification of sugarcane bagasse to fermentable sugars, biofuel, and value-added products. Ionic liquids are recognized as a relatively newly discovered promising solvent for pretreatment process because some of the ionic liquids demonstrate the rare ability to dissolve all lignocellulosic biomass components including crystalline cellulose. The data reviewed here presents prospective use of ionic liquids for biomass pretreatment.

References

Achinivu EC, Howard RM, Li G, Gracz H, Henderson WA (2014) Lignin extraction from biomass with protic ionic liquids. Green Chem 16(3):1114–1119. https://doi.org/10.1039/c3gc42306a

Aghcheh RK, Bonakdarpour B, Ashtiani FZ (2016) The influence of sugar cane bagasse type and its particle size on xylose production and xylose-to-xylitol bioconversion with the yeast *Debaryomyceshansenii*. Appl Biochem Biotechnol 180(6):1141–1151. https://doi.org/10.1007/s12010-016-2157-x

Aguilar-Rivera N (2016) Sustainable sugarcane bagasse cellulose for papermaking. Advanced engineering materials and modeling. Scrivener Publishing LLC, Beverly, pp 121–164

Al Arni S (2018) Extraction and isolation methods for lignin separation from sugarcane bagasse: a review. Ind Crop Prod 115:330–339

Bajpai P, Mishra SP, Mishra OP, Kumar S, Bajpai PK, Singh S (2004) Biochemical pulping of bagasse. Biotechnol Prog 20(4):1270–1272. https://doi.org/10.1021/bp040005q

Behera S, Arora R, Nandhagopal N, Kumar S (2014) Importance of chemical pretreatment for bioconversion of lignocellulosic biomass. Renew Sust Energ Rev 36:91–106

Brandt A, Grasvik J, Hallett JP, Welton T (2013) Deconstruction of lignocellulosic biomass with ionic liquids. Green Chem 15:550–583. https://doi.org/10.1039/C2GC36364J

Brandt-Talbot A, Gschwend FJ, Fennell PS, Lammens TM, Tan B, Weale J, Hallett JP (2017) An economically viable ionic liquid for the fractionation of lignocellulosic biomass. Green Chem 19(13):3078–3102. https://doi.org/10.1039/x0xx00000x

Chen L, Sharifzadeh M, Dowell N, Welton T, Shah N, Hallett JP (2014) Inexpensive ionic liquids: [HSO4] −−based solvent production at bulk scale. Green Chem 16:3098–3106. https://doi.org/10.1039/c4gc00016a

Chum HL, Johnson DK, Black S, Baker J, Grohmann K, Sarkanen KV, Wallace K, Schroeder HA (1988) Organosolv pretreatment for enzymatic hydrolysis of poplars: I. enzyme hydrolysis of cellulosic residues. Biotechnol Bioeng 31(7):643–649

Cooper EI, Angell CA (1983) Versatile organic iodide melts and glasses with high mole fractions of LiI: glass transition temperatures and electrical conductivities. Solid State Ionics 9:617–622

Evans DF, Chen SH, Schriver GW, Arnett EM (1981) Thermodynamics of solution of nonpolar gases in a fused salt. Hydrophobic bonding behavior in a nonaqueous system. J Am Chem Soc 103(2):481–482

George A, Brandt A, Tran K, Zahari SMNS, Klein-Marcuschamer D, Sun N, Sathitsuksanoh N, Shir J, Stavila V, Tran K, Singh S, Holmes B, Welton T, Simmons BA, Hallet JP (2015) Design of low-cost ionic liquids for lignocellulosic biomass pretreatment. Green Chem 17(3):1728–1734. https://doi.org/10.1039/C4GC01208A

Ghandi K (2014) A review of ionic liquids, their limits and applications. Green and sustainable chemistry 4(01):44–53. https://doi.org/10.4236/gsc.2014.41008

Gierer J, Norén I (1982) Oxidative pretreatment of pine wood to facilitate delignification during Kraft pulping. Holz Int J Biol Chem Phys Technol Wood 36(3):123–130

Hatakka AI (1983) Pretreatment of wheat straw by white-rot fungi for enzymicsaccharification of cellulose. Eur J Appl Microbiol Biotechnol 18(6):350–357

Heinze T, Liebert T (2001) Unconventional methods in cellulose functionalization. Prog Polym Sci 26(9):1689–1762

Hirao M, Sugimoto H, Ohno H (2000) Preparation of novel room-temperature molten salts by neutralization of amines. J Electrochem Soc 147(11):4168–4172

Ho TD, Canestraro AJ, Anderson JL (2011) Ionic liquids in solid-phase microextraction: a review. Analytica Chimica Acta 695(1–2):18–43. https://doi.org/10.1016/j.aca.2011.03.034

Hurley FH, Wier TP (1951) Electrodeposition of metals from fused quaternary ammonium salts. J Electrochem Soc 98(5):203–206

Jutz F, Andanson JM, Baiker A (2011) Ionic liquids and dense carbon dioxide: a beneficial biphasic system for catalysis. Chem Rev 111:322–353. https://doi.org/10.1021/cr100194q

Lau RM, van Rantwijk F, Seddon KR, Sheldon RA (2000) Lipase catalyzed reactions in ionic liquids. Org Lett 2:4189–4191. https://doi.org/10.1021/ol006732d

Lee SH, Doherty TV, Linhardt RJ, Dordick JS (2009) Ionic liquid-mediated selective extraction of lignin from wood leading to enhanced enzymatic cellulose hydrolysis. Biotechnol Bioeng 102(5):1368–1376. https://doi.org/10.1002/bit.22179

Li C, Cheng G, Balan V, Kent MS, Ong M, Chundawat SP, daCosta SL, Melnichenko YB, Dale BE, Simmons BA, Singh S (2011) Influence of physico-chemical changes on enzymatic digestibility of ionic liquid and AFEX pretreated corn Stover. Bioresour Technol 102(13):6928–6936. https://doi.org/10.1016/j.biortech.2011.04.005

Liu WM, Ye CF, Gong QY, Wang HZ, Wang P (2002) Tribological performance of room-temperature ionic liquids as lubricant. Tribol Lett 13:81–85

Magnuson DK, Bodley JW, Evans DF (1984) The activity and stability of alkaline phosphatase in solutions of water and the fused salt Ethylammonium nitrate. J Solut Chem 13:583–587

Mes-Hartree M, Dale BE, Craig WK (1988) Comparison of steam and ammonia pretreatment for enzymatic hydrolysis of cellulose. Appl Microbiol Biotechnol 29(5):462–468

Mosier N, Wyman C, Dale B, Elander R, Lee YY, Holtzapple M, Ladisch M (2005) Features of promising technologies for pretreatment of lignocellulosic biomass. Bioresour Technol 96(6):673–686. https://doi.org/10.1016/j.biortech.2004.06.025

Muhammad N, Man Z, Bustam MA, Mutalib MIA, Wilfred CD, Rafiq S (2011) Dissolution and delignification of bamboo biomass using amino acid-based ionic liquid. Appl Biochem Biotechnol 165(3–4):998–1009. https://doi.org/10.1007/s12010-011-9315-y

Olivier-Bourbigou H, Magna L, Morvan D (2010) Ionic liquids and catalysis: recent progress from knowledge to applications. Appl Catal A General 373(1–2):1–56. https://doi.org/10.1016/j.apcata.2009.10.008

Pacholec F, Butler HT, Poole CF (1982) Molten organic salt phase for gas-liquid chromatography. Anal Chem 54:1938–1941

Pandey A, Soccol CR, Nigam P, Soccol VT (2000) Biotechnological potential of agro-industrial residues. I: sugarcane bagasse. Bioresour Technol 74(1):69–80

Phillips BS, Zabinski JS (2004) Ionic liquid lubrication effects on ceramics in a water environment. Tribol Lett 17:533–541

Pu Y, Jiang N, Ragauskas AJ (2007) Ionic liquid as a green solvent for lignin. J Wood Chem Technol 27(1):23–33. https://doi.org/10.1080/02773810701282330

Rainey TJ (2009) A study into the permeability and compressibility properties of Australian bagasse pulp. Faculty of Built Environment and Engineering Thesis

Reddy P (2015) A critical review of ionic liquids for the pretreatment of lignocellulosic biomass. S Afr J Sci 111(11–12):1–9. https://doi.org/10.17159/sajs.2015/20150083

Robinson J, Osteryoung RA (1979) An electrochemical and spectroscopic study of some aromatic hydrocarbons in the room temperature molten salt system aluminum chloride-n-Butylpyridinium. Chlor J Am Chem Soc 101:323–327

Rocha EGA, Pin TC, Rabelo SC, Costa AC (2017) Evaluation of the use of protic ionic liquids on biomass fractionation. Fuel 206:145–154. https://doi.org/10.1016/j.fuel.2017.06.014

Salar-Garcia MJ, Ortiz-Martinez VM, Hernandez-Fernandez FJ, de los Rios AP, Quesada-Medina J (2017) Ionic liquid technology to recover volatile organic compounds (VOCs). J Hazard Mater 321:484–499

Sanjuan R, Anzaldo J, Vargas J, Turrado J, Patt R (2001) Morphological and chemical composition of pith and fibers from mexican sugarcane bagasse. Holzals Roh Werkst 59:447–450

Swatloski RP, Spear SK, Holbrey JD, Rogers RD (2002) Dissolution of cellulose with ionic liquids. J Am Chem Soc 124:4974–4975

Tan SSY, MacFarlane DR, Upfal J, Edye LA, Doherty WOS, Patti AF, Pringle JM, Scott JL (2009) Extraction of lignin from lignocellulose at atmospheric pressure using alkylbenzenesulfonate ionic liquid. Green Chem 11(3):339–345

Torimoto T, Tsuda T, Okazaki K, Kuwabata S (2010) New Frontiers in materials science opened by ionic liquids. Adv Mater 22:1196–1221. https://doi.org/10.1002/adma.200902184

Torr KM, Love KT, Çetinkol ÖP, Donaldson LA, George A, Holmes BM, Simmons BA (2012) The impact of ionic liquid pretreatment on the chemistry and enzymatic digestibility of Pinusradiata compression wood. Green Chem 14(3):778–787. https://doi.org/10.1039/c2gc16362d

Vallejos ME, Zambon MD, Area MC, da Silva Curvelo AA (2012) Low liquid−solid ratio (LSR) hot water pretreatment of sugarcane bagasse. Green Chem 14:1982–1989. https://doi.org/10.1039/c2gc35397k

Ventura SPM, Silva FAE, Quental MV, Monda D, Freire MG, Coutinho JAP (2017) Ionic-liquid-mediated extraction and separation processes for bioactive compounds: past, present, and future trends. Chem Rev 117:6984–7052. https://doi.org/10.1021/acs.chemrev.6b00550

Watanabe M, Thomas ML, Zhang SG, Ueno K, Yasuda T, Dokko K (2017) Application of ionic liquids to energy storage and conversion materials and devices. Chem Rev 117:7190–7239. https://doi.org/10.1021/acs.chemrev.6b00504

Weingaertner H, Cabrele C, Herrmann C (2012) How ionic liquids can help to stabilize native proteins. Phys Chem Chem Phys 14:415–426

Welton T (2018) Ionic liquids: a brief history. Biophys Rev 10(3):691–706

Wilkes JS, Zaworotko MJ (1992) Air and water stable 1-ethyl-3- methylimidazolium based ionic liquids. J Chem Soc Chem Commun 13:965–967

Ye CF, Liu WM, Chen YX, Yu LG (2001) Room-temperature ionic liquids: a novel versatile lubricant. Chem Commun 21:2244–2245

Yinghuai Z, Biying AO, Siwei X, Hosmane NS, Maguire JA (2011) Ionic liquids in catalytic biomass transformation. Appl Ionic Liq Sci Technol:3–26

Yu B, Zhou F, Mu Z, Liang YM, Liu WM (2006) Tribological properties of ultra-thin ionic liquid films on single-crystal silicon wafers with functionalized surfaces. Tribol Int 39:879–887

Zeng SJ, Zhang X, Bai LP, Zhang XC, Wang H, Wang JJ, Bao D, Li MD, Liu XY, Zhang SJ (2017) Ionic-liquid-based CO2 capture systems: structure, interaction and process. Chem Rev 117:9625–9673

Zhu L, O'Dwyer JP, Chang VS, Granda CB, Holtzapple MT (2008) Structural features affecting biomass enzymatic digestibility. Bioresour Technol 99(9):3817–3828. https://doi.org/10.1016/j.biortech.2007.07.033

Chapter 3
Ionic Liquids as Solvents and Catalysts for Biodiesel Production

P. Andreo-Martínez, V. M. Ortiz-Martínez, and J. Quesada-Medina

Contents

3.1 Introduction

Ionic liquids have attracted growing interest as new type of green solvents for a wide range of applications in the chemical industry over the last years. These compounds are featured with special properties that make them suitable to replace conventional organic solvents with enhanced performance. Ionic liquids can be generally defined as organic salts remaining liquid at temperatures below 100 °C, and usually at ambient temperature, in which case they are known as room temperature ionic liquids. They are typically formed by an inorganic anion (e.g., BF_4^-, Cl^-,

P. Andreo-Martínez
Department of Chemical Engineering, University of Murcia, Espinardo, Murcia, Spain

Department of Agricultural Chemistry, University of Murcia, Espinardo, Murcia, Spain

V. M. Ortiz-Martínez (✉)
Department of Chemical and Environmental Engineering, Technical University of Cartagena, Cartagena, Murcia, Spain

Department of Chemical Engineering, University of Murcia, Espinardo, Murcia, Spain
e-mail: victor.ortiz@upct.es

J. Quesada-Medina
Department of Chemical Engineering, University of Murcia, Espinardo, Murcia, Spain

Inamuddin, A. M. Asiri (eds.), *Nanotechnology-Based Industrial Applications of Ionic Liquids*, Nanotechnology in the Life Sciences,
https://doi.org/10.1007/978-3-030-44995-7_3

Table 3.1 Advantages and drawbacks of ionic liquids

Advantages	Drawbacks
• High chemical and thermal stability. • Structure and property tuning. • High catalytic activity. • Low or negligible vapor pressure and flammability. • Lower toxicity versus organic solvents. • Wide range of applications. • Liquids at room temperature. • Possible recyclability.	• High synthesis costs. • Limitations in large scale applications. • Some of them are moisture-sensitive. • Corrosiveness (depending on the ionic liquid type). • Non-biodegradability.

PF_6^-, NTf_2^-) and an organic cation (e.g., imidazolium, pyridinium, ammonium) (Earle and Seddon 2000; Hallett and Welton 2011; Lei et al. 2017).

Table 3.1 offers the main advantages and disadvantages of ionic liquids. Among the most interesting properties of ionic liquids are non-flammability, low toxicity, near zero vapor pressure, and high chemical and thermal stability. Other characteristics of ionic liquids such as solvation properties, acidity or basicity character, and polarity have promoted the application of these compounds in different chemical processes either as solvents or as catalysts (Ishak et al. 2017). Moreover, they can be properly tuned for specific applications by suitably selecting certain combinations of cations and anions (Plechkova and Seddon 2008; Ghandi 2014). In this sense, novel ionic liquids are constantly being synthesized, which offers new paths of research.

Ionic liquids have been investigated for a wide range of applications including organic synthesis, separation processes, electrochemistry, nanomaterial fabrication, or enzymatic catalysis. For such applications, imidazolium-based ionic liquids are generally the most employed species. The first industrial process involving ionic liquids dated from 2003 (BASIL process, BASF) and was employed for the removal of hydrochloric acid in the production of alkoxyphenylphosphines (Andreani and Rocha 2012).

Nowadays, the investigation of ionic liquids for biodiesel production has gained growing interest due to the advantages of these environmental-friendly compounds in order to increase the efficiency and sustainability of the process. In this sense, ionic liquids can be employed as catalysts in transesterification esterification or processes, since some ionic liquids present Brønsted acidity or basicity. Although ionic liquids have higher synthesis costs in comparison to organic conventional solvents, they offer the advantages of higher reusability, easy separation, and higher environmental safety, while minimizing the production of wastes in catalytic reactions (Muhammad et al. 2015). Ionic liquids can also be employed for the functionalization of other materials such as metallic nanoparticles or ceramic materials to be jointly employed as catalyst. Finally, these compounds can be employed in enzymatic biodiesel synthesis, with great benefits such as enzyme activity increase and reusability. Thus, this chapter reviews the recent advances related to the role of ionic liquids in biodiesel production technology.

3.2 Biodiesel Production

The depletion of worldwide oil reverses together with the environmental issues derived from the burning of fossil fuels are leading to the promotion of renewable sources of energy such as biofuels. Specially, biodiesel is an attractive alternative to petroleum-based fuels displaying several advantages, e.g., enhanced emission profile, sulfur-free content, biodegradability, and lower toxicity (Hirani et al. 2018; Ma et al. 2018).

Biodiesel typically consists of a mixture of fatty acid methyl or ethyl esters obtained through the transesterification of triglycerides coming from, e.g., vegetable, fat, and waste oils. The free fatty acids, which can also be present in these oils but in lower proportion, can be transformed into fatty acid esters through esterification. Figures 3.1 offers a schematic representation of biodiesel production based on the transesterification of triglycerides in methanol. As observed, glycerol is obtained as a sub-product in the process. Nevertheless, the separation of biodiesel from glycerol can be easily achieved and glycerol can be valorized into added-value products (Pagliaro et al. 2007).

The alcohols usually employed for these processes are ethanol and methanol although this last one is more frequently used due to its lower cost (Marchetti et al. 2007; Andreani and Rocha 2012). Edible and non-edible oils can be used for biodiesel generation; however, the second type is preferable in order to avoid the competence between food and biofuel markets. More recently, biodiesel from microalgae oil or third-generation biodiesel is also gaining growing interest.

The process can be performed in the presence of catalysts or, alternatively, at supercritical conditions (high conditions of pressure and temperature) (Lee and Saka 2010; Leung et al. 2010). In the first case, acid or base catalysts are usually employed. Sulfuric or phosphoric acids are examples of typical acid catalyst used in biodiesel generation processes. The main drawbacks related to the use of acid catalysts is the long times required to obtain high reaction yields. For its part, base catalysts, such as sodium or potassium hydroxides, although being effective, require the use of feedstocks with very low content of water; otherwise, the undesirable saponification reaction can be promoted affecting final biodiesel yield with further separation steps being needed (Talha and Sulaiman 2016). In addition, when feedstocks with high degree of free fatty acids such waste oils are employed, soap formation can occur due to the reaction between the free fatty acids and the base catalysts. The generation of biodiesel via supercritical synthesis offers promising prospects, but it

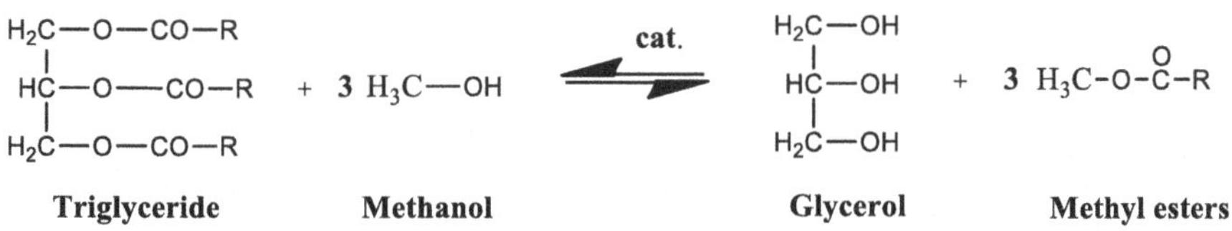

Fig. 3.1 Transesterification reaction of triglycerides in methanol

is energy-intensive due to the high conditions of temperature and pressure required. Biodiesel synthesis can also be performed through enzymatic technology. In comparison to the aforementioned chemical catalysts, enzymes need relatively low reaction temperature and less alcohol to oil molar ratios (Arumugam and Ponnusami 2017). Nevertheless, enzymes can be subjected to deactivation issues and need to be immobilized in order to enhance their recovery and reuse.

In this context, ionic liquids have been investigated as new class of effective, eco-friendly, and reusable catalysts for biodiesel production, but also as solvent and supports will be discussed in the following sections (Andreani and Rocha 2012).

3.3 Applications of Ionic Liquids for Biodiesel Production

Ionic liquids offer interesting properties as new materials for the production of biodiesel, being in line with the standards of green chemistry and showing interesting properties for industrialization. Ionic liquids can be used as single catalysts, as solvents or reaction media, or even as supports for enzymatic biodiesel generation. Ionic liquids are regarded homogeneous catalysts since they usually present the same phase as in the case of the reactants. Nevertheless, the capacity of recycling ionic liquids after the reaction and the ease of separation of products from these compounds entail the advantages of heterogeneous catalysis. Thus, it can be stated that ionic liquids combine the advantages of both homogenous and heterogeneous catalysts. Furthermore, ionic liquids have been investigated for the immobilization of metal complex compounds and conventional catalysts for biodiesel production, as well as for the functionalization, for instance, of ceramic materials. The application of ionic liquids for the enzymatic production of biodiesel are especially interesting, since they allow higher yields to be reached while avoiding enzyme deactivation. These different applications of ionic liquids in biodiesel production synthesis will be covered in the following sub-sections.

3.3.1 Ionic Liquids as Biodiesel Catalysts

Brønsted acidic type ionic liquids are among the most frequently used ionic liquids as single catalyst in transesterification and esterification processes for biodiesel production (Han et al. 2013). In this type of ionic liquids, the acidity properties are given by either the presence of the covalently attached sulfonic acid specie ($-SO_3H$) or by a Brønsted acidic counteranion, e.g., hydrogensulfate (HSO_4^-) and dihydrogen phosphate ($H_2PO_4^-$) (Vafaeezadeh and Alinezhad 2016). In comparison to other conventional acid catalysts, Brønsted acidic ionic liquids offer less equipment corrosion and present lower environmental risks. Specially, imidazole or pyridine-based ionic liquids have drawn much attention.

The use of Brønsted acidic ionic liquids are among the first attempts to directly use ionic liquids as catalysts for the transesterification of vegetable oils into biodiesel. In this sense, Wu et al. (2007) assessed several water-stable Brønsted acidic ionic liquids containing an alkane sulfonic acid group to catalyze the transesterification reaction of cottonseed in the presence of methanol. In this early study, it could be observed that the catalytic activities of the ionic liquids greatly depend on their acidic strength; as the Brønsted acidity increases, their catalytic activity also enhanced. Among the species analyzed, the ionic liquid 1-(4-sulfonic acid) butylpyridinium hydrogen sulfate offered the best performance, with a resulting biodiesel yield of over 90% after 5 h of reaction time at a fixed temperature of 170 °C. As in other catalyzed transesterification processes, there are several operational variables that can greatly influence the final performance, such as the amount of ionic liquid employed, the methanol-to-oil molar ratio, or the reaction temperature. In a further study, the performance of the ionic liquid could be enhanced by finding the proper amount ratio of ionic liquid-oil-alcohol, reducing the reaction time up to 4 h (Han et al. 2009). In line with this ionic liquid, other pyridinium-based ionic liquids such as butylpyridinium trifluoromethanesulfonate have displayed good performance (Li et al. 2010).

Brønsted acidic ionic liquids containing imidazolium cation have also shown to be effective at catalyzing the transesterification reaction of vegetable oils. Several ionic liquids derived from 1-benzyl-1H-benzimidazole have been reported in the literature as good catalysts for this reaction, obtaining conversions over 95% for optimized reaction conditions (Ghiaci et al. 2011b). Also, acidic imidazolium ionic liquids with several metal sulfates have been tested for biodiesel production; specially, the ionic liquid with formula $[BSO_3HMIm]HSO_4–Fe_2(SO_4)_3$ offered the most satisfactory catalytic activity (Li et al. 2014), with a conversion of over 95% after 1 h reaction time at a temperature of 60 °C in a microwave-assisted transesterification process. The reusability of imidazole and pyridinium-based Brønsted acidic ionic liquids can be considered high, with a slight decrease in the catalytic activity after several cycles (Li et al. 2010; Ghiaci et al. 2011b).

Other authors have highlighted the advantages of other cation species over imidazolium and pyridinium-based Brønsted ionic liquids. For example, the use of caprolactam-based iconic liquids can translate into lower synthesis costs and difficulty, together with lower toxicity issues. In this sense, Luo et al. (2017) introduced three HSO_3-functionalized ionic liquids based on caprolactam. The optimized results offered notable results with biodiesel yields over 95%, after 3 h reaction time and 140 °C.

The effect of water content in biodiesel feedstock can display an important effect on the yield of biodiesel when using Brønsted acid ionic liquids. Water content needs to be limited in order to maintain the catalytic activity of the ionic liquid. This was proved in the work by Fan et al. (2017), who observed that the reaction yield was reduced when 1% of water was present in the reaction media, and significantly dropped when the content of water reached 5%, employing imidazolium-based Brønsted ionic liquids. This was explained by the disruption of the ionic liquid structure, which is caused by the competition between the H_2O molecules and the ionic liquid anion to bind with the protons from the imidazolium cation. However, this negative effect could be reduced by increasing the temperature reaction.

Other ionic liquid species, such as surfactant-type Brønsted acids, have proven to act as amphiphilic reagents miscible with water while being capable of dissolving organic substrates. Vafaeezadeh et al. (2017) proved that the ionic liquid 1-dodecyl-3-methylimidazolium hydrogen sulfate, corresponding to the mentioned group, offered high activity as catalyst for the esterification of fatty acids with methanol and ethanol in the presence and in the absence of water. Interestingly, this ionic liquid showed good activity in both conditions, and such activity was even higher in the presence of water, with the possibility of reusing the catalyst for at least four cycles with no significant activity drop.

Another approach consists of the use of Brønsted–Lewis acidic ionic liquids, which has both Brønsted and Lewis acidic sites. A synergetic effect due to the presence of both types of catalytic sites can help to enhance selectivity and conversion yield of the biodiesel reaction (Cai et al. 2018). Several authors have shown the effectiveness of this type of ionic liquids. As representative case, Liu et al. (2013) reported that the Brønsted–Lewis acidic ionic liquid (3-sulfonic acid)-propyltriethylammonium chloroironinate was capable of catalyzing the transesterification of waste oil, achieving a biodiesel yield of 95% after 4 h, at a reaction temperature of 120 °C.

Less frequently, ionic liquids presenting Brønsted basicity have been employed for biodiesel synthesis. Recently, Wang et al. (2016) synthesized a new Brønsted basic ionic from *N*-methyl morpholine, *N*-butyl bromide, and KOH, which was used for the transesterification of soybean oil, reaching yields of over 94% at 60 °C after 6 h reaction time. Dication basic ionic liquids can even offer higher performances; this is the case of the ionic liquid bis-(3-methyl-1-imidazolium-)-ethlyene dihydroxide, with which it was possible to obtain yields over 98% after 4 h and at 55 °C. This ionic liquid also provides high reusability, with almost the same yield after being recycled for seven batch reaction cycles (Liang et al. 2010).

The use of supercritical conditions in the production of biodiesel enables to accelerate the synthesis reaction. Supercritical conditions generally refer to temperatures and pressures above the critical point of the alcohol employed, e.g., 512.6 K and 8.1 MPa for methanol (Madras et al. 2004). Under such conditions, the process is usually performed in the absence of catalysts. However, very recently, a new strategy has been introduced, consisting of the addition of ionic liquids as catalysts in supercritical conditions to reduce the temperature required to carry out the reaction while reducing reaction time. This translates into energy savings, and thanks to the high thermal stability of ionic liquids; they can be reused in this severe conditions of pressure and temperature. The ionic liquid 1-methylimidazolium hydrogen sulfate has been employed for the transesterification reaction of oil waste, providing high biodiesel yields of over 97% after very short reaction times (45 min) at low amounts of catalyst. In such conditions, even the presence of water in the reaction media, up to $3\%_{v/v}$, did not show any negative effect on the process (Caldas et al. 2016). Similarly, the ionic liquid choline dihydrogen phosphate showed to be effective at obtaining very high yields after short reaction times. In this case, the authors observed a significant interaction between the ionic liquid and the methanol molecule, which increased its reactivity and favored the transesterification reaction (Ortiz-Martínez et al. 2016).

3.3.2 Ionic Liquids in Combination with Other Materials

Ionic liquids have been studied as supports for catalysts in biodiesel synthesis processes. Specially, they have been investigated for the immobilization of metal complex compounds and conventional catalysts for biodiesel production (Andreani and Rocha 2012; Mohammad Fauzi and Amin 2012).

Several metal complexes with the formula X(3-hydroxy-2-methyl-4-pyrone)$_2$(H_2O)$_2$, in which X can be substituted by Sn, Pb, and Zn, have been used for the transesterification of vegetable oils in homogenous conditions (Abreu et al. 2003; Abreu et al. 2004). Among them, Sn-based compounds showed higher activity in comparison to conventional base and acid catalysts such as NaOH and H_2SO_4, respectively, under the same reaction conditions. In order to have the advantages of heterogeneous catalysis, the same authors immobilized this previous tin complex in the ionic liquid 1-butyl-3-methylimidazolium hexafluorophosphate, and afterward the phase was further supported onto an ion exchange resin (Abreu et al. 2005). However, the system failed in terms of reusability, due to leaching of the complex from the ionic liquid phase after the first cycle. This attempt was enhanced by the immobilization of the same tin complex catalyst in the ionic liquid 1-butyl-3-methylimidazolium tetrachloro-indate, which allowed biodiesel yield to be maintained over 80% after the third cycle (DaSilveira Neto et al. 2007).

Conventional catalysts such as acids and bases have also been immobilized in ionic liquids. In this regard, 1-n-butyl-3-methylimidazolium bis(trifluoromethylsulfonyl)imide was employed for the immobilization of K_2CO_3 and H_2SO_4, showing that the ionic liquid with the base or the acid catalyst could be reused at least in six cycles with significant biodiesel yields and selectivity for the transesterification of vegetable oil (Lapis et al. 2008). Other works have confirmed the good results obtained with the same ionic liquid and H_2SO_4 for the esterification of lauric acid (Yaacob et al. 2011), which is the main component of palm kernel oil used as feedstock for biodiesel (Lamaisri et al. 2015)

Other authors have employed ionic liquids for the modification of other catalysts for further functionalization. This is the case of clays such as bentonite modified with ionic liquid as a strategy to form new organoclays with tailored properties. For example, Ghiaci et al. (2011a) investigated the modification of bentonite (aluminum phyllosilicate clay) with a Brønsted acidic ionic liquid for the esterification of natural fatty acids with methanol, obtaining biodiesel yields of over 92% after 6 h of reaction time, at mild temperature conditions (70 °C). In terms of recyclability, the authors found that the yield was reduced down to 85% after eight cycles. Other works have followed this approached, by modifying bentonite with Brønsted ionic liquids. Interestingly, Aghabarari and Dorostkar (2014) observed that the

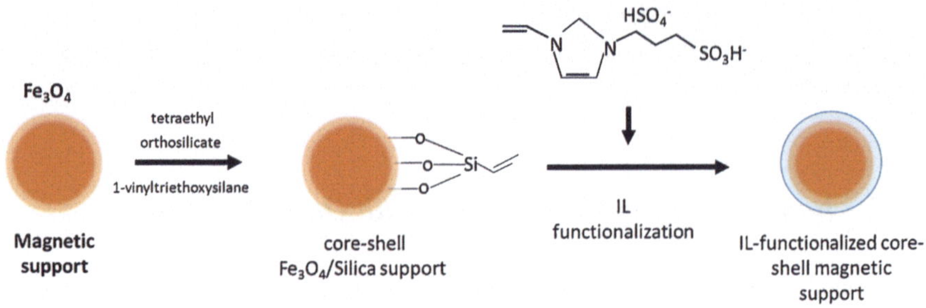

Fig. 3.2 Radical grafting copolymerization of Brønsted acidic IL, 1-vinyl-3-(3-sulfopropyl)imidazolium hydrogen sulfate (modified after Xie and Wang 2019)

combination of ionic liquids with bentonite offers better performance in comparison with that provided by bentonite modified with sulfuric acid for the esterification reaction of oleic acid in ethanol.

Other options based on the combination of ionic liquid with silica materials have been investigated (Troter et al. 2016), including the immobilizations of ionic liquids onto iron- and magnetite-silica (Zhang et al. 2012; Wu et al. 2014), ionic liquids supported on sulfhydryl group modified SiO_2 (Zhen et al. 2014) or dispersed over large surface area silica gel through physisorbed confinement (Vafaeezadeh and Hashemi 2014). These works offered effective catalysts options for biodiesel synthesis in which the ionic liquid is supported increasing the surface area between reactants and the catalyst.

Polymeric ionic liquids, which results from the polymerization of an ionic liquid monomer, can also be supported on porous matrix such as silica. Polymeric acidic Brønsted ionic liquids have been supported on a core-shell structured Fe_3O_4-SiO_2 displaying high activities for both the transesterification reaction of vegetable oils and the esterification of free fatty acids (Xie and Wang 2019). The ionic liquid 1-vinyl-3-(3-sulfopropyl)imidazolium hydrogen sulfate was immobilized through radical grafting copolymerization onto the support (see schematic representation in Fig. 3.2). In this case, the magnetic properties of the catalyst support facilitate its separation from the reaction media with a simple external and permanent magnet for further reutilization. Moreover, this immobilization method for the ionic liquid enabled a strong attachment of the polymeric ionic liquid, which in turn can favor its stability and recyclability.

3.3.3 Use of Ionic Liquids in Enzymatic Biodiesel Production

Enzymes such as lipase can act as catalysts for the production of biodiesel from lipids both through transesterification of triglycerides and esterification of fatty acids. This method can be performed at low temperatures and pressures, which translates into a reduction of energy demands for the process (Hama and Kondo

Table 3.2 Comparison of enzymatic biodiesel synthesis versus chemical biodiesel production

	Advantages	Disadvantages
Chemical process	• Wide range of operational conditions. • Availability and cost of chemicals. • Simultaneous esterification and transesterification under homogenous catalysis conditions. • Easy separation of heterogeneous catalysis.	• Sensitivity to water content. • Very slow reaction rates under homogenous acid catalyst conditions. • Saponification issues for base catalyst processes. • Catalyst recyclability.
Enzymatic synthesis	• Low temperature reactions. • Low environmental impact. • Easy catalyst recovery. • Specificity.	• Low reaction times. • High cost of biocatalysts. • Enzyme deactivation by alcohols.

2013). However, enzymatic transesterification present some limitation related to the deactivation of the enzymes, especially in the presence of methanol, and the blocking effect posed by the deposition of glycerol onto the active enzymatic sites (Andreani and Rocha 2012). Moreover, the enzymatic synthesis of biodiesel is yet to be implemented at large scale because of the high costs of the enzymes (Taher et al. 2017). Table 3.2 shows the advantages and disadvantages of enzymatic biodiesel production, which is compared to conventional chemical biodiesel synthesis (not supercritical conditions considered). (Motasemi and Ani 2012; Ribeiro et al. 2014)

The use of ionic liquids as immobilizing media for enzymes can help to overcome these issues by increasing the stability of the enzyme and ensuring its activity over time. Among the first attempts, the commercial Novozym 435 lipase (immobilized *Candida antarctica* lipase B) was used for the production of biodiesel in several ionic liquids (Ha et al. 2007; Sunitha et al. 2007), showing that the reaction yield was clearly higher in the presence of these solvents in comparison with organic solvents and ionic liquid-free conditions, with the possibility of recycling the biocatalyst several times. Later, the same feedstock was employed by Arai et al. (2010) using fungal whole cells in the presence of ionic liquids contained the anion tetrafluoroborate, in combination with the cations 1-ethyl-3-methylimidazolium and 1-butyl-3-methylimidazolium, respectively. In this case, the authors determined that it is necessary to perform the cross-linking of the biocatalyst with glutaraldehyde in order to ensure its reusability.

The type of enzyme as well as the reaction media greatly determines the final biodiesel yields and enzyme stability. For example, the enzyme stability versus methanol can vary depending on the solvent employed. An interesting work by Zhang et al. (2011) compared the performance of several lipases under different reaction conditions, namely *Penicillium expansum* lipase, the already mentioned Novozym 435 lipase and lipozyme TLIM in the ionic liquid 1-butyl-3-methylimidazolium hexafluorophosphate or hexane, respectively, for the transesterification of vegetable oil with methanol. They observed that each lipase offered a different response to methanol according to the solvent employed, and only the lipase

Penicillium expansum was stable and provided good activity when using both types of solvent (ionic liquid or hexane), while Novozym 435 was resistant to methanol in the presence of hexane, and Lipozyme TLIM in the presence of ionic liquid.

The structure of the ionic liquids is another key parameter in order to assess their viability for high-yield processes. Several authors employing short-chain cations, i.e., 1,3-dialkylimidazolium cations, have obtained low enzymatic activities due to the low solubility of lipids (triglycerides) in the ionic liquid media (Gamba et al. 2008; Yang et al. 2010). In contrast, de Diego et al. (2011) showed that the use of Novozym® 435 in ionic liquids with long alkyl-chain cations led to high synthetic activity and stability. Specifically, the ionic liquids employed were 1-hexadecyl-3-methylimidazolium triflimide and 1-octadecyl-3-methylimidazolium triflimide, providing the enzyme with suitable microenvironment for continuous reuse even at high concentrations of methanol. De los Ríos et al. (2011) also observed that, after studying different ionic liquids including imidazolium and pyridinium cations, the activity of the biocatalyst increased as the cation hydrophobicity also increased, while in the case of the anion, the activity improved as the nucleophilicity decreased.

Third-generation biodiesel obtained from microalgae has also approached through enzymatic synthesis (Xiong et al. 2008). Microalgae is gaining growing attention as a sustainable feedstock for the production of biodiesel thanks to the advantages of easy and fast cultivation and high oil content, while avoiding the competition for water and land resources (Ahmad et al. 2011; Ortiz-Martínez et al. 2019). Lai et al. (2012) compared the performances of *Penicillium expansum* lipase and Novozym® 435 for the production of biodiesel using oils from several microalgae types (*Botryococcus braunii*, *Chlorella vulgaris*, and *Chlorella pyrenoidosa*) in two solvent systems, which included the ionic liquid 1-butyl-3-methylimidazolium hexafluorophosphate and tert-butanol, respectively. On the one hand, they observed that the yield was significantly higher in ionic liquid media in comparison to the organic solvent employed; besides, the activity of the first enzyme type displayed slightly greater performance versus Novozym® 435. This findings are also in line with other works that use vegetable oil feedstock (Huang et al. 2014).

While several works have studied the use of enzymatic production of biodiesel in two stages, namely oil extraction from microalgae and subsequent transesterification, it would be preferable the development of synthesis processes in which in situ extraction and transesterification were performed with one-step procedures, in order to reduce the overall costs of the process. Ionic liquids offer the advantage of combine these two requirements, by (1) extracting algal oil without cell disruption and (2) increasing the activity of enzymes for oil transesterification. Recently, Bauer et al. (2017) studied a one-step transesterification process using a mixture of whole cells of microalgae *Chromochloris zofingiensis* with ionic liquids and Novozym® 435. For this method, the authors employed a zwitter type ionic liquid. Although the yields of the reaction was as high as 64%, one of the main advantages of this process was the possibility to directly perform the conversion of wet algal biomass into biodiesel.

3.4 Conclusions

Ionic liquids are regarded as greener solvent with advantageous properties over conventional organic solvents. Together with negligible vapor pressure, they offer chemical and thermal stability with high catalytic activity for several applications. They have proven to provide promising prospects as new type of solvents or catalysts for the synthesis of biodiesel, which in turns is a renewable fuel. Biodiesel synthesis typically involves the transesterification of oils from different sources into fatty acid methyl esters in the presence of a short-chain alcohol like methanol, or the esterification of free fatty acid. As single catalyst, they display the advantages of both homogeneous and heterogeneous catalysts. Moreover, they can be used for the modification and functionalization of other chemical catalysts to increase their activity. In addition, they have shown to be effective solvents for enzyme-catalyzed transesterification, acting as immobilization agents that enable to increase the reusability of these biocatalysts. In these applications, one of the main features of ionic liquids is their capacity of being recycled without significant loss of activity liquids. In order to enhance the implementation of ionic liquid-assisted biodiesel production processes, the cost of these materials needs to be lowered, although this issue could be partially addressed by improving its recovery and reusability.

References

Abreu FR, Lima DG, Hamú EH, Einloft S, Rubim JC, Suarez PAZ (2003) New metal catalysts for soybean oil transesterification. J Am Oil Chem Soc 80:601–604. https://doi.org/10.1007/s11746-003-0745-6

Abreu FR, Lima DG, Hamú EH, Wolf C, Suarez PAZ (2004) Utilization of metal complexes as catalysts in the transesterification of Brazilian vegetable oils with different alcohols. J Mol Catal A Chem 209:29–33. https://doi.org/10.1016/j.molcata.2003.08.003

Abreu FR, Alves MB, Macêdo CCS, Zara LF, Suarez PAZ (2005) New multi-phase catalytic systems based on tin compounds active for vegetable oil transesterificaton reaction. J Mol Catal A Chem 227:263–267. https://doi.org/10.1016/j.molcata.2004.11.001

Aghabarari B, Dorostkar N (2014) Modified bentonite as catalyst for esterification of oleic acid and ethanol. J Taiwan Inst Chem Eng 34:420–427. https://doi.org/10.1016/j.jtice.2014.03.006

Ahmad AL, Yasin NHM, Derek CJC, Lim JK (2011) Microalgae as a sustainable energy source for biodiesel production: a review. Renew. Sustain. Energy Rev 15:584–593. https://doi.org/10.1016/j.rser.2010.09.018

Andreani L, Rocha JD (2012) Use of ionic liquids in biodiesel production: a review. Brazilian J Chem Eng 29:1–13. https://doi.org/10.1590/S0104-66322012000100001

Arai S, Nakashima K, Tanino T, Ogino C, Kondo A, Fukuda H (2010) Production of biodiesel fuel from soybean oil catalyzed by fungus whole-cell biocatalysts in ionic liquids. Enzym Microb Technol 46:51–55. https://doi.org/10.1016/j.enzmictec.2009.08.008

Arumugam A, Ponnusami V (2017) Production of biodiesel by enzymatic transesterification of waste sardine oil and evaluation of its engine performance. Heliyon 3:1–18. https://doi.org/10.1016/j.heliyon.2017.e00486

Bauer G, Lima S, Chenevard J, Sugnaux M, Fischer F (2017) Biodiesel via in situ wet microalgae biotransformation: Zwitter-type ionic liquid supported extraction and transesterification. ACS Sustain Chem Eng 5:1931–1937. https://doi.org/10.1021/acssuschemeng.6b02665

Cai D, Xie Y, Li L, Ren J, Lin X, Qiu T (2018) Design and synthesis of novel BrØnsted-Lewis acidic ionic liquid and its application in biodiesel production from soapberry oil. Energy Convers Manag 166:318–327. https://doi.org/10.1016/j.enconman.2018.04.036

Caldas BS, Nunes CS, Souza PR, Rosa FA, Visentainer JV, de OS JO, Muniz EC (2016) Supercritical ethanolysis for biodiesel production from edible oil waste using ionic liquid [HMim][HSO_4] as catalyst. Appl Catal B Environ 181:289–229. https://doi.org/10.1016/j.apcatb.2015.07.047

DaSilveira Neto BA, Alves MB, Lapis AAM, Nachtigall FM, Eberlin MN, Dupont J, Suarez PAZ (2007) 1-n-Butyl-3-methylimidazolium tetrachloro-indate (BMI ṡ InCl4) as a media for the synthesis of biodiesel from vegetable oils. J Catal 68:338–343. https://doi.org/10.1016/j.jcat.2007.04.015

De Diego T, Manjón A, Lozano P, Iborra JL (2011) A recyclable enzymatic biodiesel production process in ionic liquids. Bioresour Technol 102:6336–6339. https://doi.org/10.1016/j.biortech.2011.02.071

De Los Ríos AP, Hernández Fernández FJ, Gómez D, Rubio M, Víllora G (2011) Biocatalytic transesterification of sunflower and waste cooking oils in ionic liquid media. Process Biochem 46:1475–1480. https://doi.org/10.1016/j.procbio.2011.03.021

Earle MJ, Seddon KR (2000) Ionic liquids. Green solvents for the future. Pure Appl Chem 72:1391–1398. https://doi.org/10.1351/pac200072071391

Fan P, Xing S, Wang J, Fu J, Yang L, Yang G, Miao C, Lv P (2017) Sulfonated imidazolium ionic liquid-catalyzed transesterification for biodiesel synthesis. Fuel 188:483–488. https://doi.org/10.1016/j.fuel.2016.10.068

Gamba M, Lapis AAM, Dupont J (2008) Supported ionic liquid enzymatic catalysis for the production of biodiesel. Adv Synth Catal 350:160–164. https://doi.org/10.1002/adsc.200700303

Ghandi K (2014) A review of ionic liquids, their limits and applications. Green Sustain Chem 4:44–53. https://doi.org/10.4236/gsc.2014.41008

Ghiaci M, Aghabarari B, Gil A (2011a) Production of biodiesel by esterification of natural fatty acids over modified organoclay catalysts. Fuel 90:3382–3389. https://doi.org/10.1016/j.fuel.2011.04.008

Ghiaci M, Aghabarari B, Habibollahi S, Gil A (2011b) Highly efficient Brønsted acidic ionic liquid-based catalysts for biodiesel synthesis from vegetable oils. Bioresour Technol 102:1200–1204. https://doi.org/10.1016/j.biortech.2010.09.095

Ha SH, Lan MN, Lee SH, Hwang SM, Koo YM (2007) Lipase-catalyzed biodiesel production from soybean oil in ionic liquids. Enzym Microb Technol 41:480–483. https://doi.org/10.1016/j.enzmictec.2007.03.017

Hallett JP, Welton T (2011) Room-temperature ionic liquids: solvents for synthesis and catalysis. Chem Rev 111:3508–3576. https://doi.org/10.1021/cr1003248

Hama S, Kondo A (2013) Enzymatic biodiesel production: an overview of potential feedstocks and process development. Bioresour Technol 135:386–395. https://doi.org/10.1016/j.biortech.2012.08.014

Han M, Yi W, Wu Q, Liu Y, Hong Y, Wang D (2009) Preparation of biodiesel from waste oils catalyzed by a Brønsted acidic ionic liquid. Bioresour Technol 100:2308–2310. https://doi.org/10.1016/j.biortech.2008.10.046

Han BY, Li T, Deng RQ, Xiong XF, Chen CY (2013) A review on Brønsted acid ionic liquids catalysts for biodiesel synthesis through transesterification. Appl Mech Mater 389:46–52. https://doi.org/10.4028/www.scientific.net/amm.389.46

Hirani AH, Javed N, Asif M, Basu SK, Kumar A (2018) A review on first- and second- generation biofuel productions. In: Biofuels: greenhouse gas mitigation and global warming: next generation biofuels and role of biotechnology. Springer, Berlin

Huang ZL, Yang TX, Huang JZ, Yang Z (2014) Enzymatic production of biodiesel from Millettia pinnata seed oil in ionic liquids. Bioenergy Res 7:1519–1528. https://doi.org/10.1007/s12155-014-9489-6

Ishak ZI, Sairi NA, Alias Y, Aroua MKT, Yusoff R (2017) A review of ionic liquids as catalysts for transesterification reactions of biodiesel and glycerol carbonate production. Catal Rev - Sci Eng 59:1–50. https://doi.org/10.1080/01614940.2016.1268021

Lai JQ, Hu ZL, Wang PW, Yang Z (2012) Enzymatic production of microalgal biodiesel in ionic liquid [BMIm][PF_6]. Fuel 95:329–333. https://doi.org/10.1016/j.fuel.2011.11.001

Lamaisri C, Punsuvon V, Chanprame S, Arunyanark A, Srinives P, Liangsakul P (2015) Relationship between fatty acid composition and biodiesel quality for nine commercial palm oils. Songklanakarin J Sci Technol 37:389–395

Lapis AAM, De Oliveira LF, Neto BAD, Dupont J (2008) Ionic liquid supported acid/base-catalyzed production of biodiesel. ChemSusChem 1:759–762. https://doi.org/10.1002/cssc.200800077

Lee JS, Saka S (2010) Biodiesel production by heterogeneous catalysts and supercritical technologies. Bioresour Technol 101:7191–7200. https://doi.org/10.1016/j.biortech.2010.04.071

Lei Z, Chen B, Koo YM, Macfarlane DR (2017) Introduction: Ionic Liquids. Chem Rev 117:6633–6635. https://doi.org/10.1021/acs.chemrev.7b00246

Leung DYC, Wu X, Leung MKH (2010) A review on biodiesel production using catalyzed transesterification. Appl Energy 87:1083–1095. https://doi.org/10.1016/j.apenergy.2009.10.006

Li KX, Chen L, Yan ZC, Wang HL (2010) Application of pyridinium ionic liquid as a recyclable catalyst for acid-catalyzed transesterification of jatropha oil. Catal Letters 139:151–156. https://doi.org/10.1007/s10562-010-0409-x

Li J, Peng X, Luo M, Zhao CJ, Gu CB, Zu YG, Fu YJ (2014) Biodiesel production from Camptotheca acuminata seed oil catalyzed by novel Brönsted-Lewis acidic ionic liquid. Appl Energy 115:438–444. https://doi.org/10.1016/j.apenergy.2013.10.025

Liang J, Ren X, Wang J, Jinag M, Li Z (2010) Preparation of biodiesel by transesterification from cottonseed oil using the basic dication ionic liquids as catalysts. J Fuel Chem Technol 38:275–280. https://doi.org/10.1016/s1872-5813(10)60033-3

Liu S, Wang Z, Li K, Li L, Yu S, Liu F, Song Z (2013) Brønsted-Lewis acidic ionic liquid for the "one-pot" synthesis of biodiesel from waste oil. J Renew Sustain Energy 5:023111. https://doi.org/10.1063/1.4794959

Luo H, Yin H, Wang R, Fan W, Nan G (2017) Caprolactam-based Brønsted acidic ionic liquids for biodiesel production from jatropha oil. Catalysts 7:1–10. https://doi.org/10.3390/catal7040102

Ma X, Gao M, Gao Z, Wang J, Zhang M, Ma Y, Wang Q (2018) Past, current, and future research on microalga-derived biodiesel: a critical review and bibliometric analysis. Environ. Sci Pollut Res 25:10596–10610. https://doi.org/10.1007/s11356-018-1453-0

Madras G, Kolluru C, Kumar R (2004) Synthesis of biodiesel in supercritical fluids. Fuel 83:2029–2033. https://doi.org/10.1016/j.fuel.2004.03.014

Marchetti JM, Miguel VU, Errazu AF (2007) Possible methods for biodiesel production. Renew. Sustain. Energy Rev 11:1300–1311. https://doi.org/10.1016/j.rser.2005.08.006

Mohammad Fauzi AH, Amin NAS (2012) An overview of ionic liquids as solvents in biodiesel synthesis. Renew Sust Energ Rev 16:5770–5786. https://doi.org/10.1016/j.rser.2012.06.022

Motasemi F, Ani FN (2012) A review on microwave-assisted production of biodiesel. Renew. Sustain Energy Rev 16:4719–4733. https://doi.org/10.1016/j.rser.2012.03.069

Muhammad N, Elsheikh YA, Mutalib MIA, Bazmi AA, Khan RA, Khan H, Rafiq S, Man Z, Khan I (2015) An overview of the role of ionic liquids in biodiesel reactions. J Ind Eng Chem 21:1–10. https://doi.org/10.1016/j.jiec.2014.01.046

Ortiz-Martínez VM, Salar-García MJ, Hernández-Fernández FJ, Olivares-Carrillo P, de los Ríos AP, Quesada-Medina J (2016) Ionic liquids in supercritical methanol greatly enhance transesterification reaction for high-yield biodiesel production. AICHE J 62. https://doi.org/10.1002/aic.15488

Ortiz-Martínez VM, Andreo-Martínez P, García-Martínez N, Pérez de los Ríos A, Hernández-Fernández FJ, Quesada-Medina J (2019) Approach to biodiesel production from microalgae under supercritical conditions by the PRISMA method. Fuel Process Technol 191:211–222. https://doi.org/10.1016/j.fuproc.2019.03.031

Pagliaro M, Ciriminna R, Kimura H, Rossi M, Della Pina C (2007) From glycerol to value-added products. Angew Chemie - Int Ed 46:4434–4440. https://doi.org/10.1002/anie.200604694

Plechkova NV, Seddon KR (2008) Applications of ionic liquids in the chemical industry. Chem Soc Rev 37:123–150. https://doi.org/10.1039/B006677J

Ribeiro LMO, Silva AE, Silva MCS, Almeida G, RMR (2014) A study on ethanolysis and methanolysis of coconut oil for enzymatically catalyzed production of biodiesel. J Sustain Bioenergy Syst:4. https://doi.org/10.4236/jsbs.2014.44020

Sunitha S, Kanjilal S, Reddy PS, Prasad RBN (2007) Ionic liquids as a reaction medium for lipase-catalyzed methanolysis of sunflower oil. Biotechnol Lett 29:1881–1885. https://doi.org/10.1007/s10529-007-9471-x

Taher H, Nashef E, Anvar N, Al-Zuhair S (2017) Enzymatic production of biodiesel from waste oil in ionic liquid medium. Biofuels 10:463–472. https://doi.org/10.1080/17597269.2017.1316145

Talha NS, Sulaiman S (2016) Overview of catalysts in biodiesel production. ARPN J Eng Appl Sci 11:439–448

Troter DZ, Todorović ZB, Dokić-Stojanović DR, Stamenković OS, Veljković VB (2016) Application of ionic liquids and deep eutectic solvents in biodiesel production: a review. Renew Sust Energ Rev 61:473–500. https://doi.org/10.1016/j.rser.2016.04.011

Vafaeezadeh M, Alinezhad H (2016) Brønsted acidic ionic liquids: green catalysts for essential organic reactions. J Mol Liq 218:95–105. https://doi.org/10.1016/j.molliq.2016.02.017

Vafaeezadeh M, Hashemi MM (2014) Efficient fatty acid esterification using silica supported Brønsted acidic ionic liquid catalyst: experimental study and DFT modeling. Chem Eng J 250:35–41. https://doi.org/10.1016/j.cej.2014.04.001

Vafaeezadeh M, Karbalaie-Reza M, Hashemi MM, Soleimany KQ (2017) Surfactant-like Brønsted acidic ionic liquid as an efficient catalyst for selective Mannich reaction and biodiesel production in water. J Iran Chem Soc 14:907–914. https://doi.org/10.1007/s13738-016-1036-2

Wang LL, Zhang Y, Zhang F, Feng R (2016) A bronsted basic ionic liquid as an efficient and environmentally benign catalyst for biodiesel synthesis from soybean oil and methanol. Energy Sources Part A Recover Util Environ Eff 38:2770–2776. https://doi.org/10.1080/15567036.2014.968746

Wu Q, Chen H, Han M, Wang D, Wang J (2007) Transesterification of cottonseed oil catalyzed by Brønsted acidic ionic liquids. Ind Eng Chem Res 46:7955–7960. https://doi.org/10.1021/ie070678o

Wu Z, Li Z, Wu G, Wang L, Lu S, Wang L, Wan H, Guan G (2014) Brønsted acidic ionic liquid modified magnetic nanoparticle: an efficient and green catalyst for biodiesel production. Ind Eng Chem Res 53:3040–3046. https://doi.org/10.1021/ie4040016

Xie W, Wang H (2019) Immobilized polymeric sulfonated ionic liquid on core-shell structured Fe_3O_4/SiO_2 composites: a magnetically recyclable catalyst for simultaneous transesterification and esterifications of low-cost oils to biodiesel. Renew Energy 145:1709–1719. https://doi.org/10.1016/j.renene.2019.07.092

Xiong W, Li X, Xiang J, Wu Q (2008) High-density fermentation of microalga Chlorella protothecoides in bioreactor for microbio-diesel production. Appl Microbiol Biotechnol 78:29–36. https://doi.org/10.1007/s00253-007-1285-1

Yaacob Z, Nordin AM, Yarmo MA (2011) Ionic liquid supported acid-catalysed esterification of lauric acid. Malaysian J Anal Sci Ac 15:46–53

Yang Z, Zhang KP, Huang Y, Wang Z (2010) Both hydrolytic and transesterification activities of Penicillium expansum lipase are significantly enhanced in ionic liquid [BMIm][PF6]. J Mol Catal B Enzym 63:23–30. https://doi.org/10.1016/j.molcatb.2009.11.014

Zhang KP, Lai JQ, Huang ZL, Yang Z (2011) Penicillium expansum lipase-catalyzed production of biodiesel in ionic liquids. Bioresour Technol 102:2767–2772. https://doi.org/10.1016/j.biortech.2010.11.057

Zhang L, Cui Y, Zhang C, Wang L, Wan H, Guan G (2012) Biodiesel production by esterification of oleic acid over Brønsted acidic ionic liquid supported onto Fe-incorporated SBA-15. Ind Eng Chem Res 51:16590–16596. https://doi.org/10.1021/ie302419y

Zhen B, Jiao Q, Wu Q, Li H (2014) Catalytic performance of acidic ionic liquid-functionalized silica in biodiesel production. J Energy Chem 23:97–104. https://doi.org/10.1016/S2095-4956(14)60122-4

Chapter 4
Biocatalysis in Ionic Liquids: Enzymatic Synthesis of Sugar Fatty Acid Esters

Sara Isabel da Cruz Silvério and Lígia Raquel Marona Rodrigues

Contents

Abbreviations

[BMIM][BF_4]	1-butyl-3-methylimidazolium tetrafluoroborate
[BMIM][Cl]	1-butyl-3-methylimidazolium chloride
[BMIM][dca]	1-butyl-3-methylimidazolium dicyanamide
[BMIM][$MeSO_4$]	1-butyl-3-methylimidazolium methylsulfate
[BMIM][OSO_4]	1-butyl-3-methylimidazolium octylsulfate
[BMIM][PF_6]	1-butyl-3-methylimidazolium hexafluorophosphate
[BMIM][Tf_2N]	1-butyl-3-methylimidazolium bis[(trifluoromethyl)sulfonyl]imide
[BMIM][TfO]	1-butyl-3-methylimidazolium trifluoromethanesulfonate
[BMPyr][PF_6]	1-butyl-4-methylpyridinium hexafluorophosphate
[BMPyrr][Tf_2N]	1-butyl-3-methylpyrrolidinium bis[(trifluoromethyl) sulfonyl]imide

S. I. da Cruz Silvério · L. R. M. Rodrigues (✉)
CEB-Centre of Biological Engineering, Universidade do Minho, Braga, Portugal
e-mail: sarasilverio@deb.uminho.pt; lrmr@deb.uminho.pt

Inamuddin, A. M. Asiri (eds.), *Nanotechnology-Based Industrial Applications of Ionic Liquids*, Nanotechnology in the Life Sciences,
https://doi.org/10.1007/978-3-030-44995-7_4

[BMPyrr][TfO]	1-butyl-3-methylpyrrolidinium trifluoromethanesulfonate
[DMIM][DMP]	1,3-dimethylimidazolium dimethylphosphate
[DMIM][$MeSO_4$]	1,3-dimethylimidazolium methylsulfate
[EMIM][$EtSO_4$]	1-ethyl-3-methylimidazolium ethylsulfate
[EMIM][$MeSO_3$]	1-ethyl-3-methylimidazolium methanesulfonate
[EMIM][$MeSO_4$]	1-ethyl-3-methylimidazolium methylsulfate
[EMIM][PF_6]	1-ethyl-3-methylimidazolium hexafluorophosphate
[EMIM][TfO]	1-ethyl-3-methylimidazolium trifluoromethanesulfonate
[HMIM][PF_6]	1-hexyl-3-methylimidazolium hexafluorophosphate
[HMIM][TfO]	1-hexyl-3-methylimidazolium trifluoromethanesulfonate
[HPyr][PF_6]	1-hexylpyridinium hexafluorophosphate; ILs—ionic liquids
[Me(OEt)$_3$-Et$_3$N][OAc]	Triethyl (2-(2-methoxyethoxy)ethoxy)ethylammonium acetate
[Me(OEt)$_3$-Et-IM][OAc]	3-(2-(2-methoxyethoxy)ethoxy)ethylimidazolium acetate
[MMIM][$MeSO_4$]	1,3-dimethylimidazolium methylsulfate
[MOEMIM][BF_4]	1-methoxyethyl-3-methylimidazolium tetrafluoroborate
[MOEMIM][dca]	1-methoxyethyl-3-methylimidazolium dicyanamide
[MPIM][PF_6]	1-methyl-3-propylimidazolium hexafluorophosphate
[MTOAM][Tf_2N]	Methyltrioctylammonium bis[(trifluoromethyl)sulfonyl]imide
[OMIM][Tf_2N]	1-methyl-3-octylimidazolium bis[(trifluoromethyl)sulfonyl]imide
[TBA][HSO_4]	Tetrabutylammonium bisulfate
[TBA][OAc]	Tetrabutylammonium acetate
[TBMA][Tf_2N]	Tributylmethylammonium bis[(trifluoromethyl)sulfonyl]imide
[TBP][OAc]	Tetrabutylphosphonium acetate
[TEA][OAc]	Tetraethylammonium acetate
[THTDP][PF_6]	Trihexyltetradecylphosphonium hexafluorophosphate
[TMA][OAc]	Tetramethylammonium acetate
[TMHA][Tf_2N]	Trimethylhexylammonium bis[(trifluoromethyl)sulfonyl]imide
2M2B	2-methyl-2-butanol
Ammoeng 100	Cocos alkyl pentaethoximethyl ammonium methylsulfate
Cyphos 104	Trihexyl(tetradecyl)phosphonium bis(2,4,4-trimethylpentyl)phosphinate
Cyphos 109	Trihexyl(tetradecyl)phosphonium bis[(trifluoromethyl) sulfonyl]imide
DES	Deep eutectic solvents
DMF	Dimethylformamide

DMSO	Dimethyl sulfoxide
HLB	Hydrophilic-lipophilic balance
SFAEs	Sugar fatty acid esters
THF	Tetrahydrofuran

4.1 Introduction

Sugar fatty acid esters (SFAEs) are interesting compounds containing a carbohydrate linked to one or more fatty acid chains. They are well-known non-ionic and biodegradable surfactants that can be synthesized from renewable resources (Yang and Huang 2012). Several natural sugars have been combined with different fatty acids to form unique and promising SFAEs.

The fatty acids most frequently used present saturated or unsaturated aliphatic chains containing 12 to 22 carbons, such as lauric (C12, saturated), myristic (C14, saturated), palmitic (C16, saturated), stearic (C18, saturated), oleic (C18, monounsaturated), linoleic (C18, polyunsaturated), behenic (C22, saturated), or erucic (C22, monounsaturated) (Zheng et al. 2015; Pérez et al. 2017).

The sugars involved in the synthesis of SFAEs can be mono- (fructose, glucose, or galactose), di- (sucrose, lactose, or maltose), oligo- (raffinose, maltotriose, or melezitose), polysaccharides (cellulose, dextrin or starch), or even polyols (sorbitol, xylitol or glycerol) (Pérez-Victoria et al. 2011; Zheng et al. 2015; Panpipat and Chaijan 2016). When a hydrophilic sugar head is combined with one or more lipophilic alkyl chains, amphipathic compounds with inherent hydrophilic-lipophilic balance (HLB) are generated.

Values of HLB ranging from 0 to 20 strongly determine the physicochemical properties of each type of SFAEs and consequently their specific applications. SFAEs with low HLB values (≤ 6) are oil-soluble and therefore can form water-in-oil emulsions or act as anti-foaming agents. Intermediate HLB values (7–9) means that the SFAE is equally balanced in hydrophilicity and lipophilicity and can be a good wetting and spreading agent. Higher HLB values (≥ 10) indicate that SFAEs are water-soluble and thus can form oil-in-water emulsions or act as detergents (Ye and Hayes 2014; Zheng et al. 2015; Pérez et al. 2017).

Additionally, other interesting characteristics are attributed to SFAEs such as tasteless, odorless, non-toxic and non-irritating, biodegradable, biocompatible, and several bioactive properties (Pérez et al. 2017; Ye and Hayes 2017). Therefore, SFAEs present very interesting features for industrial applications such as in the food, pharmaceutical, and cosmetics industries. SFAEs are already approved as food additives in many countries, and they can be used in the food industry as emulsifiers, fat substitutes, stabilizers, staling retarders, surfactants in colloidal delivery systems such as micro- and nanoemulsions, antimicrobial agents or to improve the organoleptic properties of the products (Neta et al. 2012; Zheng et al. 2015).

Similar applications are described for SFAEs in the pharmaceutical industry where they are important in the preparation of emulsifications and to stabilize diverse drug delivery systems. Furthermore, SFAEs can play an important role in the modification of the bioavailability of several compounds since they can affect the dissolution, absorption, and release of drugs. The antimicrobial, anti-insect, and anti-tumor properties of SFAEs are also valorized in the pharmaceutical industry (Zheng et al. 2015).

In the cosmetics industry, the non-irritating and biocompatible character of SFAEs together with all the above-mentioned features, make them suitable to be included in the preparation of several products for skin and hair treatment (Pérez et al. 2017; Zheng et al. 2015).

4.2 Conventional Synthesis of Sugar Fatty Acid Esters

4.2.1 Chemical Synthesis

The majority of marketed sugar fatty acid esters (SFAEs) are obtained by chemical synthesis and different approaches have been developed to maximize the yield and selectivity of the final products. In general, the main reactions involved in the production of SFAEs include direct esterification using fatty acid chloride or anhydride; and interesterification or transesterification using esters of the fatty acids such as methyl and vinyl esters (Nelen et al. 2015).

At an industrial level, the synthesis of the most common SFAEs (sugar esters of fructose, glucose, and sucrose) is performed by transesterification using methyl esters of the fatty acids in the presence of alkaline or metallic chemical catalysts, e.g., di-sodium hydrogen phosphate or lithium oleate (Ferrer et al. 2000; Gumel et al. 2011). The reactions are generally carried out in organic solvents at a high temperature (> 100 °C) and reduced pressure.

However, several limitations are described for these reactions, namely the lability of the sugars under high temperature; degradation of the starting material due to possible caramelization, saponification, and Maillard reactions; production of undesired by-products affecting the flavor and color of the reaction mixture hence compromising the yield; use of toxic solvents in production and purification processes; and need of a rigorous downstream processing to obtain safe products for food, pharmaceutical, and cosmetics applications (Nelen et al. 2015).

Furthermore, the severe reaction conditions lead to high energy consumption and high environmental impact due to the use of hazardous chemicals and the generation of toxic side-products (Pérez et al. 2017). Since the chemical synthesis is frequently low selective, complex mixtures of SFAEs with different degrees of esterification (e.g., mono-, di-, tri-esters, or even more substituted SFAEs, depending on the number of hydroxyl groups present in the sugar molecule used as starting material) (Zheng et al. 2015), as well as different positions of acylation can be obtained.

To overcome this limitation, the use of multiple protection and deprotection steps, associated to both primary and secondary hydroxyl groups of the sugar molecule, has been proposed. This strategy can be efficient and attractive for the synthesis of some SFAEs, but it is laborious and time-consuming, and it is not viable at an industrial scale (Faber 1995; Pérez et al. 2017).

4.2.2 Enzymatic Synthesis

The enzymatic synthesis of sugar fatty acid esters (SFAEs) is performed through a one-step process without the need of protection/deprotection reactions, and it has been proposed as a more efficient and sustainable alternative to the chemical process. Enzymatic methods are associated to lower energy consumption (temperature reaction at 35–70 °C) and environmental impact; higher degree of chem-, regio-, enantio-, and diastereo-selectivity; higher yields and productivity; and simpler product mixtures which significantly facilitate the downstream purification (Gumel et al. 2011).

Lipases and carboxylesterases are suitable biocatalysts for reactions that involve sugars and fatty acids, such as the synthesis of SFAEs. These enzymes belong to the carboxylic ester hydrolase family (EC 3.1.1) and are known by their ability to catalyze the cleavage of the ester bond of lipids. Carboxylesterases generally prefer water-soluble fatty acids shorter than C6 as substrates, while lipases have more affinity for water-insoluble and long-chain fatty acids (Ballesteros et al. 2007). Other type of enzymes reported for the synthesis of some short-chain SFAEs are proteases (Patil et al. 1991), including subtilisin (Carrea et al. 1989).

Nevertheless, lipase is the biocatalyst most commonly used in the production of SFAEs by an enzymatic route. This enzyme presents unique and interesting features such as it can catalyze heterogenous reactions at the interface of water-soluble and water-insoluble systems, it presents activity in several organic solvents, and it tolerates high temperatures (60–70 °C). Several sources of lipase have been tested, mostly from microbial origin. However, lipases from animal origin (porcine pancreas) can also be used in the synthesis of some SFAEs (Gumel et al. 2011).

In aqueous media, lipases usually catalyze hydrolysis reactions. However, in non-aqueous media these enzymes can catalyze other type of reactions, namely esterification and transesterification, which allow the synthesis of SFAEs as illustrated in Fig. 4.1. Although the reactions occur in non-aqueous media, the presence of a sparing amount of water is essential for the hydration of the enzyme and thus ensure the optimal conformation and activity (Neta et al. 2012). On the other hand, the water generated from the esterification reaction, can significantly compromise the SFAE yield since it will shift the equilibrium towards a backwards reaction (hydrolysis), and some product degradation will occur. Therefore, the water should be continuously removed from the reaction media, and several methodologies can be applied such as evaporation under reduced pressure, azeotropic distillation, or the use of molecular sieves (Yan et al. 1999; Gubicza et al. 2003; Neta et al. 2012).

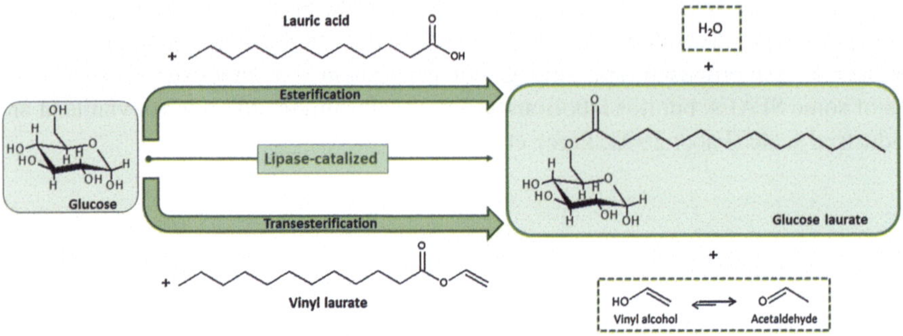

Fig. 4.1 Lipase-catalyzed reactions of esterification (lauric acid as substrate and water as by-product) and transesterification (vinyl laurate as substrate and vinyl alcohol almost irreversibly converted to acetaldehyde as by-products) in non-aqueous medium to produce glucose laurate

In the SFAEs synthesis by transesterification, methyl, ethyl, or vinyl esters of fatty acids can be used. However, vinyl ester is the most commonly chosen since its by-product, a vinyl alcohol (ethenol), is almost irreversibly converted to acetaldehyde which will contribute to shift the equilibrium to the ester formation (Adachi and Kobayashi 2005). Besides increasing the SFAE yield, this strategy presents also an additional advantage, namely the side-product acetaldehyde is easily removed from the reaction medium due to its boiling point which is around room-temperature (20.2 °C) (Panda 2002).

The SFAE yield and productivity are dependent on several parameters such as the solvent used as reaction medium, the structure of the sugar, the chain length of the fatty acid, and the source of the enzyme. Some caution should be taken when choosing the solvent since the catalytic action of lipases and solubility of the quite different substrates is strongly affected by it. Although enzymes present recognized selectivity, SFAEs with different degrees of esterification can be obtained depending on both the reaction conditions and the enzyme origin.

Generally, the enzymes have preference for the less sterically impeded fatty acyl chains and the primary hydroxyls on the sugar structure are the easiest to substitute due to its higher reactivity (Zheng et al. 2015; Pérez et al. 2017). However, some lipases can present high selectivity for long- and medium-chain fatty acids, others for short and branched fatty acids, and there are also some cases where the enzymatic conversion is apparently independent from the chain length (Zheng et al. 2015).

Although enzymatic synthesis is considered more environmentally friendly and preferable to obtain SFAEs for food, pharmaceutical, and cosmetic applications, the use of organic solvents is regarded as a negative point in the overall process. Therefore, alternative solvents and creative strategies have been proposed to generate a greener enzymatic process.

The lipase-catalyzed solid-phase synthesis of SFAEs was developed with the purpose to use minimal amount of solvent. This approach is based on a mainly solid-phase system where the acylation of the solid sugar with the fatty acid is catalyzed by immobilized lipase in the presence of a very small amount of organic solvent (usually acetone or *t*-butanol) which mostly acts as adjuvant. The presence of the solvent is important to ensure a small catalytic liquid phase for lipase. The SFAE should present low solubility in the chosen adjuvant to facilitate its removal from the reaction medium by crystallization (Cao et al. 1996; Yan et al. 1999). Besides the use of organic solvents, another important disadvantage of this strategy is the recognized effort to separate the SFAE from the unreacted substrate and immobilized catalyst (Ye and Hayes 2014).

The preparation of metastable suspensions in solvent-free reaction medium has also been proposed in the enzymatic synthesis of SFAEs. In this case, initial suspensions (2–3 micron) of sugar crystals and a mixture of the acyl donor (e.g., oleic acid) and the corresponding ester are prepared under magnetic stirring and ultrasound conditions. Subsequently, this suspension is put in contact with the immobilized enzyme under optimal temperature. The formed SFAEs will also act as co-solvent together with the acyl donor to improve the miscibility. Although laborious, this strategy provides mixtures with higher ester content (Ye et al. 2010, 2016).

The substitution of the organic solvents by supercritical carbon dioxide has also been studied in the synthesis of SFAEs. Some advantages have been described for supercritical carbon dioxide, namely it is non-toxic, non-inflammable, inexpensive, and provides higher mass transfer, low surface tension, and viscosity. Furthermore, due to the gaseous nature of the solvent, it can be easily removed without leaving any residues in the final product. Since only the fatty acid is soluble in the supercritical phase, the separation of the SFAEs from the remaining substrates and immobilized enzyme becomes easier (Šabeder et al. 2005). However, one recognized limitation of supercritical carbon dioxide is its rather nonpolar feature which can compromise the solubility of hydrophilic compounds such as sugars. To overcome this issue, the sugars can be adsorbed onto an inert solid support with high internal surface such as silica gel or cellulose (Stamatis et al. 2001).

Deep eutectic solvents (DES) which are ecologically well accepted and typically obtained from natural compounds were also reported as suitable alternatives to the hazardous organic solvents used in the enzymatic synthesis of SFAEs. Furthermore, some DES present the advantage of dissolve both hydrophilic sugars and hydrophobic fatty acids. Additionally, DES prepared with the sugar under study can act as both solvent and substrate (Pöhnlein et al. 2015; Siebenhaller et al. 2017).

Another alternative solvents, which have been widely studied in the past two decades for the synthesis of SFAEs, are the ionic liquids (ILs). The major advantages of these kind of solvents, their suitability for lipase-catalyze reactions, and the main achievements reported for the synthesis of SFAEs will be discussed in the following sections.

4.3 Potential of Ionic Liquids in Biocatalysis

In the last decades, ionic liquids (ILs) have gained increasing attention as alternative solvents for biocatalysis. There is a significant number of different enzyme-catalyzed reactions which were efficiently performed in ionic liquids (Itoh 2017).

Furthermore, ILs are widely described as green and environmentally friendly solvents with the ability to replace the hazardous, flammable, and highly volatile organic solvents.

ILs can be defined as organic salts which are liquids at room-temperature (Yang and Pan 2005) or at temperatures below 100 °C (Welton 2018), and remain liquid within a broad range of temperatures (generally below 300 °C). The successful application of ILs as alternative solvents is strictly connected with the interesting properties that come from their hybrid organic-ionic nature (Lei et al. 2017). The most important physicochemical properties generally associated to ILs include the extremely low volatility; non-flammability; high solvating potential; unique solubility for a wide range of polar and nonpolar organic, inorganic, and polymeric compounds; miscibility in several organic solvents and, in some cases, also in water; high conductivity; thermal, chemical, and electrochemical stability; and high "tunability" (Yang and Pan 2005; Dong et al. 2017; Itoh 2017).

In fact, ILs are frequently described as "designer" or "designable" solvents due to the possibility of combining different cations and anions or performing specific structural modifications in the core ions (namely the attached substituents) to obtain the desired physicochemical properties. Therefore, ILs can be tailored to meet the specific requirements of a given process or analysis which is indubitably a remarkable advantage (Vekariya 2017).

Another interesting aspect of the use of ILs as alternative solvents is their capacity to be recovered from different solutions after being used in a given application. In the last decades, several methodologies have been proposed for their recovery, recycling, and reuse. Among the strategies conventionally used, distillation of volatile compounds or distillation though specific reactions of ILs, and extraction with water, organic solvents, or supercritical carbon dioxide are the preferred techniques. However, other methodologies have also been reported such as adsorption using several adsorbent materials, membrane separation, aqueous two-phase extraction, temperature- and pressure-induced crystallization, and force field separation. Sometimes, methodologies must be combined towards a better recovery (Zhou et al. 2018).

The use of ILs as non-aqueous reaction media for biotransformation was reported for both types of catalysts, namely enzymes (free or immobilized) and whole cells. The biocatalyses have been described in different environments such as monophasic ionic liquid-based systems or in biphasic systems containing ionic liquid/water, ionic liquid/organic solvents, or ionic liquid/supercritical carbon dioxide (Mai and Koo 2018). Several advantages have been highlighted comparatively to the organic solvents, namely the enhanced solubility of the substrates and products, higher conversion rates, higher enzymatic activity and selectivity, higher thermal and operational stability (Itoh 2017; Mai and Koo 2018).

Hydrolases are the family of enzymes most widely used as biocatalysts in reaction media containing ILs. Both hydrolysis and transesterification reactions have been performed in ILs by lipases, esterases, α-chymotrypsin, or subtilisin. Oxidoreductases such as laccase, peroxidase, catalase, or tyrosinase were also successfully used as biocatalysts in oxidation and reduction reactions carried out in ILs.

The catalytic action of other types of enzymes like glucose and xylose transferase or DNA polymerase has also been reported (Itoh 2017; Mai and Koo 2018).

Recently, whole cells-based biocatalysis in systems containing ILs has gained an increased attention. The use of whole cells as biocatalysts holds recognized advantages such as high selectivity and efficiency, lower catalyst costs, and environmental safety. However, its application in ILs should be carefully considered. Although the status of non-toxic and safe generally attributed to ILs, they can exhibit a significant biological activity which can considerably compromise the integrity and efficiency of the whole cells as biocatalysts (Egorova and Ananikov 2018). Nevertheless, some successful applications in oxidation, reduction, hydrolytic, and transesterification reactions have been described using whole cells of bacteria, yeasts, and fungi (Xu et al. 2016).

4.4 Enzymatic Synthesis of Sugar Fatty Acid Esters in Ionic Liquids

The selection of the most suitable ionic liquid (IL) is a critical aspect in the enzymatic synthesis of sugar fatty acid esters (SFAEs), and it is not frequently easy. It is well-reported that ILs have great potential to solubilize compounds with different polarity, including carbohydrates (Farrán et al. 2015), proteins/enzymes (Schröder 2017), and lipids (Ventura et al. 2017). Nevertheless, in the synthesis of SFAEs these different compounds will be simultaneously present in the reaction media which could be difficult for the development of the bioprocess.

Therefore, the ideal IL should have the ability to solubilize the chemically different substrates, and eventually the SFAEs formed during the enzymatic reaction, and should not compromise the stability and activity of the biocatalyst (e.g., lipase). It is reported that the reaction rate, yield, and selectivity strongly depends on the substrates solubility and enzyme performance in the reaction medium. Hence, it is extremely important to use a suitable IL as solvent (Galonde et al. 2012).

4.4.1 Lipase Activity and Stability

Lipases, which are known for their robustness in different enzymatic processes and natural tolerance to several organic solvents, are an obvious candidate for biocataysis in ionic liquid (ILs). In fact, lipases from different sources have already showed interesting catalytic activities in ILs (Hernández-Fernández et al. 2010). For example, the esterification or transesterification activities, necessary to produce the sugar fatty acid esters (SFAEs), have been reported for lipases from *Candida antarctica*, *Candida rugosa*, *Pseudomonas cepacia*, *Thermomyces lanuginosus*, *Rhizomucor miehei*, *Bacillus subtilis*, or porcine pancreas (Nara et al. 2002; Gubicza et al. 2003;

Byun et al. 2007; de los Ríos et al. 2007; Abdulmalek et al. 2012; Gumel and Annuar 2016; Chado et al. 2018). Nevertheless, lipase from *C. antarctica* is the most widely studied among all the available sources.

The active site of lipases is characterized by the existence of a lid-region besides the substrate binding region. This lid-region is described as a mobile structure composed of amino acids with amphipathic properties covering the catalytic active site, and it may present variable length and complexity depending on the type of lipase (Secundo et al. 2006). Therefore, the amino acids composition is crucial for the so-called phenomenon of interfacial activation.

Since lipases are only active at interfaces, namely between a hydrophobic solvent and an aqueous medium, it is postulated that the phenomenon of interfacial activation results from a conformational change in the lid-region. When lipase is in the presence of those interfaces, the conformational changes may open the lid and make the active site more accessible (Verger 1997). Therefore, the lid-region seems to play an important role in modulating the activity, selectivity, and stability of lipases. Additionally, the open lid form is thermodynamically favored over the close one, as it leads to a more efficient interaction between the hydrophobic amino acids of the lid and the lipophilic medium (De María et al. 2006).

Lipases, and enzymes in general, which are surrounded by a micro-aqueous phase probably present a similar catalytic mechanism in ILs as in water and in organic solvents. Consequently, an IL may affect the enzyme action by removing the essential water from the micro-aqueous layer; by penetrating this layer, dissociating into individual cations and anions, and harmfully interacting with the enzyme through changes in the microenvironment, conformation, and/or active site; or even by negatively interacting with the substrates and products, which may occur through direct reaction with them or through changing their partition between the aqueous and non-aqueous phases (Yang 2009).

For ILs, it is believed that the effect of both cations and anions on enzyme performance may generally follow the Hofmeister series. The enzymatic activity and stability seem to be commonly favored by the presence of kosmotropic anions and chaotropic cations (Yang 2009). However, for some ILs this series is not verified or is reverted, especially for the anion effect when more kosmotropic cations are present (Lou et al. 2006). Although this unusual behavior is only observed for few enzymes, it also points out that enzyme performance in ILs may be governed by additional and complex mechanisms besides the classical approach based on the kosmotropicity/chaotropicity of the ions, which in turn is directly connected with the degree of ion hydration. These complex mechanisms, which become more emphasized for concentrated or neat ILs, may include other important effects related with the hydrophobicity of the IL, H-bond basicity and nucleophilicity of anions (Yang 2009). For all these reasons, the expression "specific ion effects" was recently proposed, instead of Hofmeister effect, to discuss the protein stabilization and enzyme activation phenomena occurring in ILs (Zhao 2016).

In some cases, the rule to predict the enzymatic performance in ILs seems to be "there is no absolute rule." This happens due to the several unknown factors involved in the enzyme–ILs interactions plus the practical observation that a specific catalytic action in a given IL may vary according to the type or source of enzyme implicated (De Diego et al. 2009).

However, it seems to exist some patterns for lipases when catalyzing esterification or transesterification reactions. The most water-miscible (hydrophilic) ILs can strongly deactivate enzyme under anhydrous conditions. On the other hand, the water-immiscible (hydrophobic) ILs can be suitable for lipase activity at low water content or anhydrous conditions (Kaar et al. 2003; De Diego et al. 2009). Additionally, the more nucleophilic anions tend to interact with the positively charged sites of lipase, causing negative conformation changes (Kaar et al. 2003; Hernández-Fernández et al. 2009). Similar negative effect is reported for anions with high H-bond basicity (Toral et al. 2007; Zhao 2012). Also, the presence of halide impurities (chloride, bromide, and iodide), resulting from the use of halide precursors in the synthesis of ILs, can significantly affect the enzymatic activity of some lipases (Lee et al. 2006; Zhao 2012).

A common strategy to improve lipase activity and stability in ILs is its immobilization through different approaches, namely solid carrier, sol-gel encapsulation, or cross-linked aggregates (Zhao 2010). Immobilized lipases are widely used in the synthesis of SFAEs in ILs (Yang and Huang 2012).

4.4.2 Substrate Solubility

Sugars and fatty acids are chemically very different and the choice of a suitable solvent which can solubilize significant amount of both substrates is not trivial. There are only few studies reporting the solubility of fatty acids in ionic liquids (ILs), and almost all are related with the application of these solvents for lipid extraction from algae (Orr and Rehmann 2016). In general, hydrophilic and water-soluble ILs, such as dialkylimidazolium cations combined with alkyl ester sulfates or phosphates, present ability to solubilize lipids, including fatty acids. Also, anions such as acetate, tetrafluoroborate, and chloride showed some potential when combined with those cations (Choi et al. 2014).

In a recent study performed with the help of a conductor-like screening model for real solvents, several combinations of cations and anions were evaluated for the extraction of a specific fatty acid, eicosapentaenoic acid. It was found that small anions with higher charge density and shorter alkyl chain cations in imidazolium-based ILs are preferred (Motlagh et al. 2019). Another study demonstrated that the IL Cocos alkyl pentaethoximethyl ammonium methylsulfate (Ammoeng 100) has great affinity for linoleic acid probably because the Cocos alkyl chain in this IL is originated from fatty acids, and it contains approximately 50% of lauric acid alkyl chain (Manic et al. 2011).

The solubility of lauric acid at 50 °C is lower in the hydrophobic ILs 1-butyl-3-methylimidazolium bis[(trifluoromethyl)sulfonyl]imide [BMIM][Tf_2N] and 1-methyl-3-octylimidazolium bis[(trifluoromethyl)sulfonyl]imide [OMIM][Tf_2N] than in the hydrophilic 1-butyl-3-methylimidazolium trifluoromethanesulfonate

[BMIM][TfO] (Ha et al. 2010a). On the other hand, the solubility of the vinyl ester of lauric acid seems to be improved in ILs with a longer alkyl chain on the cation combined with hydrophobic anions such as Tf_2N^- (Lin et al. 2015).

For carbohydrate transformation, ILs containing imidazolium-based cations combined with nucleophilic anions such as chloride, bromide, dicyanamide, formate, or acetate have been widely studied (Zhao et al. 2008; Yang and Huang 2012). Furthermore, the incorporation of an ether moiety in the side chain of the imidazolium cation can result in ILs with higher ability to dissolve different carbohydrates due to unique "sugar-philic" properties created (Kimizuka and Nakashima 2001). Sugar solubility in ILs seems to be more affected by the nature of the anion rather than the cation (Liu et al. 2005). Also, the sugar chain can affect the solubility since it generally decreases as the chain becomes longer.

The nucleophilic anions have high capacity to dissolve carbohydrates due to the formation of strong H-bond with them. For example, dicyamide-based ILs present a superior sugar solubilizing capacity due to dicyanamide's H-bond acceptor capacity. Nevertheless, high anion nucleophilicity can promote enzyme denaturation (van Rantwijk et al. 2006). On the other hand, halogenated anions, such as BF_4^-, PF_6^-, or Cl^-, may suffer hydrolysis and/or thermal decomposition during the reaction time and originate the formation of toxic compound such as hydrofluoric acid (HF), trifluorophosphate (POF_3), and hydrochloric acid (HCl), which can also deactivate the enzymes (Galonde et al. 2012).

The substrate solubility generally increases with the temperature (Yang and Huang 2012). A disadvantage often pointed out for the enzymatic synthesis is the need to work at moderate temperatures (35–70 °C), which could be inadequate to properly solubilize the hydrophilic sugar and the hydrophobic fatty acid. Besides the substrates solubility and enzyme stability, the temperature also affects the solubility of the products (which sometimes are separated by precipitation) and consequently the rate, equilibrium, and yield of the reaction (Gumel et al. 2011).

The molar ratio of sugar to fatty acid (or ester of fatty acid) significantly affects the synthesis of sugar fatty acid esters (SFAEs) since the solubility of a given substrate is dependent on the amount of the other substrate already dissolved. Generally, for moderate fatty acid chains, an excess of fatty acid favors the synthesis of SFAEs, while for longer fatty acid chains, a lower molar ratio of fatty acid to sugar is recommended (Gumel et al. 2011; Zheng et al. 2015).

4.4.3 Types of Sugar Fatty Acid Esters

Different sugar fatty acid esters (SFAEs) have been synthesized in reaction media composed of pure ionic liquids (ILs) or ILs in combination with organic solvents. Table 4.1 summarizes the most significant results reported in the literature under this topic. The most commonly synthesized SFAE is glucose laurate. However, other SFAEs have been obtained from mono- (fructose, galactose, or mannose) and disaccharides (sucrose or maltose). The catalyst typically used in the synthesis of

SFAEs is immobilized lipase B from *C. antarctica* which could yield a mixture of mono- and diesters (Ye and Hayes 2014).

4.4.3.1 Pure Ionic Liquids or Ionic Liquids Mixtures as Solvents

One of the first studies reporting the synthesis of sugar fatty acid esters (SFAEs) in ionic liquids (ILs) was performed by Liu et al. (2005) when they investigated the ability of some ILs to dissolve carbohydrates. They found that 1-butyl-3-methylimidazolium dicyanamide [BMIM][dca] have high capacity to solubilize different sugars (glucose, sucrose, lactose, and β-cyclodextrin) and successfully used this IL as reaction medium for the lipase-catalyzed esterification of sucrose and lauric acid. However, no additional information about the conversion yield, regioselectivity, or enzyme stability in this commonly denaturing IL (van Rantwijk et al. 2006) was reported.

Ganske and Bornscheuer (2005a) described the synthesis of 6-*O*-glucose laurate and 6-*O*-glucose myristate by a transesterification reaction catalyzed by poly(ethylene) glycol-modified lipase B from *C. antarctica* in 1-butyl-3-methylimidazolium tetrafluoroborate [BMIM][BF_4] and 1-butyl-3-methylimidazolium hexafluorophosphate [BMIM][PF_6] with conversion yields of 30% and 35%, respectively.

Another work described the conversion of levoglucosan (1,6-anhydroglucopyranose), an anhydrosugar derived from glucose, through lipase-catalyzed esterification and transesterification to form 4-*O*-levoglucosan laurate (Galletti et al. 2007). For this purpose, different ILs ([BMIM][BF_4], 1-methoxy ethyl-3-methylimidazolium tetrafluoroborate [MOEMIM][BF_4] and 1-methoxyethyl-3-methylimidazolium dicyanamide [MOEMIM][dca]), and biocatalysts (immobilized lipases from *P. cepacia* and *C. antarctica*) were tested. The best conversion yields were obtained with lipase from *C. antarctica* in [MOEMIM][dca] using either vinyl laurate (19%) or lauric acid (37%). Both [BMIM][BF_4] and [MOEMIM][BF_4] proved to be unsuitable solvents for the enzymatic synthesis of 4-*O*-levoglucosan laurate due to the low yields obtained (4–13%).

The synthesis of glucose palmitate was achieved by transesterification in [BMIM][TfO] using different sources of immobilized lipase as biocatalysts (*C. antarctica, T. lanuginosus,* and *M. miehei*) (Liang et al. 2012). After optimizing the reaction conditions, a 32% conversion yield was obtained for lipase from *C. antarctica* using a molar ratio of glucose:vinyl palmitate of 1:3. Glucose palmitate synthesis by lipase-catalyzed esterification was also studied in [BMIM][PF_6], trihexyl(tetradecyl) phosphonium bis(2,4,4-trimethylpentyl)phosphinate (commercial name Cyphos 104) and trihexyl(tetradecyl)phosphonium bis[(trifluoromethyl)sulfonyl]imide (commercial name Cyphos 109) (Findrik et al. 2016). The substrate conversion yield was higher in [BMIM][PF_6] (77%) when compared to Cyphos 104 (60%) and Cyphos 109 (45%). Additionally, the enzyme stability was tested using [BMIM][PF_6] as solvent and it was verified that conversion yields lowered by 13 and 19% after 4 cycles of reuse at 50 and 70 °C, respectively.

Table 4.1 Different types of sugar fatty acid esters synthesized in ionic liquids using lipase as biocatalyst

Product	Biocatalyst	Ionic liquid (IL)	Reaction conditions	References
Glucose caprylate	Lipase B from *Candida antarctica* (immobilized)	[BMIM][TfO] [OMIM][Tf_2N]	**Esterification**: Supersaturated glucose (0.33 mmol) and caprylic acid (0.33 mmol) in IL mixture (1:1) at 50 °C.	Ha et al. (2010a)
Glucose laurate	Lipase B from *Candida antarctica* (free and immobilized) Lipase B from *Candida antarctica* (immobilized) Lipase B from *Candida antarctica* (immobilized)	[BMIM][BF_4] [BMIM][PF_6] [BMIM][TfO] [BMIM][BF_4] [EMIM][TfO] [BMIM][TfO] [BMIM][PF_6]	**Transesterification**: Glucose (0.25 mmol) and vinyl laurate (0.5 mmol) in pure IL or IL/*t*-butanol (40/60% v/v) mixture at 60 °C. **Transesterification**: Supersaturated glucose (1.1 mmol) and vinyl laurate (2.2 mmol) in pure IL at 40 °C. **Esterification**: Supersaturated glucose (1.1 mmol) and lauric acid (2.2 mmol) in pure IL at 50 °C. **Transesterification**: Supersaturated glucose (1.1 mmol) and vinyl laurate (2.2 mmol) in pure IL at 40 °C with and without ultrasound radiation. **Esterification**: Supersaturated glucose (1.1 mmol) and lauric acid (2.2 mmol) in pure IL at 50 °C with and without ultrasound radiation.	Ganske and Bornscheuer (2005a), (2005b) Lee et al. (2008a) Lee et al. (2008b)
Glucose laurate	Lipase B from *Candida antarctica* (immobilized) Lipase B from *Candida antarctica* (immobilized) Lipase B from *Candida antarctica* (immobilized) Lipase B from *Candida antarctica* (immobilized)	[BMIM][TfO] [BMIM][Tf_2N] [Me(OEt)$_3$-Et-IM][OAc] [Me(OEt)$_3$-Et_3N][OAc] [BMIM][TfO] [Omim][Tf_2N] [TBA][OAc]	**Transesterification**: Supersaturated glucose (0.07–1.1 mmol) and vinyl laurate (0.5–2.2 mmol) in IL mixture (1:1) at 40 or 67 °C. **Esterification**: Supersaturated glucose (1.1 mmol) and lauric acid (2.2 mmol) in IL mixture (1:1) at 50 °C. **Transesterification**: Glucose (0.22 mmol) and vinyl laurate (0.33 mmol) in pure IL or IL/*t*-butanol mixtures at 50 °C. **Esterification**: Supersaturated glucose (0.33 mmol) and lauric acid (0.33 mmol) in IL mixture (1:1) at 50 °C. **Transesterification**: Glucose (0.3 mmol) and vinyl laurate (0.3 mmol) in pure IL at 40 °C.	Lee et al. (2008c) Mai et al. (2014) Shin et al. (2019) Zhao et al. (2009) Ha et al. (2010a) Lin et al. (2016)
Glucose laurate	Lipase B from *Candida antarctica* (immobilized)	[HMIM][TfO]	**Transesterification**: Glucose (0.05 or 0.3 mmol) and vinyl laurate (0.15 or 0.45 mmol) in pure IL or IL/2M2B (5:95) mixture at 60 or 45 °C. **Esterification**: Methyl glucoside (0.3 mmol) and laurate acid (0.3 mmol) in pure IL or IL/2M2B (5:95) at 45 °C.	Lin et al. (2015) Zhao et al. (2016)
6-O-(11-Dodecenoic)-glucose ester	Lipase B from *Candida antarctica* (immobilized) and *Thermomyces lanuginosus* (immobilized)	[BMIM][BF_4] [BMIM][Cl] [BMIM][PF_6]	**Transesterification**: Glucose (5 mmol) and 11-dodecenoic ethyl ester (15 mmol) in pure IL at 50 °C.	Yao et al. (2012)

Glucose myristate	Lipase B from *Candida antarctica* (free and immobilized)	[BMIM][BF_4] [BMIM][PF_6]	**Transesterification**: Glucose (0.25 mmol) and vinyl myristate (0.25 mmol) in pure IL or IL/*t*-butanol (40/60% v/v) mixture at 60 °C.	Ganske and Bornscheuer (2005a)
Glucose oleate	Lipase B from *Candida antarctica* (free and immobilized)	[BMIM][BF_4]	**Esterification**: Glucose (0.5 mmol) and oleic acid (0.25 mmol) in IL/*t*-butanol (1.5:1) mixture with microwave irradiation.	Rahman et al. (2012)
Glucose palmitate	Lipase B from *Candida antarctica* (free and immobilized) Lipase B from *Candida antarctica* (immobilized) Lipases from *Candida antarctica* (B immobilized), *Thermomyces lanuginosus* (immobilized) and *Mucor miehei* (immobilized) Lipases from *Candida antarctica* (B immobilized)	[BMIM][BF_4] [BMIM][PF_6] [BMIM][TfO] [Omim][Tf_2N] [BMIM][TfO] [BMIM][PF_6] Cyphos 104 Cyphos 109	**Esterification**: Glucose (0.25 mmol) and palmitic acid (0.25 mmol) in IL/*t*-butanol (40/60% v/v) mixture at 60 °C. **Esterification**: Supersaturated glucose (0.33 mmol) and palmitic acid (0.33 mmol) in IL mixture (1:1) at 50 °C. **Transesterification**: Glucose (1.1 mmol) and vinyl palmitate (3.3 mmol) in pure IL at 40 °C. **Esterification**: Glucose (5 mmol) and palmitic acid (15 mmol) in pure IL at 50 °C.	Ganske and Bornscheuer (2005a), (2005b) Ha et al. (2010a) Liang et al. (2012) Findrik et al. (2016)
Fructose laurate	Lipase B from *Candida antarctica* (immobilized)	[BMIM][BF_4] [BMIM][TfO]	**Esterification**: Fructose (0.22 mmol) and lauric acid (0.22 mmol) in IL/2M2B mixtures (20/80 and 60/40% v/v) at 50 °C.	Li et al. (2015)
Fructose laurate	Lipase B from *Candida antarctica* (immobilized)	[BMIM][TfO] [BMIM][Tf_2N]	**Transesterification**: Supersaturated fructose (178 mmol) and vinyl laurate (163 mmol) in IL mixture (1:1) at 40 °C.	Shin et al. (2019)
Fructose palmitate	Lipase B from *Candida antarctica* (immobilized)	[BMIM][TfO] [OMIM][Tf_2N]	**Esterification**: Supersaturated fructose (0.11 mmol) and palmitic acid (0.22 mmol) in IL mixture (1:1) at 60 °C with ultrasound radiation.	Ha et al. (2010b)
Mannose myristate	Lipase B from *Candida antarctica* (immobilized) Lipase B from *Candida antarctica* (immobilized)	[BMIM][TfO] [BMIM][BF_4] [BMIM][Tf_2N] [BMPyrr][TfO] [BMPyrr][TfO]	**Transesterification**: Mannose (0.05 mmol) and vinyl myristate (0.3 mmol) in pure IL at 60 °C. **esterification**: Mannose (0.05 mmol) and myristic acid (0.3 mmol) in pure IL at 60 °C. **Transesterification**: Mannose (0.024 mmol) and vinyl myristate (0.236 mmol) in pure IL at 80 °C.	Galonde et al. (2013a) Galonde et al. (2013b)

(continued)

Table 4.1 (continued)

Product	Biocatalyst	Ionic liquid (IL)	Reaction conditions	References
Galactose oleate	Lipases from *Candida antarctica* (B immobilized), *Rhizomucor miehei* (immobilized), *Thermomyces lanuginosus* (immobilized), *Candida rugosa* (free) and *Geobacillus zalihae* (free)	[BMIM][BF_4]	**Esterification**: Galactose (0.05 mmol) and oleic acid (0.1 or 0.15 mmol) in IL/DMSO (20:1) mixture at 60 °C.	Abdulmalek et al. (2012)
Maltose linoleate	Lipase B from *Candida antarctica* (immobilized)	[EMIM] [$MeSO_3$] [BMPyr][PF_6]	**Esterification**: Maltose (5 mmol) and linoleic acid (10 mmol) in pure IL, IL mixture (1:1) or IL:Organic solvent (1:1) mixture at 65 °C.	Fischer et al. (2013)
Sucrose laurate	Lipase B from *Candida antarctica* (immobilized) Lipase B from *Candida antarctica* (immobilized)	[BMIM][dca] [BMIM][TfO] [BMIM][Tf_2N]	**Esterification**: Sucrose (1.46 mmol) and lauric acid (3.0 mmol) in pure IL at 55 °C. **Transesterification**: Supersaturated sucrose (9 mmol) and vinyl laurate (163 mmol) in IL mixture (1:1) at 40 °C.	Liu et al. (2005) Shin et al. (2019)
4-*O*-levoglucosan laurate	Lipase from *Pseudomonas cepacea* (immobilized) Lipase B from *Candida antarctica* (immobilized)	[BMIM][BF_4] [MOEMIM] [BF_4] [MOEMIM] [dca]	**Transesterification**: Levoglucosan (0.25 mmol) and vinyl laurate (0.5 mmol) in pure IL at 55 °C. **esterification**: Levoglucosan (0.25 mmol) and lauric acid (0.5 mmol) in pure IL at 55 °C.	Galletti et al. (2007)

[BMIM][TfO] 1-butyl-3-methylimidazolium trifluoromethanesulfonate; [OMIM][Tf_2N] 1-methyl-3-octylimidazolium bis[(trifluoromethyl)sulfonyl]imide; [BMIM][BF_4] 1-butyl-3-methylimidazolium tetrafluoroborate; [BMIM][PF_6] 1-butyl-3-methylimidazolium hexafluorophosphate; [EMIM][TfO] 1-ethyl-3-methylimidazolium trifluoromethanesulfonate; [BMIM][Tf_2N] 1-butyl-3-methylimidazolium bis[(trifluoromethyl)sulfonyl]imide; [$Me(OEt)_3$-Et-IM][OAc] 3-(2-(2-methoxyethoxy)ethoxy)ethylimidazolium acetate; [$Me(OEt)_3$-Et_3N][OAc] triethyl (2-(2-methoxyethoxy)ethoxy)ethylammonium acetate; [TBA][OAc] tetrabutylammonium acetate; [HMIM][TfO] 1-hexyl-3-methylimidazolium trifluoromethanesulfonate; [BMIM][Cl] 1-butyl-3-methylimidazolium chloride; Cyphos 104—trihexyl(tetradecyl)phosphonium bis(2,4,4-trimethylpentyl)phosphinate; Cyphos 109—trihexyl(tetradecyl)phosphonium bis[(trifluoromethyl)sulfonyl]imide; [BMPyrr][TfO] 1-butyl-3-methylpyrrolidinium trifluoromethanesulfonate; [EMIM][$MeSO_4$] 1-ethyl-3-methylimidazolium methylsulfate; [BMPyr][PF_6] 1-butyl-4-methylpyridinium hexafluorophosphate; [BMIM][dca] 1-butyl-3-methylimidazolium dicyanamide; [MOEMIM][BF_4] 1-methoxyethyl-3-methylimidazolium tetrafluoroborate; [MOEMIM][dca] 1-methoxyethyl-3-methylimidazolium dicyanamide; 2M2B—2-methyl-2-butanol; DMSO—dimethyl sulfoxide

A different SFAE, 6-O-(11-dodecenoic)-glucose ester, was obtained in [BMIM][PF_6], [BMIM][BF_4], or 1-butyl-3-methylimidazolium chloride [BMIM][Cl] through a transesterification reaction catalyzed by an immobilized lipase from *C. antarctica* (Yao et al. 2012). The conversion yields for the different ILs followed the order [BMIM][BF_4] > [BMIM][Cl] > [BMIM][PF_6]. Under the optimal conditions (55 °C, glucose:11-dodecenoic ethyl ester ratio of 1:2, water content of 2% and 20 mg/mL of enzyme), a conversion yield of 62% was reached. However, this yield decreased to 20% after 8 cycles of enzyme reuse.

The synthesis of maltose linoleate through lipase-catalyzed esterification was studied in several ILs including [BMIM][BF_4], 1-ethyl-3-methylimidazolium methanesulfonate [EMIM][$MeSO_3$], 1-ethyl-3-methylimidazolium ethylsulfate [EMIM][$EtSO_4$], [BMIM][PF_6], 1-methyl-3-propylimidazolium hexafluorophosphate [MPIM][PF_6], 1-butyl-4-methylpyridinium hexafluorophosphate [BMPyr][PF_6], 1-hexylpyridinium hexafluorophosphate [HPyr][PF_6], trihexyltetradecylphosphonium hexafluorophosphate [THTDP][PF_6], tributylmethylammonium bis[(trifluoromethyl)sulfonyl]imide [TBMA][Tf_2N], [BMIM][Tf_2N], and methyltrioctylammonium bis[(trifluoromethyl)sulfonyl]imide [MTOAM][Tf_2N] (Fischer et al. 2013). The highest conversion yield (37%) was obtained for the water miscible [EMIM][$MeSO_3$]. Yields in the range 20–25% were obtained for [MPIM][PF_6] and [BMPyr][PF_6] and all the other ILs provided conversion yields $\leq$10%. Based on a strategy of minimal solvent, the lipase from *C. antarctica* immobilized on commercial beads was coated with the ILs under study and its performance in the synthesis of maltose linoleate was investigated without any additional solvent. Using this strategy, yields in the range 15–25% were obtained for maltose conversion (except for [TBMA][Tf_2N] which presented only 5% yield). Therefore, IL-coated lipase generally surpassed the IL solvent conditions. However, the best IL as solvent ([EMIM][$MeSO_3$]) was less efficient as coating agent. In all cases, the esterification reaction resulted in the formation of two types of monoesters, namely 6- and 6'-*O*-linoleyl-α-D-maltose.

Galonde et al. (2013a) performed a screening study involving 9 ILs to evaluate the effect of cations and anions on the synthesis of mannose myristate through esterification and transesterification using immobilized lipase from *C. antarctica*. The ILs screened were [BMIM][BF_4], [BMIM][PF_6], [BMIM][Tf_2N], [BMIM][TfO], 1-ethyl-3-methylimidazolium hexafluorophosphate [EMIM][PF_6], 1-hexyl-3-methylimidazolium hexafluorophosphate [HMIM][PF_6], 1-butyl-3-methylpyrrolidinium trifluoromethanesulfonate [BMPyrr][TfO], 1-butyl-3-methylpyrrolidinium bis[(trifluoromethyl)sulfonyl]imide [BMPyrr][Tf_2N], and trimethylhexylammonium bis[(trifluoromethyl)sulfonyl]imide [TMHA][Tf_2N]. The best results both for esterification and transesterification were obtained with the ILs containing the anion TfO^-: [BMPyrr][TfO] (44.5 and 71% conversion yield for esterification and transesterification, respectively) and [BMIM][TfO] (30% and 65% conversion yield for esterification and transesterification, respectively). The higher yields achieved for transesterification are related with the presence of better leaving groups (vinyl alcohols) when compared to water generated by esterification. On the other hand, the ILs containing the anions PF_6^- and Tf_2N^- provided the lowest conversion yields. This study also showed that lipase is directly affected by anions while the cations have only an indirect effect on its activity. Since ILs based on cation $BMPyrr^+$ were poorly explored in SFAEs synthesis, the same researchers also performed an optimization study using a full factorial design to determine the suitable conditions (80 °C, 1:10 ratio of mannose:vinyl myristate and 0.26 mmol of total substrate) for the synthesis of mannose myristate through transesterification in [BMPyrr][TFO] (Galonde et al. 2013b).

To achieve highly concentrated glucose solutions in ILs, Lee et al. (2008a) proposed a methodology based on water-mediated supersaturation. In this case, an aqueous solution of glucose was mixed with the IL followed by the water removal from the solution by vacuum evaporation. These solutions can be maintained for long periods of time probably because the high viscosity of the ILs can prevent sugar precipitation and crystallization. The synthesis of glucose laurate through esterification and transesterification was investigated using supersaturated solutions of glucose prepared in [BMIM][BF_4], 1-ethyl-3-methylimidazolium methylsulfate [EMIM][$MeSO_4$], 1-ethyl-3-methylimidazolium trifluoromethanesulfonate [EMIM][TfO], and [BMIM][TfO]. For [EMIM][$MeSO_4$], no enzymatic reaction was detected possibly due to the acidic condition of this IL which may induce enzyme inactivation. The transesterification of supersaturated glucose was performed with high initial reaction rate (15 μmol/min/g) and conversion yield (96%) in [BMIM][TfO]. It was suggested that glucose dissolved in a supersaturated solution could easily be transferred to the active site of the enzyme. The monoester formed presented low solubility in the IL and therefore it precipitated which could contribute to increase the conversion yield and avoid the formation of more complex structures such as diesters. The scale up of this reaction to 10 times higher IL volume also provided a high conversion yield (84%). For the direct esterification reaction, the conversion yield followed the order: [BMIM][TfO] > [EMIM][TfO] > [BMIM][BF_4].

The same research group also proposed an alternative approach using supersaturated glucose and ultrasound radiation to produce glucose laurate (Lee et al. 2008b). The use of ultrasound radiation allowed to improve both the dissolution rate of glucose and the lipase activity in the transesterification reaction performed in [BMIM][TfO] and [BMIM][PF_6]. In the direct esterification reaction, the enzyme activity increased about 4.7 times when [BMIM][TfO] was used as solvent under ultrasound radiation.

Glucose laurate was also synthesized in ether-functionalized ILs based on imidazolium and ammonium such as 3-(2-(2-methoxyethoxy)ethoxy)ethylimidazolium acetate [$Me(OEt)_3$-Et-IM][OAc] and triethyl (2-(2-methoxyethoxy)ethoxy)ethylammonium acetate [$Me(OEt)_3$-Et_3N][OAc] (Zhao et al. 2009). These type of ILs exhibit high lipase compatibility, low viscosity, affinity for sugars, and great potential for homogeneous enzymatic catalysis. In this case, only the transesterification reaction using immobilized lipase from *C. antarctica* was studied. A slightly higher glucose conversion was reported for $Me(OEt)_3$-Et-IM][OAc] (85%) than for [$Me(OEt)_3$-Et_3N][OAc] (71%).

Another screening study involving 16 ILs was performed by Lin et al. (2015) to evaluate the impact of different ILs on the synthesis of glucose laurate at 40 °C through transesterification using immobilized lipase from *C. antarctica*. Besides the well-studied [BMIM][PF_6] and [BMIM][BF_4], where low conversion yields were obtained (< 5%), other ILs were investigated and grouped based on the anion structure. The worst results were obtained for the methylsulfate-based ILs (1,3-dimethylimidazolium methylsulfate [MMIM][$MeSO_4$], [EMIM][$MeSO_4$], and 1-butyl-3-methylimidazolium methylsulfate [BMIM][$MeSO_4$]) for which no substrate conversion occurred. These ILs presented high ability to dissolve glucose but vinyl laurate solubility is very low (for [MMIM][$MeSO_4$] it is zero), which could help to justify the absence of conversion. The group of bis(trifluoromethylsulfonyl) imide-based ILs, whether combined with imidazolium or ammonium, provided low conversion yield (< 7%) which may be related to the hydrophobic nature of the anion that may negatively affect the enzyme. The highest conversion yields were achieved for the trifluoromethanesulfonate-based ([EMIM][TfO], [BMIM][TfO] and 1-hexyl-3-methylimidazolium trifluoromethanesulfonate [HMIM][TfO]) and acetate-based ILs (tetramethylammonium acetate [TMA][OAc] and tetrabutylammonium acetate [TBA][OAc]). When focusing on the cation nature, it was observed that the higher conversions were obtained for the ones presenting a longer alkyl chain. Additionally, the conversion yield in [HMIM][TfO] (20%) was significantly improved to 93% by optimizing the reaction temperature and substrate concentration (60 °C, 0.05 M glucose and 0.15 M vinyl laurate).

In a later study, Lin et al. (2016) reported for the first time the potential of [TBA][OAc] to act as both solvent and catalyst in the synthesis of glucose laurate. In this study, the transesterification reaction was investigated with and without the addition of immobilized lipase and it was observed that in both cases, the synthesis of glucose laurate occurred together with the formation of lauric acid as a by-product. Due to the presence of a trace amount of water in the reaction medium, lauric acid was originated from the hydrolysis of vinyl laurate and glucose laurate. Therefore, it seemed that [TBA][OAc] was able to catalyze both transesterification and hydrolysis reactions. Also, other ILs analogous to [TBA][OAc] were investigated for this dual role, namely [TMA][OAc], tetraethylammonium acetate [TEA][OAc], tetrabutylammonium bisulfate [TBA][HSO_4], and tetrabutylphosphonium acetate [TBP][OAc]. Neither [TBA][HSO4] nor [TBP][Ac] were able to promote the synthesis of glucose laurate. On the other hand, in the tetraalkylammonium acetate ILs, glucose laurate was formed, and the conversion yield followed the order [TBA][Ac] > [TEA][Ac] > [TMA][Ac]. Additionally, a plausible mechanism based on the ions dissociation in the presence of water and the nucleophilic attack of the acetate ion to the hydroxyl group of glucose is proposed.

Besides pure ILs, also some mixtures have been prepared to be used as reaction medium in the enzymatic synthesis of SFAEs. However, it is important to point out that the combination of two ionic liquids results in a quaternary solvent such as a salt in salt mixture. Therefore, these “binary ionic liquids” are slightly different from binary organic solvent mixtures. The evidence that immobilized lipase from *C. antarctica* presented higher activity in the hydrophilic ILs ([BMIM][TfO] and

[BMIM][BF_4]) and higher stability in the hydrophobic ILs ([BMIM][Tf_2N] and [BMIM][PF_6]) led to the preparation of ILs mixtures (Lee et al. 2008c). Therefore, the synthesis of glucose laurate by esterification or transesterification was performed in a mixture of [BMIM][Tf_2N]:[BMIM][TfO] (1:1, v/v) using a supersaturated solution of glucose prepared by the water-mediated methodology. Although some reduction in enzyme activity was detected, its stability was significantly improved and the conversion yield (51–59%) was maintained for 5 cycles of reuse in transesterification reaction and for 24 h in the esterification reaction.

Similarly, mixtures of hydrophilic ([BMIM][TfO]) and hydrophobic ([BMIM][Tf_2N] and [OMIM][Tf_2N]) ILs were prepared to optimize lipase activity and stability in the synthesis of glucose laurate through direct esterification using a supersaturated glucose solution (Ha et al. 2010a). The activity of immobilized lipase was significantly enhanced in a mixture of [BMIM][TfO]:[OMIM][Tf_2N] (9:1, v/v). However, the highest enzyme stability was obtained in the mixture [BMIM][TfO]:[OMIM][Tf_2N] (1:1, v/v) which allowed a 78% of residual activity after 5 cycles of both enzyme and ILs reuse. Also, the synthesis of other glucose esters, namely glucose caprylate and glucose palmitate, through direct esterification was studied. The conversion was higher for the shorter chain fatty acids (60, 54, and 40% conversion for caprylic, lauric, and palmitic acids, respectively) probably due to the lower solubility of longer chain fatty acids in the ILs mixture.

The optimization of glucose laurate synthesis through transesterification was performed by response surface methodology using immobilized lipase from C. *antarctica*, supersaturated glucose solution, and the mixture [BMIM][TfO]/[BMIM][Tf_2N] (1:1 v/v) as reaction medium (Mai et al. 2014). It was verified that the significance of the parameters under study followed the order: vinyl laurate/glucose molar ratio > temperature > enzyme load. Under optimal conditions (67 °C, 73 g/L lipase and substrate molar ratio of 7.6), a maximal conversion yield of 96% was achieved. The synthesis of glucose laurate was scaled up to 2.5 L with very similar conversion yields. Furthermore, the immobilized enzyme and IL mixture could be effectively recycled and reused for up 10 cycles with only 25% reduction of the conversion yield.

To increase sugar solubility in the mixture [BMIM][TfO]/[BMIM][Tf_2N] (1:1 v/v), a solvent-mediated method was developed (Shin et al. 2019). Therefore, glucose, fructose, or sucrose were fist solubilized in a methanol:water (1:1, v/v) solution which was further added to the ILs mixture. Methanol and water were removed from the medium by vacuum evaporation. Glucose, fructose, or sucrose laurate were synthesized through transesterification using this reaction medium and immobilized lipase from *C. antarctica*. This method allowed to significantly increase the amount of sugar dissolved and consequently the productivity of the three esters of laurate.

The mixture of ILs ([BMIM][TfO]:[OMIM][Tf_2N] (1:1, v/v)) was used to synthesize fructose palmitate through esterification under ultrasound radiation with immobilized lipase, supersaturated fructose solution, and initial water content of 0.5% (w/w) (Ha et al. 2010b). The ultrasound irradiation was reported to be a suitable strategy to increase the enzyme activity by increasing the mass transfer rate of the product from the active site in viscous reaction media. The enzyme stability was also enhanced under these conditions and 84% activity remained after 5 cycles of enzyme and ILs reuse.

Fischer et al. (2013) used a mixture of [EMIM][$MeSO_3$]:[BMPyr][PF_6] (1:1, v/v) to obtain maltose linoleate via esterification reaction and the conversion yield (60%) significantly increased compared with the use of each IL alone as solvent (37% for [EMIM][$MeSO_3$] and 25% for [BMPyr][PF_6]).

4.4.3.2 Ionic Liquids/Organic Solvents Mixtures as Solvents

Different mixtures of ionic liquids (ILs) and organic solvents have been studied with the aim to lower the viscosity of the medium, increase substrate solubility and consequently the reaction yield. Co-solvent systems of IL/organic solvent are described as a balance approach between lipase selectivity and mass transfer limitation of pure ILs (Li et al. 2015). The addition of the co-solvent may result either in the formation of a homogeneous or a biphasic system.

As mentioned before, Ganske and Bornscheuer (2005a) used pure [BMIM][BF_4] and [BMIM][PF_6] as solvents in the synthesis of 6-*O*-glucose laurate and 6-*O*-glucose myristate. However, the conversion yield significantly increased from 30–35% to 90% when a mixture of ILs and *t*-butanol (60/40% v/v) was used as solvent and non-modified enzyme as catalyst. In this case, also the synthesis of 6-*O*-glucose palmitate, which did not occur in pure ILs, was achieved by direct esterification. The same authors also studied the performance of free and immobilized lipases from different sources (*P. cepacia*, *Aspergillus* sp., *C. antarctica*, *C. rugosa*, *Rhyzomucor miehei,* and *T. lanuginosa*) and optimized the synthesis of 6-*O*-glucose laurate and 6-*O*-glucose palmitate in IL/*t*-butanol mixtures of several ILs (Ganske and Bornscheuer 2005b). Although lipases from *T. lanuginosa* and *R. miehei* provided some substrate conversion (8–33%), the highest conversion yields (around 60%) were obtained with immobilized lipase from *C. antarctica* in mixtures containing 40% (v/v) *t*-butanol and 60% (v/v) [BMIM][BF_4] or [BMIM][PF_6] at 60 °C. For all the other ILs studied (1-butyl-3-methylimidazolium octylsulfate [BMIM][OSO_4], 1,3-dimethylimidazolium methylsulfate [DMIM][$MeSO_4$] and 1,3-dimethylimidazolium dimethylphosphate [DMIM][DMP]) no conversion was detected. It was also found that immobilized lipase from *C. antarctica* has preference for fatty acids with moderate chain length and the synthesis of sugar fatty acid esters (SFAEs) is faster when using vinyl esters of fatty acids. The IL/*t*-butanol mixture resulted in a two-phase system (IL in the bottom phase and *t*-butanol in the top phase) with the immobilized enzyme at the interface which seems to favor the substrate conversion.

The ether-functionalized IL [$Me(OEt)_3$-Et-IM][OAc] was also mixed with *t*-butanol in different proportions (60/40 and 40/60% v/v) and the synthesis of glucose laurate by transesterification was evaluated (Zhao et al. 2009). Better conversion yield (82%) was obtained when using lower amount of *t*-butanol, suggesting that [$Me(OEt)_3$-Et-IM][OAc] is more enzyme stabilizing than *t*-butanol.

Another co-solvent system used in the synthesis of glucose laurate by lipase-catalyzed transesterification was 2-methyl-2-butanol (2M2B) in combination with [HMIM][TfO] (Lin et al. 2015). For a mixture composed mainly of organic solvent ([HMIM][TfO]/2M2B volume ratio of 0.05/0.95), the substrate conversion yield was significantly improved (62%) when compared with that obtained for pure IL (13%) or pure 2M2B (34%). However, after optimizing the synthesis of glucose laurate by response surface methodology this yield was further enhanced to 76%. The same mixture was also used in the synthesis of glucose laurate through esterification using immobilized lipase as biocatalyst and methyl glucoside and palmitic acid as substrates (Zhao et al. 2016). Using this approach, a conversion yield of 62% was obtained, which was higher than that achieved through transesterification using glucose and vinyl laurate under the optimal conditions (46%). Therefore, it was suggested that the use of a more hydrophobic sugar favored the SFAEs synthesis probably due to an improvement in substrate solubility and/or a better compatibly between the substrate and the active site of lipase.

The use of dimethyl sulfoxide (DMSO) as co-solvent in IL:DMSO (20:1, v/v) mixtures containing [BMIM][BF_4], [BMIM][TfO], [BMIM][Tf_2N], or [BMIM][PF_6] was studied in the synthesis of galactose oleate through esterification catalyzed by several lipases (free and immobilized) (Abdulmalek et al. 2012). The best conversion yield (87%) was achieved at 60 °C after 3 h, using immobilized lipase from *R. miehei* (2% w/w) and a molar ratio of galactose:oleic acid of 1:3 in a reaction medium composed of [BMIM][BF_4]:DMSO (20:1, v/v).This hydrophilic IL is miscible with DMSO, and it seemed to contribute to a better solubilization of galactose. Among the lipases tested, the immobilized ones showed higher potential for the synthesis of galactose oleate. Free lipases were probably more exposed to the environment which may result in a quick denaturation. The short reaction time was a consequence of the complete solubilization of galactose in the reaction medium.

The application of microwave irradiation in the synthesis of glucose oleate performed in a biphasic mixture of [BMIM][BF_4]:*t*-butanol (1.5:1, v/v) allowed to increase lipase activity and remarkably reduce the reaction time to 30 min (Rahman et al. 2012). Although no conversion was detected when pure [BMIM][BF_4] was used, the application of microwave heating at 60 °C in the mixture [BMIM][BF_4]:*t*-butanol resulted in a conversion yield of 90%. The immobilized lipase was probably activated by the superheating of the water layer which surrounds the enzyme particles.

The synthesis of maltose linoleate through lipase-catalyzed esterification was performed in several mixtures of IL:organic solvent (1:1, v/v), namely [EMIM][$MeSO_3$]:tetrahydrofuran (THF), [EMIM][$MeSO_3$]:dimethylformamide (DMF), [EMIM][$MeSO_3$]:acetone, [BMPyr][PF_6]:THF, [BMPyr][PF_6]:DMF, and [BMPyr][PF_6]:acetone (Fischer et al. 2013). The conversion yields obtained (50–67%) followed the order: [EMIM][$MeSO_3$]:THF > [BMPyr][PF_6]:THF > [EMIM][$MeSO_3$]:acetone > [BMPyr][PF_6]:DMF > [BMPyr][PF_6]:acetone > [EMIM][$MeSO_3$]:DMF. In all cases, the conversion yields were higher than those obtained for the pure ILs.

Co-solvent mixtures containing [BMIM][TfO]/2M2B (60/40% v/v) or [BMIM][BF_4]/2M2B (20/80% v/v) were used in the synthesis of fructose laurate (mono- and diesters) through lipase-catalyzed esterification (Li et al. 2015). The substrate solubility, limitations of mass transfer, and interactions of lipase-IL were pointed out as the most relevant factors in the formation of fructose esters. Also, mass transfer can be related with the dissolution and diffusion of substrates. Therefore, several experimental parameters were studied with the aim to improve the monoester yield. The formation of diesters seems to be associated with the low amount of soluble fructose, the depletion of substrates, and the synthesis of a considerable amount of mono-acylated products. Kinetic features and conformational changes caused in lipase structure by the co-solvent mixtures were also referred as determinants for the selectivity in product formation. It was found that monoester fraction was constituted by a mixture of 1-α-fructofuranose laurate, 1-β-fructopyranose laurate, 6-lauroyl-α-d-fructofuranose, and 6-lauroyl-β-d-fructofuranose. On the other hand, the diesters fraction were composed of 1,6-dilauroyl-α-d-fructofuranose laurate and 1,6-dilauroyl-β-d-fructofuranose laurate.

4.5 Conclusion

Ionic liquids (ILs) proved to have an enormous potential to be used as solvents, coating agents, or even catalysts in the synthesis of sugar fatty acid esters (SFAEs). Furthermore, they can be used pure or in mixtures obtained from the combination of different ILs or ILs and organic solvents. The application of these ILs mixtures generally results in a considerable increase of the substrate solubility, enzyme activity and/or stability. ILs also demonstrated potential to be combined with different methodologies such as ultrasound and microwaves radiation to improve the synthesis of SFAEs through the enhancement of the mass transfer rate in viscous reaction media.

Nevertheless, there are still some aspects that limit the wide scale application of ILs as solvents in the synthesis of interesting compounds such as SFAEs. Those aspects comprise the high cost of the ILs, unfavorable physicochemical and functional properties (high viscosity, low melting point, enzyme deactivation), and in some cases their high toxicity or low biodegradability. Therefore, it is important to properly use the accurate available tools to analyze and predict IL properties in order to proficiently design and develop more suitable and cost-effective ionic liquids. Besides the already proven potential of ILs as substitutes of the hazardous and highly volatile organic solvents, efforts should be taken to develop truly green IL-based bioprocesses for the synthesis of SFAEs. These bioprocesses are expected to include the application of less (eco)toxic and more biodegradable ILs such as the ones presenting linear alkyl chains, oxygen containing functionalities, and ether substitution or even ILs derived from natural origin (e.g., choline, betaine, phenylalanine, or levulinate). Furthermore, the toxicity mechanisms of ILs have been barely investigated. Thus, a more complete understanding of this mechanism is required to discuss the toxicity issue at a biological level.

Another important aspect in the development of greener and cost-effective biocatalytic processes is the recyclability and reuse of both solvents and catalysts. The application of immobilized biocatalysts truly facilitates this task and ILs already demonstrated their great potential to be recycled and reused in several bioconversion cycles. However, it is important to ensure that alternative recycling methodologies not based on the use of organic solvents will be gradually adopt.

Since the accumulation of the product in the reaction media frequently suppresses its rate of formation and compromises the yields, a continuous removal of the synthesized SFAE is desired. Therefore, the design of alternative reactors and/or processes of synthesis involving ILs and the simultaneous reaction and extraction of the SFAEs are considered of utmost importance. Non-conventional methods such as microwave, ultrasound, and supercritical carbon dioxide can be combined with ILs to improve the enzyme performance and to create a unique bioprocess of product synthesis and extraction.

The improvement of the enzymatic synthesis of SFAEs also includes the potential tuning of the molecular interactions between enzymes and their environment (solvent, substrates, and products). Simple chemical changes such as the preparation of polymer- or detergent-modified lipases can contribute to increase the activity and stability. Another interesting strategy to enhance enzyme performance could be the rational design of lipases. Specifically engineered lipases can present improved activity, stability, and notable selectivity. Also, the affinity of lipases to more complex substrates such as branched chain fatty acids can be achieved by genetic modifications.

The synthesis of SFAEs by combining different sugars and fatty acids presents enormous versatility and potential to generate new and interesting compounds with innovative properties. Therefore, it is expected that the number of different SFAEs produced via enzymatic synthesis could greatly increase in the future.

Acknowledgements This work received financial support from Portuguese Foundation for Science and Technology (FCT) under the scope of the strategic funding of UID/BIO/04469/2013 unit; COMPETE 2020 (POCI-01-0145-FEDER-006684); the project Lignozymes (POCI-01-0145-FEDER-029773). The authors also acknowledge BioTecNorte operation (NORTE-01-0145-FEDER-000004) and the project MultiBiorefinery (POCI-01-0145-FEDER-016403) funded by the European Regional Development Fund under the scope of Norte2020—Programa Operacional Regional do Norte.

References

Abdulmalek E, Mohd Saupi HS, Tejo BA, Basri M, Salleh AB, Raja Abd Rahman RNZ, Abdul Rahman MB (2012) Improved enzymatic galactose oleate ester synthesis in ionic liquids. J Mol Catal B Enzym 76:37–43. https://doi.org/10.1016/j.molcatb.2011.12.004

Adachi S, Kobayashi T (2005) Synthesis of esters by immobilized-lipase-catalyzed condensation reaction of sugars and fatty acids in water-miscible organic solvent. J Biosci Bioeng 99:87–94. https://doi.org/10.1263/jbb.99.087

Ballesteros A, Plou FJ, Alcalde M, Ferrer M, García-Arellano H, Reyes-Duarte D, Ghazi I (2007) Enzymatic synthesis of sugar esters and oligosaccharides from renewable resources. In: Patel RN (ed) Biocatalysis in the pharmaceutical and biotechnology industries. CRC Press, Boca Raton, pp 463–500

Byun H-G, Eom T-K, Jung W-K, Kim S-K (2007) Lipase catalyzed production of monoacylglycerols by the esterification of fish oil fatty acids with glycerol. Biotechnol Bioprocess Eng 12:491–496. https://doi.org/10.1007/BF02931345

Cao L, Fischer A, Bornscheuer UT, Schmid RD (1996) Lipase-catalyzed solid phase synthesis of sugar fatty acid esters. Biocatal Biotransformation 14:269–283. https://doi.org/10.3109/10242429609110280

Carrea G, Riva S, Secundo F (1989) Enzymatic synthesis of various 1'-*O*-sucrose and 1-*O*-fructose esters. J Chem Soc Perkin Trans 1:1057–1061. https://doi.org/10.1039/P19890001057

Chado GR, Holland EN, Tice AK, Stoykovich MP, Kaar JL (2018) Modification of lipase with poly(4-acryloylmorpholine) enhances solubility and transesterification activity in anhydrous ionic liquids. Biomacromolecules 19:1324–1332. https://doi.org/10.1021/acs.biomac.8b00176

Choi S-A, Oh Y-K, Jeong M-J, Kim SW, Lee J-S, Park J-Y (2014) Effects of ionic liquid mixtures on lipid extraction from *Chlorella vulgaris*. Renew Energy 65:169–174. https://doi.org/10.1016/j.renene.2013.08.015

De Diego T, Lozano P, Abad MA, Steffensky K, Vaultier M, Iborra JL (2009) On the nature of ionic liquids and their effects on lipases that catalyze ester synthesis. J Biotechnol 140:234–241. https://doi.org/10.1016/j.jbiotec.2009.01.012

De los Ríos AP, Hernández-Fernández FJ, Martínez FA, Rubio M, Víllora G (2007) The effect of ionic liquid media on activity, selectivity ans stability of *Candida antarctica* lipase B in transesterification reactions. Biocatal Biotransformation 25:151–156. https://doi.org/10.1080/10242420701379213

De María PD, Sánchez-Montero JM, Sinisterra JV, Alcántara AR (2006) Understanding *Candida rugosa* lipases: an overview. Biotechnol Adv 24:180–196. https://doi.org/10.1016/j.biotechadv.2005.09.003

Dong K, Liu X, Dong H, Zhang X, Zhang S (2017) Multiscale studies on ionic liquids. Chem Rev 117:6636–6695. https://doi.org/10.1021/acs.chemrev.6b00776

Egorova KS, Ananikov VP (2018) Ionic liquids in whole-cell biocatalysis: a compromise between toxicity and efficiency. Biophys Rev 10:881–900. https://doi.org/10.1007/s12551-017-0389-9

Faber K (1995) Biotransformations in organic chemistry – a textbook, 2nd edn. Springer, Heidelberg, pp 278–295

Farrán A, Cai C, Sandoval M, Xu Y, Liu J, Hernáiz MJ, Linhardt RJ (2015) Green solvents in carbohydrate chemistry: from raw materials to fine chemicals. Chem Rev 115:6811–6853. https://doi.org/10.1021/cr500719h

Ferrer M, Cruces MA, Plou FJ, Pastor E, Fuentes G, Barnabé M, Parra JL, Ballesteros A (2000) Chemical versus enzymatic synthesis catalysis for the regioselective synthesis of sucrose esters of fatty acids. Stud Surf Sci Catal 130A:509–514. https://doi.org/10.1016/S0167-2991(00)81008-3

Findrik Z, Megyeri G, Gubicza L, Bélafi-Bakó K, Nemestóthy N, Sudar M (2016) Lipase catalyzed synthesis of glucose palmiate in ionic liquid. J Clean Prod 112:1106–1111. https://doi.org/10.1016/j.jclepro.2015.07.098

Fischer F, Happe M, Emery J, Fornage A, Schütz R (2013) Enzymatic synthesis of 6- and 6'-*O*-linoleyl-α-D-maltose: from solvent-free to binary ionic liquid reaction media. J Mol Catal B Enzym 90:98–106. https://doi.org/10.1016/j.molcatb.2013.01.019

Galletti P, Moretti F, Samorì C, Tagliavini E (2007) Enzymatic acylation of levoglucosan in acetonitrile and ionic liquids. Green Chem 9:987–991. https://doi.org/10.1039/b702031g

Galonde N, Nott K, Debuigne A, Deleu M, Jerôme C, Paquot M, Wathelet J-P (2012) Use of ionic liquids for biocatalytic synthesis of sugar derivatives. J Chem Technol Biotechnol 87:451–471. https://doi.org/10.1002/jctb.3745

Galonde N, Nott K, Richard G, Debuigne A, Nicks F, Jerôme C, Fauconnier M-L (2013a) Study of the influence of pure ionic liquids on the lipase-catalyzed (trans)esterification of mannose based on their anion and cation nature. Curr Org Chem 17:763–770. https://doi.org/10.2174/1385272811317070010

Galonde N, Brostaux Y, Richard G, Nott K, Jerôme C, Fauconnier M-L (2013b) Use of response surface methodology for the optimization of the lipase-catalyzed synthesis of mannosyl myristate in pure ionic liquid. Process Biochem 48:1914–1920. https://doi.org/10.1016/j.procbio.2013.08.023

Ganske F, Bornscheuer UT (2005a) Lipase-catalyzed glucose fatty acid ester synthesis in ionic liquids. Org Lett 7:3097–3098. https://doi.org/10.1021/ol0511169

Ganske F, Bornscheuer UT (2005b) Optimization of lipase-catalyzed glucose fatty acid ester synthesis in a two-phase system containing ionic liquids and *t*-BuOH. J Mol Catal B Enzym 36:40–42. https://doi.org/10.1016/j.molcatb.2005.08.004

Gubicza L, Nemestóthy N, Fráter T, Bélafi-Bakó K (2003) Enzymatic esterification in ionic liquid integrated with pervaporation for water removal. Green Chem 5:236–239. https://doi.org/10.1039/b211342m

Gumel AM, Annuar MSM (2016) *Thermomyces lanuginosus* lipase-catalyzed synthesis of natural flavor esters in a continuous flow microreactor. 3 biotech 6:24. https://doi.org/10.1007/s13205-015-0355-9

Gumel AM, Annuar MSM, Heidelberg T, Chisti Y (2011) Lipase mediated synthesis of sugar fatty acid esters. Process Biochem 46:2079–2090. https://doi.org/10.1016/j.procbio.2011.07.021

Ha SH, Hiep NM, Lee SH, Koo Y-M (2010a) Optimization of lipase-catalyzed glucose ester synthesis in ionic liquids. Bioprocess Biosyst Eng 33:63–70. https://doi.org/10.1007/s00449-009-0363-4

Ha SH, Hiep NM, Koo Y-M (2010b) Enhanced production of fructose palmitate by lipase-catalyzed esterification in ionic liquids. Biotechnol Bioprocess Eng 15:126–130. https://doi.org/10.1007/s12257-009-3091-3

Hernández-Fernández FJ, de los Ríos AP, Tomás-Alonso F, Gómez D, Víllora G (2009) Stability of hydrolases enzymes in ionic liquids. Can J Chem Eng 87:910–914. https://doi.org/10.1002/cjce.20227

Hernández-Fernández FJ, de los Ríos AP, Lozano-Blanco LJ, Godínez C (2010) Biocatalytic ester synthesis in ionic liquid media. J Chem Technol Biotechnol 85:1423–1435. https://doi.org/10.1002/jctb.2453

Itoh T (2017) Ionic liquids as tool to improve enzymatic organic synthesis. Chem Rev 117:10567–10607. https://doi.org/10.1021/acs.chemrev.7b00158

Kaar JL, Jesionowski AM, Berberich JA, Moulton R, Russel AJ (2003) Impact of ionic liquid physical properties on lipase activity and stability. J Am Chem Soc 125:4125–4131. https://doi.org/10.1021/ja028557x

Kimizuka N, Nakashima T (2001) Spontaneous self-assembly of glycolipid bilayer membranes in sugar-philic ionic liquids and formation of ionogels. Langmuir 17:6759–6761. https://doi.org/10.1021/la015523

Lee SH, Ha SH, Lee SB, Koo Y-M (2006) Adverse effect of chloride impurities on lipase-catalyzed transesterifications in ionic liquids. Biotechnol Lett 28:1335–1339. https://doi.org/10.1007/s10529-006-9095-6

Lee SH, Dang DT, Ha SH, Chang W-J, Koo Y-M (2008a) Lipase-catalyzed synthesis of fatty acid sugar ester using extremely supersaturated sugar solution in ionic liquids. Biotechnol Bioeng 99:1. https://doi.org/10.1002/bit.21534

Lee SH, Nguyen HM, Koo Y-M, Ha SH (2008b) Ultrasound-enhanced lipase activity in the synthesis of sugar ester using ionic liquids. Process Biochem 43:1009–1012. https://doi.org/10.1016/j.procbio.2008.05.001

Lee SH, Ha SH, Hiep NM, Chang W-J, Koo YM (2008c) Lipase-catalyzed synthesis of glucose fatty acid ester using ionic liquids mixtures. J Biotechnol 133:486–498. https://doi.org/10.1016/j.jbiotec.2007.11.001

Lei Z, Chen B, Koo Y-M, MacFarlane DR (2017) Introduction: ionic liquids. Chem Rev 117:6633–6635. https://doi.org/10.1021/acs.chemrev.7b00246

Li L, Ji F, Wang J, Jiang B, Li Y, Bao Y (2015) Efficient mono-acylation of fructose by lipase-catalyzes esterification in ionic liquid co-solvents. Carbohydr Res 416:51–58. https://doi.org/10.1016/j.carres.2015.08.009

Liang J, Zeng W, Yao P, Wei Y (2012) Lipase-catalyzed regioselective synthesis of palmitolyglucose ester in ionic liquids. Adv Biol Chem 2:226–232. https://doi.org/10.4236/abc.2012.23027

Lin X-S, Wen Q, Huang Z-L, Cai Y-Z, Halling PJ, Yang Z (2015) Impacts of ionic liquids on enzymatic synthesis of glucose laurate and optimization with superior productivity by response surface methodology. Process Biochem 50:1852–1858. https://doi.org/10.1016/j.procbio.2015.07.019

Lin X-S, Zou Y, Zhao K-H, Yang T-X, Halling P, Yang Z (2016) Tetraalkylammonium ionic liquids as dual solvents-catalysts for direct synthesis of sugar fatty acid esters. J Surf Deter 19:511–517. https://doi.org/10.1007/s11743-016-1798-7

Liu Q, Janssen MHA, van Rantwijk F, Sheldon RA (2005) Room-temperature ionic liquids that dissolve carbohydrates in high concentrations. Green Chem 7:39–42. https://doi.org/10.1039/b412848f

Lou W-Y, Zong M-H, Liu Y-Y, Wang J-F (2006) Efficient enantioselective hydrolysis of D,L-phenylglycine methyl ester catalyzed by immobilized *Candida antarctica* lipase B in ionic liquid containing systems. J Biotechnol 125:64–74. https://doi.org/10.1016/j.jbiotec.2006.01.017

Mai NL, Koo Y-M (2018) Enzymatic reaction in ionic liquids. In: Chang HN (ed) Emerging areas in bioengineering, vol 1. Wiley-VCH, Weinheim, pp 35–65. https://doi.org/10.1002/9783527803293.ch3

Mai NL, Ahn K, Bae SW, Shin DW, Morya VK, Koo Y-M (2014) Ionic liquids as novel solvents for the synthesis of sugar fatty acid ester. Biotechnol J 9:1565–1572. https://doi.org/10.1002/biot.201400099

Manic MS, Najdanovic-Visak V, da Ponte MN (2011) Ectraction of free fatty acids from soybean oil using ionic liquids or poly(ethyleneglycol)s. AICHE J 57:1344–1355. https://doi.org/10.1002/aic

Motlagh SR, Harun R, Biak DRA, Hussain SA, Ghani WAWAK, Khezri R, Wilfred CD, Elgharbawy AAM (2019) Screening of suitable ionic liquids as green solvents for extraction of eicosapentanoic acid (EPA) from microalgae biomass using COSMO-RS model. Molecules 24:317. https://doi.org/10.3390/molecules24040713

Nara SJ, Harjani JR, Salunkhe MM (2002) Lipase-catalysed transesterification in ionic liquids and organic solvents: a comparative study. Tetrahedron Lett 43:2979–2982. https://doi.org/10.1016/S0040-4039(02)00420-3

Nelen BAP, Bax L, Cooper JM (2015) Sucrose esters. In: Norn V (ed) Emulsifiers in food technology, 2nd edn. John Wiley & Sons, Ltd, United Kingdom, pp 147–180

Neta NS, Teixeira JA, Rodrigues LR (2012) Sugar ester surfactants: enzymatic synthesis and applications in food industry. Crit Rev Food Sci Nutr 55:595–610. https://doi.org/10.1080/10408398.2012667461

Orr VCA, Rehmann L (2016) Ionic liquids for the fractionation of microalgae biomass. Curr Opin Green Sustain Chem 2:22–27. https://doi.org/10.1016/j.cogsc.2016.09.006

Panda H (2002) Handbook on chemical industries (alcohol based). Asia Pacific Business Press Inc., Delhi, p 37

Panpipat W, Chaijan M (2016) Ionic liquids in the synthesis of sugar/carbohydrate and lipid conjugates. In: Xu X, Guo Z, Cheong L-Z (eds) Ionic liquids in lipid processing and analysis: opportunities and challenges. OACS Press, London, pp 347–371. https://doi.org/10.1016/B978-1-63067-047-4.00011-8

Patil DR, Rethwisch DG, Dordick JS (1991) Enzymatic synthesis of a sucrose-containing linear polyester in nearly anhydrous organic media. Biotechnol Bioeng 37:639–646. https://doi.org/10.1002/bit.260370706

Pérez B, Anankanbil S, Guo Z (2017) Synthesis of sugar fatty acids esters and their industrial utilizations. In: Ahmad MU (ed) Fatty acids: chemistry, synthesis, and applications. AOCS Press, London, pp 329–354. https://doi.org/10.1016/B978-0-12-809521-8.00010-6

Pérez-Victoria I, Pérez-Victoria FJ, Roldán-Vargas S, García-Hernández R, Carvalho L, Castanys S, Gamarro F, Morales JC, Pérez-Victoria JM (2011) Non-reducing trisaccharide fatty acid monoesters: novel detergents in membrane biochemistry. Biochim Biophys Acta 1808:717–726. https://doi.org/10.1016/j.bbamem.2010.11.031

Pöhnlein M, Ulrich J, Kirschhöfer F, Nusser M, Muhle-Goll C, Kannengiesser B, Brenner-Weiß G, Luy B, Liese A, Syldatk C, Hausmann R (2015) Lipase-catalyzed synthesis of glucose-6-*O*-hexanoate in deep eutectic solvents. Eur J Lip Sci Technol 117:161–166. https://doi.org/10.1002/ejlt.201400459

Rahman MBA, Arumugam M, Khairuddin NSK, Abdulmalek E, Basri M, Salleh AB (2012) Microwave assisted enzymatic synthesis of fatty acid sugar ester in ionic liquid-*tert*-butanol biphasic solvent system. Asian J Chem 24:5058–5062

van Rantwijk F, Secundo F, Sheldon RA (2006) Structure and activity of *Candida antarctica* lipase B in ionic liquids. Green Chem 8:282–286. https://doi.org/10.1039/b513062j

Šabeder S, Habulin N, Knez Ž (2005) Comparison of the esterification of fructose and palmitic acid in organic solvent and in supercritical carbon dioxide. Ind Eng Chem Res 44:9631–9635. https://doi.org/10.1021/ie050266k

Schröder C (2017) Proteins in ionic liquids: current status of experiments and simulations. Top Curr Chem 375:25. https://doi.org/10.1007/s41061-017-0110-2

Secundo F, Carrea G, Tarabiono C, Gatti-Lafranconi P, Brocca S, Lotti M, Jaeger K-E, Puls M, Eggert T (2006) The lid is a structural and functional determinant of lipase activity and selectivity. J Mol Catal B Enzym 39:166–170. https://doi.org/10.1016/j.molcatb.2006.01.018

Shin DW, Mai NL, Bae S-W, Koo Y-M (2019) Enhanced lipase-catalyzed synthesis of sugar fatty acid esters using supersaturated sugar solution in ionic liquids. Enzym Microb Technol 126:18–23. https://doi.org/10.1016/j.enzmictec.2019.03.004

Siebenhaller S, Hajek T, Muhle-Goll C, Himmelsbach M, Luy B, Kirschhöfer F, Brenner-Weiß G, Hahn T, Zibek S, Sydatk C (2017) Beechwood carbohydrates for enzymatic synthesis of sustainable glycolipids. Bioresour Bioprocess 4:25. https://doi.org/10.1186/s40643-017-0155-7

Stamatis H, Sereti V, Kolisis FN (2001) Lipase-catalyzed synthesis of sugar fatty acid esters in supercritical carbon dioxide. In: Vulfson EN, Holland HL (Eds) Enzymes in nonaqueous solvents: methods and protocols, volume 15. Humana Press Inc., New Jersey, pp 517–522

Toral AR, de los Ríos AP, Hernández FJ, MHA J, Schoevaart R, van Rantwijk F, Sheldon RA (2007) Cross-linked *Candida antarctica* lipase B is active in denaturating ionic liquids. Enzym Microb Technol 40:1095–1099. https://doi.org/10.1016/j.enzmictec.2006.08.027

Vekariya RL (2017) A review of ionic liquids: applications towards catalytic organic trnasformations. J Mol Liq 227:44–60. https://doi.org/10.1016/j.molliq.2016.11.123

Ventura SPM, de Silva FA, Quental MV, Mondal D, Freire MG, JAP C (2017) Ionic-liquid-mediated extraction and separation processes for bioactive compounds: past, present and future trends. Chem Rev 117:6984–7052. https://doi.org/10.1021/acs.chemrev.6b00550

Verger R (1997) "Interfacial activation" of lipases: facts and artifacts. Trends Biotechnol 15:32–38. https://doi.org/10.1016/S0167-7799(96)10064-0

Welton T (2018) Ionic liquids: a brief history. Biophys Rev 10:691–706. https://doi.org/10.1007/s12551-018-0419-2

Xu P, Zheng G-W, Du P-X, Zhong M-H, Lou W-Y (2016) Whole-cell biocatalytic processes with ionic liquids. ACS Sustain Chem Eng 4:371–386. https://doi.org/10.1021/acssuschemeng.5b00965

Yan Y, Bornscheuer UT, Cao L, Schmid RD (1999) Lipase-catalyzed solid-phase synthesis of sugar fatty acid esters: removal of by-products by azeotropic distillation. Enzym Microb Technol 25:725–728. https://doi.org/10.1016/S0141-0229(99)00106-4

Yang Z (2009) Hofmeister effects: an explanation for the impact of ionic liquids on biocatalysis. J Biotechnol 144:12–22. https://doi.org/10.1016/j.jbiotec.2009.04.011

Yang Z, Huang Z-L (2012) Enzymatic synthesis of sugar fatty acid esters in ionic liquids. Cat Sci Technol 2:1767–1775. https://doi.org/10.1039/c2cy20109g

Yang Z, Pan W (2005) Ionic liquids: green solvents for nonaqueous catalysis. Enzym Microb Technol 37:19–28. https://doi.org/10.1016/j.enzmictec.2005.02.014

Yao P-J, Huang G-J, Yan W, Zhang X, Li Q, Wei Y-A (2012) Lipase-catalyzed synthesis and characterization of 6-O-(11-dodecenoic)-glucose ester in ionic liquids. Adv Chem Eng Sci 2:204–211. https://doi.org/10.4236/aces.2012.22025

Ye R, Hayes DG (2014) Recent progress for lipase-catalysed synthesis of sugar fatty acid esters. J Oil Palm Res 26:355–365

Ye R, Hayes DG (2017) Bioactive properties of sugar fatty acid esters. In: Aguilar MV, Otero C (eds) Frontiers in bioactive compounds, At the crossroads between nutrition and pharmacology, vol 2. Bentham eBooks, Sharjah, pp 124–145. https://doi.org/10.2174/97816810842991170 20009

Ye R, Pyo S-H, Hayes DG (2010) Lipase-catalyzed synthesis of saccharide-fatty acid esters using suspensions of saccharide crystals in solvent-free media. J Am Oil Chem Soc 87:281–293. https://doi.org/10.1007/s11746-009-1504-2

Ye R, Hayes DG, Burton R, Liu A, Harte FM, Wang Y (2016) Solvent-free lipase-catalyzed synthesis of technical-grade sugar esters and evaluation of their physicochemical and bioactive properties. Catalysts 6:78. https://doi.org/10.3390/catal6060078

Zhao H (2010) Methods for stabilizing and activating enzymes in ionic liquids – a review. J Chem Technol Biotechnol 85:891–907. https://doi.org/10.1002/jctb.2375

Zhao H (2012) Ionic liquids as (co-)solvents for hydrolytic enzymes. In: de María PD (ed) Ionic liquids in biotransformations and organocatalysis: solvents and beyond. John Wiley & Sons, Inc., New Jersey, pp 151–227

Zhao H (2016) Protein stabilization and enzyme activation in ionic liquids: specific ion effects. J Chem Technol Biotechnol 91:25–50. https://doi.org/10.1002/jctb.4837

Zhao H, Baker GA, Song Z, Olubajo O, Crittle T, Peters D (2008) Designing enzyme-compatible ionic liquids that can dissolve carbohydrates. Green Chem 10:696–705. https://doi.org/10.1039/b801489b

Zhao H, Jones CL, Cowins JV (2009) Lipase dissolution and stabilization in ether-functionalized ionic liquids. Green Chem 11:1128–1138. https://doi.org/10.1039/b905388c

Zhao K-H, Cai Y-Z, Lin X-S, Xiong J, Halling PJ, Yang Z (2016) Enzymatic synthesis of glucose-based fatty acid esters in bisolvent systems containing ionic liquids or deep eutectic solvents. Molecules 21:1294. https://doi.org/10.3390/molecules21101294

Zheng Y, Zheng M, Ma Z, Xin B, Guo R, Xu X (2015) Sugar fatty acid esters. In: Ahmad MU, Xu X (eds) Polar lipids: biology, chemistry, and technology. AOCS Press, Urbana, Illinois, pp 215–243. https://doi.org/10.1016/B978-1-63067-044-3.50012-1

Zhou J, Sui H, Jia Z, Yang Z, He L, Li X (2018) Recovery and purification of ionic liquids from solutions: a review. RSC Adv 8:32832–32864. https://doi.org/10.1039/c8ra06384b

Chapter 5
Ionic Liquid for the Extraction of Plant Phenolics

Muhammad Mushtaq and Sumia Akram

Contents

Abbreviations

$[C_4MIM]Cl$	1-butyl-3-methylimidazolium chloride
GHz	Gigahertz
GRAS	Generally regarded as safe
ILs	Ionic liquids
KHz	Kilohertz
MAE	Microwave-assisted extraction
MHz	Megahertz
MIM	Methylimidazolium

M. Mushtaq (✉)
Department of Chemistry, Government College University Lahore, Lahore, Pakistan

S. Akram
Division of Science and Technology, University of Education Lahore, Lahore, Pakistan
e-mail: sumia.akram@ue.edu.pk

Inamuddin, A. M. Asiri (eds.), *Nanotechnology-Based Industrial Applications of Ionic Liquids*, Nanotechnology in the Life Sciences,
https://doi.org/10.1007/978-3-030-44995-7_5

5.1 Introduction

Separation and purification are indispensable processes for every walk of science to study the composition and behavior of matter. These processes are often used to reduce the complexity of the system under investigation. In a wider sense, the separation may be carried out by applying energy (evaporation, distillation, centrifugation), mass (extraction, adsorption, flotation), barrier (filtration), or fields (electrophoresis, mass spectrometer). During any of these applications, chemical constituents of matter separate out on the basis of difference in chemical or physical properties such as melting points, masses, ionization energies, and solubility.

Extraction, especially solvent extraction, falls among the most frequently applied sample preparation technique for its easiness, effectiveness, simplicity, and versatility. In extraction, the analytical chemist applies matter, energy, or both to isolate/concentrate a particular analyte or reduce the complexity of the analytical sample. In the simplest sense, extraction involves the distribution of sample components between two or more solvent systems of counterpolarities or limited mutual solubilities. This can separate out the component of interest or at least reduce the complexity of the matter for further qualitative or quantitative analysis. There are millions of such biphasic solvent combinations and approaches and every researcher at one or other stage of their studies has to go for extraction to separate out analyte, reduce the complexity of sample for chromatographic/spectroscopic analysis, or simply enrich the traces of toxins or pollutants.

The phenolic bioactives are metabolites mainly produced during shikimic acid and malonate/acetate pathways in plant chloroplast and utilized as pigments, flavoring agents, antioxidants against stress and diseases, signaling molecules, pheromones, and an integral part of structural elements and defense system. Similar to those for plants, the phenolic compounds regulate a diversity of physiological process and play a vital role in the proper functioning of the human and animal body. Although these are non-nutrient xenobiotics of the human diet, their deficiency may break out in partial or complete failure of any organ. Another important role of phenolic compounds in the human body is their ability to clean up/neutralize various toxic and highly reactive-free radicals produced during various physiological changes.

Chemically phenolic compounds are characterized as compounds having at least one aromatic ring and one or more hydroxyl groups and normally cover flavonoids (C_6–C_3–C_6) like flavan-3-ols, flavones, flavonols, flavanones, and isoflavones, anthocyanins, dihydrochalcones, phenolic acids, hydroxycinnamates, stilbenes, and tannins (Fig. 5.1). The structural diversity of phenolic compounds endorse that a single solvent combination will not be equally suitable for all the classes of compounds. Moreover, the selection of an extraction system or technique is not so simple as it seems because of matrix effects, structural interferances, nature of bonding a particular phenolic compounds bear with cellulosic microfibrils, and above all potential use of the extracts produced.

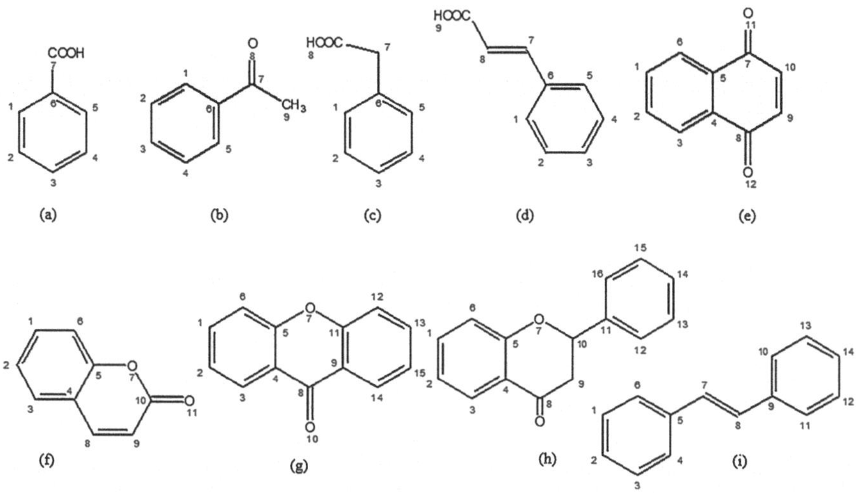

Fig. 5.1 Basic carbon structure of various classes of phenolic compounds (**a**) phenolic acid, (**b**) acetophenone, (**c**) phenylacetic acids, (**d**) hydroxyl cinnamic acid, (**e**) naphthoquinones, (**f**) coumarins, (**g**) xanthanones, (**h**) flavonoids, and (**i**) stilbenes (at least one numbered position must have a hydroxyl group)

5.1.1 Ionic Liquids

The term ionic liquids (ILs) stand for liquids salts or ionic glasses which contains plenty of ions and neutral molecules so there exist strong ionic interactions (ILs) in addition to van der Waals forces normally responsible for the existence of the liquid state. These extraordinary strong ionic interactions produce the inflammable liquids of high thermal stability, tunable viscosity and miscibility, and extraordinary low vapor pressure, or melting points. The term ionic liquid was introduced by Walden (1914) for ethyl ammonium nitrate $(C_2H_5)NH_3^+NO_3^-$ having melting point below 50 °C. The ionic liquids of reasonably low melting points (about room temperature) generally consist of bulky organic cations such as 1-alkyl pyridinium and 1-alkyl-3-methylimidazolium.

5.1.1.1 Preparation of Ionic Liquids

Until the end of twentieth century, the ILs were rare but now there are trillions of large asymmetric organic cations or mixture of asymmetric organic salts with inorganic compounds that behave like liquids at room temperature as asymmetry lowers the lattice energy and the melting point of subsequent melt. Normally, cation controls the thermodynamic behavior of melt except in the cases when anions are large enough to decrease lattice energy and lower the melting point of final melt. It is generally believed that more than 10^{18} binary and ternary ionic combinations are possible and preparation or synthesis of anions and cations shown in Fig. 5.1 vary accordingly. In general, IL synthesis can be carried out in a variety of ways depending upon the characteristics someone looking for and nature of precursors or chemicals present in the shelf. Practically, ILs have extremely low vapor pressure so their purification, particularly via conventional methods, is quiter difficult than their synthesis. Before somebody shot to hunt a new or existing IL combination care must be taken while dealing with certain alkylating agents like bromoalkanes which may react so exothermally to cause safety issues or deterioration of final IL. Next to safety, purity of the synthesized IL matters at the user application level. Fortunately, the majority of raw materials used for IL synthesis are water-soluble and synthesis need no additional solvents still to improve reaction or extraction dynamics certain organic solvents are often applied. Moreover, a certain type of raw materials or precursors may contain impurities or solvents from previous states, for example, pyridines or alkylimidazoles may contain water or oxygen which should be distilled by calcium hydrides or sodium hydroxide and stored under dry argon or nitrogen at 0 °C. Similarly, to obtain colorless halides, haloalkanes need sulfuric acid washing and neutralization prior to distillation with $CaCl_2$ under protective environment.

The majority of researchers dealing with 1-alkyl-3-methylimidazolium [C_nMIM]-based cations have used diazoles (aromatic compounds bearing non-adjacent nitrogen) as raw materials and proceeded synthesis and purification of ILs via a large number of pathways (Fig. 5.2). The most frequently used pathway (Pathway 1) involves the metathesis exchange of anions starting from alkylimidazole. The harmful halides like AgCl produced during the reaction are hard to take out especially during the preparation of more hydrophilic ILs. The second pathway involves neutralization of alkylimidazole base with bronsted acid (HX) to produce halides free ILs but the traces of acid or base often challenge the purity of final products. Another quite simple pathway, i.e., (pathway 3 of Fig. 5.3) direct alkylation of imidazole bases is useful for the synthesis of phosphate, sulfate, or sulfonate-based ILs but limited due to availability and reactivity of alkylating agents.

Pathway 4 (Fig. 5.3) provides an efficient and clean class of ILs while applying dimethyl carbonates as an alkylating agent instead of alkyl halides, but this route again requires acids or NH_4 of anions. A single-step reaction of sultones with alkylimidazoles (pathway 5) leads to the formation of high melting zwitterions which on subsequent reaction with an acid generate a new class of functionalized ILs.

5.1.1.2 Analytical Potential of Ionic Liquids

In general, neoteric ILs are salts of nitrogen-containing organic cations of large unsymmetrically substituted molecules of ammonium, phosphonium, sulphonium, imidazole, pyrrolidine, or pyridine families and organic (alkyl sulfates) or inorganic (halides) anions. A recent report by Gavhane et al. (2019) indicates that nucleobases purine and pyrimidine can also be used for the synthesis of ILs. Unlike inorganic salts produced by electron transfer between acids and bases, the ILs are pure ionic compounds formed by virtue of proton transfer between two species and bear low

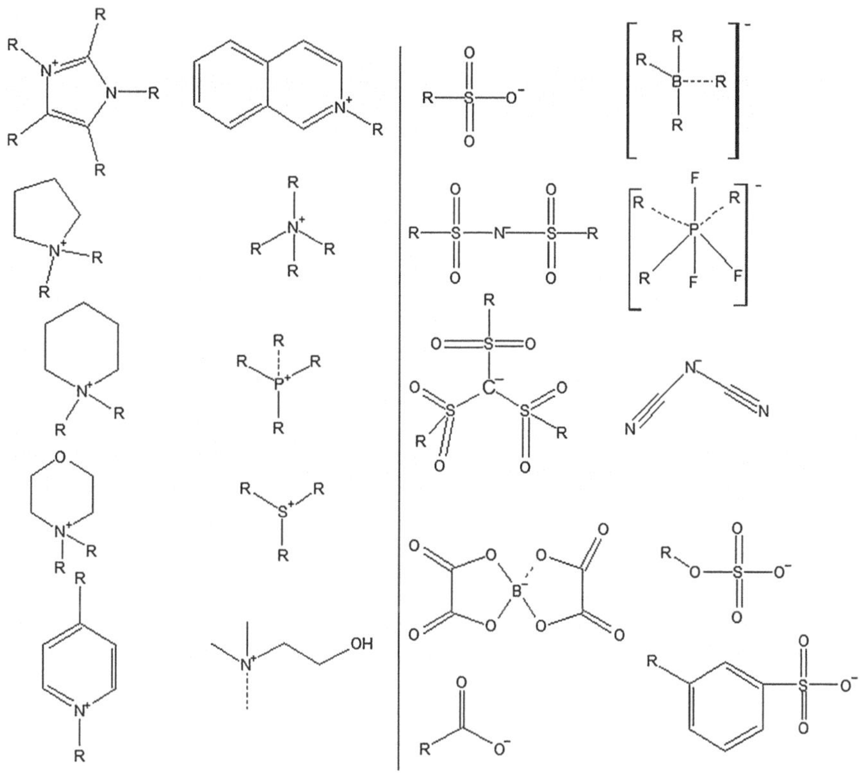

Fig. 5.2 Lewis structure of well-known cations and ions for the synthesis of ionic liquids

melting points (below room temperature), high thermal stability, tunable viscosity, polarity, and aqueous miscibility. The most interesting features of these liquids is their solvation behavior with replacement or modification of functional groups and substituents. In this context, the forecasts show that there can be 10^{18} combinations of non-volatile ILs of different solubility, viscosity, miscibility, and other physicochemical characteristics. These combinations in addition to being the solvents of choice are increasing focused in all areas of analytical chemistry like chromatography, electrochemistry, electrophoresis, and above all mass spectrometry.

In chromatographic separation, ILs have been investigated as a water-immiscible stationary phase. Armstrong et al. (1999) observed that room-temperature ILs can act as nonpolar stationary phases to separate nonpolar analytes which do not act as proton-donor or -acceptor. ILs-based stationary phases are highly interactive and retentive towards the compounds containing acidic or basic functional groups. It was observed that chloride-based ILs interacted more strongly with proton-donor or acceptor molecules whereas hexafluorophosphate-containing ILs retained nonpolar solutes. So, ILs might be an attractive stationary phase to examine differences in ions or their interactions towards organic molecules. Moreover, ILs have the ability to solubilize complex macrocyclic compounds like cyclodextrins or their derivatives.

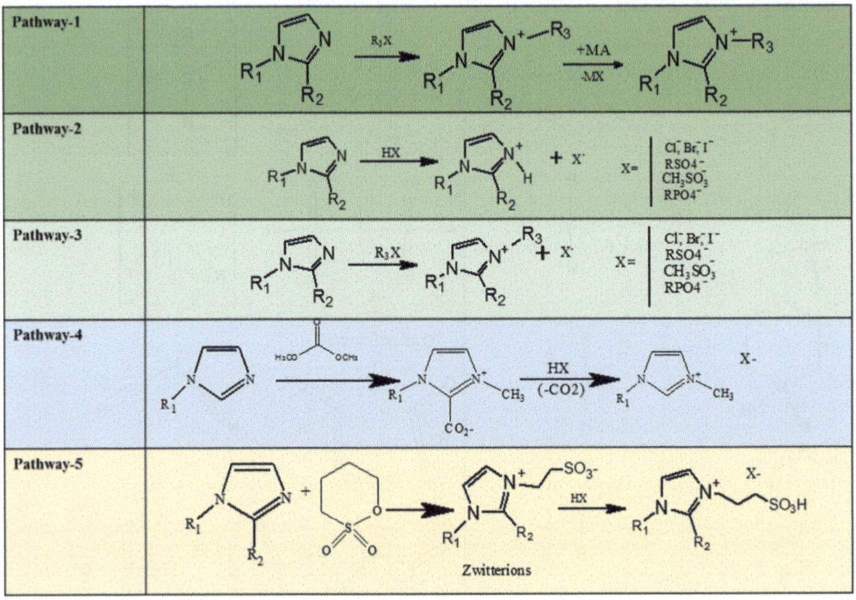

Fig. 5.3 A comparison of pathways frequently adopted for the synthesis of ionic liquids

Ding et al. (2004) observed that chiral ILs-based stationary phases can effectively separate out enantiomers in gas chromatography.

ILs-based solvent have also shown outstanding stability under vacuum and extremely low vapor pressure which have offered opportunities for their use in mass spectrometers especially those based on matrix-assisted laser desorption ionization (MALDI). The ideal matrixes for MALDI are those which can (a) dissolve a liquid sample or co-crystallized solid sample, (b) strongly absorb the laser light, (c) remain condensed under high-vacuum (do not produce vapors), (d) retard chemical or thermal degradation of the analytes in sample, and (e) induce the sample ionization. ILs qualify the majority of these characteristics and have been increasingly used to make useful MALDI matrixes. ILs-based matrixes offer a more homogenous solution of liquid and solid samples with greater vacuum stability. Another interesting feature of ILs is their ability to readily dissolve biological samples (proteins) and produce more intense spectral intensities as compared to classical solid matrixes. For example, ILs matrix for low molecular weight basic compounds will produce a single molecular ion of protonated analyte which in turn intensifies the mass spectrometric identification of analytes.

5.1.1.3 Ionic Liquids Green Solvents

The ionic liquids definitely cover advanced, non-volatile, and technological tunable solvents which are eventually going to replace all volatile organic solvent. When applied as a reaction solvent, an IL would not produce volatile organic compounds under a wide range of thermodynamic conditions. Moreover, many of ILs are easy and safe to dispose of as compared to highly flammable organic solvents. However, it is too early to declare these solvents as greener extraction choices because a majority of these solvents are either toxic or the environmental fate and potential toxicity are not known. In particular, alkyl-imidazolium-based ILs should be treated as other research chemicals of unknown toxicity and should not be discharged to water bodies unless sufficient data becomes available regarding their toxicity.

It would be reasonable to suggest that concepts of generally regarded as safe (GRAS) and pharmaceutically acceptable ions should be applied along with general regulations set by Joint FAO/WHO Expert Committee on Food Additives (JECFA) and Codex Alimentarius Commission. In this context, inorganic and organic anions like chlorides (Cl_2), bromides (Br_2), sulfates (SO_4), phosphates (PO_4), nitrogen trioxide (NO_3), acetates, succinates, glycolates, lactates, malates, tartrates, citrates, ascorbates, glutamates, benzoates, salicylates, and methanesulfonates have been declared pharmaceutically acceptable non-toxic additives, and many of these can be used for the synthesis of ILs. Finally, it would be fair to conclude that ILs are not going to contribute by design such as do volatile organic compounds still their handling or imitated discharge can cause severe damages to human health, property, and environment.

5.1.2 Ionic Liquids as an Extraction Solvent

The tunable viscosity and polarity of ionic liquids via anion replacement or change in cationic substituents make the ILs potential extraction solvents. However, the selection of appropriate ionic pair for a particular class of biomolecules would be an exhausting task and need comprehensive research regarding physicochemical and thermodynamic behavior of ILs. At the analyte side, especially for biomolecules like phenolics, the size of a biomolecule or molecular mass, its surface characteristics, charge, pH, and nature of substituents may affect partitioning coefficient. At present, ILs based on alkyl-imidazolium cations are potential candidates to replace organic solvent used during liquid–liquid type extraction because, these ILs are (1) air and water-soluble, (2) liquids at room temperature and over a wider range of thermodynamic conditions, (3) less viscous, and (4) less dense. Another potential but unexplored area related to the use of ILs as solvents involves the separation of ILs from the phenolic rich phase. The majority of researchers undertaken IL-based extraction have evaluated extracts (IL + phenolic compounds) for in vitro antioxidant activities or quantified the phenolic compound by various chromatographic or spectroscopic techniques. In this context, da Costa Lopes et al. (2016) adopted a methodology for the pretreatment and fractionation of wheat straw using [C_2MIM][OAc]. Briefly, the powdered cellulosic sample (wheat straw) was mixed IL and heated up at 120 °C for 6 h under constant stirring. The cellulose was fractionated by alkali and filtrate after neutralization and evaporation was further mixed with an excess of ethanol 96% (v/v) to separate out hemicellulose fraction, and 4 M HCl to precipitate out lignin. The aliquot left behind was neutralized with NaOH pellets and evaporated to solid residues (phenolic compounds, NaCl salt, IL entrapped in NaCl, and impurities) which was mixed with plenty of acetonitrile to extract IL and phenolic compounds.

5.1.3 Ionic Liquids as Macerating Agent

An additional mentionable feature of these ILs towards the extraction of natural bioactives is their solvating power or the ability to dissolve biopolymers. For example, Kilpeläinen et al. (2007) have demonstrated both soft and hard type wood are soluble in imidazolium-based ionic liquids. However, during the dissolution process, the wood components may remain intact, which can open an array of opportunities like production of glucose, chemicals, and fuels. Another brief report by Usuki et al. (2011) indicates that incubation of *Ginkgo biloba* leaves with 1-butyl-3-methylimidazolium chloride [C_4MIM]Cl at 150 °C enhanced the extraction of shikimic acid 2–3 times as compared to conventional methanol solvent. The increase in extraction yield was attributed to cellulose dissolving power of [C_4MIM]Cl. In addition to imidazolium-based ILs, certain choline-based ILs like choline chlorides (Ribeiro et al. 2013) have been also used for the extraction of polyphenols and saponins. The latter class of ILs are considered to be more economical, biodegradable, and less toxic; nevertheless, there are very limited number of evidence to establish their suitability regarding extraction of phenolic compounds.

5.1.4 Ionic Liquid-Based Liquid–Liquid Extraction

Bogel-Łukasik et al. (2010) investigated phase equilibrium phenomena of C_4MIM-based ILs bearing different anions and observed that solubilities of phenolic compounds like gallic acid, tannins, and quercetin can be tuned via an appropriate selection of anion and thermodynamic conditions. The authors claimed [C_4MIM][BF_4] as best IL solvent for tannic acid whose solubility in IL followed E^N_T polarity scale. One more representative phenolic antioxidant vanillin was extracted by Cláudio et al. (2010) using ILS, and buffers-based aqueous two-phase system. The authors tried a large number of C_nMIM cations (n = 2, 4, 6, 7, 10 and OHC2) with halogen, acetate, methyl sulfate, phosphates, and dicyanides anions to design an extraction process for standard antioxidant vanillin without compromising its functional characteristics. It is worth mentioning that vanillin preferably migrates towards ILs-rich phase but partition depends upon cation and anion structure, concentration of vanillin, temperature, and viscosities of two phases. For the ILs system investigated by these researchers, 1-butyl-3-methylimidazolium chloride $[C_4mim]Cl$ offered reasonably higher affinities for overall concentration ranges followed by 1-benzyl-3-methylimidazolium chloride, $[C_7H_7mim]Cl$ which produced comparable results when applied at higher concentration (<5 g/dm^3).

Gallic acid (3,4,5-trihydroxybenzoic acid) falls among the most abundant and frequently studied representative phenolic bioactive of high antioxidant, antitumor, and antifungal properties. It also finds application in the analytical assays followed for the estimation of total phenolic compounds as a reference standard. The majority of agricultural commodities especially fruits, teas, and herbs contain a reasonable amount of free and hydrolyzable (tannins) gallic acid. Cláudio et al. (2012) used ILs along with some inorganic salt to partition gallic acid. A large number of ILs combinations to develop a two-phase aqueous system for the purification of gallic acid and observed that under moderately acidic conditions the neutral form of gallic acid, and its homologies prefer ILs rich phase while its conjugate forms prefer salt-rich phase. The outcomes indicated that ILs-based liquids under acidic conditions can be used for the extraction of gallic acid from natural biomasses.

5.1.5 Ionic Liquid-Assisted Microwave Extraction

Microwaves consist of two electric and magnetic fields oscillating perpendicular to each other within the frequency range of 300 megahertz (MHz) to 300 gigahertz (GHz) which interact with polar compounds to induce ionic conduction and continuous rotations of dipoles. The heat energy is mainly generated due to the resistance of matrix towards ionic migrations and rotations which often rupture cell vacuoles and liberate polar compounds (solutes) from a matrix. At this stage, the presence of a compatible solvent may cause selective diffusion of analytes, or they may be allowed to trickle down under the effect of gravity.

Ionic liquids being polar in nature interact with microwave without undergoing any structural changes and may induce more heat within the same irradiation time. This property of ILs has attracted analytical chemist to execute simultaneous matrix-energy-assisted extraction of phenolic compounds. Lou et al. (2012) carefully investigated the effect of carbon chain length of imidazolium-based cations, anions, and microwave parameters on the recovery of caffeic acid, chlorogenic acid, and quercetin from burdock (*Arctium lappa* L.) leaves. It was observed that an increase in carbon number of alkyl substituent (Table 5.1) from C-2 (ethyl) to C-4 (butyl) sharply increased the recovery of caffeic and chlorogenic acid but further increase in carbon number (C-5 to C-8) does not cause a parallel increase in extraction yield. This kind of behavior indicates that C-4 alkyl chain, i.e., 1-butyl-3-methylimidazolium may work as a compromising point between polarity and viscosity of proposed ILs system. Similarly, the extraction efficiency of 1-butyl-3-methylimidazolium-based ILs depends upon of anions. In general, anions offering good interaction (π-π, ionic

Table 5.1 Summary of ionic liquids used for the extraction of phenolic compounds from plant matrix

ILS	Acronym	Characteristics/conditions density (ρ; g/mL), viscosity (*η;* mPa s)	Analytes	Matrix	Recovery (percent)	References
Liquid liquid extraction						
1-butyl-3-methylimidazolium chloride	$[C_4MIM]Cl$	Non-volatile, water-miscible, $\rho = 1.10$, MP = 65 °C	Caffeine	guaraná (*Paullinia cupana)*	9	Cláudio et al. (2013)
do	do	do	Shikimic acid	*Ginkgo biloba* leaves	2–3	Usuki et al. (2011)
1-Butyl-3-methylimidazolium tetrafluoroborate	$[C_4MIM]$ [BF4]	$\rho = 1.21$, $\eta = 233$, −80 °C, water-miscible	Piperine	White pepper	98.70	Cao et al. (2009)
1-Alkyl-3-methylimidazolium Bromides, chlorides, and methylsulfates	$[C_nMIM]Br$ $[C_nMIM]Cl$ $[C_nMIM]$ CH_3SO_4 $n = 2–8$	A biphasic system consisting ILs and inorganic salts (K_3PO_4, Na_2SO_4, KH_2PO_4,K_2HPO_4) was formed	Gallic acid	Spiked	<50	Cláudio et al. (2012)
Imidazolium cations with halogens, methanesulfonate, acetate, and dicyanides anions	$[C_nMIM]X$ $[C_nMIM]$ CH_3CO_2 $[C_nMIM]$ $N(CN)_2$ $[C_nMIM]$ CH_3SO_3 $n = 2, 4, 6, 7, 10$, OHC_2	do	Vanillin	Spiked	<50 for $[C_4MIM]Cl$ & $[C_7MIM]Cl$	Cláudio et al. (2010)
Chlorides of 1-butyl-3-methylimidazolium 1-butyl-1-methylpyrrolidinium 1-butyl-3-methylpiperidinium Tetrabutylammonium Cholinium	$[C_4MIM]Cl$ $[C_4MPYR]$ Cl $[C_4MPIP]Cl$ $[N_{4444}]Cl$ $[Ch]Cl$	A ternary system consisting 5% (by weight) of ILs in PEG 400+ $(NH_4)_2SO_4$ was formed to study salting out and partition coefficients	Vanillic acid Gallic acid and eugenol	Spiked	nd	Neves et al. (2019)

1-Butyl-3-methylimidazolium bromide	$[C_4MIM]Br$	0.50 g sample + 10 mL of 1.2 Mol/L $[C_4MIM]Br$ subjected to microwave irradiation for 8 min at 70 °C.	Isoflavones	Radix puerariae		Zhang et al. (2014)
Solid-phase extraction						
Chlorides, bromides, tetrafluoroborates, and tosylates of 1-butyl-3-methylimidazolium	$[C_4MIM]Cl$ $[C_4MIM]Br$ $[C_4MIM][BF_4]$ $[C_4MIM][Tso]$	Solid-phase extraction	Protocatechuic, ferulic, and caffeic acid	*Saliconia herbaces* L.	79–94	Bi et al. (2012)
1-methylimidazole+blank polymer	–	Solid-phase extraction	Caffeine and theophylline	Green tea	87.2–91.03	Tian et al. (2009)
Microwave/ultrasound-assisted extraction						
Chlorides, bromides, tetrafluoroborates, and dihydrogen phosphate of 1-butyl-3-methylimidazolium	$[C_4MIM]Cl$ $[C_4MIM]Br$ $[C_4MIM][BF_4]$ $[C_4MIM][H_2PO_4]$	5 g powder + ILs processed under variable microwave condition while ultrasound power was fixed to 50 W	Caffeic acid, Chlorogenic acid, and Quercetin	Burdock (*Arctium lappa* L.) leaves	96.1–105.3	Lou et al. (2012)
1-Alkyl-3-methylimidazolium Bromides, chlorides, tetrafluoroborates, and hexafluorophosphates	$[C_nMIM]X$ ($n = 2, 4, 6$) $X = Cl^{-1}$, $Br^{-1}, BF_4^{-1}, PF_6^-$	$[C_4MIM][BF_4]$: $\rho = 1.16$; $\eta = 248$ water-miscible, $[C_6MIM][BF_4]$: $\rho = 1.23$; $\eta = 380$ Partly soluble 1 g of plant +10 mL IL subjected to microwave irradiation	Phenolic alkaloids	Medicinal plant *Nelumbo nucifera Gaertn*	98–105% for $[C_4MIM][BF_4]$ & $[C_6MIM][BF_4]$	Lu et al. (2008)

(continued)

Table 5.1 (continued)

ILS	Acronym	Characteristics/conditions density (ρ; g/mL), viscosity (η; mPa s)	Analytes	Matrix	Recovery (percent)	References
1-butyl-3-methylimidazolium bromide	[C_4MIM]Br	3.0 g sample + 30 mL of 0.75 M ILs at 60 °C, solvent to solid ratio 12:1 mL/g, and irradiation time 7 min	Phenolic compounds	Pigeon pea leaves	97–103	Wei et al. (2013)
1-butyl-3-methylimidazolium methylsulfate	[C_4mim] CH_3SO_4	1.0 Mol L^{-1} [C_4mim]CH_3SO_4 at pH 1.0 was applied at L/S of 1:10 and ultrasonic power of 200 W	Luteolin and apigenin	Celery	72.7–89.5	Han and Row (2011)
1-butyl-3-methylimidazolium hexafluorophosphate	[C_4MIM] [PF6]	1.0 g powdered sample+3 mL of IL mixture was subjected to ultrasound treatment	Chlorogenic acid, gallic acid, psoralen, rutin, and bergapten	*Ficus carica* L	95–104	Qin et al. (2015)
1-ethyl-3-methylimidazolium acetate	[C_2MIM] [OAc]	3.5 g wheat straw+ [C_2MIM][OAc] kept in a 100 mL flask at 120 °C for 6 h under constant stirring.	Phenolic compounds	Wheat straw	No data available	da Costa Lopes et al. (2016)
1-butyl-3-methylimidazolium chloride	[C_4MIM]Cl	0.5 g of sample (60-mesh) mixed with 25 mL of 1.5 Mol L – 1 aqueous IL for 8 min under 136 W of microwave power	Paeonol	Roots of *Cynanchum paniculatum*	85.2–126.7	Jin et al. (2011)
1-Octy-3-methylimidazolium bromide	[C_8MIM]Br	0.5 g sample was added to 0.75 M 1-octyl-3-methylimidazolium bromide followed by ultrasound treatment	Alkaloids	*Camptotheca acuminata samara.*	100	Ma et al. (2012)
Tetramethyl guanidium lactate	TMGL	5 g of sample was dispersed in 50 mL of0.25 M IL and extraction was carried out under 58 W of ultrasound treatment	Forskolin	*Coleus forskohlii* roots	87.4	Harde et al. (2014)
1-Butyl-3-methylimidazolium bromide	[C_4MIM]Br	0.50 g sample + 10 mL of 1.2 Mol/L [C_4MIM]Br subjected to microwave irradiation for 8 min at 70 °C.	Isoflavones	Radix puerariae	No data available	Zhang et al. (2014)

interaction or hydrogen bonding) like chlorides and bromides offered good recovery rates of phenolic acids. For less polar phenolic like quercetin reverse might be true. The authors also observed that an increase in the concentration of [C_4MIM][Br] up to 1.5 mMol increased the extraction of phenolic compounds particularly quercetin but further increase in concentration of IL. The cite worthy outcomes of this study indicates that incorporation of microwave and ultrasound treatments enhance the recovery of liberated phenolic compounds due to increased heat flux and ILs absorbs more heat radiations as compared to conventional solvents. Within imidazolium-based ionic liquids, longer chain cations may favor the recovery of less polar compounds. Similarly, small size anions will favor the liberation of polar phenolic compounds.

Cláudio et al. (2013) reported the use of 1-butyl-3-methylimidazolium (C_4mim), methylimidazolium (C_2mim), 1-hydroxyethyl-3-methylimidazolium (OHC_2mim) cations-based ILs for the selective extraction of caffeine from guaraná (Table 5.1). The authors investigated a wide range of anions like chloride, acetate, and tosylate and extraction conditions including, extraction times, ILs concentration, liquid–solid ratio, and temperature. Finally, it was observed that 1-butyl-3-methylimidazolium chloride (C_4mimCl) produced an outstanding recovery of caffeine (9% by dry weight of sample) under moderate temperature and short withholding time. The ILs was found to be recyclable and reusable up to three consecutive treatments.

Hibiscus (*Hibiscus sabdariffa* L.), an edible flower of *Malvaceae* family has been ranked as the national flower of many countries like Malaysia, South Korea, and Hawaii for its importance. The fusion made from the calyces of hibiscus flower has become a popular beverage (available under different traditional names) for the presence of phenolic antioxidants. Cassol et al. (2019) compared the phenolic compounds recovered from the calyx of hibiscus during exhaustive extraction (methanol for 25 min), simple microwave-assisted extraction (MAE), treatment with citric acid prior to MAE, and MAE followed by citric acid extraction. The outcomes indicate that MAE before extraction with citric (acidic aqueous solution) was more effective for the recovery of hibiscus phenolics as compared to exhaustive extraction.

Likewise et al. (2008) used ILs of C_nMIM (n = 2, 4 and 6) cations with four anions (Cl^{-1}, Br^{-1}, BF_4^{-1}, and PF_6^-) for microwave-assisted extraction of phenolic alkaloids from *Nelumbo nucifera Gaertn*. It was observed that among selected anions ILs containing tetrafluoroborates (BF_4^-) of selected cations were more efficient. Within the similar anions (BF_4^-) an increase in carbon chain could not cause significant changes in extraction. This kind of behavior might be a matter of concern because an increase in the number of the alkyl chain of C_nMIM increases its viscosity and hydrophobicity. The similar kind of behavior was also raised by these authors for anion PF_6^{-1} when [C_4MIM][PF_6] could not furnish parallel extraction yield. Overall, research of present investigation claimed that incorporation of [C_4MIM][BF_4] and [C_6MIM][BF_4] during microwave-assisted extraction not only enhance the recovery rates but also reduced the extraction time from 2 h to 90 s. A similar kind of results have been reported by Zeng et al. (2010) during microwave-ionic liquid-based extraction of rutin from medicinal plants. Zhang et al. (2014) found recovery of total isoflavonoids from *Radix pueraria* increases steadily with an increase in pH values during microwave-assisted extraction aqueous solution of [C_4MIM][Br] IL until pH reaches 9.0 because the isoflavonoids are highly O-glycosylated and their O-linked saccharides are covalently linked to other cell wall components. Under alkaline conditions, O-chains tend to be cleaved off, and thus the O-glycosylated isoflavonoids are liberated. Most importantly, the cell wall breakage can also be attributed to weakening and/or disruption of inter- and intramolecular hydrogen bonds in the polysaccharide chains. Flavonoids are normally solubilize under weak alkaline solutions. A further increase in pH can hydrolyze flavonoid backbone and reduce the extraction yield.

5.1.6 Ionic Liquid-Assisted Ultrasound Extraction

Likewise, microwave, ultrasounds covers sound waves which fall beyond human hearing, i.e., 20 kilohertz (KHz) to 100 megahertz (MHz). The passage of ultrasound during sample medium causes cavitation (bubbles) via nonstop compressions and expansions and a sufficient amount of energy is liberated during kinetic motions liquids or gases entrapped. A typical cavity generated inside liquid may have temperature as high as 5000 K and grow to the pressure of 1000 atm which on collapse cools suddenly at 1010 k/s and liberated energy may facilitate solutes from plant material. Ultrasound energy irradiation has been coupled with almost all classical extraction techniques to improve their efficiency, and the same trend has been applied in the case of IL-based extraction. Nevertheless, what kind of interaction/change sample may undergo when dissolved in IL the majority of researchers have claimed the progressive effect of ultrasound irradiations on the extraction of phenolics.

Cao et al. (2009) investigated 1-alkyl-3-methylimidazolium (CnMIM) cations where $n = 2$, 4 and 6 and anions including tetrafluoroborate (BF_4^-), bromide (Br^-), dihydrogphosphate ($H_2PO_4^-$), and hexafluorophosphate (PF_6^-) as anions for the extraction of from white pepper piperine during ultrasound-assisted extraction. In addition to extraction conditions like temperature, irradiation time, and ultrasound intensity, the nature of ILs influenced the recovery of piperine. The water miscibility of series of methylimidazolium (MIM)-based ionic liquids decreased with the increase in carbon chain length of alkyl substituents and anions following order $BF_4^- > Br^- > H_2PO_4^- > PF_6^-$. Similar was the case of extraction efficiencies, i.e., extraction efficiency decreased with the decrease in hydrophilic character of anion or increase in alkyl chain length of cations. Under the optimum extraction conditions for 10 mL experiment (IL concentration, ultrasonic power 500 W), the authors recovered up to 98.70% of piperine within 30 min of extraction time whereas simple ultrasound-assisted extraction could recovery up to 5% of piperine under same conditions. It should be mentioned here that the application of ILs for the recovery of piperine during ultrasound-assisted extraction does not change the absorbance of retention characteristics of analyte.

Neves et al. (2019) tested the effect of chloride-based ILs with cholinium (Ch), 1-butyl-3-methylimidazolium (C_4MIM), 1-butyl-1-methylpyrrolidinium (C_4MPYR), 1-butyl-1-methylpiperidinium (C_4MPIP), tetrabutylammonium (N_{4444}), and tetrabutylphosphonium (P_{4444}) at 5% by weight on polyethylene glycol (PEG 400) and $(NH_4)_2SO_4$-based aqueous two-phase system. The resultant pseudo-ternary liquids were used for the extraction of phenolic compounds vanillic acid, gallic acid, and eugenol). It was observed that mixture of IL-PEG behave more hydrophobic character and ILs prefer PEG-rich part of phase following an overall of partition coefficients as Ch < C_4MIM < C_4MPYR < C_4MPIP < N_{4444} < P_{4444} thus can be easily salted out from aqueous media. ILs can work as adjuvants in the ternary phase systems of water + PEG + salt and improve the extraction of biomolecules including phenolics; however, the outcomes are inconsistent. Overall, the addition of IL as an adjuvant at very low concentration affect the partition coefficient of phenolics, an IL bearing more hydrophobic cation transfer more hydrophobic phenolics towards PEG-rich phase and vice versa.

5.1.7 Conclusions and Future Prospectus

The above-cited scientific reports and features of ionic liquids (ILs) indicate that these solvents may become green designer's choice solvents within next couple of decades. In a short time, ILs will become non-volatile digestive and extraction solvent for plant phenolics. Obviously, this requires the synthesis of more symmetric cations and anions. Similarly, there needs a great deal of research regarding structural features of ILs and phenolic compounds to design selective liquid or solid-phase extraction of particular class of phenolic compounds. As stated above, ILs have been frequently linked with green chemistry when used as catalyst; however, their application as extraction solvent needs more comprehensive and guided evaluation of cations and anions before calling them "green extraction solvent." Finally, lot of combinations and opportunities are there regarding new ionic liquids and many of these may provide more green and selective choice for the extraction of phenolic bioactives.

Acknowledgements We would like to express our deepest regard for the motivation, encourage, and guidance of our senior supervisors, especially Dr. Bushra Sultana, Assistant Professor, Department of Chemistry, University of Agriculture, Faisalabad, Pakistan.

References

Armstrong DW, He L, Liu Y-S (1999) Examination of ionic liquids and their interaction with molecules, when used as stationary phases in gas chromatography. Anal Chem 71(17):3873–3876. https://doi.org/10.1021/ac990443p

Bi W, Zhou J, Row KH (2012) Solid phase extraction of three phenolic acids from Saliconia herbacea l. by different ionic liquid-based silicas. J Liq Chromatogr Relat Technol 35(5):723–736. https://doi.org/10.1080/10826076.2011.608230

Bogel-Łukasik R, Nobre Gonçalves LM, Bogel-Łukasik E (2010) Phase equilibrium phenomena in solutions involving tannins, flavonoids and ionic liquids. Green Chem 12(11):1947–1953. https://doi.org/10.1039/C0GC00308E

Cao X, Ye X, Lu Y, Yu Y, Mo W (2009) Ionic liquid-based ultrasonic-assisted extraction of piperine from white pepper. Anal Chim Acta 640(1):47–51. https://doi.org/10.1016/j.aca.2009.03.029

Cassol L, Rodrigues E, Zapata Noreña CP (2019) Extracting phenolic compounds from Hibiscus sabdariffa L. calyx using microwave assisted extraction. Ind Crop Prod 133:168–177. https://doi.org/10.1016/j.indcrop.2019.03.023

Cláudio AFM, Freire MG, Freire CSR, Silvestre AJD, Coutinho JAP (2010) Extraction of vanillin using ionic-liquid-based aqueous two-phase systems. Sep Purif Technol 75(1):39–47. https://doi.org/10.1016/j.seppur.2010.07.007

Cláudio AFM, Ferreira AM, Freire CSR, Silvestre AJD, Freire MG, Coutinho JAP (2012) Optimization of the gallic acid extraction using ionic-liquid-based aqueous two-phase systems. Sep Purif Technol 97:142–149. https://doi.org/10.1016/j.seppur.2012.02.036

Cláudio AFM, Ferreira AM, Freire MG, Coutinho JAP (2013) Enhanced extraction of caffeine from guaraná seeds using aqueous solutions of ionic liquids. Green Chem 15(7):2002–2010. https://doi.org/10.1039/C3GC40437D

da Costa Lopes AM, Brenner M, Falé P, Roseiro LB, Bogel-Łukasik R (2016) Extraction and purification of phenolic compounds from lignocellulosic biomass assisted by ionic liquid, polymeric resins, and supercritical CO2. ACS Sustain Chem Eng 4(6):3357–3367. https://doi.org/10.1021/acssuschemeng.6b00429

Ding J, Welton T, Armstrong DW (2004) Chiral ionic liquids as stationary phases in gas chromatography. Anal Chem 76(22):6819–6822. https://doi.org/10.1021/ac049144c

Gavhane RJ, Madkar KR, Kurhe DN, Dagade DH (2019) Room temperature ionic liquids from purine and pyrimidine nucleobases. Chemistry Select 4(19):5823–5827. https://doi.org/10.1002/slct.201900626

Han D, Row KH (2011) Determination of luteolin and apigenin in celery using ultrasonic-assisted extraction based on aqueous solution of ionic liquid coupled with HPLC quantification. J Sci Food Agric 91(15):2888–2892. https://doi.org/10.1002/jsfa.4553

Harde SM, Lonkar SL, Degani MS, Singhal RS (2014) Ionic liquid based ultrasonic-assisted extraction of forskolin from Coleus forskohlii roots. Ind Crop Prod 61:258–264. https://doi.org/10.1016/j.indcrop.2014.07.016

Jin R, Fan L, An X (2011) Microwave assisted ionic liquid pretreatment of medicinal plants for fast solvent extraction of active ingredients. Sep Purif Technol 83:45–49. https://doi.org/10.1016/j.seppur.2011.09.005

Kilpeläinen I, Xie H, King A, Granstrom M, Heikkinen S, Argyropoulos DS (2007) Dissolution of wood in ionic liquids. J Agric Food Chem 55(22):9142–9148

Lou Z, Wang H, Zhu S, Chen S, Zhang M, Wang Z (2012) Ionic liquids based simultaneous ultrasonic and microwave assisted extraction of phenolic compounds from burdock leaves. Anal Chim Acta 716:28–33. https://doi.org/10.1016/j.aca.2011.03.012

Lu Y, Ma W, Hu R, Dai X, Pan Y (2008) Ionic liquid-based microwave-assisted extraction of phenolic alkaloids from the medicinal plant Nelumbo nucifera Gaertn. J Chromatogr A 1208(1):42–46. https://doi.org/10.1016/j.chroma.2008.08.070

Ma CH, Wang SY, Yang L, Zu YG, Yang FJ, Zhao CJ et al (2012) Ionic liquid-aqueous solution ultrasonic-assisted extraction of camptothecin and 10-hydroxycamptothecin from Camptotheca acuminata samara. Chem Eng Process Process Intensif 57-58:59–64. https://doi.org/10.1016/j.cep.2012.03.008

Neves CMSS, de Cássia S, Sousa R, Pereira MM, Freire MG, Coutinho JAP (2019) Understanding the effect of ionic liquids as adjuvants in the partition of biomolecules in aqueous two-phase systems formed by polymers and weak salting-out agents. Biochem Eng J 141:239–246. https://doi.org/10.1016/j.bej.2018.10.022

Qin H, Zhou G, Peng G, Li J, Chen J (2015) Application of ionic liquid-based ultrasound-assisted extraction of five phenolic compounds from fig (Ficus carica L.) for HPLC-UV. Food Anal Methods 8(7):1673–1681. https://doi.org/10.1007/s12161-014-0047-9

Ribeiro BD, Coelho MAZ, Rebelo LPN, Marrucho IM (2013) Ionic liquids as additives for extraction of Saponins and polyphenols from mate (Ilex paraguariensis) and tea (Camellia sinensis). Ind Eng Chem Res 52(34):12146–12153. https://doi.org/10.1021/ie400529h

Tian M, Yan H, Row KH (2009) Solid-phase extraction of caffeine and theophylline from green tea by a new ionic liquid-modified functional polymer sorbent. Anal Lett 43(1):110–118. https://doi.org/10.1080/00032710903276554

Usuki T, Yasuda N, Yoshizawa-Fujita M, Rikukawa M (2011) Extraction and isolation of shikimic acid from Ginkgo biloba leaves utilizing an ionic liquid that dissolves cellulose. Chem Commun 47(38):10560–10562. https://doi.org/10.1039/C1CC13306C

Walden, P. (1914) Molecular weights and electrical conductivity of several fused salts. Bull Acad Imper Sci 1800

Wei Z, Zu Y, Fu Y, Wang W, Luo M, Zhao C, Pan Y (2013) Ionic liquids-based microwave-assisted extraction of active components from pigeon pea leaves for quantitative analysis. Sep Purif Technol 102:75–81. https://doi.org/10.1016/j.seppur.2012.09.031

Zeng H, Wang Y, Kong J, Nie C, Yuan Y (2010) Ionic liquid-based microwave-assisted extraction of rutin from Chinese medicinal plants. Talanta 83(2):582–590. https://doi.org/10.1016/j.talanta.2010.10.006

Zhang Y, Liu Z, Li Y, Chi R (2014) Optimization of ionic liquid-based microwave-assisted extraction of isoflavones from Radix puerariae by response surface methodology. Sep Purif Technol 129:71–79. https://doi.org/10.1016/j.seppur.2014.03.022

Chapter 6
Ionic Liquids for the Sustainable Development of Chemistry

Haydar Göksu, Nursefa Zengin, Hilal Acıdereli, Ayşenur Aygün, Kemal Cellat, and Fatih Şen

Contents

6.1 Introduction

Analytical research has a very important role in daily life. For this reason, multidimensional studies are conducted in order to increase the speed, safety, and quality of analytical processes. Ionic liquids (ILs) are one of the examples of these studies. For more than a decade, various researches have been carried out studies on ionic liquids. The melting point of ionic liquids is below 100 °C. Ionic liquids are composed by two different components as cations and anions. The anions are organic or inorganic, while the cations are organic. Cations are larger components than anions. They have high thermal stability and can be characterized in liquid form with a wide temperature range (Nawała et al. 2018). Ionic liquids, which are generally in liquid form under room conditions and at low temperatures, have an ionic-covalent crystalline structure. They have high polarity, low vapor pressure, and are resistant to high temperatures. These properties are the results of the coulomb effect, hydrogen bonds, and van der Waals interactions (Wasserscheid and Schröer 2014). Ionic liquids are considered green solvents, regarding to this property there is increasing

H. Göksu (✉) · N. Zengin
Kaynasli Vocational College, Duzce University, Duzce, Turkey
e-mail: haydargoksu@duzce.edu.tr

H. Acıdereli · A. Aygün · K. Cellat · F. Şen (✉)
Sen Research Group, Department of Biochemistry, Dumlupınar University, Kütahya, Turkey
e-mail: fatih.sen@dpu.edu.tr

Inamuddin, A. M. Asiri (eds.), *Nanotechnology-Based Industrial Applications of Ionic Liquids*, Nanotechnology in the Life Sciences,
https://doi.org/10.1007/978-3-030-44995-7_6

interest. Ionic liquids are mainly used as media in chemical processes. It is desirable to improve the use of ionic liquids as both solvent and catalyst. Extensive studies are carried out to increase efficiency. It is noteworthy that it is clean and environmentally beneficial, and cost of ionic liquids are considerably low in industrial technology (Holbrey and Seddon 1999).

Furthermore, the need for volatile organic solvents is eliminated thanks to IL solvents. Due to the high polarity of ILs, they do not mix with most hydrocarbons. Ionic liquids are also used as a very good solvent for inorganic complexes such as catalysts. In ionic liquids, the catalyst is immobilized, and biphasic catalysis is formed. Thus, ionic liquids are easily separated from the product. Since the two-phase catalysis is composed and the product is readily separated, the recycling rate is high. (Welton 1999; Wasserscheid and Keim 2000; Earle and Seddon 2000). Therefore, ionic liquids often used as a catalyst or solvent in chemical reactions as an environmentally friendly alternative. On the other hand, ionic liquids can be reused many times due to the easily removing properties of them from the reaction medium without creating any waste (Sasikumar et al. 2018; Kunz and Häckl 2016; Dobras and Orlińska 2018; Weerachanchai and Lee 2017). Besides, due to their eco-friendly properties, ionic liquids are preferred according to other organic solvents fully meet the concept of "green chemistry." Paul Walden synthesized the first ionic liquid in the literature in 1914. This ionic liquid is ethyl ammonium nitrate [$EtNH_3 + NO_3^-$] salt with a melting point of 120 °C (Walden 1914). With each passing year, the demand for ionic liquids is increasing day by day. According to the anion and cation types, IL's physical and thermal properties vary. In addition, these properties also vary according to the length of the alkyl groups on the cation (Kim et al. 2004) Protic ionic liquids have very different properties than aprotic ionic liquids due to the change of proton (Del Pópolo et al. 2006; Dupont and Suarez 2006; Huang et al. 2006; Ogihara et al. 2006; Markusson et al. 2007; Martinelli et al. 2007; Nakamoto and Watanabe 2007; Nakamoto et al. 2007; Nuthakki et al. 2007; Greaves and Drummond 2008; Greaves et al. 2008; Hernández-Fernández et al. 2015; Rika and Lee 2007). In theory, ionic liquids are formed purely ions (Noda et al. 2003) The liquid must be highly ionized in order to be considered as pure salt (Johansson et al. 2008). There is a wide range of solvent properties due to the wide variety of ionic liquids that are completely formed of ions. This allows for more workspaces to be used (Fernández et al. 2015). As the structure of ions in ionic liquids is changed, ionic liquids are obtained in different properties and different application areas (Chen et al. 2018). Physical properties of ionic liquids such as viscosity, solubility, melting point, density, and hydrophobicity can be modified depending on the requirements for a specific task.

As a consequence of this ability, ionic liquids called designer solvents, and reaction products can be separated by using conventional solvents. Owing to these properties, ionic liquids are more preferable than traditional volatile solvents and catalysts in the various chemical process such as electrochemistry, organic synthesis, solid support, chemical separation, nanoparticle formation, catalysis, and biocatalysis. (Vekariya 2017; Zafarani-Moattar and Majdan-Cegincara 2007). Currently, the use of ionic liquids has increased. Ionic liquids can be applied not only in the field of chemistry but also in nanotechnology, physics, and other sciences (Khare et al. 2012; Bennett and Leo 2006). In recent years, the using of ionic liquids have become widespread with the Montreal Protocol in 1989 and have taken its place as an alternative "green" and "designable" solvents (Plechkova and Seddon 2008) for sustainable and green chemistry.

6.2 The Applications of Ionic Liquids

Ionic liquids are used in different fields such as physics, chemistry, and material science (Beichel et al. 2014). Previously mentioned properties of ionic liquids have attracted interest with a varied range of applications such as inorganic (Freudenmann et al. 2011; Lei et al. 2014), catalysis, and organometallic or organic synthesis (Hallett and Welton 2011; Olivier-Bourbigou et al. 2010; Conrad Zhang 2006; Wasserscheid and Keim 2000), biocatalysts (van Rantwijk and Sheldon 2007; Kragl et al. 2002), chemical analysis (Shamsi and Danielson 2007; Koel 2008; Pandey 2006), electrochemical devices (Hagiwara and Lee 2007), media for electrodeposition of metals (Endres et al. 2008; Endres and El Abedin 2006), pharmaceutical ingredients (Hough and Rogers 2007), batteries (Osada et al. 2016), ionogels (Mitra et al. 2017), sensors (Kuberský et al. 2015), fuel cells (Hernández-Fernández et al. 2015), capacitors (Lin et al. 2016; Li et al. 2015), plasticizers (Li et al. 2015), thermal fluids (Chernikova et al. 2015), extractants (Zhang et al. 2015), and lubricants (Bermúdez et al. 2009). Many species of ionic liquids have been developed for specific targets such as organic synthesis, catalysis, and separation. Ionic liquids derived from imidazolium and aluminate are mostly used in the production of batteries as electrolytes (Wilkes et al. 1982; Chum et al. 1975).

In recent years, studies on chiral ionic liquids instead of organic chiral solvents have been reported in the literature. Because of the specific properties of chiral ionic liquids and their easy synthesis compared to the chiral organic molecules, their importance in organic chemistry as enantiomer selector are becoming increasingly. In general, chiral ionic liquids are used as chromatographic fillers in the separation of racemic mixtures, as a solvent in the determination of enantiomeric mixtures in NMR spectroscopy, as both solvent and catalyst in both industrial and synthetic chemical reactions.

On the other hand, with the increasing population in the north of the world, also there is an increase in global pollution and decrease in fossil fuels. These issues compel people to find new sustainable energy resources. Biodiesel, a product of vegetable and animal origin, is an alternative fuel source for oil, and it is also a renewable and environmentally friendly energy source. However, there is a problem with the resources of biodiesel. Developed countries pursue alternative methods for the production of biodiesel, especially in the scientific community. In the literature, ionic liquids are used as catalysts or solvents in biodiesel production. Using of ionic liquids in the catalyst of biodiesel are also taken attention. Some disadvantages of homogeneous or heterogeneous catalysts are reported in the literature, especially in the case of biodiesel. They are often used as solvents or catalysts in the transesterification of vegetable oils or the esterification of fatty acids (Ullah et al. 2018;

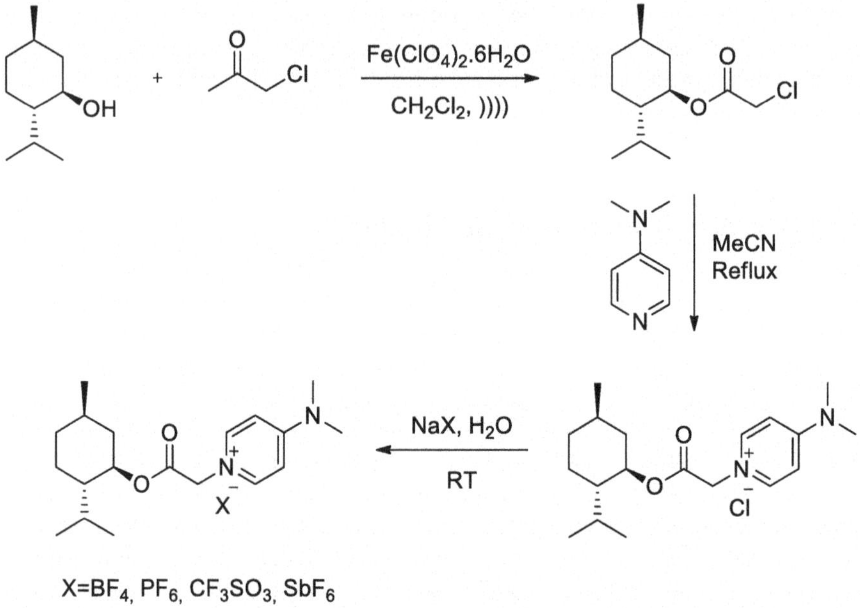

Scheme 6.1 Synthesis of some chiral ionic liquids

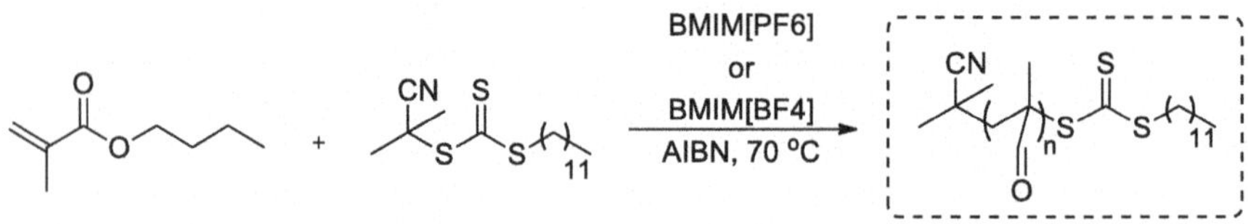

Scheme 6.2 Reversible addition-fragmentation chain transfer polymerization of butyl methacrylate with BMIM [PF_6] or BMIM [BF_4]

Wasserscheid and Keim 2000). By the way, homogeneous and heterogeneous catalysts have also been used in the synthesis of organic molecules in recent years (Eris et al. 2018a; Eris et al. 2018b; Şen et al. 2018c; Günbatar et al. 2018; Ayranci et al. 2017; Şen et al. 2018b). In particular, chiral catalysts or solvents used in the synthesis of chiral molecules have become popular in the literature (Şen et al. 2018a). However, the high cost of these catalysts has led scientists to alternative catalysts. For this purpose, the tendency for effective and easily synthesized chiral/achiral ionic liquids is increasing day by day. Singh and Chopra used chiral ionic liquids obtained by using 4-dimethyl aminopyridine as the Mosher's acid. Their research indicated that an effective catalyst used in the reduction of carbonyl compounds as given in Scheme 6.1. (Singh and Chopra 2018).

Ionic liquids are also used in the polymerization reactions for solvent, catalyst, or other purposes. In the literature, reversible addition-fragmentation chain transfer polymerization of butyl methacrylate was carried out using ionic liquids such as

BMIM [BF_4] (1-butyl-3-methylimidazolium tetrafluoroborate), BMIM [PF_6] (1-butyl-3-methylimidazolium hexafluorophosphate), EMIM [$EtOSO_3$] (1-ethyl-3-methylimidazolium ethyl sulfate). In addition to the use of ionic liquids as solvents, it also carried out a mass-controlled polymerization, as shown in Scheme 6.2 (Kumar et al. 2018).

6.3 Ionic Liquids to Carry out Isomerization and Dimerization Reactions

Many studies were reported on synthesis and applications of ionic liquids. Ionic liquids are mainly used as co-catalyst and inert solvents due to their ability to dissolve organometallic compounds by maintaining polar medium (Vekariya 2017). Some of the transition metal-catalyzed reactions are oxidation reactions (Owens and Abu-Omar 2000), alkoxycarbonylation reactions (Zim et al. 1998), Heck reactions (Calo et al. 2000), hydrogenation reactions (Melo et al. 2016), hydroformylation reactions (Walter et al. 2015), Suzuki cross-coupling reactions (Jiao et al. 2015), Trost–Tsuji coupling reactions (W. Chen et al. 1999), ring-closing metathesis, olefin dimerization, and oligomerization reactions. Ionic liquids can also be used to carry out the isomerization reactions in a more controlled manner. In particular, isomerization can be controlled in metal-catalyzed dimerization reactions that carried out in the ionic liquid. For instance, in the literature, the nickel-catalyzed dimerization reaction of propylene was carried out in a controlled manner in organochloroaluminate IL. In this study, not only the IL used for isomer-controlled dimerization but also the temperature and pressure of dimerization is examined in detail. As shown in the study, active C_1 and C_2 carbons of propylene reacted with nickel separately to form active species which may react with propylene in the next step. The dimerization products formed by the elimination reactions are isomerized to form more stable olefin products, as shown in Scheme 6.3 (de Souza et al. 2007).

In another study, Fabry et al. focused on catalytic (E)- to (Z)-isomerization of olefins in ionic liquids using a photoredox catalyst. They develop a technique which consists of two-phase system. The photosensitizer was absorbed in IL phase while the olefin in an apolar solvent phase to be separated easily. They used toluene/[BMIM/BF_4] and stilbene/[Ir(ppy)$_2$(bpy)]PF_6 as two-phase photocatalyst ionic liquid system. Results showed that (Z)-olefins can be obtained efficiently compared to typical methods, and no-loss of reactivity was observed after repeated cycles (Fabry et al. 2015).

In the dimerization of organic molecules, new molecules are introduced into the literature by the use of ILs, which are defined as an environmentally friendly catalyst. In the literature, the IL acting as Bronsted–Lewis acid is used as a catalyst in the dimerization of the resin molecule. (3-sulfonic acid)-propyl-triethylammonium chlorozincinate [$HSO_3(CH_2)_3NEt_3$] Cl_nZCl_2 was used as Bronsted–Lewis acid. Herein, this ionic liquid was synthesized and characterized, as shown in Scheme 6.4. The activity in the chemical reactions of this synthesized catalyst is quite high.

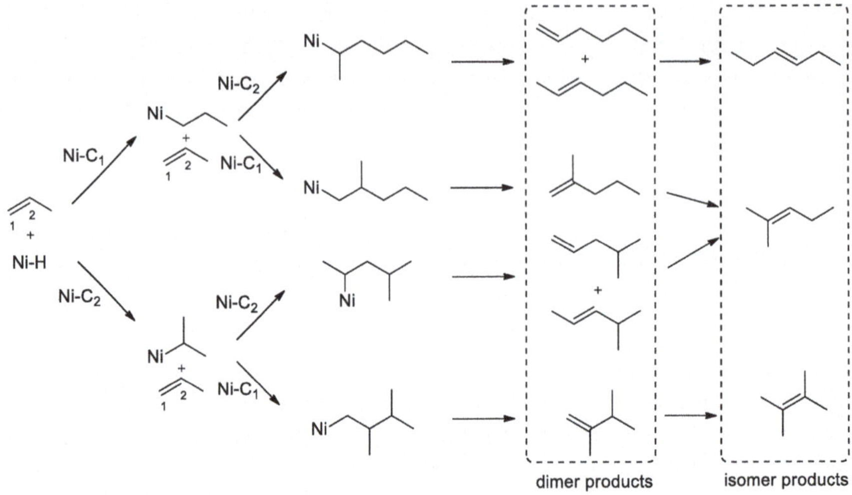

Scheme 6.3 Proposed mechanism for dimerization of propylene

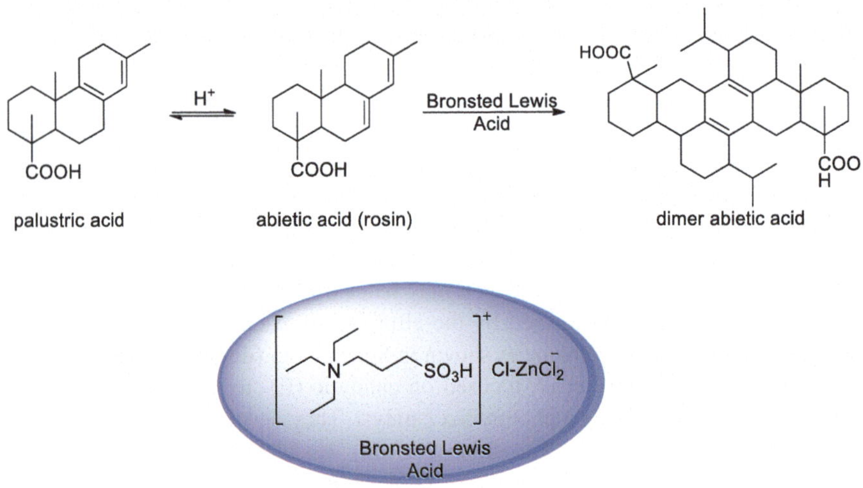

Scheme 6.4 Dimerization of abietic acid

Ionic liquids can also be used repeatedly after being easily removed from the reaction medium (Liu et al. 2008).

Oligomerization can generally be used for a variety of chemicals in academia and industry. Among those materials, chlorinated ionic liquids also have acid-base properties (Morton and Hamer 2018; Atkins et al. 2011; Olivier-Bourbigou et al. 2009; Welton 1999). Due to the continuous increase in energy demand for transportation, the demand for oligomerization is increasing.

Oligomers obtained using olefins in the chemical industry and science community as excipients, lubricants, surfactants, and prepolymers are still significant.

Oligomers, also important sources of petroleum products such as ethylene, propylene, and butene, are obtained by catalytic reactions of olefins. In the literature, oligomers have been achieved under various reaction conditions such as acidic media, zeolites, clays, resins, homogeneous, and heterogeneous catalysts (Forestière et al. 2009; Antunes et al. 2015; Thiele and de Souza 2007; Azizov et al. 2010). However, it is desirable to use ionic liquids as solvents or catalysts in oligomerization reactions, considering both the controllability of the reactions and the environmental factors. For the oligomerization of 1-butene, metal ionic liquids obtained from [NiCl $(PMe_3)_2$] [$AlCl_4$] with 1-butyl-3-methylimidazolium chloride have been used, and effective results have been obtained (Li et al. 2018; Behr et al. 2016). In the literature, Bronsted–acid ionic liquid-like 1-(4-sulfonic acid) butyl-3-hexylimidazolium hydrogen sulfate ([HIMBs]HSO_4) was also synthesized and used in oligomerization reactions, as shown in Scheme 6.5.

This ionic liquid ([HIMBs] HSO_4), which can be easily removed from the reaction medium and has the potential to be reused, provides the conversion of isobutene with 83.27%, resulting in 52.02% dimer and 35.80% trimer formation as given in Table 6.1.

In the oligomerization of isobutene, a process starting with the formation of a reactive intermediate such as carbocation under acidic conditions, followed by a process terminating with dimer, trimer, and tetramer formation is seen as shown in Scheme 6.6 (Wang et al. 2018).

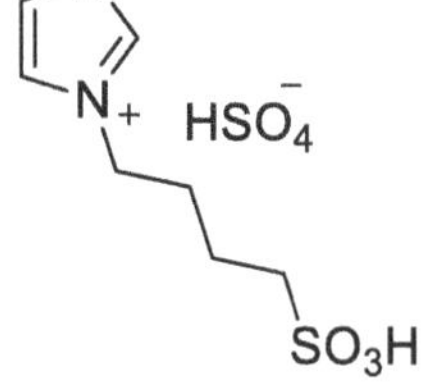

Scheme 6.5 Bronsted–acid ionic liquid for oligomerization reactions

Table 6.1 Oligomers derived from some olefins

			Oligomers	
Entry	Olefins	Conversion yield, %	Dimer, %	Trimer, %
1	Ethylene	59.01	15.96	38.96
2	Propylene	43.87	67.01	9.44
3	1-butene	4.66	31.57	3.99
4	Isobutene	83.27	52.02	35.80

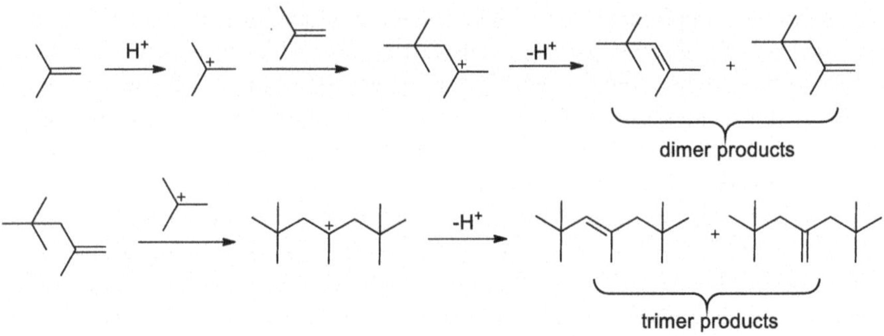

Scheme 6.6 Oligomerization of isobutene in the presence of [HIMBs]HSO_4 as an ionic liquid

Scheme 6.7 [1-butyl-3-methylimidazolium] [tetrachloroaluminate] as ionic liquid

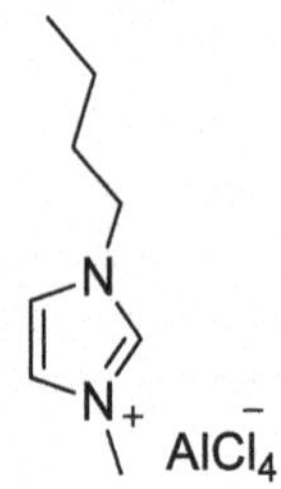

Scheme 6.8 Nickel complex for ethylene oligomerization

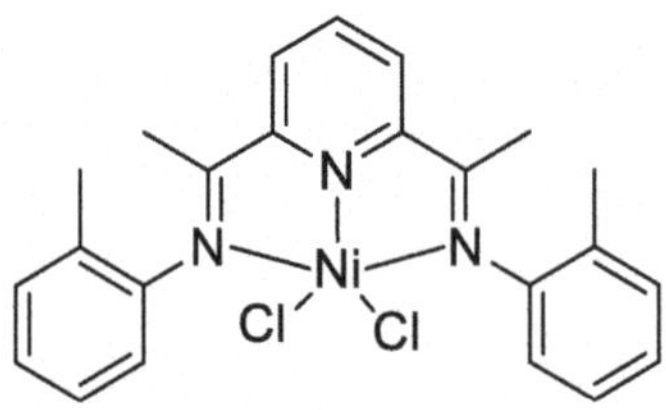

In another study, oligomerization of straight-chain olefins was carried out using chloroaluminate ILs. For this purpose, [1-butyl-3-methylimidazolium] [tetrachloroaluminate] as the IL was synthesized as shown in Scheme 6.7. Dimer, trimer, and tetramer formations were provided from olefins in the presence of a Lewis acid such as $AlCl_3$. The results show that the best conversion in the oligomerization reaction is observed in the ethene molecule at 60 °C. As the chain lengthens, the conversion rates in the reactions decrease, even the tetramer formation is not seen after the propene molecule. This indicates the selectivity of the ionic liquid used (Stenzel et al. 2003).

Another ionic liquid used in oligomerization reactions is the 1-methyl-3-butylimidazolium tetrachloroaluminate (BMI·$AlCl_4$). In this study, the ionic liquid used for oligomerization of ethylene served as co-catalyst. Bis (imino) pyridine nickel (II) complexes were used as the primary catalyst, as shown in Scheme 6.8.

The activity and selectivity of ionic liquid supported catalysts are quite high. With the dimerization of ethylene, over 90% of butene is observed, and 87% is seen as 1-butene (Thiele and De Souza 2011).

6.4 Conclusions

As a conclusion, in recent years, new ionic liquids have been developing owing to the unique properties of them, and applications in this field are increasing rapidly. Nowadays, there are many applications related to ionic liquids. They can be used as catalysts in the processing of biomass, in metal processing, batteries, solar panels, fuel cells, electro-optics, heat transfer, and in the physical properties of the material such as diffusion viscosity. Although the structure of ionic liquids is an irregular structure, a short and medium distance arrangement can be mentioned when an ion reference is taken. Short and medium distances can be examined by determining the structure of distribution functions. This section shows that ionic liquids are used as catalysts, co-catalysts, or solvents in dimerization and oligomerization reactions. In particular, the cation and anion structures in the ionic liquids have an important effect on the conversion and the reaction mechanism. The formation of structures such as a dimer, trimer, and tetramer with the controlled reaction of olefins are essential for the chemical industry. ILs, which are used for different purposes in the flow of science and continue to develop primarily in the plastic sector, have to renew themselves continuously.

References

Antunes BM, Rodrigues AE, Lin Z, Portugal I, Silva CM (2015) Alkenes Oligomerization with resin catalysts. Fuel Process Technol. https://doi.org/10.1016/j.fuproc.2015.04.031

Atkins MP, Seddon KR, Swadźba-Kwaśny M (2011) Oligomerisation of linear 1-olefins using a Chlorogallate(III) ionic liquid. Pure Appl Chem. https://doi.org/10.1351/pac-con-10-12-03

Ayranci R, Başkaya G, Güzel M, Bozkurt S, Şen F, Ak M (2017) Carbon based nanomaterials for high performance optoelectrochemical systems. Chem Select 2(4):1548–1555. https://doi.org/10.1002/slct.201601632

Azizov AG, Kalbalieva ES, Alieva RV, Bektashi NR, Aliev BM (2010) Oligomerization of hexene-1 in the presence of catalytic systems on the basis of ionic liquids. Pet Chem. https://doi.org/10.1134/s0965544110010081

Behr A, Bayrak Z, Samli G, Yildiz D, Peitz S, Stochniol G (2016) Oligomerization of N-butenes in a two-phase reaction system with homogeneous Ni/Al-catalysts. Chem Eng Technol. https://doi.org/10.1002/ceat.201500317

Beichel W, Preiss UP, Verevkin SP, Koslowski T, Krossing I (2014) Empirical description and prediction of ionic liquids' properties with augmented volume-based thermodynamics. J Mol Liq 192(April):3–8. https://doi.org/10.1016/J.MOLLIQ.2013.07.006

Bennett M, Leo D (2006). Physics of transduction in ionic liquid-swollen nafion membranes. Smart structures and materials 2006: active materials: behavior and mechanics. https://doi.org/10.1117/12.658480

Bermúdez MD, Jiménez AE, Sanes J, Carrión FJ (2009) Ionic liquids as advanced lubricant fluids. Molecules. https://doi.org/10.3390/molecules14082888

Calo V, Nacci A, Lopez L, Mannarini N (2000) Heck reaction in ionic liquids catalyzed by a Pd-Benzothiazole Carbene complex. Tetrahedron Lett. https://doi.org/10.1016/S0040-4039(00)01592-6

Chen W, Xu L, Chatterton C, Xiao J (1999) Palladium catalysed Allylation reactions in ionic liquids. Chem Commun. https://doi.org/10.1039/a903323h

Chen M, White BT, Kasprzak CR, Long TE (2018) Advances in Phosphonium-based ionic liquids and poly(ionic liquid)s as conductive materials. Eur Polym J. https://doi.org/10.1016/j.eurpolymj.2018.08.015

Chernikova EA, Glukhov LM, Krasovskiy VG, Kustov LM, Vorobyeva MG, Koroteev AA (2015) Ionic liquids as heat transfer fluids: comparison with known systems, possible applications, advantages and disadvantages. Russ Chem Rev. https://doi.org/10.1070/rcr4510

Chum L, Helena V, Koch R, Miller LL, Osteryoung RA (1975) An electrochemical scrutiny of organometallic Iron complexes and Hexamethylbenzene in a room temperature molten salt. J Am Chem Soc. https://doi.org/10.1021/ja00844a081

Conrad Zhang Z (2006) Catalysis in ionic liquids. https://doi.org/10.1016/s0360-0564(05)49003-3

Del Pópolo MG, Kohanoff J, Lynden-Bell RM (2006) Solvation structure and transport of acidic protons in ionic liquids: a first-principles simulation study. J Phys Chem B. https://doi.org/10.1021/jp0602326

Dobras G, Orlińska B (2018) Aerobic oxidation of Alkylaromatic hydrocarbons to Hydroperoxides catalysed by N-Hydroxyimides in ionic liquids as solvents. Appl Catal A Gen. https://doi.org/10.1016/j.apcata.2018.05.012

Dupont J, Suarez PAZ (2006) Physico-chemical processes in Imidazolium ionic liquids. Phys Chem Chem Phys. https://doi.org/10.1039/b602046a

Earle MJ, Seddon KR (2000) Ionic liquids. Green solvents for the future. Pure Appl Chem 72(7):1391–1398. https://doi.org/10.1351/pac200072071391

Endres F, El Abedin SZ (2006) Air and water stable ionic liquids in physical chemistry. Phys Chem Chem Phys. https://doi.org/10.1039/b600519p

Endres F, MacFarlane DR, Abbott A (2008) Electrodeposition from ionic liquids. Wiley-VCH, Weinheim

Eris S, Daşdelen Z, Sen F (2018a) Enhanced Electrocatalytic activity and stability of Monodisperse Pt Nanocomposites for direct methanol fuel cells. J Colloid Interface Sci 513(March):767–773. https://doi.org/10.1016/j.jcis.2017.11.085

Eris S, Daşdelen Z, Yıldız Y, Sen F (2018b) Nanostructured Polyaniline-RGO decorated platinum catalyst with enhanced activity and durability for methanol oxidation. Int J Hydrog Energy 43(3):1337–1343. https://doi.org/10.1016/J.IJHYDENE.2017.11.051

Fabry DC, Ronge MA, Rueping M (2015) Immobilization and continuous recycling of Photoredox catalysts in ionic liquids for applications in batch reactions and flow systems: catalytic alkene isomerization by using visible light. Chem Eur J. https://doi.org/10.1002/chem.201406653

Fernández H, José F, de los Ríos AP, Quesada-Medina J, Sánchez-Segado S (2015) Ionic liquids as extractor agents and reaction media in ester synthesis. Chem Bio Eng Reviews. https://doi.org/10.1002/cben.201400019

Forestière A, Olivier-Bourbigou H, Saussine L (2009) Oligomerization of monoolefins by homogeneous catalysts. Oil & Gas Science and Technology - Revue de l'IFP. https://doi.org/10.2516/ogst/2009027

Freudenmann D, Wolf S, Wolff M, Feldmann C (2011) Ionic liquids: new perspectives for inorganic synthesis? Angewandte Chemie - International Edition. https://doi.org/10.1002/anie.201100904

Greaves TL, Drummond CJ (2008) Protic ionic liquids: properties and applications. Chem Rev. https://doi.org/10.1021/cr068040u

Greaves TL, Weerawardena A, Krodkiewska I, Drummond CJ (2008) Protic ionic liquids: physicochemical properties and behavior as amphiphile self-assembly solvents. J Phys Chem B. https://doi.org/10.1021/jp0767819

Günbatar S, Aygun A, Karataş Y, Gülcan M, Şen F (2018) Carbon-nanotube-based rhodium nanoparticles as highly-active catalyst for hydrolytic dehydrogenation of dimethylamineborane at room temperature. J Colloid Interface Sci 530(November):321–327. https://doi.org/10.1016/j.jcis.2018.06.100

Hallett JP, Welton T (2011) Room-temperature ionic liquids. Solvents for synthesis and catalysis - chemical reviews (ACS publications). Chem Rev. https://doi.org/10.1021/cr1003248

Hernández-Fernández FJ, de los Ríos AP, Mateo-Ramírez F, Godínez C, Lozano-Blanco LJ, Moreno JI, Tomás-Alonso F (2015) New application of supported ionic liquids membranes as proton exchange membranes in microbial fuel cell for waste water treatment. Chem Eng J. https://doi.org/10.1016/j.cej.2015.04.036

Holbrey JD, Seddon KR (1999) Ionic liquids. Clean Techn Environ Policy 1(4):223–236. https://doi.org/10.1007/s100980050036

Hough WL, Rogers RD (2007) Ionic liquids then and now: from solvents to materials to active pharmaceutical ingredients. Bull Chem Soc Jpn. https://doi.org/10.1246/bcsj.80.2262

Huang JF, Baker GA, Luo H, Hong K, Li QF, Bjerrum NJ, Dai S (2006) Brønsted acidic room temperature ionic liquids derived from N,N-Dimethylformamide and similar Protophilic amides. Green Chem. https://doi.org/10.1039/b604777g

Jiao N, Li Z, Wang Y, Liu J, Xia C (2015) Palladium nanoparticles immobilized onto supported ionic liquid-like phases (SILLPs) for the carbonylative Suzuki coupling reaction. RSC Adv. https://doi.org/10.1039/c5ra00443h

Johansson KM, Izgorodina EI, Forsyth M, MacFarlane DR, Seddon KR (2008) Protic ionic liquids based on the dimeric and oligomeric anions: [(AcO)XHx−1]−. Phys Chem Chem Phys 10(20):2972. https://doi.org/10.1039/b801405a

Khare V, Ruby C, Sonkaria S, Taubert A (2012) A green and sustainable nanotechnology: role of ionic liquids. Int J Precis Eng Manuf 13(7):1207–1213. https://doi.org/10.1007/s12541-012-0160-x

Kim KS, Shin BK, Lee H (2004) Physical and electrochemical properties of 1-butyl-3-methylimidazolium bromide, 1-butyl-3-methylimidazolium iodide, and 1-butyl-3-methylimidazolium tetrafluoroborate. Korean J Chem Eng. https://doi.org/10.1007/BF02705586

Koel M (2008) Ionic liquids in chemical analysis. https://doi.org/10.1201/9781420046472

Kragl U, Eckstein M, Kaftzik N (2002) Enzyme catalysis in ionic liquids. Curr Opin Biotechnol. https://doi.org/10.1016/S0958-1669(02)00353-1

Kuberský P, Altšmíd J, Hamáček A, Nešpůrek S, Zmeškal O (2015) An electrochemical NO2 sensor based on ionic liquid: influence of the morphology of the polymer electrolyte on sensor sensitivity. Sensors. https://doi.org/10.3390/s151128421

Kumar S, Arunjunai RS, Roy M, Singha NK (2018) Effect of ionic liquids on the RAFT polymerization of butyl methacrylate. Eur Polym J. https://doi.org/10.1016/j.eurpolymj.2018.08.018

Kunz W, Häckl K (2016) The hype with ionic liquids as solvents. Chem Phys Lett. https://doi.org/10.1016/j.cplett.2016.07.044

Lei Z, Dai C, Wang W, Chen B (2014) UNIFAC model for ionic liquid-CO2 systems. AICHE J. https://doi.org/10.1002/aic.14294

Li Z, Liu N, Yao Y, Chen S, Wang H, Wang H (2015) Thermal behavior of cellulose Diacetate melt using ionic liquids as plasticizers. RSC Adv. https://doi.org/10.1039/c4ra10618k

Li K, Choudhary H, Rogers RD (2018) Ionic Liquids for Sustainable Processes: Liquid Metal Catalysis. Current Opinion in Green and Sustainable Chemistry. https://doi.org/10.1016/j.cogsc.2018.02.011

Lin Z, Taberna PL, Simon P (2016) Graphene-based Supercapacitors using eutectic ionic liquid mixture electrolyte. Electrochim Acta. https://doi.org/10.1016/j.electacta.2015.12.097

Liu S, Xie C, Yu S, Liu F (2008) Dimerization of rosin using Brønsted-Lewis acidic ionic liquid as catalyst. Catal Commun. https://doi.org/10.1016/j.catcom.2008.03.045

Markusson H, Belières JP, Patrik J, Austen Angell C, Jacobsson P (2007) Prediction of macroscopic properties of Protic ionic liquids by Ab initio calculations. J Phys Chem A. https://doi.org/10.1021/jp072036k

Martinelli A, Matic A, Jacobsson P, Börjesson L, Fernicola A, Panero S, Scrosati B, Ohno H (2007) Physical properties of proton conducting membranes based on a Protic ionic liquid. J Phys Chem B. https://doi.org/10.1021/jp0735029

Melo CI, Szczepańska A, Bogel-Łukasik E, Da Ponte MN, Branco LC (2016) Hydrogenation of carbon dioxide to methane by ruthenium nanoparticles in ionic liquid. ChemSusChem. https://doi.org/10.1002/cssc.201600203

Mitra S, Cerclier C, Berrod Q, Ferdeghini F, de Oliveira-Silva R, Judeinstein P, le Bideau J, Zanotti JM (2017) Ionic liquids confined in silica Ionogels: structural, thermal, and dynamical behaviors. Entropy. https://doi.org/10.3390/e19040140

Morton MD, Hamer CK (2018) Ionic liquids – the beginning of the end or the end of the beginning? – a look at the life of ionic liquids through patent claims. Sep Purif Technol. https://doi.org/10.1016/j.seppur.2017.11.023

Nakamoto H, Watanabe M (2007) Brønsted Acid-Base ionic liquids for fuel cell electrolytes. Chem Commun. https://doi.org/10.1039/b618953a

Nakamoto H, Noda A, Hayamizu K, Hayashi S, Hamaguchi HO, Watanabe M (2007) Proton-conducting properties of a Brønsted acid-base ionic liquid and ionic melts consisting of bis(trifluoromethanesulfonyl)imide and benzimidazole for fuel cell electrolytes. J Phys Chem C. https://doi.org/10.1021/jp0661286

Nawała J, Dawidziuk B, Dziedzic D, Gordon D, Popiel S (2018) Applications of ionic liquids in analytical chemistry with a particular emphasis on their use in solid-phase microextraction. TrAC Trends Anal Chem 105(August):18–36. https://doi.org/10.1016/J.TRAC.2018.04.010

Noda A, Susan MABH, Kudo K, Mitsushima S, Hayamizu K, Watanabe M (2003) Brønsted acid-base ionic liquids as proton-conducting nonaqueous electrolytes. J Phys Chem B. https://doi.org/10.1021/jp022347p

Nuthakki B, Greaves TL, Krodkiewska I, Asoka W, Iko Burgar M, Mulder RJ, Drummond CJ (2007) Protic ionic liquids and Ionicity. Aust J Chem. https://doi.org/10.1071/ch06363

Ogihara W, Kosukegawa H, Ohno H (2006) Proton-conducting ionic liquids based upon multivalent anions and alkylimidazolium cations. Chem Commun. https://doi.org/10.1039/b606186a

Olivier-Bourbigou H, Forestire A, Saussine L, Magna L, Chodorge JA, Pucci A, Debuisschert Q (2009) Olefin oligomerization for the production of fuels and petrochemicals. DGMK Tagungsbericht

Olivier-Bourbigou H, Magna L, Morvan D (2010) Ionic liquids and catalysis: recent progress from knowledge to applications. Appl Catal A Gen. https://doi.org/10.1016/j.apcata.2009.10.008

Osada I, De Vries H, Scrosati B, Passerini S (2016) Ionic-liquid-based polymer electrolytes for battery applications. Angewandte Chemie - International Edition. https://doi.org/10.1002/anie.201504971

Owens GS, Abu-Omar MM (2000) Methyltrioxorhenium-catalyzed Epoxidations in ionic liquids. Chem Commun. https://doi.org/10.1039/b001661f

Pandey S (2006) Analytical applications of room-temperature ionic liquids: a review of recent efforts. Anal Chim Acta. https://doi.org/10.1016/j.aca.2005.06.038

Plechkova NV, Seddon KR (2008) Applications of ionic liquids in the chemical industry. Chem Soc Rev 37(1):123–150. https://doi.org/10.1039/B006677J

van Rantwijk F, Sheldon RA (2007) Biocatalysis in ionic liquids. Chem Rev 107(6):2757–2785. https://doi.org/10.1021/cr050946x

Rika H, Lee JS (2007) Ionic liquids for electrochemical devices. Electrochemistry 75(1):23–34. https://doi.org/10.5796/electrochemistry.75.23

Sasikumar B, Arthanareeswaran G, Ismail AF (2018) Recent progress in ionic liquid membranes for gas separation. J Mol Liq. https://doi.org/10.1016/j.molliq.2018.06.081

Şen B, Akdere EH, Şavk A, Gültekin E, Paralı Ö, Göksu H, Şen F (2018a) A novel thiocarbamide functionalized graphene oxide supported bimetallic monodisperse Rh-Pt nanoparticles (RhPt/TC@GO NPs) for knoevenagel condensation of aryl aldehydes together with malononitrile. Appl Catal B Environ 225(November 2017):148–153. https://doi.org/10.1016/j.apcatb.2017.11.067

Şen B, Aygün A, Okyay TO, Şavk A, Kartop R, Şen F (2018b) Monodisperse palladium nanoparticles assembled on graphene oxide with the high catalytic activity and reusability in the dehydrogenation of dimethylamine-borane. Int J Hydrog Energy 43(44):20176–20182. https://doi.org/10.1016/J.IJHYDENE.2018.03.175

Şen B, Aygün A, Şavk A, Akocak S, Şen F (2018c) Bimetallic palladium–iridium alloy nanoparticles as highly efficient and stable catalyst for the hydrogen evolution reaction. Int J Hydrog Energy 43(44):20183–20191. https://doi.org/10.1016/J.IJHYDENE.2018.07.081

Shamsi SA, Danielson ND (2007) Utility of ionic liquids in analytical separations. J Sep Sci. https://doi.org/10.1002/jssc.200700136

Singh A, Chopra HK (2018) Synthesis, characterization and applications of some novel DMAP-based chiral ionic liquids. J Mol Liq. https://doi.org/10.1016/j.molliq.2018.06.058

de Souza RF, Leal BC, de Souza MO, Thiele D (2007) Nickel-catalyzed propylene dimerization in Organochloroaluminate ionic liquids: control of the isomerization reaction. J Mol Catal A Chem. https://doi.org/10.1016/j.molcata.2007.03.003

Stenzel O, Brüll R, Wahner UM, Sanderson RD, Raubenheimer HG (2003) Oligomerization of olefins in a Chloroaluminate ionic liquid. J Mol Catal A Chem. https://doi.org/10.1016/S1381-1169(02)00450-8

Thiele D, de Souza RF (2007) The role of aluminum species in biphasic Butene dimerization catalyzed by nickel complexes. J Mol Catal A Chem. https://doi.org/10.1016/j.molcata.2006.09.026

Thiele D, de Souza RF (2011) Biphasic ethylene oligomerization using Bis(Imino)pyridine cobalt complexes in methyl-Butylimidazolium organochloroaluminate ionic liquids. J Mol Catal A Chem. https://doi.org/10.1016/j.molcata.2011.03.015

Ullah Z, Khan AS, Muhammad N, Ullah R, Alqahtani AS, Shah SN, Ghanem OB, Bustam MA, Man Z (2018) A review on ionic liquids as perspective catalysts in transesterification of different feedstock oil into biodiesel. J Mol Liq. https://doi.org/10.1016/j.molliq.2018.06.024

Vekariya RL (2017) A review of ionic liquids: applications towards catalytic organic transformations. J Mol Liq. https://doi.org/10.1016/j.molliq.2016.11.123

Walden P (1914) Molecular eights and electrical conductivity of several fused salts. Bull Acad Imp Sci St-Petersbourg

Walter S, Haumann M, Wasserscheid P, Hahn H, Franke R (2015) N-butane carbonylation to n-pentanal using a Cascade reaction of dehydrogenation and SILP-catalyzed hydroformylation. AICHE J. https://doi.org/10.1002/aic.14676

Wang G, Song H, Li R, Li Z, Chen J (2018) Olefin oligomerization via new and efficient Brönsted acidic ionic liquid catalyst systems. Cuihua Xuebao/Chinese Journal of Catalysis. https://doi.org/10.1016/S1872-2067(18)63071-2

Wasserscheid, Keim (2000) Ionic liquids-new "solutions" for transition metal catalysis. Angew Chem

Wasserscheid P, Schröer W (2014) Towards a better fundamental understanding of ionic liquids. J Mol Liq. https://doi.org/10.1016/j.molliq.2014.04.014

Weerachanchai P, Lee JM (2017) Recovery of lignin and ionic liquid by using organic solvents. J Ind Eng Chem. https://doi.org/10.1016/j.jiec.2017.01.018

Welton T (1999) Room-temperature ionic liquids. Solvents for synthesis and catalysis. Chem Rev. https://doi.org/10.1021/cr980032t

Wilkes JS, Levisky JA, Wilson RA, Hussey CL (1982) Dialkylimidazolium chloroaluminate melts: a new class of room-temperature ionic liquids for electrochemistry, spectroscopy, and synthesis. Inorg Chem. https://doi.org/10.1021/ic00133a078

Zafarani-Moattar MT, Majdan-Cegincara R (2007) Viscosity, density, speed of sound, and refractive index of binary mixtures of organic solvent + ionic liquid, 1-butyl-3-methylimidazolium hexafluorophosphate at 298.15 K. Journal of Chemical and Engineering Data. https://doi.org/10.1021/je700338t

Zhang J, Yu K, Yang Q, Bao Z, Zhang Z, Yang Y, Ren Q, Xing H (2015) Selective separation of zwitterionic phospholipid homologues with functional ionic liquids as Extractants. RSC Adv. https://doi.org/10.1039/c5ra14202d

Zim D, De Souza RF, Dupont J, Monteiro AL (1998) Regioselective synthesis of 2-arylpropionic esters by palladium- catalyzed hydroesterification of styrene derivatives in molten salt media. Tetrahedron Lett. https://doi.org/10.1016/S0040-4039(98)01551-2

Chapter 7
Ionic Liquids for Enhanced Enzymatic Saccharification of Cellulose-Based Materials

Carmen-Alice Teacă, Cristina-Magdalena Stanciu, Fulga Tanasă, and Mărioara Nechifor

Contents

C.-A. Teacă (✉)
Advanced Research Center for Bionanoconjugates and Biopolymers, "Petru Poni" Institute of Macromolecular Chemistry, Iasi, Romania

C.-M. Stanciu
Natural Polymers, Bioactive and Biocompatible Materials Department, "Petru Poni" Institute of Macromolecular Chemistry, Iasi, Romania

F. Tanasă · M. Nechifor
Polyaddition and Photochemistry Department, "Petru Poni" Institute of Macromolecular Chemistry, Iasi, Romania

Inamuddin, A. M. Asiri (eds.), *Nanotechnology-Based Industrial Applications of Ionic Liquids*, Nanotechnology in the Life Sciences,
https://doi.org/10.1007/978-3-030-44995-7_7

7.1 Introduction

The continuous diminution of fossil resources and the increasing environmental issues related to their utilization on a global scale require the identification of sustainable alternatives for the manufacture of commodities (gas, chemical substances, and goods), which are essential to the societal development. Lignocellulosic biomass is a widespread natural resource presenting a net zero carbon footprint and significant renewability. Together with lignin and hemicelluloses, cellulose develops a very complex polymer matrix, known as lignin-carbohydrate complex in lignocellulose biomass.

The emerging biorefinery concept which involves the efficient transformation of renewable biomass resources into biofuels and chemicals is potently linked to the hydrolysis process involving cellulose, the main polymer component of lignocellulose materials (namely, wood, straw, crop residues), to monosaccharides, especially glucose. This process represents the most important part for a biorefinery strategy based on carbohydrate polymers (Huber et al. 2006; Gallezot 2007; Corma et al. 2007; Cherubini and Strømman 2011; Tuck et al. 2012; Davison et al. 2014).

Efficient saccharification of cellulose from different renewable sources to glucose is very useful when considering obtainment of energy and chemicals. Saccharification of cellulose substrates in aqueous enzyme solutions is characterized by a reduced rate, cellulose being a semi-crystalline biopolymer insoluble in water. The significant crystalline part of cellulose structure makes different substrates comprising it less accessible to hydrolytic enzymes of cellulase type. Total biodegradation of cellulose is a natural process catalyzed by cellulolytic microorganisms, namely cellulases, which releases organic carbon in the air as gases. There are some kind of enzymes [namely, glucanases—both of endo- and exo-type, and β-glucosidase] involved in the complete saccharification of crystalline part of cellulose in the presence of enzymes (Zhang and Lynd 2004; Zhang 2008).

Total cellulase activities are invariably determined using insoluble cellulosic materials (Zhang et al. 2006; Zhang et al. 2009), such as pure cellulose substrates (e.g., cellulose powder, microbial cellulose, cellulose from algae, Whatman No. 1 filter paper, cotton linter), as well as other sources of cellulose (colored cellulose, insoluble form of cellulose derived from cotton or wood pulp, and pre-treated biomass materials).

Usually, chemical or physical treatments are applied to lignocellulose materials in order to disrupt or weaken the strong bonds present between their polymer components, namely cellulose, lignin, and hemicelluloses, an easier degradation process occurring with conversion of materials into useful bio-fuel intermediates. A comprehensive overview on the applications and breakthrough of a novel category of solvents derived from organic salts with low melting points for biomass pre-treatment and disintegration strategies with focus upon environmental impact and advantageous tunable properties of cellulosic materials resulted from their sustainable valorization processes was presented (Tadesse and Luque 2011).

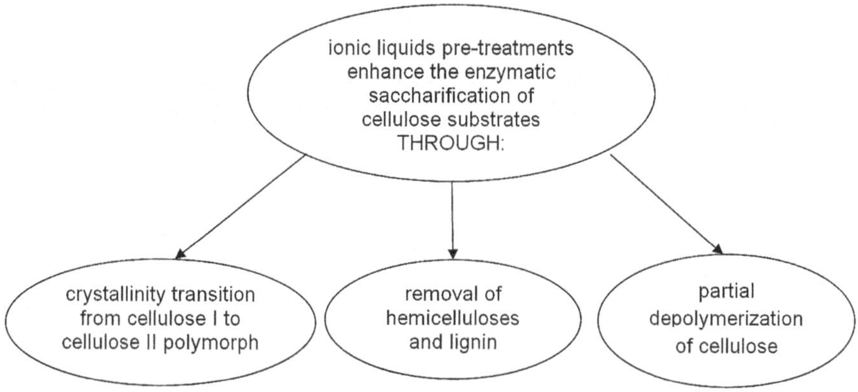

Fig. 7.1 Correlation effects between pre-treatments with ionic liquids and cellulose substrates accessibility to enzymes action

In this context, the preliminary methodologies aiming to intensify the cellulose availability to enzymes attack are essential for the hydrolysis process of this polysaccharide with $\beta(1 \rightarrow 4)$ linked D-glucose units (Swatloski et al. 2002; Zhu et al. 2006; Bodîrlău et al. 2010a; Bodîrlău et al. 2010b; Spiridon et al. 2011). A modern pre-treatment method of cellulose substrates involves use of ionic liquids (IL's), a novel category of solvents for an efficient dissolution of cellulose (Swatloski et al. 2002; Kosan et al. 2008; Tadesse and Luque 2011; Hou et al. 2017), and biocatalysis reactions (Van Rantwijk and Sheldon 2007), which present also effective utility and good properties as media for the enzymatic reactions of such polysaccharide (Dadi et al. 2006; Kamiya et al. 2008; Zhao et al. 2010; Xu et al. 2015).

In Fig. 7.1, there are presented the main effects on the cellulose substrates during pre-treatment with ionic liquids.

Cellulose substrates used in different studies for investigating the influence of ionic liquids pre-treatment upon enzymatic saccharification include corn stover (Li et al. 2011; He et al. 2015; Jin et al. 2018), rice straw (Xu et al. 2015; Cheenkachorn et al. 2016), wheat straw (Ren et al. 2016), kenaf (Ninomiya et al. 2012), switchgrass (Li et al. 2010; Zhao et al. 2010; Samayam and Schall 2010; Singh et al. 2009), bagasse (Uju et al. 2012; Uju et al. 2013), wood species—such as poplar (Samayam and Schall 2010; Wang et al. 2011), and eucalyptus (Uju et al. 2012; Li et al. 2016; Xu et al. 2017), and cellulose-enriched substrates, namely milkweed floss, and white poplar floss (Bodîrlău et al. 2010a; Bodîrlău et al. 2010b; Spiridon et al. 2011).

Crystallinity changes during cellulose pre-treatment with ionic liquids include significant polymorph transitions from crystal form I to crystal form II, as well as a consequently decrease of total crystallinity in cellulose substrates (Dadi et al. 2006; Dadi et al. 2007; Cheng et al. 2011; Bian et al. 2014). A mild treatment of lignocellulose materials with diluted solutions of ionic liquids (Hou et al. 2012; Hou et al. 2013a; Hou et al. 2013b) results in hemicelluloses removal. Ionic liquids can also fractionate the lignin-polysaccharide networks through particular cleavage of lignin-carbohydrate bonds present in lignocelluloses-based substrates (Singh et al. 2009; Brandt et al. 2013). An enhanced dissolving of carbohydrates, mainly those non-tied to lignin or containing a small number of lignin bonds in the presence of suitable catalysts (Sun et al. 2011) can be achieved. The dissolving of the polymeric complex constituted from cellulose, lignin, and hemicelluloses, through intricate entangled and covalent linkages in wood, is also possible (Fort et al. 2007; Kilpeläinen et al. 2007; Miyafuji 2015). The complete dissolution of both hardwoods and softwoods can be obtained by decreasing the particle size of wood samples, but is limited in the presence of water.

An effective enzymatic saccharification of cellulose substrates is closely linked to the chain length of cellulose (Zhang and Lynd 2004), its diminution being observed during pre-treatments with ionic liquids (Uju et al. 2013; Bian et al. 2014). Our previous investigations (Bodîrlău et al. 2010a; Bodîrlău et al. 2010b; Spiridon et al. 2011) have evidenced that dissolving with IL's changed the crystalline orderly pattern of cellulose substrates (namely milkweed floss, and white poplar floss) with a favorable influence on the enzymatic hydrolysis course by increasing the available part of cellulose. An enhanced deterioration process of above-mentioned cellulose substrates by dissolving with 1-ethyl-3-methyl-imidazolium-tetrachloroaluminate [EMIM]Cl-$AlCl_3$ was evidenced (Spiridon et al. 2011) this being caused by the increased acidity of reaction medium which can contribute to enhanced depolymerization of polysaccharides to saccharides.

Chloride anion-based IL's dissolve cellulose more efficiently than other IL systems because chloride anions can readily interact with the hydroxyl groups on cellulose due to the higher basicity of hydrogen bonds (Torimoto et al. 2010). As presented in a previous paper (Spiridon et al. 2011), cellulose can be efficiently dissolved in 1-alkyl-3-methylimidazolium based ionic liquids with chloride anion in the chemical structure, such as [BMIM]Cl (Swatloski et al. 2002; Remsing et al. 2006; Mazza et al. 2009) and [AMIM]Cl (Wu et al. 2004; Zhang et al. 2005). Other useful solvent systems based on ionic liquids are represented by the ammonium salt like benzyldimethyl(tetradecyl)ammonium chloride ([BDTA]Cl) and pyridinium salt like 1-butyl-3-methylpyridinium chloride ([BMPy]Cl) (Heinze et al. 2005; El Seoud et al. 2007).

In all these mentioned above, the interaction between the halogen anions and the OH groups from cellulose causes dissolution of the cellulose materials through disruption of the hydrogen bonds network from polymer chains (Swatlovski et al. 2002; Moulthrop et al. 2005). Such interplays can be revealed by NMR spectroscopy method in solutions of cellulose in 1-butyl-3-methylimidazolium based ionic liquid (Moulthrop et al. 2005; Remsing et al. 2006). Ionic liquids with good transport properties and being liquid at room temperature are represented by those based on 1-alkyl-3-methylimidazolium chloride aluminum chloride type (Wilkes 2002; Brown et al. 2017). Such system with acidic compositions may prove to be an efficient mixture of dissolution medium and catalytic agent for Friedel-Crafts reactions (Boon et al. 1986; Wilkes 2002).

In this paper, different cellulose substrates including pure cellulose, namely microcrystalline cellulose (Avicel), and Whatman filter paper (WFP), respectively, as well as an industrial source of cellulose, namely bleached softwood Kraft pulp, were treated with ionic liquids prior to enzymatic hydrolysis. After pre-treatment with ionic liquid (cellulose/ionic liquids ratio 1/10, at 100 °C for 60 min), cellulose materials were immersed in deionized water under agitation and further filtrated to obtain a product containing amorphous and partial crystalline polymer compounds.

Changes in crystalline structure for the different types of cellulose substrates after dissolution in solutions of ionic liquids (ILs) and saccharification in the presence of enzyme were investigated through Fourier transform infrared (FTIR) spectroscopy and wide-angle X-ray diffraction (WAXD) methods and compared.

Whatman filter papers are well-known as the standards for laboratory general filtration purposes (when exhibit particle retention levels down to 2.5 μm). These pure cellulose substrates are characterized by guaranteed grade, dependability, reproducibility, and uniformity being obtained from raw materials of higher quality, namely cotton linters which contain a minimum alpha cellulose content of 98%. Microcrystalline cellulose, as a very crystalline type compound, is the most used cellulose substrate for different pre-treatment strategies and hydrolysis investigations. It contains 97% of carbohydrate calculated as cellulose on the dried basis.

Pulp is the fibrous material that generally results when its processing through delignification methods disrupts the chemical linkages in the lignocellulosic substrates (wood, fiber crops, or waste paper), separating predominantly cellulose fibers. Delignification methods include mechanical, chemical, or semi-chemical (various combinations of the first two mentioned) approaches. Pulp is an air-dried product comprising 90% oven-dry pulp and 10% water. Kraft pulp results from a dominant industrial chemical process, namely Kraft (sulfate) process which represents 91% of chemical delignification and 75% of all pulp production. In chemical delignification, the wood fibers fibrillation occurs by removing the lignin and mannan (a significant hemicelluloses component in softwood, this being glucomannan-rich), when result fibers with most of its cellulose component and few hemicelluloses (e.g., xylan component). Its main advantages are represented by chemical recovery and pulp strength. For bleached softwood kraft pulp, one of the main chemical pulp grades, a cellulose yield, as pulp component, of 35% can be obtained. Kraft softwood process makes xylan very susceptible to allow the enzymatic hydrolysis which occurs more effectively in the absence of lignin.

Two different ionic liquids (ILs), based on 1-alkyl-3-methylimidazolium chloride ([BMIM]Cl; [EMIM]Cl-$AlCl_3$]), were applied to cellulose substrates as pre-treatment agents for increasing accessibility to further enzymatic hydrolysis with specific enzyme, namely endo-1,4-beta-D-glucanase (cellulase).

Enzymatic saccharification kinetics of the IL-treated cellulose substrates was considerably improved as evidenced by reducing sugars yields. The initial enzymatic reaction rates for cellulose substrates dissolved in ILs were higher than those for initial samples, this evolution being caused by possible *interactive* influence of dissolving solvent and hydrolytic enzyme upon polysaccharide substrates. Comparative with usual cellulose dissolution processes, the ionic liquid solutions were applied under mild parameters in order to ensure an optimum activity of enzyme. Crystallinity index values for cellulose substrates, initial, pre-treated with

ILs, enzyme hydrolysis, as well as after combined pre-treatment with ILs and enzyme hydrolysis, were evaluated from spectral and diffraction data and compared.

7.2 Experimental Part

7.2.1 *Materials*

Microcrystalline cellulose as powder (coded as MC) was obtained from Sigma-Aldrich (USA). Industrial pulp cellulose from softwood species (coded as ISC), as sheets, was supplied by EcoPaper S.A. Zarnesti, Brasov, Romania, while Whatman filter paper was obtained from GE Healthcare UK Ltd. The two last-mentioned cellulose substrates were previously ground on a Retsch PM 200 planetary ball mill at 200 rpm for 30 min, and further sieved to achieve uniform lengths under powder form. Ionic liquids (ILs) were purchased from Acros Organics (NY, USA), namely 1-butyl-3-methylimidazolium chloride ([BMIM]Cl-—coded as IL1), and from Fluka (BASF, Germany), namely 1-ethyl-3-methyl-imidazolium-tetrachloroaluminate ([EMIM]Cl-$AlCl_3$—coded as IL2), respectively. Both ILs were used per se. These have the chemical structures shown in Fig. 7.2.

Cellulase used for enzymatic hydrolysis is derived from *Aspergillus niger* species, being supplied by BioChemika Fluka, Germany. Sodium acetate dihydrate, sodium potassium tartrate (Rochelle salt), 3,5-dinitrosalicylic acid (DNS), sodium hydroxide, phenol, sodium metabisulfite, acetic acid, and ethanol were obtained from Sigma-Aldrich Ltd.(Germany). All reagents and chemicals were analytical grade, being used as received.

7.2.2 *Cellulose Substrates Pre-Treatment with Ionic Liquids*

The pre-treatment of microcrystalline cellulose (MC), Whatman filter paper, and industrial softwood pulp cellulose (ISC) with different ionic liquids (as presented in Fig. 7.2, coded as IL1, and IL2, respectively) was performed in hermetically closed

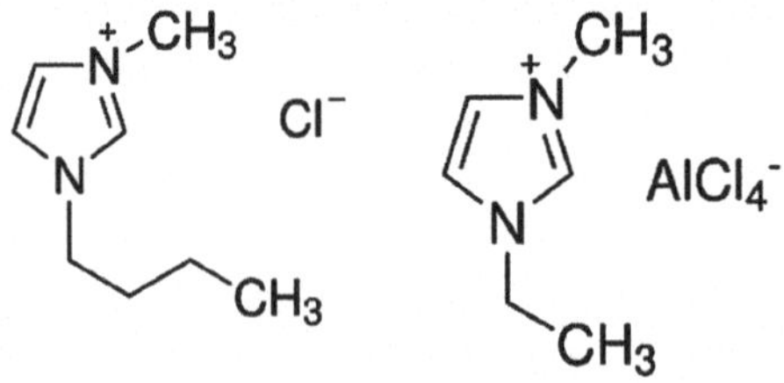

[BMIM]Cl [EMIM]$AlCl_4$

Fig. 7.2 Ionic liquids used for pre-treatment of different cellulose substrates

Table 7.1 Code names for cellulose substrates under study

Cellulose substrates	Initial	After enzymatic hydrolysis	After pre-treatment with ILs	After pre-treatment with ILs and enzymatic hydrolysis
Microcrystalline cellulose	MC	MC(EH)	MC-IL1 MC-IL2	MC-IL1(EH) MC-IL2(EH)
Whatman filter paper	WFP	WFP(EH)	WFP-IL1 WFP-IL2	WFP-IL1(EH) WFP-IL2(EH)
Industrial softwood pulp cellulose	ISC	ISC(EH)	ISC-IL1 ISC-IL2	ISC-IL1(EH) ISC-IL2(EH)

flasks under the following conditions: cellulose substrate/IL ratio 1/10, for 60 min immersion in an oil bath at 100 °C. After pre-treatment, cellulose substrates were immersed in deionized water under agitation (200 rpm, 30 min) resulting a product formed from amorphous and partial crystalline polymer components (regeneration process). The cellulose substrates were subsequently dried for 24 h, at 70 °C in a vacuum-oven and then subjected to enzymatic hydrolysis with cellulase. The cellulose substrates, without and with treatments, were coded as presented in Table 7.1.

7.2.3 *Cellulase Catalyzed Enzymatic Saccharification of Cellulose Substrates*

Cellulose substrates under study (50 mg, non-treated or treated with ILs) were incubated with cellulase (14 IU/g cellulose) in vials containing 15 mL acetic acid–sodium acetate buffer solution (50 mM, pH 4.8) at 50 °C for 4 h using an without shaker water-bath WNB 7–45. After centrifugation and removing samples from the supernatant periodically, the concentration of released soluble reducing sugars was measured and calculated by the dinitrosalicylic acid assay (DNS) using D-glucose as a standard (Miller 1959; Ghose 1987). All cellulose substrates were hydrolyzed using the same enzyme stock solution. The control enzymatic hydrolysis reactions were performed simultaneously with those conducted through dissolution in solutions of ILs in order to exclude possible variable background as temperature or enzyme loading. Experiments were run in triplicate, the error bars showing the standard deviation.

Yield of soluble released sugars from cellulose substrates saccharification was determined as follows:

$$\text{Yield}(\%) = \frac{\text{reducing sugars weight}}{\text{cellulose substrate weight}} \times 100$$

A schematic representation of the pre-treatment stages applied to cellulose substrates is given below, in Scheme 7.1.

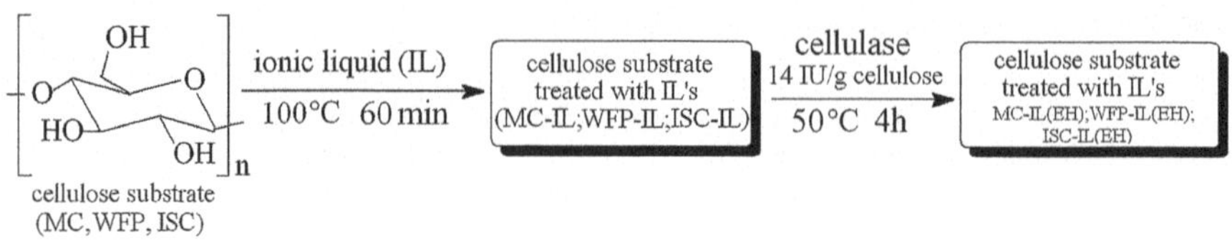

Scheme 7.1 Pre-treatment stages for cellulose substrates under study

7.2.4 Investigation Methods Used for Structural Characterization of Cellulose Substrates

A Bruker Vertex 70 spectrophotometer was used for FTIR spectroscopy investigation on the range 4000–400 cm^{-1}, at 4 cm^{-1} resolution and applying 64 scans for each sample.

A Bruker AD8 ADVANCE X-ray diffractometer with CuKa radiation (60 kV, 50 mA) was used for WAXD measurements on cellulose substrates, at room temperature. The angle of diffraction (2θ) ranged from 10 to 30° using a rate of 2° min^{-1}.

7.3 Experimental Results and Discussion

7.3.1 Enzymatic Saccharification of Cellulose Substrates

The effect of ILs used for dissolution stage of cellulose substrates (MC, WFP, ISC) on their further enzymatic saccharification with cellulase was determined and compared. As represented in Fig. 7.3a-c, the rate of enzymatic hydrolysis under cellulase action presents an upward trend for all cellulose substrates under study. There are significant differences between the substrates and their pre-treatment stages with ionic liquids.

For microcrystalline cellulose (MC), the rate of enzymatic saccharification is upper level for initial substrate without pre-treatment using ionic liquid, as expected, MC being a crystalline compound with high pure cellulose content. For the Whatman filter paper substrate, an enhanced rate of enzymatic hydrolysis is observed for sample pre-treated with IL1—see Fig. 7.3b. For both MC and WFP substrates, the chloroaluminate-based ionic liquid IL2 has a less influence upon release of soluble sugars given the increased acidity of the reaction medium which causes significant degradation processes through depolymerization of cellulose by hydrolysis to smaller fractions.

An enhanced rate of enzymatic hydrolysis is observed for the industrial softwood pulp cellulose (ISC), mainly when using pre-treatment with ionic liquids, IL2 having a greater influence than IL1 upon cellulose substrate during hydrolysis reaction. This is due to the particular chemical composition of cellulose substrate which

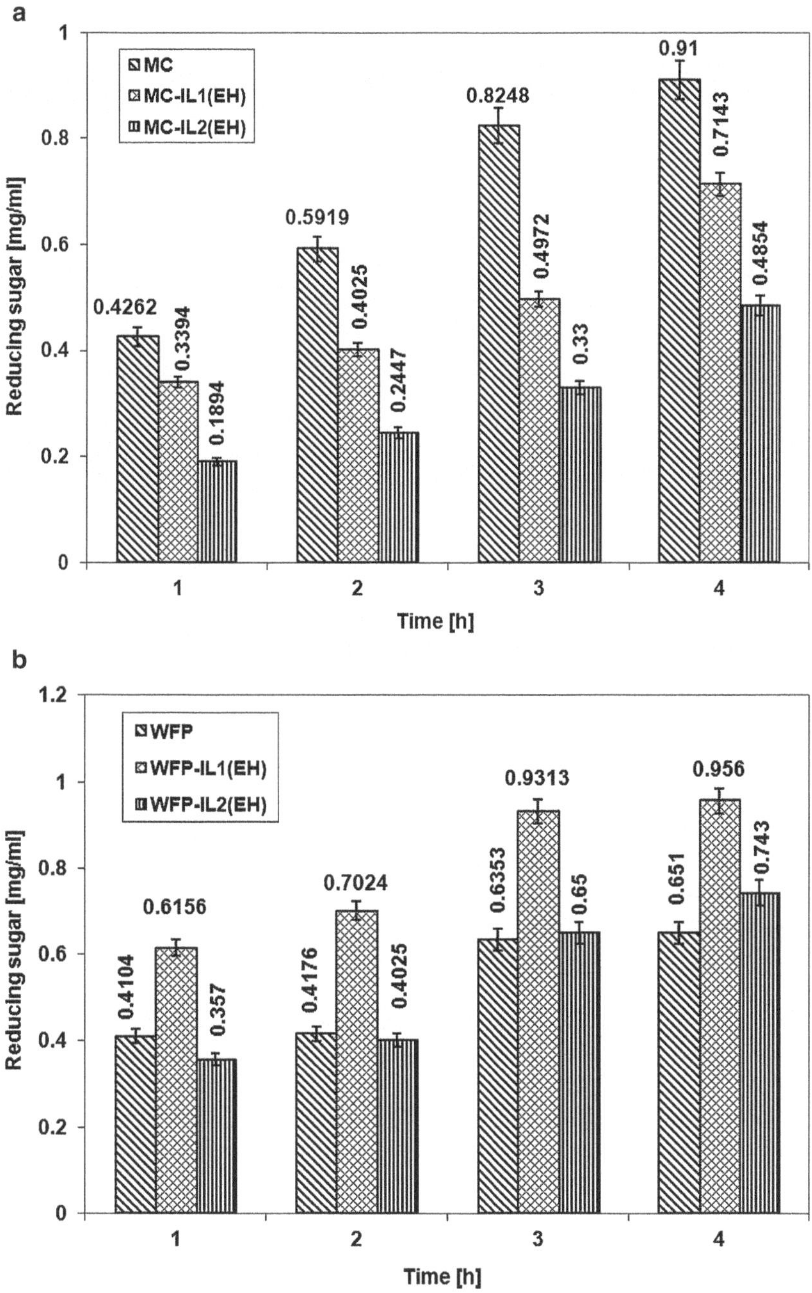

Fig. 7.3 Comparative reducing sugar yields of non-treated and treated different cellulose substrates with ILs under action of cellulase: MC (**a**), WFP (**b**), and ISC (**c**). (Data shown are the mean of three experiments. Error bars indicate standard error mean)

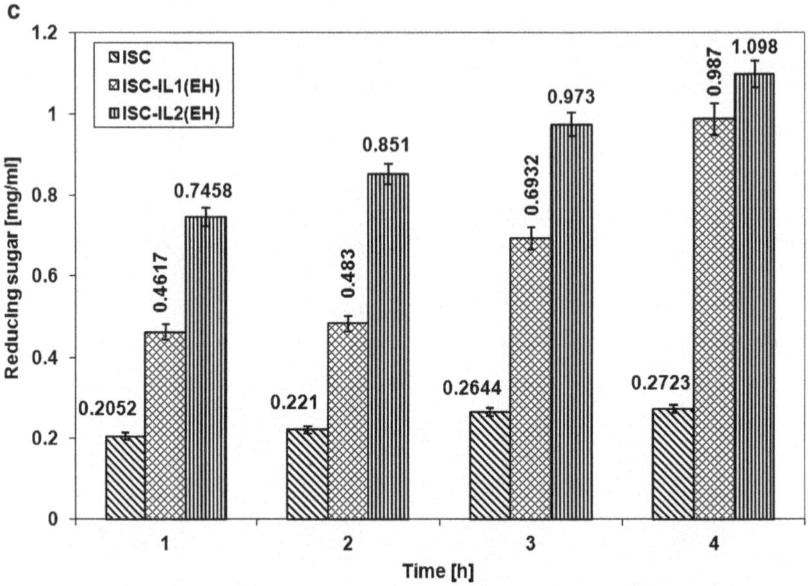

Fig. 7.3 (continued)

contains 80% cellulose, 18% hemicelluloses (xylan as predominant component, and glucomannan), and 2% residual lignin (Sjöström and Westermark 1999). Hemicelluloses component present in ISC chemical composition contain –COOH groups that can enhance decay reactions. Under acidic conditions, both ILs can exhibit some catalytic action. As a consequence, the –OH groups of the polysaccharides dissolved in ILs are activated (Moulthrop et al. 2005; Mazza et al. 2009). The increased acidity of reaction medium during pre-treatment with IL2 can enhance the depolymerization processes for polysaccharides components by hydrolysis to saccharides.

The cellulose substrates pre-treatment with ionic liquids modified their ordered structure with a positive impact on the resulted rate values for enzymatic saccharification. When compared with ILs used here, we can notice that IL1([BMIM]Cl) is a good solvent, while IL2 ([EMIM]Cl-$AlCl_3$) increases the acidic character of the reaction medium which intensifies the degradation processes in cellulose substrates, as also evidenced in a previous study (Spiridon et al. 2011).

All treated cellulose substrates contain reduced crystalline parts and consequently, become more susceptible to enzyme attack than those without treatment. The ionic liquids used in our experiments influenced the rate of enzymatic hydrolysis as follows: [BMIM]Cl > [EMIM]$AlCl_4$.

7.3.2 FTIR Spectroscopy Investigation of Cellulose Substrates During Dissolution in Ionic Liquids and Saccharification Under Enzyme Attack

FTIR spectroscopy investigation represents an effective useful method for structural changes evidence in cellulose materials (Hinterstoisser and Salmén 2000). In this study, it was applied to evidence the changes in the structure of cellulose substrates resulted after pre-treatment with ionic liquids. Figures 7.4, 7.5, and 7.6 illustrate the FTIR spectra recorded for cellulose samples (MC, WFP, ISC) and those treated with IL1 with/no further enzymatic hydrolysis (EH) for two domains 4000–800 cm^{-1} (a) and 1400–800 cm^{-1} (b), respectively.

The FTIR spectra recorded for initial cellulose and cellulose treated cu IL1 are alike, indicating that the structure for both substrates is identically with slight differences as shifted wavenumber values. It can also be concluded that IL1 is a good solvent for cellulose.

In general, the large peaks at 3330 cm^{-1} are assigned to the mixed OH stretching vibration mainly from polysaccharides in cellulose substrates. The spectral vibrations appeared at 2900 cm^{-1} are specific to the C–H stretching in methyl and methylene groups. The peaks indicated at 1360 cm^{-1} (WFP), 1368 cm^{-1} (MC), and 1427 cm^{-1} for all cellulose substrates is assigned to the CH_2 symmetric bending/scissoring movement in cellulose. The peak appeared at 1020 cm^{-1} peak is attributed to the C–O stretching in cellulose. The band evidenced at 1370 cm^{-1} for industrial softwood pulp cellulose is ascribed to the symmetric C–H deforming in cellulose and hemicelluloses. The peak located at 897 cm^{-1} is specific to the C–O–C stretching at the β-(1 → 4)-glycosidic linkage in cellulose (Nelson and O'Connor 1964), C–C=O and C–CH deformation and stretching, as well as to the C1–H bending in cellulose and hemicelluloses.

However, there are still some differences in the spectral bands evidenced in the FTIR spectra recorded for cellulose substrates after dissolution in ionic liquid IL1 and enzymatic saccharification with cellulase, as presented in Figs. 7.4, 7.5, and 7.6.

The absorption spectral band from O–H stretching increased as wavenumber value from 3330 cm^{-1} to 3436 cm^{-1} after pre-treatment with IL1 and enzymatic hydrolysis. This evolution indicates the broken hydrogen bonds in cellulose. An additional some evidence of hydrogen bonds disruption in C6–OH is given by shifting the spectral band from 1427 cm^{-1} to 1419 cm^{-1} (Hinterstoisser and Salmén 2000).

The FTIR spectra recorded in the 800–1800 cm^{-1} region were useful to evidence the structural changes occurred in cellulose substrates under study. The spectral data evidence the cellulose I polymorph form for all initial cellulose substrates without pre-treatments, in accordance with the knowledge that native celluloses sources have the crystal structure of cellulose I (O'Sullivan 1997). After dissolution in solution of ionic liquid and enzymatic saccharification, the crystalline cellulose I polymorph was changed into an amorphous polymorph, namely cellulose II.

The spectral values noticed at 1427 cm^{-1} and 897 cm^{-1} are related to the crystalline part in cellulose substrates. The presence of cellulose I polymorph can be

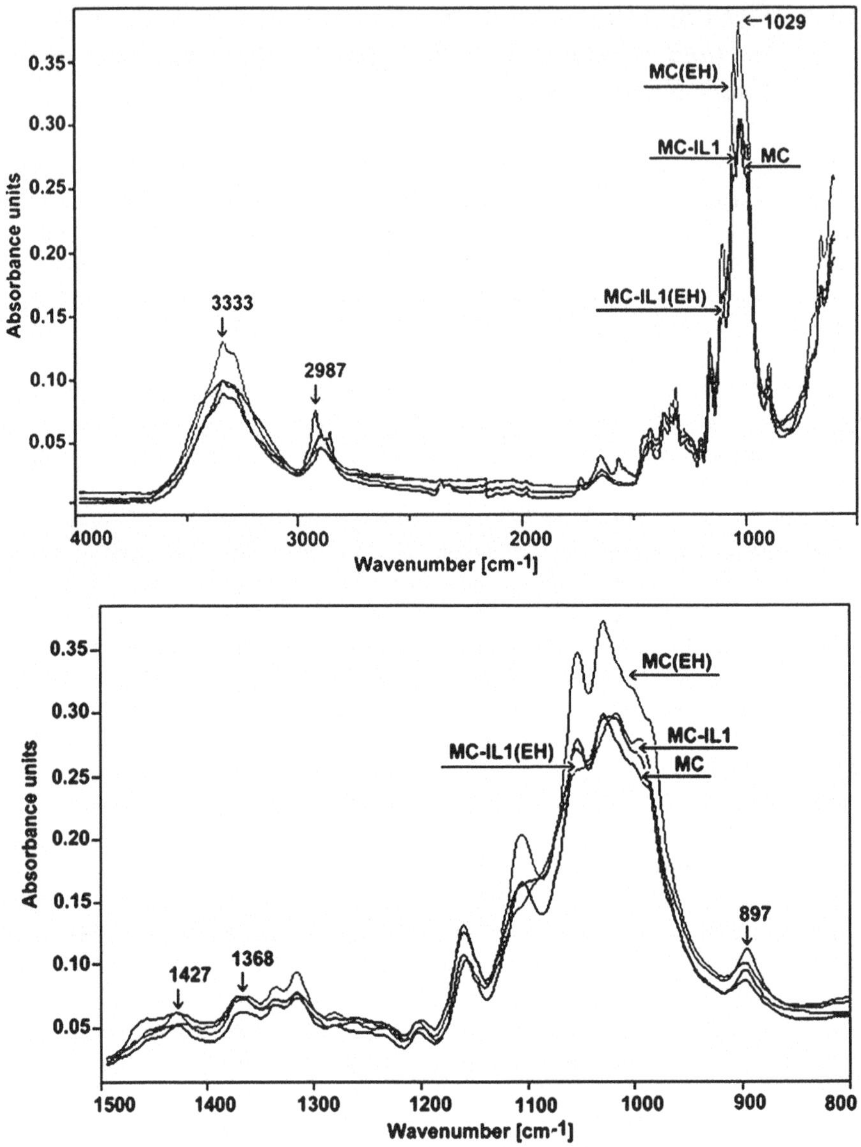

Fig. 7.4 FTIR spectra recorded for microcrystalline cellulose (MC) before and after treatment(s) with ionic liquids and enzymatic hydrolysis

evidenced by calculation of crystallinity index through absorbance ratio A_{1427}/A_{897} (O'Connor et al. 1958; Oh et al. 2005; Široký et al. 2010) which is also known as lateral order index (Hurtubise and Krassig 1960). The spectral evidence of crystalline part in cellulose substrates is given in the region around 1000–1200 cm^{-1} (Zhbankov et al. 2000). The spectral band evidenced close to

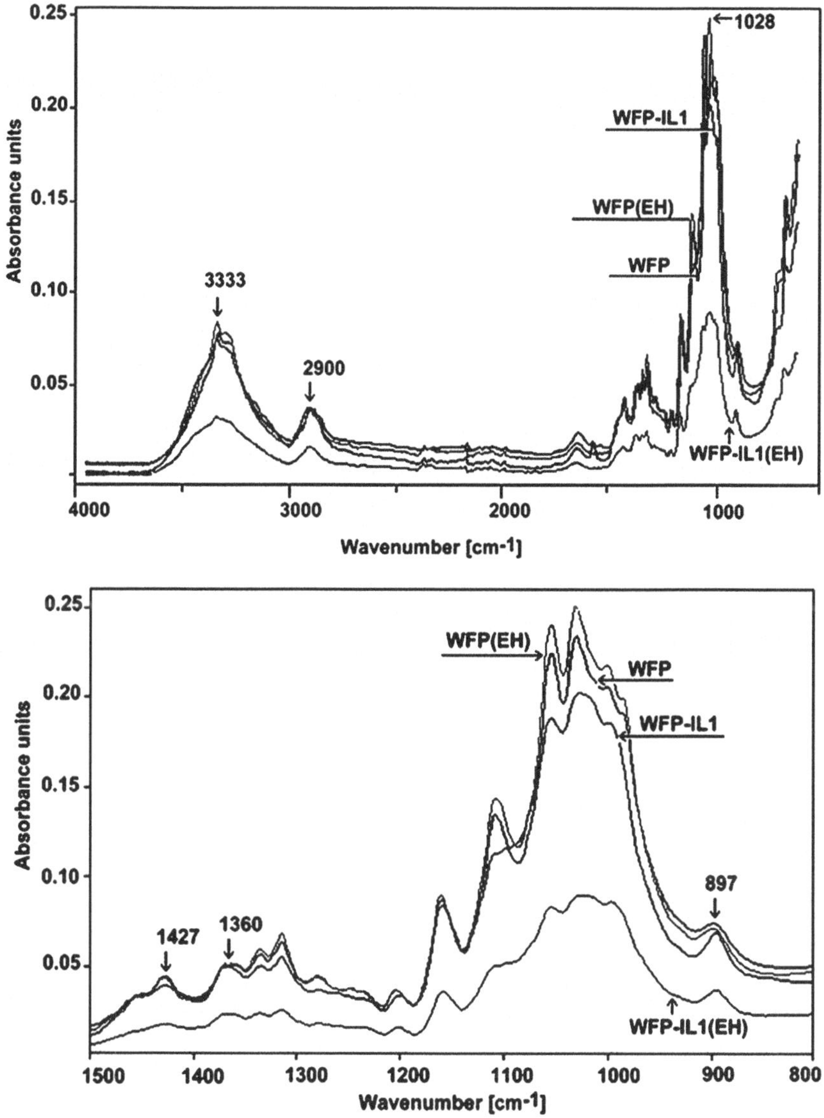

Fig. 7.5 FTIR spectra recorded for Whatman filter paper (WFP) before and after treatment(s) with ionic liquids and enzymatic hydrolysis

1160 cm^{-1} is attributed to the anti-symmetric bridge stretching of C-O-C groups from cellulose, while that around 1318 cm^{-1} can be assigned to CH_2 wagging type vibrations in cellulose.

The results of FTIR spectroscopy investigation evidenced some degradation processes in cellulose as shown by the lower intensity for spectral bands of cellulose substrates after saccharification in enzyme solutions and mainly after dissolution

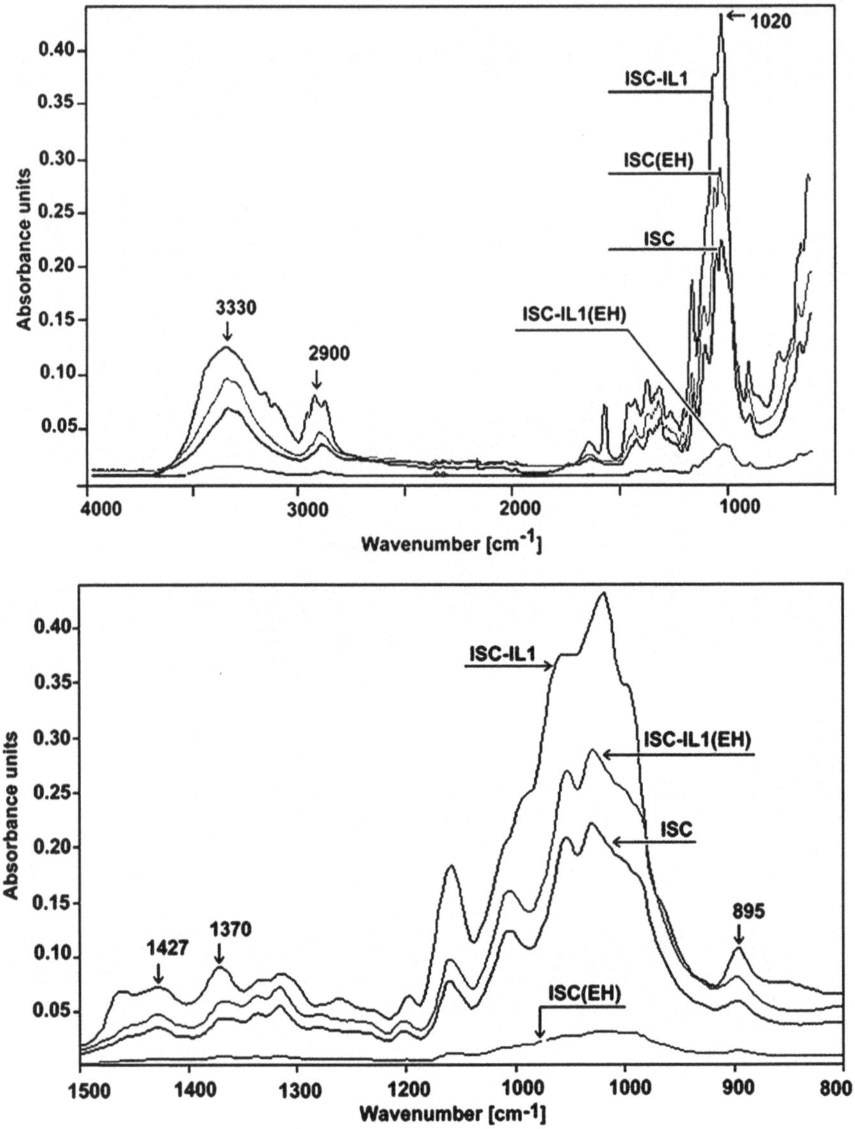

Fig. 7.6 FTIR spectra recorded for industrial softwood cellulose pulp (ISC) before and after treatment(s) with ionic liquids and enzymatic hydrolysis

stage using ionic liquids. The bending vibrations specific to the absorbed water significantly decreased as intensity in cellulose substrates after enzyme attack and pretreatment with ionic liquids, as evidenced by the spectral band at 1635–1640 cm^{-1}. Another infrared ratio A_{1370}/A_{2900}, known as the total crystallinity index (TCI), was calculated (Nelson and O'Connor 1964). These spectral indices, TCI and LOI, which are nearly connected to the crystalline form and level of the intermolecular uniformity in cellulose part from cellulosic substrates, are presented in Table 7.2.

Table 7.2 Crystallinity index of cellulose substrates evaluated from FTIR-ATR spectral data: total crystallinity index (TCI) and lateral order index (LOI) values

Cellulose substrate	Total crystallinity index TCI ($A_{1370\ cm^{-1}}/A_{2900\ cm^{-1}}$)	Lateral order index LOI ($A_{1427\ cm^{-1}}/A_{897\ cm^{-1}}$)
MC	1.2965	0.5909
MC(EH)	1.3117	0.6071
MC-IL1	0.9607	0.6419
MC-IL1(EH)	1.2932	0.4533
WFP	1.3363	0.5975
WFP(EH)	1.3596	0.6155
WFP-IL1	1.2934	0.5726
WFP-IL1(EH)	1.3635	0.4739
ISC	1.4102	0.5991
ISC(EH)	1.4114	0.5861
ISC-IL1	1.2904	0.6787
ISC-IL1(EH)	1.3138	0.4600

A higher index value, both for TCI and LOI respectively, indicates that the cellulose substrate has a dominant crystalline orderly pattern. This means that some of the crystalline part of cellulose became a non-crystalline part after dissolution in ILs (Li et al. 2010). In this case, the resulted cellulose substrates present a more fragmented and porous morphology given the highly content of amorphous part of cellulose polymer, and such structures could become more susceptible to enzyme attack, as also evidenced by the rate of saccharification in the presence of enzyme of cellulose substrates after dissolution in ionic liquids.

LOI decrease for all cellulose substrates after pre-treatment stage with IL1 and enzymatic saccharification by comparison with those without pre-treatment (see Table 7.2), indicating a more accessible surface in cellulose component for efficient cellulase enzymatic hydrolysis. After pre-treatment with ILs, a significant diminution of crystalline part in cellulose substrates may occur with their further restructuration into mostly amorphous parts.

The specific action of IL2 ([EMIM]Cl-$AlCl_3$) is evidenced by more intense cellulose degradation processes during dissolution (data are not presented here).

7.3.3 WAXD Investigation of Cellulose Substrates During Dissolution in Ionic Liquids and Saccharification Under Enzyme Attack

The cellulose substrates treated with ILs presented a reduced crystalline part and enhanced susceptibility to enzyme attack in comparison with those without treatment. As a result, the cellulose substrates were attacked by enzyme more rapidly after pre-treatment with ionic liquids than those without treatment.

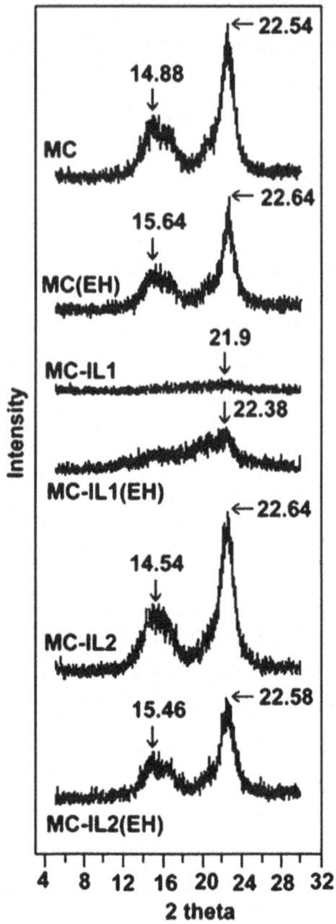

Fig. 7.7 WAXD data recorded for microcrystalline cellulose (MC) before and after pre-treatment(s) with ionic liquids and enzymatic hydrolysis

Behavior of cellulose substrates treated with IL1 is quite different from that noticed for those pre-treated with IL2 due to the ILs different chemical structure. After treatment with IL1 (e.g., MC and ISC) less cellulose has a crystalline structure, this fact being evidenced by a large peak at 21.9° in the corresponding diffractograms, as shown in Figs. 7.7, 7.8, and 7.9.

Pre-treatment of cellulose substrates with IL1 resulted in subsequent decrease of the crystallinity index, fact showed by the significant change in the X-ray diffraction spectra (see Figs. 7.7, 7.8, and 7.9, and Table 7.3).

7.4 Concluding Remarks

The total enzymatic saccharification of cellulose substrates involving methodologies based on ionic liquid pre-treatment stages can be achieved through two distinct process routes as follows: regeneration procedure (when the pre-treatment with IL is followed by its removal through washing before enzymatic saccharification—as

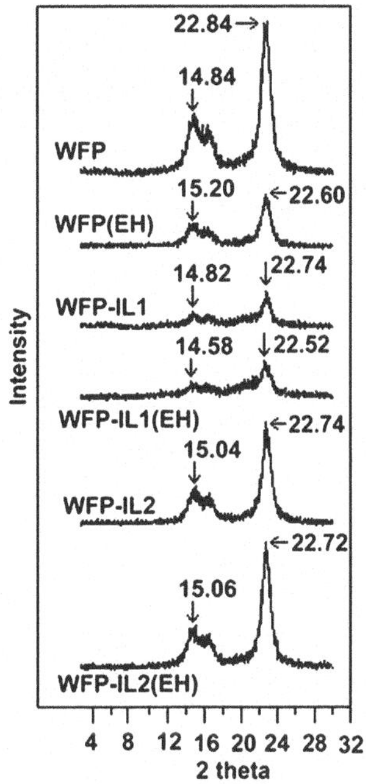

Fig. 7.8 WAXD data recorded for Whatman filter paper (WFP) before and after pre-treatment(s) with ionic liquids and enzymatic hydrolysis

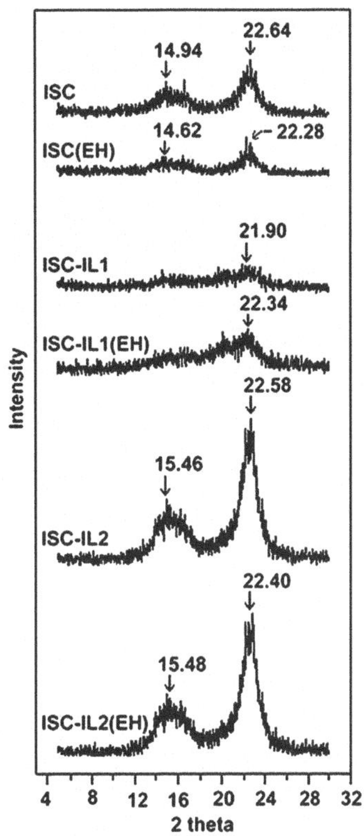

Fig. 7.9 WAXD data recorded for industrial softwood cellulose pulp (ISC) before and after pre-treatment(s) with ionic liquids and enzymatic hydrolysis

Table 7.3 Crystallinity index of cellulose substrates (evaluated from wide-angle X-ray diffraction data by Segal method—Segal et al. 1959)

Cellulose substrate	Cellulose substrate without/after pre-treatment					
	Untreated	Enzymatic hydrolysis (EH)	IL1	IL1 and EH	IL2	IL2 and EH
MC	0.7222	0.7149	0.2592	0.4045	0.7308	0.6611
WFP	0.7099	0.4964	0.3852	0.2523	0.6322	0.6746
ISC	0.3899	0.5555	0.1975	0.4791	0.7470	0.7398

we have presented here), and the one-pot method (when the enzyme solution and IL solvent are put together in the same vessel). Even considering the main advantages of the last-mentioned procedure (less washing, better recycling of IL), the further efficient enzyme attack is hindered given the increased level of ILs concentration. Enzymatic saccharification of cellulose substrates is limited not only by the cellulase activity in ILs, but also by the interaction of ILs with substrates through coating effects. There are still some discrepancies between reduced enzymatic saccharification yields and optimum activity of enzyme in ILs. ILs are relatively expensive and their complete recyclability, as well as biocompatibility, toxicology, and biodegradability issues must also be considered and efficiently solved if ILs are integrated at large-scale industry processes for biorefinery strategies involving biomass, including polysaccharides hydrolysis and fermentation processes. Nevertheless, the aforementioned problem related to the enzyme inactivation in cellulose-dissolving ILs may be technically solved through design of such ILs to support efficiently enzyme activity or to become tolerant to the enzymes presence, even if their applications at large-scale processing industry are quite far to be feasible. Cellulose saccharification under enzyme action in ILs represents only one of the most interesting ways to modify its structure. Oxidation, acylation, and other derivatization processes may represent other potentially feasible routes to modify cellulose dissolved in ILs solutions through enzymatic hydrolysis, all these leading to novel cellulose products or best obtainment technologies at industry scale for present cellulose products.

In this chapter, the changes in crystalline structure for the different types of cellulose substrates during dissolution in ionic liquids (ILs) and saccharification in the presence of enzyme were analyzed through Fourier transform infrared (FTIR) spectroscopy and wide-angle X-ray diffraction (WAXD) methods. There is a noticeable evidence of increased reducing sugars yields after pre-treatment of cellulose substrates with ionic liquids (mainly after treatment with [BMIM]Cl—IL1). There are obviously differences between cellulose substrates interacting with ILs and cellulase, considering their different chemical composition. For both MC and WFP substrates, the chloroaluminate-based ionic liquid IL2 has a less influence upon release of soluble sugars given the increased acidity of the reaction medium which causes significant degradation processes through depolymerization of cellulose.

Cellulose substrates pre-treatment with ionic liquids resulted in subsequent decrease of their crystallinity index, as evidenced from spectral and diffraction data. Slight differences in the FTIR spectra of cellulose samples subjected to dissolution in ILs and further saccharification under cellulase action were evidenced. The absorption spectral band specific to OH groups stretching shifted to increased wavenumber values after dissolution in ILs and saccharification under cellulase action, indicating the disruption of hydrogen bonds in cellulose structure. The spectral data evidence the cellulose I polymorph form for all initial cellulose substrates without pre-treatments. During the pre-treatment stages involving dissolution in ionic liquid and addition of enzyme solution, the crystalline polymorph type cellulose I was transformed into amorphous polymorph, namely cellulose II.

Acknowledgements We dedicate this work to the memory of our colleague and friend, Dr. Ruxanda Bodîrlău, and thank her for working together with such discretion and enthusiasm.

We are grateful to Mrs. Elena Marlică for assistance on the WAXD analysis.

References

Bian J, Peng F, Peng X, Xiao X, Peng P, Xu F, Sun R (2014) Effect of [Emim]ac pretreatment on the structure and enzymatic hydrolysis of sugarcane bagasse cellulose. Carbohydr Polym 100:211–217. https://doi.org/10.1016/j.carbpol.2013.02.059

Bodîrlău R, Teacă C-A, Spiridon I (2010a) Influence of ionic liquid on hydrolyzed cellulose material: FT-IR spectroscopy and TG-DTG-DSC analysis. Int J Polym Anal Charact 15:460–469. https://doi.org/10.1080/1023666X.2010.510112

Bodîrlău R, Teacă C-A, Spiridon I (2010b) Enzymatic hydrolysis of *Asclepias syriaca* fibers in the presence of ionic liquids. Monatsh Chem 141:1043–1048. https://doi.org/10.1007/s00706-010-0348-7

Boon JA, Levisky JA, Pflug JL, Wilkes JS (1986) Friedel-crafts reactions in ambient-temperature molten salts. J Org Chem 51:480–483. https://doi.org/10.1021/jo00354a013

Brandt A, Gräsvik J, Hallett JP, Welton T (2013) Deconstruction of lignocellulosic biomass with ionic liquids. Green Chem 15:550–583. https://doi.org/10.1039/c2gc36364j

Brown LC, Hogg JM, Swadźba-Kwaśny M (2017) Lewis acidic ionic liquids. Top Curr Chem 375:78–118. https://doi.org/10.1007/s41061-017-0166-z

Cheenkachorn K, Douzou T, Roddecha S, Tantayotai P, Sriariyanun M (2016) Enzymatic saccharification of rice straw under influence of recycled ionic liquid pretreatments. Energy Procedia 100:160–165. https://doi.org/10.1016/j.egypro.2016.10.159

Cheng G, Varanasi P, Li C, Liu H, Melnichenko YB, Simmons BA, Kent MS, Singh S (2011) Transition of cellulose crystalline structure and surface morphology of biomass as a function of ionic liquid pretreatment and its relation to enzymatic hydrolysis. Biomacromolecules 12:933–941. https://doi.org/10.1021/bm101240z

Cherubini F, Strømman AH (2011) Chemicals from lignocellulosic biomass: opportunities perspectives, and potential of biorefinery systems. Biofuels Bioprod Biorefin 5:548–561. https://doi.org/10.1002/bbb.297

Corma A, Iborra S, Velty A (2007) Chemical routes for the transformation of biomass into chemicals. Chem Rev 107:2411–2502. https://doi.org/10.1021/cr050989d

Dadi AP, Varanasi S, Schall CA (2006) Enhancement of cellulose saccharification kinetics using an ionic liquid pretreatment step. Biotechnol Bioeng 95:904–910. https://doi.org/10.1002/bit.21047

Dadi AP, Schall CA, Varanasi S (2007) Mitigation of cellulose recalcitrance to enzymatic hydrolysis by ionic liquid pretreatment. Appl Biochem Biotechnol 137:407–421. https://doi.org/10.1007/s12010-007-9068-9

Davison RA, Dixon P, Gilna M, Keller P, Langan AK, Naskar JN, Saddler TJ, Tuskan GA, Wyman CE (2014) Lignin valorization: improving lignin processing in the biorefinery. Science 344:1246843-(1-10). https://doi.org/10.1126/science.1246843

El Seoud OA, Koschella A, Fidale LC, Dorn S, Heinze T (2007) Applications of ionic liquids in carbohydrate chemistry: a window of opportunities. Biomacromolecules 8:2629–2647. https://doi.org/10.1021/bm070062i

Fort DA, Remsing RC, Swatloski RP, Moyna P, Moyna G, Rogers RD (2007) Can ionic liquids dissolve wood? Processing and analysis of lignocellulosic materials with 1-N-butyl-3-methylimidazolium chloride. Green Chem 9:63–69. https://doi.org/10.1002/cssc.201000272

Gallezot P (2007) Process options for converting renewable feedstocks to bioproducts. Green Chem 9:295–302. https://doi.org/10.1039/b615413a

Ghose TK (1987) Measurement of cellulase activities. Pure Appl Chem 59:257–268. https://doi.org/10.1351/pac198759020257

He Y-C, Liu F, Gong L, Zhu Z-Z, Ding Y, Wang C, Xue Y-F, Rui H, Tao Z-C, Zhang D-P, Ma C-L (2015) Significantly improving enzymatic saccharification of high crystallinity index's corn Stover by combining ionic liquid [Bmim]Cl–HCl–water media with dilute NaOH pretreatment. Bioresour Technol 189:421–425. https://doi.org/10.1016/j.biortech.2015.04.047

Heinze T, Schwikal K, Barthel S (2005) Ionic liquids as reaction medium in cellulose functionalization. Macromol Biosci 5:520–525. https://doi.org/10.1002/mabi.200500039

Hinterstoisser B, Salmén L (2000) Application of dynamic 2D FTIR to cellulose. Vib Spectrosc 22(1–2):111–118. https://doi.org/10.1016/S0924-2031(99)00063-6

Hou X, Smith TJ, Li N, Zong M (2012) Novel renewable ionic liquids as highly effective solvents for pretreatment of rice straw biomass by selective removal of lignin. Biotechnol Bioeng 109:2484–2493. https://doi.org/10.1002/bit.24522

Hou X, Li N, Zong M (2013a) Renewable bio ionic liquids-water mixtures-mediated selective removal of lignin from rice straw: visualization of changes in composition and cell wall structure. Biotechnol Bioeng 110:1895–1902. https://doi.org/10.1002/bit.24862

Hou X, Li N, Zong M (2013b) Significantly enhancing enzymatic hydrolysis of rice straw after pretreatment using renewable ionic liquid–water mixtures. Bioresour Technol 136:469–474. https://doi.org/10.1016/j.biortech.2013.02.118

Hou Q, Ju M, Li W, Liu L, Chen Y, Yang Q (2017) Pretreatment of lignocellulosic biomass with ionic liquids and ionic liquid-based solvent systems. Molecules 22(3):490–514. https://doi.org/10.3390/molecules22030490

Huber GW, Iborra S, Corma A (2006) Synthesis of transportation fuels from biomass: chemistry, catalysts, and engineering. Chem Rev 106:4044–4098. https://doi.org/10.1021/cr068360d

Hurtubise FG, Krassig H (1960) Classification of fine structural characteristics in cellulose by infared spectroscopy. Use of potassium bromide pellet technique. Anal Chem 32:177–181. https://doi.org/10.1021/ac60158a010

Jin L, Yu X, Peng C, Guo Y, Zhang L, Xu Q, Zhao ZK, Liu Y, Xie H (2018) Fast dissolution pretreatment of the corn Stover in gamma-valerolactone promoted by ionic liquids: selective delignification and enhanced enzymatic saccharification. Bioresour Technol 270:537–544. https://doi.org/10.1016/j.biortech.2018.09.083

Kamiya N, Matsushita Y, Hanaki M, Nakashima K, Narita M, Goto M, Takahashi H (2008) Enzymatic in situ saccharification of cellulose in aqueous-ionic liquid media. Biotechnol Lett 30:1037–1040. https://doi.org/10.1007/s10529-008-9638-0

Kilpeläinen I, Xie H, King A, Granstrom M, Heikkinen S, Argyropoulos DS (2007) Dissolution of wood in ionic liquids. J Agric Food Chem 55(22):9142–9148. https://doi.org/10.1021/jf071692e

Kosan B, Michels C, Meister F (2008) Dissolution and forming of cellulose with ionic liquids. Cellulose 15:59–66. https://doi.org/10.1007/s10570-007-9160-x

Li C, Knierim B, Manisseri C, Arora R, Scheller HV, Auer M, Vogel KP, Simmons BA, Singh S (2010) Comparison of dilute acid and ionic liquid pretreatment of switchgrass: biomass recalcitrance, delignification and enzymatic saccharification. Bioresour Technol 101:4900–4906. https://doi.org/10.1016/j.biortech.2009.10.066

Li C, Cheng G, Balan V, Kent MS, Ong M, Chundawat S, daCosta SPS, Melnichenko YB, Dale BE, Simmons BA, Singh S (2011) Influence of physico-chemical changes on enzymatic digestibility of ionic liquid and AFEX pretreated corn Stover. Bioresour Technol 102:6928–6936. https://doi.org/10.1016/j.biortech.2011.04.005

Li HY, Chen X, Wang CZ, Sun SN, Sun RC (2016) Evaluation of the two-step treatment with ionic liquids and alkali for enhancing enzymatic hydrolysis of Eucalyptus: chemical and anatomical changes. Biotechnol Biofuels 9:166–179. https://doi.org/10.1186/s13068-016-0578-y

Mazza M, Catana DA, Vaca-Garcia C, Cecutti C (2009) Influence of water on the dissolution of cellulose in selected ionic liquids. Cellulose 16:207–215. https://doi.org/10.1007/s10570-008-9257-x

Miller GL (1959) Use of dinitrosalicylic acid reagent for determination of reducing sugar. Anal Chem 31:426–428. https://doi.org/10.1021/ac60147a030

Miyafuji H (2015) Application of ionic liquids for effective use of woody biomass. J Wood Sci 61:343–350. https://doi.org/10.1007/s10086-015-1489-4

Moulthrop JS, Swatloski RP, Moyna G, Rogers RD (2005) High-resolution ^{13}C NMR studies of cellulose and cellulose oligomers in ionic liquid solutions. Chem Commun 12:1557–1559. https://doi.org/10.1039/b417745b

Nelson ML, O'Connor RT (1964) Relation of certain infrared bands to cellulose crystallinity and crystal lattice type. Part II. A new infrared ratio for estimation of crystallinity in celluloses I and II. J Appl Polym Sci 8:1325–1341. https://doi.org/10.1002/app.1964.070080323

Ninomiya K, Kamide K, Takahashi K, Shimizu N (2012) Enhanced enzymatic saccharification of kenaf powder after ultrasonic pretreatment in ionic liquids at room temperature. Bioresour Technol 103(1):259–265. https://doi.org/10.1016/j.biortech.2011.10.019

O'Connor RT, DuPré EF, Mitcham D (1958) Applications of infrared absorption spectroscopy to investigations of cotton and modified cottons. Part I: physical and crystalline modifications and oxidation. Textile Res J 28:382–392. https://doi.org/10.1177/004051755802800503

O'Sullivan AC (1997) Cellulose: the structure slowly unravels. Cellulose 4:173–207. https://doi.org/10.1023/A:1018431705579

Oh SY, Dong IY, Shin Y, Hwan CK, Hak YK, Yong SC, Won HP, Ji HY (2005) Crystalline structure analysis of cellulose treated with sodium hydroxide and carbon dioxide by means of X-ray diffraction and FTIR spectroscopy. Carbohydr Res 340:2376–2391. https://doi.org/10.1016/j.carres.2005.08.007

Remsing RC, Swatloski RP, Rogers RD, Moyna G (2006) Mechanism of cellulose dissolution in the ionic liquid 1-N-butyl-3-methylimidazolium chloride: a ^{13}C and $^{35/37}Cl$ NMR relaxation study on model systems. Chem Commun 12:1271–1273. https://doi.org/10.1039/b600586c

Ren H, Zong M-H, Wu H, Li N (2016) Efficient pretreatment of wheat straw using novel renewable cholinium ionic liquids to improve enzymatic saccharification. Ind Eng Chem Res 55(6):1788–1795. https://doi.org/10.1021/acs.iecr.5b03729

Samayam IP, Schall CA (2010) Saccharification of ionic liquid pretreated biomass with commercial enzyme mixtures. Bioresour Technol 101(10):3561–3566. https://doi.org/10.1016/j.biortech.2009.12.066

Segal L, Creely JJ, Martin AE Jr, Conrad CM (1959) An empirical method for estimating the degree of crystallinity of native cellulose using the X-ray diffractometer. Textile Res J 29(10):786–794. https://doi.org/10.1177/004051755902901003

Singh S, Simmons BA, Vogel KP (2009) Visualization of biomass solubilization and cellulose regeneration during ionic liquid pretreatment of switchgrass. Biotechnol Bioeng 104:68–75. https://doi.org/10.1002/bit.22386

Široký J, Blackburn RS, Bechtold T, Taylor J, White P (2010) Attenuated total reflectance Fourier-transform infrared spectroscopy analysis of crystallinity changes in lyocell following continuous treatment with sodium hydroxide. Cellulose 17:103–115. https://doi.org/10.1007/s10570-009-9378-x

Sjöström E, Westermark U (1999) Chemical composition of wood and pulps: basic constituents and their distribution. In: Sjöström E, Alén R (eds) Analytical methods in wood chemistry, pulping, and papermaking. Springer series in wood science. Springer, Berlin, pp 1–19. https://doi.org/10.1007/978-3-662-03898-7_1

Spiridon I, Teacă C-A, Bodîrlău R (2011) Structural changes evidenced by FTIR spectroscopy in cellulosic materials after pre-treatment with ionic liquid and enzymatic hydrolysis. Bio Resources 6(1):400–413. https://doi.org/10.15376/biores.6.1.400-413

Sun N, Jiang X, Maxim ML, Metlen A, Rogers RD (2011) Use of polyoxometalate catalysts in ionic liquids to enhance the dissolution and delignification of woody biomass. Chem Sus Chem 4(1):65–73. https://doi.org/10.1002/cssc.201000272

Swatloski RP, Spear SK, Holbrey JD, Rogers RD (2002) Dissolution of cellulose with ionic liquids. J Am Chem Soc 124:4974–4975. https://doi.org/10.1021/acs.jpcb.7b09073

Tadesse H, Luque R (2011) Advances on biomass pretreatment using ionic liquids: an overview. Energy Environ Sci 4:3913–3929. https://doi.org/10.1039/c0ee00667j

Torimoto T, Tsuda T, Okazaki K-I, Kuwabata S (2010) New frontiers in materials science opened by ionic liquids. Adv Mater 22:1196–1221. https://doi.org/10.1002/adma.200902184

Tuck CO, Perez E, Horvath IT, Sheldon RA, Poliakoff M (2012) Valorization of biomass: deriving more value from waste. Science 337:695–699. https://doi.org/10.1126/science.1218930

Uju SY, Nakamoto A, Goto M, Tokuhara W, Noritake Y, Katahira S, Ishida N, Nakashima K, Ogino C, Kamiya N (2012) Short time ionic liquids pretreatment on lignocellulosic biomass to enhance enzymatic saccharification. Bioresour Technol 103(1):446–452. https://doi.org/10.1016/j.biortech.2011.10.003

Uju NA, Shoda Y, Goto M, Tokuhara W, Noritake Y, Katahira S, Ishida N, Ogino C, Kamiya N (2013) Low melting point pyridinium ionic liquid pretreatment for enhancing enzymatic saccharification of cellulosic biomass. Bioresour Technol 135:103–108. https://doi.org/10.1016/j.biortech.2012.06.096

Van Rantwijk F, Sheldon RA (2007) Biocatalysis in ionic liquids. Chem Rev 107:2757–2785. https://doi.org/10.1021/cr050946x

Wang Y, Radosevich M, Hayes D, Labbé N (2011) Compatible ionic liquid-cellulases system for hydrolysis of lignocellulosic biomass. Biotechnol Bioeng 108(5):1042–1048. https://doi.org/10.1002/bit.23045

Wilkes JS (2002) A short history of ionic liquids-from molten salts to neoteric solvents. Green Chem 4:73–80. https://doi.org/10.1039/b110838g

Wu J, Zhang J, He J, Ren Q, Guo M (2004) Homogeneous acetylation of cellulose in a new ionic liquid. Biomacromolecules 5:266–268. https://doi.org/10.1021/bm034398d

Xu J, Liu X, He J, Hu L, Dai B, Wu B (2015) Enzymatic in situ saccharification of rice straw in aqueous-ionic liquid media using encapsulated Trichoderma aureoviride cellulose. J Chem Technol Biotechnol 90:57–63. https://doi.org/10.1002/jctb.4458

Xu J, Liu B, Hou H, Hu J (2017) Pretreatment of eucalyptus with recycled ionic liquids for low-cost biorefinery. Bioresour Technol 234:406–414. https://doi.org/10.1016/j.biortech.2017.03.081

Zhang Y-HP (2008) Reviving the carbohydrate economy via multi-product biorefineries. J Ind Microbiol Biotechnol 35(5):367–375. https://doi.org/10.1007/s10295-007-0293-6

Zhang Y-HP, Lynd LR (2004) Toward an aggregated understanding of enzymatic hydrolysis of cellulose: noncomplexed cellulose systems. Biotechnol Bioeng 88:797–824. https://doi.org/10.1002/bit.20282

Zhang H, Wu J, Zhang J, He J (2005) 1-Allyl-3-methylimidazolium chloride room temperature ionic liquid: a new and powerful non-derivatizing solvent for cellulose. Macromolecules 38:8272–8277. https://doi.org/10.1021/ma0505676

Zhang Y-HP, Himmel M, Mielenz JR (2006) Outlook for cellulase improvement: screening and selection strategies. Biotechnol Adv 24(5):452–481. https://doi.org/10.1016/j.biotechadv.2006.03.003

Zhang Y-HP, Hong J, Ye X (2009) Cellulase Assays. In: Mielenz J (ed) Biofuels, Methods in molecular biology (methods and protocols), vol 581. Humana Press, Totowa, NJ. https://doi.org/10.1007/978-1-60761-214-8_14

Zhao H, Baker GA, Cowins JV (2010) Fast enzymatic saccharification of switchgrass after pretreatment with ionic liquids. Biotechnol Prog 26(1):127–133. https://doi.org/10.1002/btpr.331

Zhbankov RG, Firsov SP, Korolik EV, Petrov PT, Lapkovski MP, Tsarenkov VM, Marchewka MK, Ratajczak H (2000) Vibrational spectra and the structure of medical biopolymers. J Mol Structure 555:85–96. https://doi.org/10.1016/S0022-2860(00)00590-1

Zhu SD, Wu YX, Chen QM, Yu ZN, Wang CW, Jin SW, Ding YG, Wu G (2006) Dissolution of cellulose with ionic liquids and its application: a mini-review. Green Chem 8:325–327. https://doi.org/10.1039/b601395c

Chapter 8
Biological Applications of Ionic Liquids-Based Surfactants: A Review of the Current Scenario

Magaret Sivapragasam and Cecilia Devi Wilfred

Contents

Nomenclature

$[C_4mim]$ $[C_{12}SO_4]$	1-butyl-3-methylimidazolium dodecyl sulfate
$[C_4MP][C_{12}SO_4]$	N-butyl-*N*-methylpyrrolidinium dodecyl sulfate
SDS	Sodium dodecyl sulfate
$[C_{12}mim][Br]$	1-dodecyl-3-methylimidazolium bromide

M. Sivapragasam (✉)
Centre of Research on Ionic Liquid, Universiti Teknologi PETRONAS, Bandar Seri Iskandar, Perak, Malaysia

School of Biological Science, Faculty of Science and Technology, QUEST International University Perak, (QIUP), Ipoh, Perak, Malaysia
e-mail: magaret.sivapragasam@qiup.edu.my

C. D. Wilfred
Centre of Research on Ionic Liquid, Universiti Teknologi PETRONAS, Bandar Seri Iskandar, Perak, Malaysia

Faculty of Fundamental and Applied Science, University Teknologi PETRONAS, Bandar Seri Iskandar, Perak, Malaysia

Inamuddin, A. M. Asiri (eds.), *Nanotechnology-Based Industrial Applications of Ionic Liquids*, Nanotechnology in the Life Sciences,
https://doi.org/10.1007/978-3-030-44995-7_8

$[C_{12}MP][Br]$	N-dodecyl-*N*-methylpyrrolidinium bromide
$[C_{12}mim][Br]$	1-dodecyl-3-methylimidazolium bromide
$[C_{14}mim][Br]$	1-tetradecyl-3-methylimidazolium bromide
$[C_{16}mim][Br]$	1-hexadecyl-3-methylimidazolium bromide
$[C_4mim][(C_{12}H_{2)5}SO_3]$	Dodecyl sulfonate
SDDS	Sodium dodecyl sulfonate
$[C_4mim][(C_{12}H_{2)5}SO_4]$	1-butyl-3-methylimidazolium dodecyl sulfonate
[Na][DBS]	Sodium dodecyl benzene sulfonate
$[N_{1114}][DBS]$	n-butyltrimethylammonium dodecyl benzene sulfonate
$[C_4mim][DBS]$	1-butyl-3-methylimidazolium dodecyl benzene sulfonate
$[C_4Py]DBS$	*N*-butylpyridinium dodecyl benzene sulfonate
$[C_4mim][BF4]$	1-butyl-3-methylimidazolium tetrafluoroborate
DH	Dopamine hydrochloride
AC	Acetylcholine chloride
TTAB	Tetradecyltrimethylammonium bromide
NaC	Sodium cholate
NaDC	Sodium deoxycholate
ME	Microemulsion
IL/o	Ionic liquid-in-oil
$[C_1mim][(MeO)_2PO_2]$	1-methyl-3-methylimidazolium dimethyl phosphate
o/w	Oil-in-water
w/o	Water-in-oil
$[C_6mim][Cl]$	1-hexyl-3-methylimidazolium chloride
$[C_4mim][PF_6]$	1-butyl-3-methylimidazolium hexafluorophosphate
5-FU	5-fluorouracil
$[C_4mim][Br]$	1-butyl-3-methylimidazolium bromide
DMBA	Dimethylbenz anthracene
TPA	12-*O*-tetradecanoylphorbol-13-acetate
ETO	Etodolac
$[C_{10}mim][Br]$	1-decyl-3-methylimidazolium bromide
$[C_4mim][PF_6]$	1-butyl-3-methylimidazolium hexafluorophosphate
$[C_{14}mim][Br]$	1-tetradecyl-3-methylimidazolium bromide
$[C_8mim][C_{12}OSO_3]$	1-methyl-3-octylimidazolium dodecylsulfate
$[C_{10}mim][Cl]$	1-decyl-3-methylimidazolium chloride
CaLB	Candida antarctica lipase B
rhL	Rhizopus oryzae lipase
$[C_{16}mim][Br]$	1-hexadecyl-3-methylimidazolium bromide
CAC	Critical aggregation concentration

CMC Critical micellar concentration

8.1 Introduction

Simply said Ionic Liquids-Based Surfactants (ILBS) are known to be ionic liquids with long-chain "tails." ILs which contain long alkyl chains in their structures are surface active, but they have melting points exceeding 100 °C. Thus, the term "ionic liquids" is not valid for many of these surfactants. Therefore, the term "ionic liquid-based surfactants" is deemed to describe them better. These "technology enablers" are said to possess the same remarkable features of ionic liquids (ILs), which include high chemical and thermal stability, negligible vapor pressure, polarity, as well as

Box 8.1 Advantages of Ionic Liquids in Comparison to Conventional Surfactants

- Ionic liquids have melting points or glass transition temperatures below 100 °C, making them user friendly.
- Ionic liquids are tunable molecules with a varied number of cations and anions combination.
- Some ionic liquids have a relatively higher viscosity that surfactant solution, which is favorable in a change of mobility ratio.
- Thermally and chemically stable in wide liquid ranges.
- In the process of micellar flooding, co-surfactants are used with surfactants. However in the case of ionic liquids, they have strong cohesive forces which enables them to form stable micelles without the need of additional cosurfactants.
- Typically non-flammable and have a wide range of solubility and miscibility.
- Properties of the micellar solution can be easily modified by changing the structural make up of the cation-anion pair.

molecular structural flexibility. Typically, there are four classes of surfactants, namely non-ionic, zwitterionic, anionic, and cationic. ILs hold much in common with the cationic surfactants which are derived from quaternized nitrogen salts, and the anionic derivatives derived from sulfonate salts (Jordan and Gathergood 2015). Box 8.1 shows the advantages of ILs in comparison to conventional solvents.

Due to it characteristics of having long alkyl side chains, ILBS are said to be toxic to a certain extent. Hence, the environmental, health, and safety impact of ILs has been a growing concern. Information on its biological activity, ecotoxicity, and cytotoxicity are of great interest due to their booming number of biological applica-

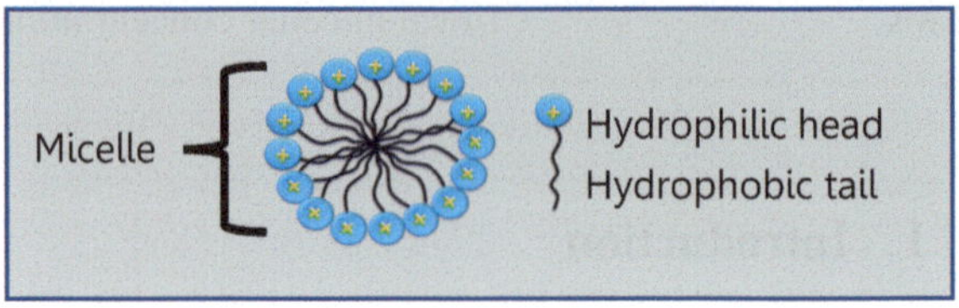

Fig. 8.1 Micelle formation above CMC level

tions in recent years; (or as reviewed by Ying 2006 and Mehta and Kushwinder 2010). This has been an area of study by many research groups (Tawfik 2015; Negm et al. 2015; Egorova et al. 2015; Zaki and Tawfik 2014; Mukherjee et al. 2013; Negm et al. 2011). This then led them to be an outstanding green media for catalysis, synthesis, and separation processes (Klein et al. 2013; Brown et al. 2013; Li et al. 2012).

Due to the variety of structurally diversed ILBS that can be synthesized, it generates the development of a huge field of surfactants with great structural diversity affecting overall aggregation behavior. It was found that the aggregation behavior of ILs can be structured by designing the cation/anion and alkyl chain length. They exhibit the surfactant-like behavior when $n > 8$ due to their inherent amphiphilicity (Chen et al. 2010). These long-chain analogues of the common ILs can self-assemble forming micelles in an aqueous solution. Micellar aggregates are formed when ILs that possess long carbon chain are substituents within the cationic head group are dissolved in water (Fig. 8.1). It is based on the concept of microemulsions, but without a surfactant to promote the aggregates formation (Ventura et al. 2012; Brown et al. 2011). Instead, it uses the capacity of long alkyl chain ILs. Imidazolium (1-alkyl-3-methylimidazolium) salts were the first ILBS used to study ILs with surfactant properties. This was due to its easily tunable nature. For example, while increasing the alkyl chain length and symmetry of the cation, the melting points decrease, whereas while changing the anion hydrophobicity, the hydrogen-bond-accepting ability differs (Brown et al. 2011).

The impetus for studying ILBSs is great potential of their applications, for example, in the biological field. The progress in these applications rests on a clear understanding of the relationship between surfactant molecular structure and properties of its solution. For example, ILBS have been successfully employed as substitutes of conventional organic solvents in extraction processes or as modifiers of the chemical structure of sorbents with better reported performance compared to conventional organic solvents and conventional cationic surfactants (Heidarizadeh and Majdi-nasab 2015). ILBS has also received great attention for drug delivery whereby IL-based microemulsions overcame the inability of conventional microemulsions to dissolve a number of chemicals which includes hydrophilic and hydrophobic substances for dermal delivery water-soluble drugs (Goindi et al. 2015; Mahajan et al. 2012; Zhou et al. 2014; Tourné-Péteilh et al. 2014; Rao et al. 2012). ILBS are also seen promising for the extraction of proteins or other biomacromolecules (Mao et al. 2014; Sun et al. 2014; Wang et al. 2015) and the separation of protein (Hasan et al. 2015; Ventura et al. 2012) due to their structure which can be readily designed to offer adjustable properties. This review puts forth the current scenario of ILBS, covering the biological applications of ILBS in the last decade. It is believed that our present review will provide a platform for new researchers to get an overview of the biological applications of ILBS in for biological applications.

8.2 ILBS Self-Assembly Features

Like conventional surfactants, ILBS are amphiphilic molecules compromised of a hydrophilic charged head group and one or more hydrophobic alkyl chain tails. The ability of ILBS to self-assemble into various aggregations form were governed by its dual characteristic of polar and apolar regions (Bowers et al. 2004; Vanyur et al. 2007; Blesic et al. 2007; Miskolczy et al. 2004; Jungnickel et al. 2008; Geng et al. 2010). These amphiphilic substances (with pronounced hydrophilic and lipophilic fragments) are able to self-organize at an individual state or in solutions. Some studies in the past have looked at the self-organizing pattern of ILs via computational simulations. Simulation studies (Smirnova and Safonova 2010; Jiang et al. 2007) denote the formation of two types of domains—positively charged imidazolium rings and anions arranged in three-dimensional network supported by strong electrostatic interactions, whereas alkyl groups form non-polar domains with short-ranged *van der Waals* interactions. Herein, a brief review on various self-assembly of the ILBS system such as micellar formation, IL-microemulsion, and vesicle/gel IL cooperative system will be further elaborated.

8.2.1 Micellar Formation in ILBS

The micellar formation in ILBS plays a vital role in understanding ILs behavior in aqueous phases. Micelles are defined as a self-aggregation of surfactant in solutions in which it only occurs when the concentration of surfactant reaches its critical micelle concentration (CMC). In principal, below the CMC levels surfactant molecules appear as a single monomer. Above its CMC point, the hydrophobic alkyl chain region starts to orient itself pointing towards the center of micelle with the head group exposed to the aqueous phase and vice versa in inverse micelles (Jungnickel et al. 2008; Vanyur et al. 2007; Blesic et al. 2007; Geng et al. 2010). Determination of CMCs for surface tension measurement and aggregation behavior is typically investigated using dynamic light scattering (DLS) and force tensiometer.

Jingjing et al. (2012) reported values of γcmc and cmc at given temperature for two candidates, namely $[C_4mim][C_{12}SO_4]$ and $[C_4MP][C_{12}SO_4]$ which were recorded over traditional ionic surfactant, SDS. It showed that studied imidazolium and pyrrolidinium ILs have superior surface activity as compared to SDS. The micelle formation in both ILs systems was aided by the weaker hydration of the bulky imidazolium or pyrrolidinium cations. This diminishes the electrostatic repulsion between ILs head groups, which in turn resulted in lower cmc and γcmc values. In a recent study, Rao et al. (2015) observed the same trend on ILs with three different cationic groups with dodecylbenzenesulfonate anion specifically $[N_{1114}]$DBS $[C_4mim]$DBS $[C_4Py]$DBS in comparison with anionic surfactant [Na]DBS. It was again proved that due to the electrostatic repulsion reduction of charged ILs head groups that led to the lowering of surface tension and CMC. Gemini dicationic ILBS presented lower CMC values than monocationic ILs. This was seemingly interesting as micelles were formed with only a little amount of ILs as gemini-based ILBS possessed very low CMC as well as a strong efficacy in decreasing surface tension. Studies related to the formation micelle in ILBS is summarized in Table 8.1.

8.2.2 ILBS—Microemulsion System

Typically, microemulsions are often referred to as a thermodynamically stable, transparent, and a micro-heterogenous system. They possess two immiscible polar and non-polar solvents which is stabilized by an interfacial film of surfactants; sometimes in the presence of a co-surfactant. In the Winsor's classification, there are four types of microemulsions systems; they are oil-in-water (O/W, Winsor I), water-in-oil (W/O Winsor II), and bi-continuous (BC, Winsor III) structure with excess water and oil or both (Winsor 1948). Winsor IV type microemulsion, on the other hand, can be obtained where the components are not in excess. Microemulsions have the capacity to simultaneously co-solubilize huge volumes of polar and non-polar compounds, which overcomes the solubilization barriers of immiscible liquids.

A lot of studies have been steered towards increasing water solubilization capacity, stability and the interfacial curvature of microemulsion systems, and this is where the emergence of ILs in microemulsion systems came about (Singh et al. 2017). In fact, in the past, several reviews have elegantly discussed the unique properties of IL-based microemulsions (Qiu and Texter 2008 and Kunz et al. 2012). In recent years, many studies associated to microemulsion system dealing with ionic liquids as substituents as oil or water phase has emerged (Tadros et al. 2004; Cheng et al. 2007; Qiu and Texter 2008; McClements 2012). Microemulsions are two or more immiscible liquids phase that are stabilized thermodynamically by surfactants. Formation of ionic liquid-in-oil microemulsions was first studied by Gao and co-workers (2004) between three component consisting of [C4mim][BF4] and cyclohexane, aided by the T-X100 non-ionic surfactant. The formation of dispersed nano-droplets with ILs (core assisted by non-ionic surfactant T-X100) were clearly demonstrated using freeze-fracture electron microscopy (FFEM) and DLS for its size, shape of aggregates, and distribution. Wei and co-workers found that the water solubilization capacity of W/AOT/O (water/sodium bis-ethylhexyl sulphosuccinate/oil) microemulsions reduced at high concentrations of imidazolium-based ILs but

Table 8.1 Critical Micelle Concentration (CMC) (mmol/L) and Surface Tension γ_{cmc} (mN/m) measurement in aqueous solution at $T = 298$ K

No	ILBS	Structure	cmc (mmol/L)	γ_{cmc} (mN/m)	References
1	[C_4mim] [$C_{12}SO_4$]	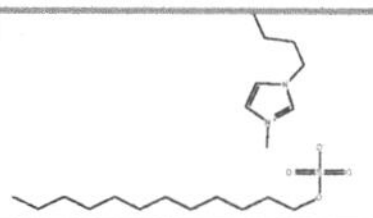	1.8	31.9	Jingjing et al. (2012)
2	[C_4MP] [$C_{12}SO_4$]	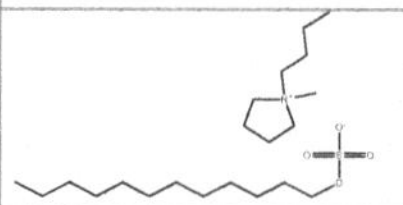	2.7	34.3	
3	SDS	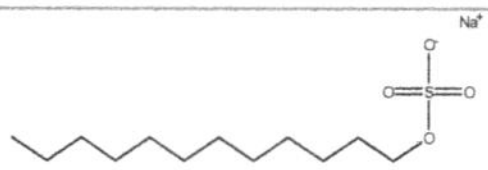	7.8	39.6	
4	[C_{12}mim]Br	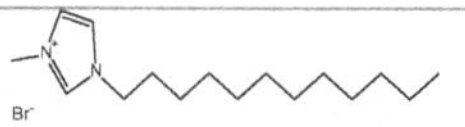	10.9	39.4	
5	[C_{12}MP]Br	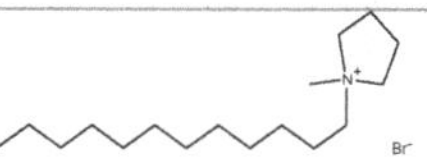	13.5	42.4	
6	[C_{12}mim]Br	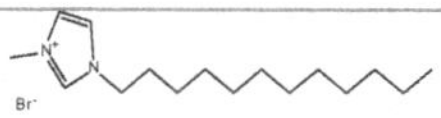	10.9	39.4	Bin et al. (2008)
7	[C_{14}mim]Br	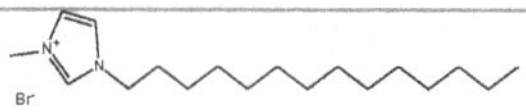	2.8	39.2	
8	[C_{16}mim]Br	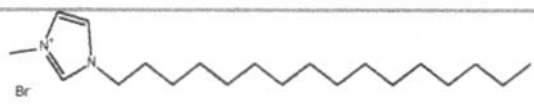	0.55	39.1	
9	[C_4mim] [$C_{12}H_{25}SO_3$]	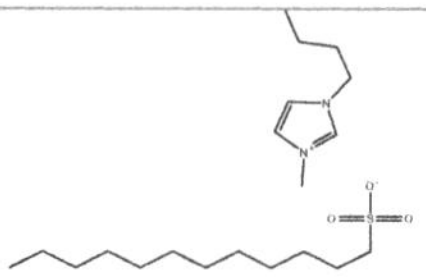	4.4	36.9	Yanxin et al. (2015)
10	SDDS	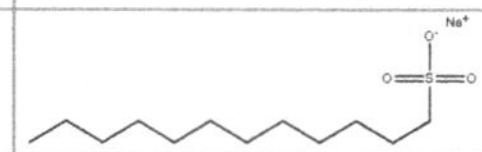	7.4	37.6	
11	[C_4mim] [$C_{12}H_{25}SO_4$]	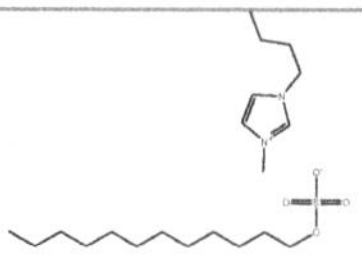	1.8	31.9	
12	[Na]DBS	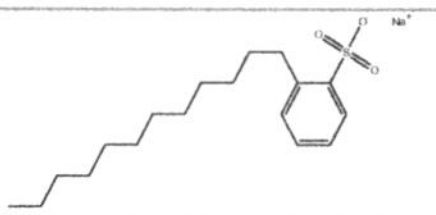	2.75	32.49	Rao et al. (2015)
13	[N_{1114}] DBS	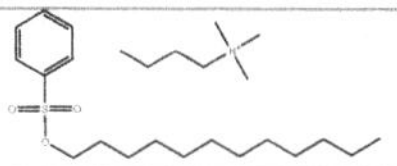	1.34	29.11	
14	[C_4mim]DBS	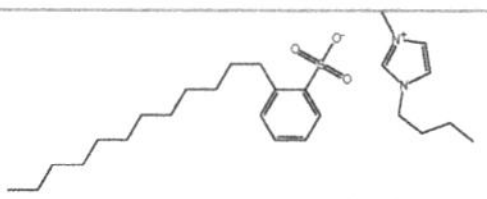	1.08	29.18	
15	[C_4Py] DBS	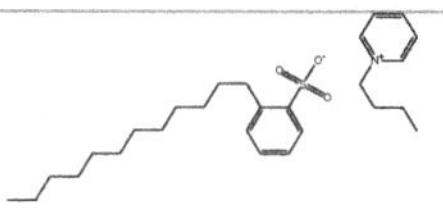	0.92	29.69	

(continued)

Table 8.1 (continued)

No	ILBS	Structure	cmc (mmol/L)	γ_{cmc} (mN/m)	References
16	SALSIL 1 SALSIL 2 SALSIL 3 SALSIL 4 SALSIL 5 SALSIL 6 SALSIL 7 SALSIL 8 SALSIL 9	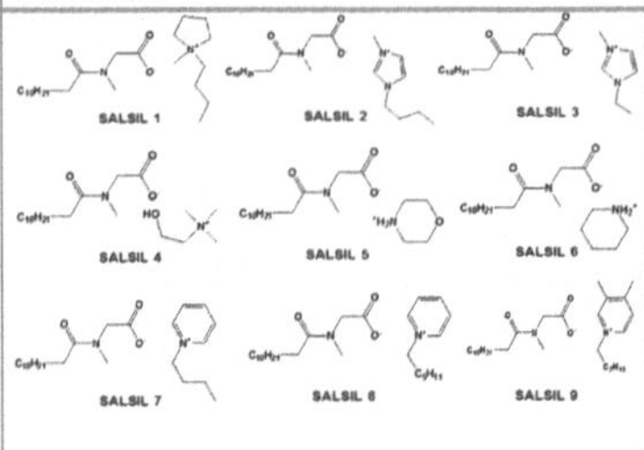	2.42 2.44 2.62 2.67 2.79 2.8 2.46 1.15 1.02	39.6 37.1 36.31 36.24 31.72 33.40 27.3 29.81 30.63	Mustahil et al. (2019)
17	[Cho][Lin] [Cho][Ole] [Cho][Eru]	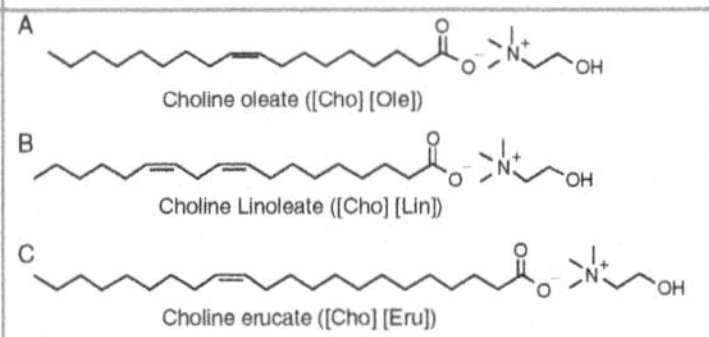	2.0 1.7 0.8	26.6 24.4 23.2	Ali et al. (2019)

increased at low concentrations of the aforementioned ILs (Wei et al. 2011). Another study by Kang et al. (2015) showed that the SANS data was in the agreement with traditional water-in-oil (w/o) microemulsions and with the addition of ILs, a progressive swollen of micelle was observed through an increase in droplet volume (Kang et al. 2015). In another study, Kundu et al. 2014 showed that the ions of imidazolium cation-based ILs influenced the water solubilization capacity of W/O microemulsions. All the above-mentioned studies dealt with the incorporation of imidazolium-based ILs into their microemulsion systems. Recently, researchers have been venturing into the potentials of protic ILs as a formulation for IL-in-oil microemulsions. One such example is a pioneering study done by Jiu and team whereby they incorporated 1,1,3,3-tetramethylguanidinium (TMG)-based protic ILs to formulate IL-in-CO_2 microemulsions (Liu et al. 2007). Later, Chandran et al. 2010 explained the mechanism of formation of these microemulsions and predicted greater stability of protic TMG-based microemulsions compared to aprotic imidazolium-based systems. The key property that distinguishes pILs from aILs is their ability to transfer protons, which often promotes the formation of an extended hydrogen bonding network in the system. In lieu with the benefits of pILs, Singh et al. 2017 studied the physicochemical behavior of a series of protic guanidinium-based ionic liquid-based water/oil microemulsions in order to increase the stability and water uptake capacity of water-in-oil (W/O) microemulsions. They found a new approach to tune the formulation and properties of the W/O microemulsions by the addition of task-specific amounts of the ILs. This was credited to the excellent proton transfer abilities of pILs which promotes the formation of extended hydrogen bonding networks thus stabilizing microemulsion systems.

8.2.3 Self-Assembly of ILBS Vesicles

Vesicles have been considered as a mimic of biological membranes and have been widely applied in pharmaceuticals especially in drug delivery system and drug encapsulation (Huiyong et al. 2013; Rao et al. 2015; Qi and Shiping 2015). Unprecedented outcome from spontaneous assembly of vesicle in ILs had been observed by Kimizuka and Nakashima (2001). These opened a new horizon where spontaneous formation and growth of ILs vesicles being explored further. Due to large variety and unlimited number of combinations between cation and anion, this so-called designer solvent offers encapsulation of pharmaceutical active molecules or enabling encapsulation of targeted drug in the micellar phase prior to formation of vesicle (Rao et al. 2015). The self-assembly behavior of IL surfactants is less understood from the viewpoint of scientific research and practical application. Typically, there are two categories of ILs as self-assembly media, these include those of pILs and aILs. PILs are similar to water due to their protic nature as they form a hydrogen bond between ion. However, the situation is not alike for aILs as self-assembly in the bulk aILs is weak/absent (Pei et al. 2018).

8.3 Application of ILBS in Biological Systems

8.3.1 ILBS in Pharmaceuticals

Many reports on drug–surfactant interactions have been discussed in the past, they include those of (Enache and Volanschi 2011; Sharma and Mahajan 2012). Surfactants helped in the solubilization of poorly water-soluble drugs by reducing the interfacial tension between the solution and the drug, hence increased the drugs solubility. Solubilization capacity of surfactants had led to better drugs bioavailability upon oral administration. Besides, surfactants have been employed as permeation enhancer in transdermal drug delivery of low lipid solubility drugs. They acted as one of the important ingredients of emulsion formulation to improve the drugs absorption and penetration.

For example, the drug–surfactant association in the incidence of [C4mim][BF_4] IL as a modulator has been studied by Patra and Barakat (2011). There was an increase in the association of anionic surfactant; sodium dodecyl sulfate (SDS) with curcumin as compared to SDS without ILs. In the absence of ILs, there was a repulsion between the negative charge of the head group (sulfate ion) of SDS and the negative charge of the deprotonated form of curcumin. Upon the IL addition, its positive charge head group acted as a stabilizer between the negatively charged SDS and negatively charged curcumin (deprotonated form), consequently assisting the association of curcumin with SDS. On the other hand, the association of curcumin with neutral Triton X-100 (TX100) increased suggestively in the presence of IL. This might be due to the induction of hydrogen bonding and dipole–dipole forces by the positive charge of the head group of the IL with TX100 (Behera et al. 2007) which helped interactions or strong associations of curcumin with the neutral surfactant solution.

Study by Mahajan et al. (2012), described the interactions of a surface active IL, $[C_{14}mim][Br]$ with dopamine hydrochloride (DH) and acetylcholine chloride (AC) which are drugs widely used in several heart diseases. The results were linked with the structurally similar cationic surfactant, tetradecyltrimethylammonium bromide (TTAB) and found that $[C_{14}mim][Br]$ IL acted as a better drug carrier for DH and AC, whereby DH bound strongly with the $[C_{14}mim][Br]$ and TTAB surfactants compared to AC. This was due to the cation-π interactions between the positive charge of the surfactant moieties and the aromatic region of DH. Furthermore, the binding capacity of $[C_{14}mim][Br]$ with DH was greater than TTAB and DH due to the π-π interactions of the π system of DH and imidazolium ring of $[C_{14}mim][Br]$. Besides, $[C_{14}mim][Br]$ gave lower critical micelle concentration (cmc) value than TTAB, due to the difference in their head groups. The positive charge in the imidazolium ring of $[C_{14}mim][Br]$ is delocalized, whereas in TTAB, it exists as a point charge. Hence, the head group–head group repulsions are more prevalent among quaternary ammonium molecules, leading to steric hindrance to form aggregates.

Vashishat et al. (2015) discussed on surface active ionic liquid. They reported synergism between $[C_{12}mim][Br]$ and bile salts; sodium cholate (NaC) and sodium deoxycholate (NaDC) in the solubilization of a poorly water-soluble drug phenothiazine, a tricyclic aromatic compound where its derivatives were largely employed as tranquilizers and neuroleptics in psychiatry (Karpinska et al. 1996). The binary aqueous mixtures of NaC/NaDC and $[C_{12}mim][Br]$ exhibited strong synergistic activities in the mixed micelles and at the air–water interface. Among the two bile salts used, NaDC exhibited better synergy.

Recently, ILs-based microemulsions have been explored as a beneficial tool in the solubilization and permeation enhancement for drug molecules. Moniruzzaman et al. (2010) developed ionic liquid-in-oil (IL/o) microemulsion (ME) and evaluated its ability to dissolve drug molecules (acyclovir, methotrexate, dantrolene sodium) that were insoluble and/or sparingly soluble in water. The solubility study indicated a very significant degree of solubilization of such drugs in ILs microemulsions, while IL-free systems could solubilize small amounts of drugs. The study also suggested the useful influence of IL in the formation of stable microemulsion with small droplets size (8–34 nm) in diameter which were stable for 45 days. Among various ILs used, hydrophilic IL $[Cmim][(MeO)_2PO_2]$ was the most effective for the formation of IL/o ME.

In a related study, Dobler et al. (2013) described the influence of ILs; $[C_6mim][Cl]$ and $[C_4mim][PF_6]$ on the conventional oil-in-water (o/w) and water-in-oil (w/o) emulsions. ILs were found to be successfully incorporated into the emulsions structure and resulted in stable formulations. Dobler and co-workers deduced that the incorporation of hydrophilic IL, $[C_6mim][Cl]$ as a component of the water phase in o/w emulsion led to a decrease in droplet sizes and the viscosity of the formulation while maintaining long-term stability. Moreover, $[C_6mim][Cl]$ IL led to a significant decrease of surface tension of the formulations depending on its concentration in the water. The presence of the hydrophobic [C_4mim][PF_6] resulted in efficient penetration of drug into the skin.

Later, Goindi et al. (2014) reported an IL/o ME system for dermal delivery of a poorly permeating drug, called 5-fluorouracil (5-FU). Result depicted a noteworthy enhancement in the solubility of 5-FU in the IL [C_4mim][Br] compared to water. The IL/o ME consisted of isopropyl myristate, Tween 80, Span 20, and [C_4mim][Br] which have been evaluated on ex vivo skin permeation studies through mice skin exhibited a fourfold enhancement in percent drug permeation compared to an aqueous solution. The selected IL/o ME also showed 2.3-fold enhanced permeation as compared to hydrophilic ointment and 1.6-fold permeation than water-in-oil (w/o) ME. In addition, the side effects such as erythema and irritation associated with the conventional topical 5-FU cream preparation were not observed. In the in vivo studies against dimethylbenzanthracene (DMBA)/12-O-tetradecanoylphorbol-13-acetate (TPA)-induced mice skin carcinogenesis, IL/o ME was successful in treating skin tumors in a period of 4 weeks, with other formulations including the commercial cream were ineffective. Further studies by Goindi et al. (2015) on IL/w ME formulation focused on the solubilization ability of etodolac (ETO), a poorly water-soluble anti-inflammatory drug. The formulation consisted of [C_4mim][PF_6] as an IL whereas Tween 80 and ethanol were used as a surfactant and co-surfactant, respectively. Goindi and co-workers emphasized that an IL/w-based ME can efficiently enhance the solubility and permeability of ETO for its topical delivery with better skin permeation in comparison with other formulations studied. In vivo anti-arthritic and anti-inflammatory activities of the prepared formulations discovered that ETO loaded IL/w-based ME was found to be more profound in controlling inflammation than oily solutions, o/w ME and marketed formulation of ETO. Histopathological studies also confirmed that IL/w-based ME caused no anatomical and pathological variations to the skin.

8.3.2 ILBS in Protein and Enzyme-Based Applications

ILs have the ability to dissolve various materials such as proteins, fats, salts, and other wide range of organic molecules. This ability makes them suitable for liquid–liquid extraction which is one of the applications for ILBS in protein extraction. In a study by Mao et al. 2014, a dual-ionic liquid microemulsion system was developed using 1-decyl-3-methylimidazolium bromide (DmimBr) as a surfactant and 1-butyl-3-methylimidazolium hexafluorophosphate (BmimPF_6) as a replacement for organic solvents. This dual—IL microemulsion system was proven to be effective for protein extraction with an increase in extraction efficiency of hemoglobin compared to pure bmimPF_6. The composition of the microemulsion became the main factor of the stability of the microemulsion system and hemoglobin extraction capacity. The extraction efficiency of hemoglobin increased with an increase of DmimBr. This is due to the increasing of DmimBr aggregates formed which was used to isolate the hemoglobin from human whole blood.

Studies by Geng et al. (2009) explained the binding of a long chained C_{14}mimBr to BSA. Based on the surface tension plots, critical micellar concentration (CMC), and critical aggregation concentration (CAC) and were investigated under different concentration of BSA with different temperatures. From the surface tension results, C_{14}mimBr bound to BSA through electrostatic attraction when CAC is near which occurs when C_{14}mimBr /BSA complexes begun to form. At low concentrations, C_{14}mimBr reduced the burial of tryptophan residue in a hydrophobic microenvironment with the binding constant at 1.55×10^3 L/mol, 4.90×10^3 L/mol for 2×10^{-6} mol/L BSA and 5×10^{-5} mol/L BSA, respectively. In conclusion, at CAC, C_{14}mimBr bound to BSA by electrostatic interaction while at CMC C_{14}mimBr bound to BSA by hydrophobic interaction.

Mester et al. 2012 and team conducted a study regarding the combination of $[C_n\text{mim}]^+$ cation and chaotropic anions on the casein solubility in aqueous solution. ILs used in the study are ($[C_n\text{mim}]^+$ cation with Cl^- and SCN^- anions where $n = 2$, 4, 6). The objective of this study was to investigate the effect of chaotrope–surfactant combination on solubility of protein. As the ILs concentration increases, the solubility of casein also increased. From the anion prospect, SCN^--based salts dissolved more casein than Cl^- as anion, and this can be proved by Hofmeister series, where the SCN^- anion is considered more chaotropic than Cl^- anion. This research proved that $[C_n\text{mim}]$-based ionic liquids with $n = 6$ possessed surfactant-like properties.

In a review by Bharmoria et al. (2014), the structural and functional stability of cellulose was investigated with the interaction to a biamphiphilic ILs, (3-methyl-1-octylimidazolium dodecylsulfate $[C_8\text{mim}][C_{12}OSO_3]$). $[C_8\text{mim}][C_{12}OSO_3]$ bound sequentially to cellulase from the monomeric to vesicular regime in which the electrostatic interactions were reinforced by hydrophobic interactions. From the result obtained, cellulase was active in the presence of $[C_8\text{mim}][C_{12}OSO_3]$ at pH 4.8. As the critical vesicular concentration (CVC) of $[C_8\text{mim}][C_{12}OSO_3]$ reduced, the ionic interactions decreased, leading to cellulase functional stability in $[C_8\text{mim}][C_{12}OSO_3]$ solution.

Laccase, a copper-containing enzyme exhibited catalytic activity after solubilizing in water/ionic liquid (W/IL) microemulsion. Sun et al. (2014) investigated a mixed surfactant for the formation of microemulsions. The research used a mixed surfactant consisting of 1-tetradecyl-3-methylimidazolium bromide ([$(C_{14}$mim]Br, Triton X-100, H_2O and 1-butyl-3-methylimidazolium hexafluorophosphate ([Bmim] PF_6). From the electrical conductivity measurement, the presence of W/IL microemulsion could be detected when the electrical conductivity of the system increased linearly with the increase of water.

In a study by Ventura et al. (2012), 1-decyl-3-methylimidazolium chloride $[C_{10}\text{mim}]$Cl was used in the formation of micelles by self-aggregation in order to increase the enzyme activity of *Candida antarctica* lipase B (CaLB). Previously, organic solvents were used to improve the solubility of hydrophobic reagents but due to its disadvantageous characteristics such as the significant vapor pressure, toxicity and flammability, ionic liquid were employed. $[C_{10}\text{mim}]$Cl IL was used to promote the formation of micelles via self-aggregation thus increasing the activity of the CaLB. The effect of various molar concentrations of [C_{10}mim]Cl on the relative activity of lipase was investigated. As the molar concentration of [C_{10}mim]Cl increased, the activity of lipase also increased. From the dynamic light scattering measurements, the longest alkyl chains tested, [C_{10}mim]Cl showed the presence of aggregates in the 1–10 nm size range. The self-aggregation of the [C_{10}mim]Cl led to the increase of the IL–water interface, thus increasing enzyme activity.

Adak et al. (2015) studied on how imidazolium ILBS improved the activity and thermal stability of lipase from *Rhizopus oryzae* (NRRL 3562). This study focuses on the effect of 1-hexadecyl-3-methylimidazolium bromide (C_{16}mim)Br IL on *Rhizopus oryzae* lipase (rhL). When (C_{16}mim)Br was introduced into the system, the deactivation temperature of the rhL increased and the enzyme thermal stability increased. In terms of rhL activity, (C_{16}mim)Br enhanced the enzyme activity by 80% at low concentrations of ILs (100 μM).

8.4 Conclusion

ILBS are known to have great potential and applications especially for biomolecules. The progress in these applications is based on their designer like properties which make them easily tunable according to their functions. However, long chained ILs are known to carry high toxicological values. Much can still be studied on creating "green" and "environmentally-friendly" ILBS systems.

Acknowledgements The present work has been supported by the Centre of Research in Ionic Liquids (CORIL), Universiti Teknologi PETRONAS.

References

Adak, S., Datta, S., Bhattacharya, S., & Banerjee, R. (2015). Imidazolium based ionic liquid type surfactant improves activity and thermal stability of lipase of Rhizopus oryzae. J Mol Catal B Enzym 119:12–17

Ali MK, Moshikur RM, Wakabayashi R, Tahara Y, Moniruzzaman M, Kamiya N, Goto M (2019) Synthesis and characterization of choline–fatty-acid-based ionic liquids: a new biocompatible surfactant. J Coll Int Sci 551:72–80

Behera K, Pandey MD, Porel M, Pandey SJ (2007) Unique role of hydrophilic ionic liquid in modifying properties of aqueous triton X-100. J Chem Phys 127:184501

Bharmoria P, Mehta MJ, Pancha I, Kumar A (2014) Structural and functional stability of cellulase in aqueous-biamphiphilic ionic liquid surfactant solution. J Phy Chem B 118(33):9890–9899

Bin D, Xueyan Z, Liqiang Z, Jin Z, Na L, Tohru I (2008) Aggregation behavior of long-chain imidazolium ionic liquids in aqueous solution: Micellization and characterization of micelle microenvironment. Colloids Surf A Physicochem Eng Asp 317:666–672

Blesic M, Marques MH, Plechkova NV, Seddon KR, Rebelo LPN, Lopes A (2007) Self-aggregation of ionic liquids: micelle formation in aqueous solution. Green Chem 9:481–490

Bowers JP, Butts CJ, Martin PC, Vergara-Gutierrez M (2004) Aggregation behaviour of aqueous solutions in ionic liquids. Langmuir 20:2191–2198

Brown P, Butts C, Dyer R, Eastoe J, Grillo I, Guittard F, Sarah R, Heenan R (2011) Anionic surfactants and surfactant ionic liquids with quaternary ammonium counterions. Langmuir 27:4563–4571
Brown P, Butts CP, Eastoe J, Grillo I, James C, Khan A (2013) New catanionic surfactants with ionic liquid properties. J Coll Inter Sci 395:185–189
Chandran A, Prakash K, Senapati S (2010) Self-assembled inverted micelles stabilize ionic liquid domains in supercritical CO_2. J Am Chem Soc 132:12511–12516
Chen X, Liu J, Wang J (2010) Ionic liquids in the assay of proteins. Anal Meth:1222–1226
Cheng S, Zhang J, Zhang Z, Han B (2007) Novel microemulsions: ionic liquid-in-ionic liquid. Chem Comm 24:2497–2499
Dobler D, Schmidts T, Klingenhofer I, Runkel F (2013) Ionic liquids as ingredients in topical drug delivery systems. Int J Pharma 441:620–627
Egorova KS, Seitkalieva MM, Posvyatenko AV, Khrustalev VN, Ananikov VP (2015) Cytotoxic activity of salicylic acid-containing drug models with ionic and covalent binding. ACS Med Chem Lett 11:1099–1104
Enache M, Volanschi E (2011) Spectral studies on the molecular interaction of anticancer drug mitoxantrone with CTAB micelles. J Pharm Sc 100:558–565
Gao, H., Li, J., Han, B., Chen, W., Zhang, J., Zhang, R., & Yan, D. (2004). Microemulsions with ionic liquid polar domains. Phy Chem Chem Phy 6(11):2914–2916
Geng, F., Zheng, L., Liu, J., Yu, L., & Tung, C. (2009). Interactions between a surface active imidazolium ionic liquid and BSA. Colloid Polym Sci 287(11):1253
Geng F, Liu J, Zheng L, Yu L, Li Z, Li G, Tung C (2010) Micelle formation of long-chain imidazolium ionic liquids in aqueous solution measured by isothermal titration microcalorimetry. J Chem Eng Data 55:147–151
Goindi S, Arora P, Kumar N, Puri A (2014) Development of novel ionic liquid-based microemulsion formulation for dermal delivery of 5-fluorouracil AAPS. Pharm Sci Tech 15:810–821
Goindi S, Kaur R, Kaur R (2015) An ionic liquid-in-water microemulsion as a potential carrier for topical delivery of poorly water soluble drug: development, ex-vivo and in-vivo evaluation. Int J Pharma 495:913–923
Hasan F, Vidanapathirana P, Das S, Fernand VE, Siraj N, Losso JN, Warner IM (2015) Ionic liquids as buffer additives in ionic liquid-polyacrylamide gel electrophoresis separation of mixtures of low and high molecular weight proteins. RSC Adv 5:69229–69237
Heidarizadeh F, Majdi-nasab A (2015) A green, homogeneous and reusable surfactant/copper based ionic liquid for the N-arylation of indoles, pyrazoles and imidazoles. Tetra Lett 56:6360–6363
Huiyong W, Lamei Z, Jianji W, Zhiyong L, Suojiang Z (2013) The first evidence for unilamellar vesicle formation of ionic liquids in aqueous solutions. Chem Comm 49:5222–5224
Jiang W, Wang Y, Voth GA (2007) Molecular dynamics simulation of nanostructural organization in ionic liquid/water mixtures. J Phy Chem B 111(18):4812–4818
Jingjing J, Bin D, Huina Z, Yingyuan Z, Xiaoqing W, Rui W, Li Y (2012) Aggregation behaviors of dodecyl sulfate-based anionic surface active ionic liquids in water. J Phys Chem B 16:958–965
Jordan A, Gathergood N (2015) Biodegradation of ionic liquids–a critical review. Chem Soc Rev 44:8200–8237
Jungnickel C, Łuczak J, Ranke J, Fernandez JF, Muller A, Thoing J (2008) Micelle formation of imidazolium ionic liquids in aqueous solution. Colloids Surf A Physicochem Eng Asp 316:278–284
Kang TH, Jeon Y, Kim MW (2015) Structure study of a microemulsion system with an ionic liquid. Phy Chem Chem Phy 17(41):27833–27839
Karpinska J, Starczewska B, Tarasiewicz HP (1996) Analytical properties of 2- and 10-disubstituted phenothiazine derivatives. Anal Sci 2:161–170
Kimizuka N, Nakashima T (2001) Spontaneous self-assembly of glycolipid bilayer membranes in sugar-philic ionic liquids and formation of ionogels. Langmuir 17(22):6759–6761
Klein R, Kellermeier M, Taoraud D, Müller E, Kunz W (2013) Choline alkylsulfates : new promising green surfactant. J Coll Inter Sci 392:274–280

Kundu K, Bardhan S, Saha SK, Paul BK (2014) Water solubilization, conductivity and structural characteristics of single and mixed surfactant water-in-oil microemulsions in absence and presence of ionic liquids. Fluid Phase Equilib 361:237–249

Kunz W, Zemb T, Harrar A (2012) Using ionic liquids to formulate microemulsions: current state of affairs. Curr Opi Coll Inter Sci 17(4):205–211

Li H, Yu C, Chen R, Li J, Li J (2012) Novel ionic liquid-type Gemini surfactants: synthesis, surface property and antimicrobial activity. Colloid Surf A Phy Eng Asp 395:116–124

Liu J, Cheng S, Zhang J, Feng X, Fu X, Han B (2007) Reverse micelles in carbon dioxide with ionic-liquid domains. Ange Chem Int Ed 46(18):3313–3315

Mahajan S, Sharma R, Mahajan RK (2012) An investigation of drug binding ability of a surface active ionic liquid: micellization, electrochemical and spectroscopic studies. Langmuir 28:17238–17246

Mao QX, Wang H, Shu Y, Chen XW, Wang JH (2014) A dual-ionic liquid microemulsion system for the selective isolation of hemoglobin. RSC Adv 4:8177–8182

McClements DJ (2012) Nanoemulsions versus microemulsions: terminology, differences, and similarities. Soft Matt 8:1719–1729

Mehta SK, Kushwinder K (2010) Ionic liquid microemulsions and their technlogical applications. Ind J Chem 49A:662–684

Mester P, Wagner M, Rossmanith P (2012) Ionic liquids designed as chaotrope and surfactant for use in protein chemistry. Sep Purif Tech 97:211–215

Miskolczy Z, Sebok-Nagy K, Biczok L, Gokturk S (2004) Aggregation and micelle formation of ionic liquids in aqueous solution. Chem Phys Lett 400:296–300

Moniruzzaman M, Kamiya N, Goto M (2010) Ionic liquid based microemulsion with pharmaceutically accepted components: formulation and potential applications. J Colloid Int Sci 352:136–142

Mukherjee I, Mukherjee S, Naskar B, Ghosh S, Moulik SP (2013) Amphillic behaviour of two phosphonium based ionic liquids. J Colloid Inter Sci 395:135–144

Mustahil NA, Baharuddin SH, Abdullah AA, Reddy AVB, Mutalib MIA, Moniruzzaman M (2019) Synthesis, characterization, ecotoxicity and biodegradability evaluations of novel biocompatible surface active lauroyl sarcosinate ionic liquids. Chemosphere 229:349–357

Negm NA, Kandile NG, Mohamad MA (2011) Synthesis, characterization and surface activity of new eco-friendly Schiff bases vanillin derived cationic surfactants. J Surf Deter 14:325–331

Negm NA, Badr EA, Zakaria K, El-Raouf MA (2015) Environmentally friendly nonionic surfactants derived from jatropha oil fatty acids as inhibitors for carbon steel corrosion in acidic medium. J Surfac Deter 18:1011–1024

Patra D, Barakat C (2011) Unique role of ionic liquid [bmim][BF4] during curcumin-surfactant association and micellization of cationic, anionic and non-ionic surfactant solutions. Spectro Acta Mol Biomol Spectro 79:1823–1828

Pei Y, Hao L, Ru J, Zhao Y, Wang H, Bai G, Wang J (2018) The self-assembly of ionic liquids surfactants in ethanolammonium nitrate ionic liquid. J Mol Liq 254:130–136

Qi Z, Shiping Z (2015) Ionic liquids: versatile Media for Preparation of vesicles from polymerization-induced self-assembly. ACS Macro Lett 4:755–758

Qiu Z, Texter J (2008) Ionic liquids in microemulsions. Curr Opin Coll Inter Sci 13:252–262

Rao VG, Mandal S, Ghosh S, Banerjee C, Sarkar N (2012) Ionic liquid-in-oil microemulsions composed of double chain surface active ionic liquid as a surfactant: temperature dependent solvent and rotational relaxation dynamics of Coumarin-153 in [Py][TF_2N]/[C_4mim][AOT]/benzene microemulsions. J Phy Chem B 116:8210–8221

Rao KS, Gehlot PS, Gupta H, Markus D, Arvind K (2015) Sodium bromide induced micelle to vesicle transitions of newly synthesized anionic surface-active ionic liquids based on dodecylbenzenesulfonate. J Phys Chem B 119:4263–4274

Sharma R, Mahajan RK (2012) An investigation of binding ability of ionic surfactants with trifluoperazine dihydrochloride: insights from surface tension, electronic absorption and fluorescence measurements. RSC Adv 2:9571–9583

Singh AP, Kundu K, Singh V, Gardas RL, Senapati S (2017) Enhanced stability and water solubilizing capacity of water-in-oil microemulsions based on protic ionic liquids. Phy Chem Chem Phy 19(38):26132–26144

Smirnova NA, Safonova EA (2010) Ionic liquids as surfactants. Russ J Phy Chem A 84(10):1695–1704

Sun Y, Yan K, Huang X (2014) Formation, characterization and enzyme activity in water-in-hydrophobic ionic liquid microemulsion stabilized by mixed cationic/nonionic surfactants. Colloids Surf B: Biointer 122:66–71

Tadros T, Zquierd P, Esquena J, Solans C (2004) Formation and stability of nanoemulsions. Adv Coll Int Sci 108-109:303–318

Tawfik SM (2015) Simple one step synthesis of gemini cationic surfactant-based ionic liquids: physicochemical, surface properties and biological activity. J Mol Liq 209:320–326

Tourné-Péteilh C, Coasne B, In M, Brevet D, Devoisselle JM, Vioux A, Viau L (2014) Surfactant behavior of ionic liquids involving a drug: from molecular interactions to self-assembly. Langmuir 30:1229–1238

Vanyur R, Biczok L, Miskolczy Z (2007) Micelle formation of 1-alkyl-3-methylimidazolium bromide ionic liquids in aqueous solution. Coll Surf A Physico Eng Asp 299:256–261

Vashishat R, Sanan R, Mahajan RK (2015) Bile salt-surface active ionic liquid mixtures: mixed micellization and solubilization of phenothiazine. RSC Adv 5:72132–72141

Ventura SP, Santos LD, Saraiva JA, Coutinho JA (2012) Ionic liquids microemulsions: the key to Candida antarctica lipase B superactivity. Green Chem 14:1620–1625

Wang A, Chen L, Xu F, Yan Z (2015) Phase behavior of epoxidized soybean oil-based ionic liquid microemulsions: effects of ionic liquids, surfactants, and co-surfactants. Colloid Surf A: Phy Eng 481:500–505

Wei J, Su B, Yang J, Xing H, Bao Z, Yang YY, Ren Q (2011) Water solubilization capacity and volume-induced percolation of sodium bis (2-ethylhexyl) sulfosuccinate microemulsions in the presence of 1-alkyl-3-methylimidazolium chloride ionic liquids. J Chem Eng Data 56:3698–3702

Winsor PA (1948) Hydrotropy, solubilisation and related emulsification processes. Trans Faraday Soc 44:376

Yanxin J, Lushan W, Tao W, Peng C, Yanhui B, Li Y (2015) Aggregation behavior of dodecylsulfonate-based surface active ionic liquids in water. J Mol Liq 212:23–29

Ying GG (2006) Fate, behavior and effects of surfactants and their degradation products in the environment. Environ Int 32:417–431

Zaki MF, Tawfik SM (2014) Synthesis, surface properties and antimicrobial activity of some germanium nonionic surfactants. J Oleo Sci 63:921–931

Zhou G, Luo Z, Fu X (2014) Preparation of starch nanoparticles in a water-in-ionic liquid microemulsion system and their drug loading and releasing properties. J Agri Food Chem 62:8214–8220

Chapter 9
Ionic Liquid for Water Purification

Syeda Mariam Hasany, Sumia Akram, Muhammad Mushtaq, and Ahmad Adnan

Contents

Abbreviations

*P	Permeate side pressure
A	Area
ILs	Ionic liquids
P	Pressure
PDMS	Polydimethylsiloxane
rpm	Revolutions per minute
RTILs	Room-temperature ionic liquids
SILMs	Supported ionic liquid membranes
T	Temperature
t	Time
VOCs	Volatile organic compounds

S. M. Hasany
Kinnaird College for Women University, Lahore, Pakistan

S. Akram
Division of Science and Technology, University of Education Lahore, Lahore, Pakistan
e-mail: sumia.akram@ue.edu.pk

M. Mushtaq (✉) · A. Adnan
Department of Chemistry, Government College University Lahore, Lahore, Pakistan
e-mail: muhammad.mushtaq@gcu.edu.pk; ahmadadnan@gcu.edu.pk

Inamuddin, A. M. Asiri (eds.), *Nanotechnology-Based Industrial Applications of Ionic Liquids*, Nanotechnology in the Life Sciences,
https://doi.org/10.1007/978-3-030-44995-7_9

9.1 Introduction

9.1.1 Water

Water (dihydrogen oxide; H_2O) is one of the most abundant and important chemical substances for any kind of life prevalence on earth. The presence of hydrogen bonding and other short-range forces in water molecules makes it to exist in all three states and interact with majority of other chemical species (universal solvent) under all its physical states. The extraordinary higher solvation power of water makes it fit for the transport (supply or dispose) of nutrients and energy to every cell or organ of living organism and to regulate body function. In the case of humans, our body normally encloses more than 70% water and an adult man and woman have to drink 3.7–2.7 L of water depending upon the age, physical activities, health, and climate conditions. Fortunately, almost 71% of global population has access to safe drinking water (a water that does not cause significant adverse effect over lifetime consumption); still we had to taste more than 400 deaths per day due to lack of safe drinking water (WHO 2019). Although water covers almost 70% of earth crust about 97% of water resources are saline whereas 3% remains hard to access so only 0.014% of water resources are considered to be fresh and easily accessible. This small fraction of water resources, i.e., freshwaters including water in lakes, rivers, glaciers, streams, and underground, have acceptable range of dissolved salts or solids that may be fit for drinking purposes. The ground or lake water is normally considered as an accessible from of freshwater. The abovementioned resource of freshwater is globally sufficient to meet the drinking water demand but an increase in world population, bulk urbanization, change in lifestyle, rural-urban migration, modern consumption patterns, and elevated level of human interference with water cycle have increased the threats of water dearth. Moreover, temporal and spatial variations and change in climate conditions may lead to physical or actual water scarcity in different parts of the world. This whole scenario demands more efficient, sustainable, and green water purification strategies.

9.1.2 Water Purification

Water purification focuses on the set of techniques used to remove suspended solids, undesirable chemicals, toxins, biotic contaminants, and gases from water to make it fit for human consumption. The absolute nature and extent of purification strategy mainly depend upon the type of contamination water has and objective of its final use. The most universal and frequently adopted approaches include (a) physical separation like filtration, adsorption, distillation, and sedimentation; (b) biological techniques which use a variety of simple and biologically active membranes; and (c) chemical processes like chlorination and flocculation. In addition to these methods, ozone, ultraviolet radiations, and ionizig radiations have also been used to disinfect the water but these methods are not the part of commercial/industrial scale purification of water. In this chapter, the authors will try to explain how the use of recently introduced ionic liquids (ILs) in water purification methods can improve the efficiency of the process or reduce the effluent volume to make the process more sustainable and green.

9.1.3 Ionic Liquids

Ionic liquids (ILs) are liquids or molten salts of organic compounds which possess exceptionally nominal vapor pressure above the surface of liquid. Another term room-temperature ionic liquids (RTILs) has been introduced for the liquid organic salts/mixtures which melt below 100 °C or less. The ILs for their nonvolatile character and high thermal stability have surfaced as a class of soft materials with very unique and beneficial physicochemical properties which can be tailored by a change in structure and class of cations and anions (Amde et al. 2015; Ghandi 2014; Han and Armstrong 2007; Jordan and Gathergood 2015; Nulwala et al. 2018).

9.1.3.1 Tunable Character of Ionic Liquids

The ILs being salts have two species which together give rise to the unique configuration of ionic liquids. One part is the cationic species and the other is the anionic species (Fig. 5.2). There are many other ways and basis upon which ILs may be classified but the most important way to categorize these solvents is the cationic part of the organic liquid salt. The ILs have four major groups when they are categorized on the basis of their cationic part; these classes are as follows: dialkylimidazolium-, alkylammonium-, N-alkylpyridinium, and lastly phosphonium-based ILs (Ghandi 2014). The key features of ILs are their tunable structure, solvent characteristics (Nulwala et al. 2018), tailor-made materials, and "designer solvents". Moreover, ILs can be designed to fit for any function (Han and Armstrong 2007; Stepnowski 2007) like separation of organic matter, disinfectants, and gases.

9.1.3.2 Polymerization in Ionic Liquids

Another special characteristic of ILs is their ability to connect with other monomers that give rise to a macromolecule called as poly(ionic liquid)s or polymeric ILs. This phenomenon is known as salt-responsive polymerization because their hydrophilic and hydrophobic character of the cationic part of ILs can be modified by replacing its anionic counterpart (Nulwala et al. 2018). The salt-responsive polymerization has opened new horizons in the preparation of exchange resins and other ion-selective membranes (Isik et al. 2016).

9.1.3.3 Non-volatile Character of Ionic Liquids

A considerable volume of pollutants and greenhouse gases are contributed to the environment by researchers or industrialists during production, transport, use, storage, or dispose of volatile organic solvent like *n*-hexane, pyridine, petroleum ether, dichloromethane, and many more. These toxic solvents either seep into the ground or end up in the form of toxic vapors in the environment (Sheldon 2012). In my opinion, the property of non-volatility is the one which has earned these organic liquid salts (ILs) the title of green solvents. Ionic liquids are non-volatile; hence these can be employed as a much safer alternative for the volatile solvents in various reactions. The best thing about the use of ILs is that these will not cause damage and toxicity for atmosphere (Akbari et al. 2019). As ILs have negligible vapor pressure, this property makes them eligible to be employed in harsh conditions of pressure and temperature; therefore Dahi et al. (2014) designated them the title of green solvents. The ILs are found in the liquid state because of the peculiar chemical structure of cationic and anionic species involved so as to destabilize the crystal structure and the ILs acquire a liquid state at ambient temperature (Yan et al. 2019).

9.1.3.4 Solvation Power

ILs usually have higher density as compared to that of water, but tetraalkylborate-based ILs are an anomaly to this general trend as they are lighter in density. Furthermore, if the cation alkyl chain length changes then the density of the ILs may change; similar will be the case with the size and nature of anions. In general, an increase in the cationic alkyl chain length increases the viscosity of the respective ILs (Endres and Zein El Abedin 2006). ILs are known to be polar solvents, and their polarity is similar to that of alcohols having short alkyl chain length (Baker et al. 2002). As of recently, there has been a shift in the use of ILs, and scientists are now coupling ILs with membrane processes and membranes. This type of utilization of ILs by immobilizing them on a membrane not only facilitates the reusability and recovery of ILs but also minimizes the loss of ILs out of the system. Moreover, using ionic liquids in such a way ensures the reduction of active phase required for a given operation. Also, using a matrix to immobilize ILs allows a barrier to be formed between the receiving and feeding phases. In addition to good solvation behavior, the ILs also have the property of high ionic conductivity, because they themselves are ionic in nature and hence have the ability to conduct ions (Ali Ayati et al. 2019; Ghandi 2014; Uragami et al. 2016b; Yan et al. 2019).

9.1.3.5 Mesoscale Structure

The versatility of IL properties and their applications in almost all domains can be attributed to the extraordinary large number of IL structures. The most widely existing structures, which are commonly found in ILs, are the "ionic clusters." Ionic clusters are a very characteristic mesoscale phenomenon; these ionic clusters are found to be stable under particular conditions but tend to change with slight change in experimental conditions (Liu et al. 2019). The IL system usually possesses two kinds of mesoscale structures viz., "ionic cluster" and "hydrogen bond." The strength of hydrogen bond usually affects the viscosity of IL system (Ma et al. 2018) while ionic clusters are responsible for properties like dissolution, viscosity, alkalinity, and acidity (Liu et al. 2019).

According to Yan et al. (2019) there are no prescribed regulations and rules for the preparation of ionic liquids, because all that needs to be done is the balancing out of ion-ion interactions. An IL system can be prepared by combining bulky organic cations with a vast range of anions and there are countless number of arrangements through which this can be done. This is the reason why ionic liquids are considered as "designer solvents" (Nulwala et al. 2018; Stepnowski 2007).

9.1.4 Ionic Liquids in Water Purification

Ionic liquids were initially used to hydrolyze or catalyze different various reactions particularly taking place in organic and aqueous phases for their extraordinary thermal stability and nonvolatile character. Until recently, these liquids are rapidly replacing organic solvents and catalysts in synthesis, stationary phases or adsorbents of chromatographic instruments, and many conventional filtration membranes. Ionic liquids may find their application in water desalination, integrated filtration, adsorption, removal of toxins, and volatile organic compounds. In the same way, the integration of ionic liquid in membrane filtration technologies can open new areas of research for water purification and desalination. Table 9.1 provides a bird's-eye view of various techniques where ILs have been incorporated to get rid of water contaminates.

9.1.4.1 Ionic Liquid-Mediated Solvent Extraction

The solvent extraction falls among the most frequently adopted classical approaches for water purification nevertheless to its effect on the final quality of water; this technique efficiently removes organic pollutants. The traditional application of solvent extraction involves the use of some water-immiscible solvents which can be effectively portioned into organic impurities like aromatic compounds, azo dyes, phenols, and its derivative. The efficiency of extractive-type water purification depends upon the polarity of extraction solvent and impurities. In general, water purification via solvent extraction requires nonpolar solvents or media; however, introduction of ILs has created a lot of more prospects.

The volatile organic compounds (VOCs) are substances of carbon except carbon dioxide, carbon monoxide, metallic carbides, carbonic acid, carbonates, and ammonium carbonate that take part in atmospheric photochemical reactions (Tisserand and

Table 9.1 Types of ionic liquids, target contaminant in water, and conditions for the removal of contaminants from water

Solvent/support/ adsorbent	Ionic liquids (ILs)	Characteristics	Target impurity	Conditions	Adsorption capacity (%)	References
Ionic liquid-mediated solvent extraction of organic contaminants						
Tributyl phosphate	1-Butyl-3-methyl imidazolium tetrafluoroborate	Water miscible	Chlorophenols	pH 3, T 30 °C, t 20 min	99.2–99.7	Brinda Lakshmi et al. (2013)
None	1-Butyl-3-methyl imidazolium bis-trifluoromethanesulfonimide	Hydrophobic	1-Nonanol, toluene, hexanoic acid	pH 3.5, T 22 °C, t 2 min	NA	McFarlane et al. (2005)
None	1-Butyl-3 methyl imidazolium hexafluorophosphate	Hydrophobic	1-Naphthol, 2-naphthol, catechol, resorcinol, 4-chlorophenol, 4-nitrophenol, 2,4-dinitrophenol, 2,6-dinitrophenol, 2,4,6-trinitrophenol (picrate anion)	pH 1–7, T 22 °C, t 30 min	20- >90	Khachatryan et al. (2005)
None	Tetrahexylammoniumdihexyl sulfosuccinate	Hydrophobic	Phenol, 4-nitrophenol, 2,4-dinitrophenol, 2,6-dinitrophenol, picric acid, 1-naphthol, 2-naphthol, aniline hydrochloride, p-toluidine, 3-nitroaniline, tryptamine hydrochloride	pH NA, T 20 °C, t 15 min	94%: NA	Egorov et al. (2008)
None	1-Methyl-3-alkyl imidazolium hexafluorophosphate 1-Methyl-3-alkyl imidazolium tetrafluoroborate	Water immiscible	Phenol, bisphenol A, pentachlorophenol, 4-octylphenol, 4-nonylphenol	pH 7, T 24 °C, t 30 min	NA	Fan et al. (2008)
None	1-Octyl-3-methyl imidazolium hexafluorophosphate	Hydrophobic	Phenol, 4-nitrophenol, aniline, nitrobenzene	pH 2–8, T NA°C, t 40 min	NA	Zhang et al. (2010)
None	1-Hexyl-2-phenyl-imidazolium bis(trifluoromethanesulfonyl)amide	Recyclable ILs	Methylene blue, methyl orange	pH >6, T ambient, t 15 min	89–96%	Kermanioryani et al. (2016)

Triton X-114	1-Butyl-3-methyl imidazolium hexafluorophosphate	Hydrophobic	Methylene blue	pH 8, T 35 °C, t 120 min	66–97.8%	Talbi et al. (2014)
Sodium chloride	Tricaprylmethylammonium thiocyanate	Immiscible in hexane and water, soluble in ethyl acetate and ethanol	Methylene blue, methyl orange	pH NA, T 20 °C, t 30 min	64–89%	Chen et al. (2013)
Triton X-114	1-Butyl-3-methyl imidazolium hexafluorophosphate	Hydrophobic	Methylene blue	pH 8, T 35 °C, t 120 min	66–97.8%	Talbi et al. (2014)
Ionic liquid-mediated pervaporation for the removal of volatile organic compounds						
Poly(methyl methacrylate)-graft-poly(dimethylsiloxane) membrane	1-Allyl-3-butylimidazolium bis(trifluoromethane sulfonyl) imide	Hydrophobic, immiscible with water	Volatile organic compounds	*P 1.33 pa, Pt 40 °C, A 13.8cm^2.	Weight % of benzene is approx. 22%	Uragami et al. (2016a)
Polystyrene and poly(dimethylsiloxane) membranes	1-Allyl-3-butylimidazolium bis(trifluoromethane sulfonyl) imide	Hydrophobic water-repellent membranes	VOCs like toluene, chloroform, and benzene	Pt 40 °C, *P 1.33 pa, A 13.8cm^2	% weights of benzene, toluene, and chloroform were 42.3, 30.5, and 51.4%, respectively	Uragami et al. (2016b)
Ceramic ultrafiltration module	(1-Ethenyl-3-ethyl-imidazolium hexafluorophosphate ionic liquid (IL1) Tetrapropylammonium tetracyano-borate ionic liquid (IL2)	Appreciable thermal stability Selective membranes	Butan-1-ol, acetone, water	Pt 23 °C, pH 7	βi of acetone is 3.2 and 7.9 for IL1 and IL2, respectively βi of butan-1-ol was 3.1 and 10.9 for IL1 and 2, respectively	Pavel Izák et al. (2008)

(continued)

Table 9.1 (continued)

Solvent/support/ adsorbent	Ionic liquids (ILs)	Characteristics	Target impurity	Conditions	Adsorption capacity (%)	References
Polypropylene (PP) and nylon membranes used as porous support	1-Decyl-3-methylimidazolium tetracyanoborate, trihexyltetradecylphosphonium tetracyanoborate, 1-decyl 3-methylimidazolium tris(pentafluoroethyl) trifluorophosphate	Hydrophobic and porous membranes coated with silica	*n*-Butanol	Pt 37 °C, P 10 mbar, feed flow rate 35 L/h	n-Butanol in permeate 55 wt.%	Heitmann et al. (2012)
PDMS membranes	1-Octyl-3-methylimidazolium bis(trifluoromethylsulfonyl) imide	Hydrophobic membranes, high permeate flux, high selectivity, $[Tf_2N]^-$-based ionic liquid	Acetone, butanol, ethanol	Pt 25 °C, vacuum pressure 0.8 bar, A 13.4cm^2	βi of butanol, acetone, and ethanol for SILM was 5.26,1.61, and 1.24, respectively	Mai et al. (2013)
PDMS membranes, ultrafiltration ceramic module	1-Ethenyl-3-ethyl-imidazolium hexafluorophosphate (IL1) Tetrapropyl-ammonium tetracyanoborate (IL2)	Hydrophobic membranes, very stable throughout, IL2 was found to be more efficient	Butan-1-ol and water binary mixture	37°C, pH 7, P 20 pa	βi approx. 12 for butan-1-ol	Izák et al. (2009)
Ionic liquid-mediated adsorption-based water purification						
None	1-Hexadecyl-3-methylimidazolium chloride monohydrate-crafted zeolite	Efficacious adsorbent	Methyl orange	pH 2–11, shaking 150 rpm (24 h), centrifugation 4500 rpm (20 min)	45–116 mmol/kg	Xing et al. (2016)

None	Poly-divinylbenzene hydroxyl-functionalized ionic liquids	Efficient adsorbent	Sunset Yellow and Orange II	2 mg Adsorbent, 150 rpm agitation, centrifugation 8000 r/min (10 min)	2095.80 mg/g for Orange II 2100.98 mg/g for Sunset Yellow	Gao et al. (2013)
ZnO nanoparticles	1-Butyl-3-methylimidazolium tetrafluoroborate	IL-functionalized nanoparticles (nanoadsorbent)	Brilliant Blue R 250	pH 2–6, t 5–10 min	>80% (53.73 mg/g)	Chaudhary et al. (2016)
Organosilica	1,3-Bis(3-trimethoxysilylpropyl) imidazolium chloride ionic liquid	Self-assembled ionic liquid-based organosilica (SAILBO)	Malachite green	pH 7, T 40 °C, t 30 min	96% (19.23 mg/g)	Elhamifar et al. (2016)
Layered double hydroxides (LDH)	2-Hydroxyethylammonium acetate	Intercalation of the anionic portion of IL	Reactive orange 5		300.9 mg/g	Zhou et al. (2016)
Chitosan	Trioctyl methylammonium chloride (Aliquat-336)	Efficient adsorbent	Palladium	pH 3.5–4.0	187.61 mg/g	Kumar et al. (2015)
Copolymer hydrogel	1-Allyl-3-methylimidazolium chloride	Efficient IL-modified hydrogel-based adsorbents	Chromium (VI)	pH 1–10, T 20–55 °C	74.5 mg L^{-1}	Jiang et al. (2015)
D-Galactose	1-Methyl-3-(((3aS,5S,5aR,8aR,8bS)-2,2,7,7-tetramethyltetrahydro-3aH-bis([1,3]dioxolo)[4,5-b:4′,5′-d]pyran-5-yl)methyl)-1H-imidazol-3-ium hexafluorophosphate(V)	Efficient adsorbents for lead ions, five times recyclable, hydrophobic	Lead (II)		374.9 mg/g	Jayachandra et al. (2016)
Palm shell-activated carbon	Trioctyl methylammonium thiosalicylate	Task-specific ionic liquids as adsorbents	Hg(II)	pH 8, T room temp, t 180 min	83.33 mg/g	Ismaiel et al. (2013)

Where ∗*P* permeate-side pressure, *P* pressure, *T* temperature, *t* time, *rpm* revolutions per minute, *A* area

Young 2014). The most common VOCs include xylene, toluene, acetaldehyde, formaldehyde, chloroform, and benzene; these are released into the environment by various anthropogenic activities like burning of fuel and/or production and application of industrial chemicals, paints, solvents, biogenics, cosmetics, furnishings, insulation, and adhesives. The majority of VOCs disturb neverious and digestive systems to cause diarrhea, headache, respiratory problems, sensory irritation, organ dysfunction, and oxidative stress (Schubert and Miekisch 2013; Thurston 2017; Tisserand and Young 2014). There are many ways reported in literature which can be employed to remove VOCs from aqueous media, for example, pervaporation, adsorption, and sorption (Ali Ayati et al. 2019; Noble and Gin 2011; Uragami et al. 2016a; Uragami et al. 2016b).

Brinda Lakshmi et al. (2013) found that chlorophenols from aqueous media can be completely removed by applying appropriate IL and stripping media. The authors applied 1-butyl-3-methyl imidazoliumtetrafluoroborate $[C_4mim][BF_4]$ in tributyl phosphate (TBP) to remove chlorophenols from water under different pH, mixing speed, temperature, IL/TBP, and contamination level. It was observed that 0.5% of IL $[C_4mim][BF_4]$ solution in TBP can extract more than 90% of chlorophenols (Table 9.1) from aqueous solutions when applied at a water-to-solvent ratio of 5.0. Previously, McFarlane et al. (2005) have removed polar organic compounds like alcohols, organic acids, and phenols using a wide range of hydrophobic ILs of imidazolium, phosphonium, and pyrrolidinium families. It was observed that partitioning of organic contaminants in water varied depending upon salinity, pH, and nature of pollutants. In general the organic contaminants exhibited higher order solubilities (distribution coefficients >100 for 1-nonanol and toluene), except for acetic acid which portioned sulfonate anion-based ionic liquids and hexanoic acid whose distribution was high in imidazolium cation-based ILs under mild acidic conditions. The distribution coefficients for toluene, 1-nonanol, cyclohexanone, and hexanoic acid were independent of ionic liquid-to-water ratio over the range from 0.02 to 1.0.

The discovery of ionic liquid has opened a new but highly complicated discussion regarding water purification. The direct application of hydrophobic ILs might be limited to the recovery of more polar contaminants into immiscible IL phase. The widely practiced methods involve dissolution of ILs in a particular solvent like tributyl phosphate or octanol so solvent/water partitioning coefficient of an analyte does not affect the IL/water partitioning. The IL hydrophobicity usually increases with the increase in alkyl chain and decreases in anion counterpart; for example within imidazolium cation, IL-containing PF_6^- anion would be more hydrophobic as compared to those having BF_4^- or Cl^- anions. In the case of hydrophilic ILs like imidazolium based, water may also partition into IL phase and vice versa which means that recovery of purification of IL will also be a troublesome task.

9.1.4.2 Ionic Liquid Membranes

There are various ways (Fig. 9.1) the ILs have been integrated with membranes like ionic liquid membrane contactors (ILMCs), ionic liquid composite polymer membranes (ILPMs), supported ionic liquid membranes (SILMs), poly(ionic liquid)s

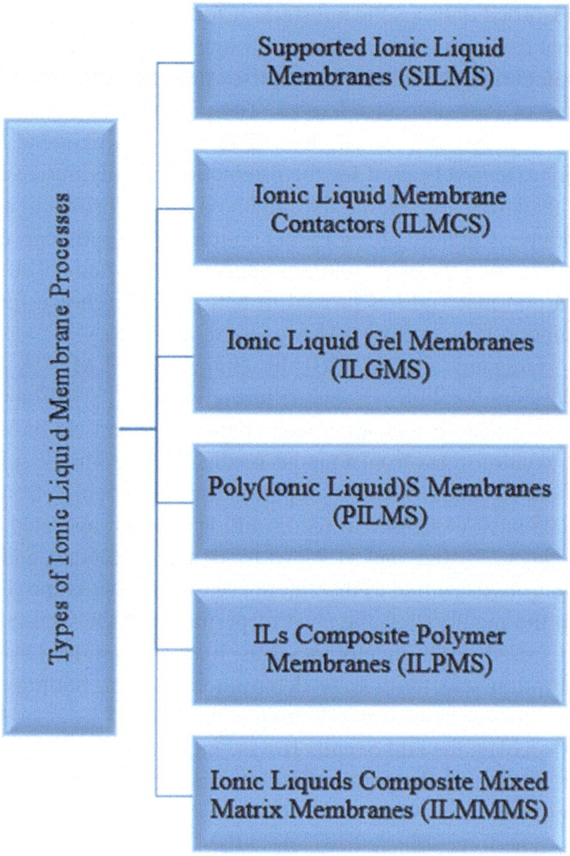

Fig. 9.1 Types of ionic liquid membranes often applied in separation science

membranes (PILMs), ionic liquid gel membranes (ILGMs), and ionic liquid composite mixed matrix membranes (ILMMMs) and out of these SILMs were found to be highly selective and effective for water purification.

9.1.4.3 Supported Ionic Liquid Membranes (SILMs)

The traditional water purification assemblies consist of a series of filters or membranes for sequential or continuous removal of impurities but these membranes are nonselective and hard to reuse. In contrast, supported ionic liquid membranes (SILMs) contain ionic liquid-based molecular components assembled like multifunctional composites also called supported ionic liquid phases (SILPs). The nature of SILPs and its concentration in the membrane control the nature and extent of interactions between analytes and support. In practice, each configuration of SILPs has the ability to remove a single type of impurity so their continuous and sequential design may provide a good water purification choice. when an ionic liquid is adsorbed onto a solid substrate it forms a solid composite material with properties that can be selectively tuned through chemical modification. The SILMs fall in two categories; the first one comprises polymeric membranes and the second one consist of inorganic membranes. The polymers like polysulfone (PS), polypropylene (PP), polytetrafluoroethylene (PTFE), polyvinylidene fluoride (PVDF), and nylon are extensively used for the production of polymeric support membranes. Ceramic substances are commonly employed for the production of inorganic support membranes. Ceramic offers a good option for the synthesis of these membranes because it has good thermal endurance and great mechanical strength. Apart from ceramics, carbon material has proved to be a suitable choice for the synthesis of inorganic support base in SILMs (Karousos et al. 2018).

Uragami et al. (2016b) prepared SILMs by embedding IL 1-allyl-3-butylimidazolium bis(trifluoromethane sulfonyl)imide ([ABIM]TFSI). Pervaporation technique was employed to get rid of VOCs like toluene, benzene, and chloroform from aqueous media. The membrane selectivity partitions the VOCs from aqueous media to separate them efficiently from water. In the experiment, 0.05 wt% VOC used as feed solution was pervaporated and the permeate was found to be 42.3, 30.5, and 51.4%wt for benzene, toluene, and chloroform, respectively. It was observed that as the amount of dimethylsiloxane (DMS) and ILs was increased in the membranes, the %wt of permeated VOCs raised. The maximum amount of ILs used in the membranes was 30 % by weight. Lastly, diffusion selectivity and sorption selectivity were determined, and the findings proved that the VOC perm selectivity from water is mainly influenced by "sorption" rather than diffusion. Moreover, the author calls ILs as "molecular recognition devices" which play a role in increasing the perm selectivity and permeability, hence aiding in effective VOC removal from water.

The force which drives the SILM system, i.e., the pressure/concentration gradient among the feeding and the collecting phase, makes the SILMs an energy-efficient system of separation. Three fundamental methodologies like vacuum, pressure, and immersion exist to prepare SILMs (Wang et al. 2016). The major advantages of SILMs lie in their selectivity, but some studies show that the ILs may bleed out of the support base after a prolonged and heavy use of the SILMs. So, for industrial application scientists are working to enhance the membrane stability by focusing on three areas, viscosity of ILs, preparation method of SILMs, and properties of the support membranes.

The selectivity of SILMs was seen to be dependent on the length of alkyl chain. According to Hernández-Fernández et al. (2009a) the selectivity of the membranes increased as a result of increase in the length of alkyl chain, but opposite scenario was observed when certain anions were used and alkyl chain length on the imidazolium ring was increased; that is, when the alkyl chain length of such ILs is increased it results in the decrease in the selectivity of membranes. In order to embed ILs strongly in the membrane systems to ensure its good performance, low IL bleeding and appreciable operational robustness ionic liquids must be immobilized onto the desired membranes employing the pressure (Hernández-Fernández et al. 2009b). This not only offers the greatest impregnation of ionic liquids onto membrane system but also provides maximum stability to the supported ionic liquid membranes formed. In the same study, Hernández et al. also utilized the "vacuum method" for the synthesis of SILMs and impregnation of ILs and studied the stability of the synthesized membrane in order to establish a comparison of the two methods, and it was observed that the "pressure method" is much efficacious as compared to the vacuum method.

New studies have shown that the membranes can be treated with "plasma" to prevent the loss/bleeding of ILs from the system. The cold plasma treatment method strengthens the connection among ILs and support membrane, hence making the SILMs more long-lasting (Dahi et al. 2017). The problem of membrane stability can be solved by the use of polymeric membranes, by having accommodated the chains or clusters of polymers with the ILs, hence providing increased mechanical strength (Dai et al. 2016; Lu et al. 2009; Shalu et al. 2013). Unfortunately, the increased membrane stability of polymeric membranes comes with a compromise between high selectivity and greater permeability during the separation of gaseous mixtures. The polymer matrix is amorphous which greatly increases the permeability of the gaseous mixtures.

These membranes find applications in separation of organic substances, gaseous mixtures, and vapor permeation. Izák et al. (2008) prepared ultrafiltration SILMs and impregnated two ILs 1-ethenyl-3-ethyl-imidazolium hexafluorophosphate $[C_2MIM]PF_6$ and tetrapropylammonium tetracyanoborate $TPA[(CN)]_4$. These IL-impregnated membranes were further supported using a titanium oxide ceramic in an ultrafiltration module. It was observed that the membranes and the ceramic support were stable in aqueous media for more than 5 months at 23 °C temperature, pH 7, and a minimal pressure of 20 Pa. The membranes were subsequently used to separate the ternary mixtures of acetone, 1-butanol, and water. The application of IL-based membranes caused an increase in the enrichment factor of butan-1-ol from 2.2 to 3.1 in case of $[C_2MIM]PF_6$ and 10.9 in case of $TPA[(CN)]_4$ ILs. For acetone the enrichment factors raised from 2.3 to 3.2 and 2.3 to 7.9 times, respectively. Moreover, these SILMs were found to be very efficient and durable for a period of 5 months, but the process of separation can be a bit time consuming. Recently, Dahi et al. (2014) prepared SILMs for the separation of ethanol from water. Similarly, Uragami et al. (2016a) focused on the addition of hydrophobic ionic liquid to poly(methyl methacrylate)-graft-poly(dimethylsiloxane) (PMMA-graft-PDMS) membranes for the removal of dilute benzene from water.

9.1.4.4 Ionic Liquid-Mediated Pervaporation

Pervaporation or "pervaporative separation" involves partial evaporation of liquids through a porous or nonporous membrane into vacuum (McKeen 2012) and generally depends on "polymer-solute interaction" rather than vapor-liquid equilibrium (Böddeker et al. 1997). The pervaporation apparatus (Fig. 9.2) consists of multiple round separator elements fixed on an axial, and a central permeate removal tube. The separator elements possess an axial permeate transport plate and a permeation membrane surrounding the permeate transport plate along with the annular sealing rings located among adjoining membranes; hence the liquid must move through the

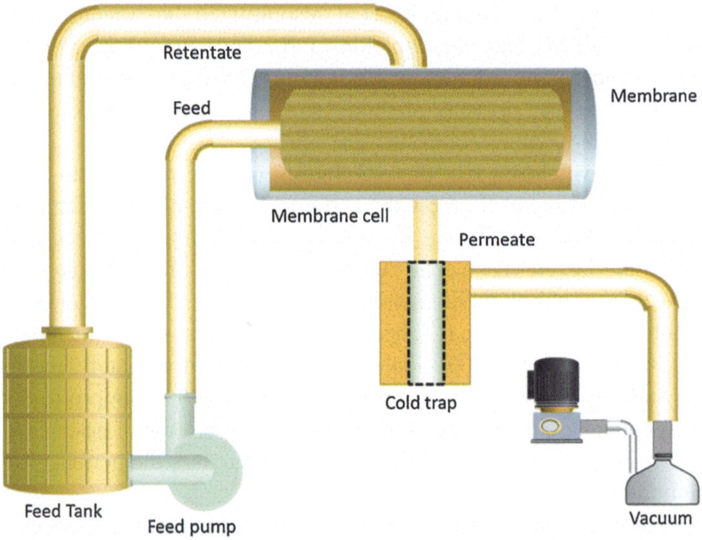

Fig. 9.2 Process instrumentation diagram for pervaporation-based water purification

membranes in order to reach the permeate removal tube. When vacuum is applied to the permeate the removal tube sucks/extracts the permeate past the permeate transport plates and pervaporation membranes inside the permeate removal tube; later on this collected permeate could be condensed within the same tube via direct contact condensation using chilled liquid permeate. However, an external condenser may be used for the same purpose. Streams of bubbles or solution of permeate vapor passing through the separator elements diminish or avert concentration polarization (McGregor et al. 2001).

Pervaporation has been announced a useful technique to cater a variety of contaminants. For instance, Li et al. (2018) followed pervaporation to recycle phenol from aqueous media through the Zeolite (ZSM-5)/polydimethylsiloxane (PDMS)/polyvinylidene fluoride (PVDF) hollow fiber composite membrane. Similarly, Zhou et al. (2014) also used pervaporation via composite membranes for the separation of dimethyl carbonate from a methanolic solution. Recently, Uragami et al. (2016a) applied pervaporation technique to remove benzene from water. For this purpose, the authors used poly(methyl methacrylate)-graft-poly(dimethylsiloxane) (PMMA-g-PDMS) membranes and enhanced its capacity by coupling/associating ionic liquids with them. Due to the unique properties of ionic liquids 1-allyl-3-butylimidazolium bis(trifluoromethane sulfonyl)imide [ABIM]TFSI, the permeability and perm selectivity of water and benzene were improved appreciably. The incorporation of ILs into the membrane increased the attraction of membranes with benzene because the ILs incorporated were hydrophobic in nature and had a great affinity for benzene and less affinity for water, hence causing its easy removal. The ILs were imbedded in the PMMA layer of the membrane. The PMMA-g-PDMS membrane acts as an absorbent which critically partitions the contaminant benzene from the benzene-water mixture. Moreover, it was observed that benzene permeability and perm selectivity increased with the increase in the amount of DMS in the membrane and its permeability. At an amount of 40 mol % DMS the normalized rate of permeation and perm selectivity of benzene raised significantly. Typically, the physical and chemical structure of the polymer membrane influences its perm selectivity, which later on governs the permeate diffusivity through the membrane and permeate solubility into the membrane. Contact angle study/determination showed that the membranes having higher DMS proportion had greater contact angles; this attributes to an enhanced hydrophobic character of the membrane. It was also seen that the permeation rate is dependent upon the extent of packing of molecules in a membrane.

A comparison of sorption selectivity and diffusion selectivity depicts that the VOC (benzene) perm selectivity is mainly influenced by "sorption" rather than diffusion. Scientists reached this solution because the sorption selectivity was higher than the diffusion selectivity. Moreover, Differential Scanning Calorimetery analysis proved that the [ABIM]TFSI was preferentially dispersed in the PMMA phase as compared to the PDMS phase, therefore causing a higher benzene absorption into the PMMA phase due to greater benzene-[ABIM]TFSI attraction. In this chapter, the ionic liquids [ABIM]TFSI are termed as "molecular recognition devices" because these ILs serve to enhance the perm selectivity and permeability of the membranes facilitating effective pervaporation of benzene from aqueous media.

Uragami et al. (2016b) performed a similar study in which the membranes impregnated with ILs were used for the removal of volatile organic compounds (VOCs) like toluene, chloroform, and benzene. The aqueous solution containing 0.05% by weight of VOCs was permeated through the IL-modified membranes to selectively remove the VOCs from the solution. The ILs played a significant role as absorbents which selectively partition the target contaminants (VOCs) from the aqueous media. It was also observed that by increasing the amount of ILs and DMS content in the membranes the perm selectivity of the desired adulterants increased (by increasing the amount of DMS the hydrophobic nature of the membranes was enhanced, hence attracting the VOCs towards themselves and repelling the water away). After the synthesis of IL membranes, they were dried and submerged in a vessel containing 0.05 wt% benzene aqueous solution. The temperature of the system was maintained at 40 °C and the membranes were kept like this until equilibrium state was reached. After equilibrium was attained the membranes were removed from the vessel and were blotted with filter paper (at this stage the solution has absorbed into the membranes hence swelling them). This swollen membrane was subjected to low pressure in order to desorb the solution, which was later collected in a cold trap. The gas chromatographic analysis of the collected solution was performed in order to determine the concentration and composition of the collected VOCs. The benzene concentrations in permeate are greater as compared to those found in feed solution, indicating the high perm selectivity and benzene permeability.

The major benefit of using this membrane for the VOC separation was the enhanced perm selectivity of benzene without compromising its permeability. It was seen that the normalized permeation rate was almost 14 (10^{-7} kgm/m^{2}h) for [ABIM]TFSI membranes at 30% wt of ILs. The % weights of benzene, toluene, and chloroform in the permeate obtained after pervaporation through a membrane having 30% by weight ILs were found to be 42.3, 30.5, and 51.4%, respectively.

Mai et al. (2013) used pervaporation technique employing immobilized IL-PDMS membranes for the selective separation of acetone, ethanol, and butanol from fermentation broth. Here the immobilized IL-PDMS membranes were found to be more efficacious and possessed greater operational stability in the separation process as compared to the regular and common supported IL-PDMS membranes in which the ILs are absorbed physically on to the supported membranes; hence their long-term stability must be ameliorated. The [Omim][Tf_2N] (5.2% was the concentration of this IL in the membranes) was proved to be the unexampled IL for the recovery of butanol from water. This research is quite comprehensive in a sense that it presents a comparative analysis of permeate flux, enrichment factors, and nature of all three membranes, i.e., PDMS membrane, supported IL membranes, as well as immobilized IL membranes for the recovery of acetone, ethanol, and butanol from water. Enrichment factors of butanol in immobilized IL-PDMS membrane, supported IL-PDMS membrane, and PDMS membrane were found to be 5.26, 6.20, and 1.75, respectively. Moreover, the enrichment factor of butanol for immobilized IL membrane and supported IL membranes was found to be 3.0 and 3.5 times greater than that of simple PDMS membranes, respectively. Apart from that immobilized IL-PDMS membranes also depicted a greater permeate flux which ensured increased recovery of butanol by pervaporative separation; another benefit of using the IL-modified membranes (supported and immobilized) was that they could also enrich ethanol and acetone while the simple PDMS membranes lacked this feature.

P. Izák et al. (2009) used 1-ethenyl-3-ethyl-imidazolium hexafluorophosphate-PDMS (IL1-membranes) and tetrapropyl-ammonium tetracyanoborate-PDMS membranes (IL2-membranes) for the pervaporative separation of butan-1-ol from aqueous media. The experimental parameters employed are given in Table 9.1. It was observed that at low concentration of butan-1-ol in the feed solution the maximum enrichment factor was achieved which is a very unusual finding as compared to the rest of the studies on pervaporative separation of VOCs from water using IL membranes because typically it is higher flux with which low selectivity is observed. The reason behind this deviation from the normal trend was that high permeation rate was caused by a rise in the diffusion coefficient of butan-1-ol, because scientists believed that ILs embedded in PDMS membrane create small passages and channels via which the desired contaminant is conveyed from the feed to the permeate side, hence separating it from the aqueous media. And this is the reason as to why diffusion coefficient and permeation flux were found to be greater for IL membranes as compared to simple PDMS ones. It was concluded that IL2 membranes were found to be more efficient in the separation process as compared to the IL1 membranes, regardless of their miscibility deficiency with PDMS.

9.1.4.5 Ionic Liquid-Mediated Adsorption

The metals which have specific gravity greater than 5.0 are classified as heavy metals. But it must be noted that all heavy metals are not hazardous; for instance silver and gold are considered nontoxic heavy metals, while the metals like mercury, lead, chromium, cadmium, and arsenic are generally considered toxic heavy metals. On the other hand, there are heavy metals like copper, iron, and zinc which possess noted levels of toxicity but it is also known that these metals are available in our body in trace amounts. And these trace metals are required for the daily functioning of our body and are crucial to our survival. But overexposure to these heavy metals can cause serious health effects (Becker 2016). Nowadays heavy metal pollution of water is one of the most critical environmental issues. Heavy metals are known to contaminate water via natural as well as anthropogenic sources. Hence removal of heavy metals from water is the number one focus of scientists these days (Yadav et al. 2019).

Azo dyes form the biggest class of synthetic dyes that possess an azo group [N=N]/chromophore; hence it can be available in synthetic dyes accompanied with an aromatic ring (Rovina et al. 2016). Azo dyes comprise one of the most expansively used chromophores in the dye chemistry, and they generally show absorption in the range of 350–650 nm (Kim and Son 2011). Since azo dyes comprise almost 70% of all the dyes used in the dyeing industry their importance cannot be understated, but their extensive use comes with a drawback. If these dyes come in contact with human body and get absorbed, they get converted into carcinogenic amines via reduction decomposition reaction catalyzed by reductive enzymes in our body. Cancer-causing amines which are produced by azo dyes due to reduction decomposition are based on regulations in the European Union, and the German Government, via the German Goods Ordinance (ETAD, 1998) (Wakelyn 2007). As mentioned above, the azo dyes are carcingenicncerious. It is crucial to remove these harmful and detrimental compounds from water. The following paragraphs will elaborate the use of ILs in the removal of azo dyes from aqueous media.

Xing et al. (2016) used clinoptilolite zeolite as a support and modified it with the ionic liquid 1-hexadecyl-3-methylimidazolium chloride monohydrate to form the adsorbent for the extraction of methyl orange, an anionic dye, from aqueous media. The adsorption process was quite rapid and the solution pH was maintained between 2 and 11. The adsorption capacity was found to be 45 mmol/kg for ILCZ at 1.0 (external cation exchange capacity) ECEC and 116 mmol/kg for ILCZ at 2.0 ECEC. Gao et al. (2013) developed a special IL-based adsorbent for the elimination of two carcinogenic dyes: Orange II and Sunset Yellow. The hydroxyl-functionalized ionic liquid-based cross-linked polymer [PDVB-IL-OH] was used as an adsorbent to remove these two hazardous anionic dyes from water. The adsorbents were prepared by the method of refluxing. The thermic stability was determined by doing thermogravimetric analysis (TGA). The adsorption capacity for the removal of Orange II was 2095.80 mg/g and Sunset Yellow was 2100.98 mg/g.

Jiang et al. (2015) have devised a method to modify hydrogels by using ionic liquids and then employed these IL-modified hydrogels as adsorbents for the removal of hexavalent chromium from aqueous media. The ionic liquids used in the study were 1-allyl-3-methylimidazolium chloride. Copolymerization of monomers acrylamide (AM) and dimethyldiallylammonium chloride (DADMAC) was done to get the desired hydrogel. The adsorbents proved to be highly effective against the Cr (VI) impurity in water because the removal rate was 95.9% in just 10 min of contact time. Moreover, the highest adsorption capability for Cr (VI) ion removal from water was found to be 74.5 mg L^{-1}.

Jayachandra et al. (2016) made 1-methyl-3-(((3aS,5S,5aR,8aR,8bS)-2,2,7,7-tetramethyltetrahydro-3aH-bis([1,3]dioxolo)[4,5-b:4′,5′-d]pyran-5-yl)methyl)-1H-imidazol-3-ium hexafluorophosphate(V) ionic liquids from d-galactose and used them for the adsorptive removal of lead ions from water. Thermodynamic studies proved that this lead ion (Pb^{+2}) sequestering process in water is exothermic and spontaneous in nature. These ionic liquids were efficient for the complete sorption and hence elimination of lead ions from aqueous media. The maximum adsorption/loading capacity was found to be 374.9 mg/g, while the most economically beneficial feature of this study is that these ionic liquids could be reused about five times at an industrial level. To study the reusability and recyclability of these ILs various desorption-regeneration studies were performed.

Ismaiel et al. (2013) devised a unique method to eliminate toxic mercury from aqueous media. In this study trioctylmethylammonium thiosalicylate ionic liquids were embedded on to palm shell-activated carbon to develop task-specific ionic liquids which effectively scavenge hazardous and carcinogenic mercury ions from polluted water. The loading capability of these task-specific ionic liquids was found to be 83.33 mg/g, which proved the efficacy of these modified ILs in the Hg(II) removal. The optimum pH for this mercury ion adsorption process was 8, but there is one limitation of this method which was long contact time of about 3 h, because most of the other IL-based adsorbents that have been studied for the removal of dyes or metals from water have reaction time of 5–15 min. Apart from this shortcoming, this adsorbent is good in eliminating mercury ions from water.

Kumar et al. (2015) used a biopolymer "chitosan" as an adsorbent and modified it simply with trioctylmethylammonium chloride ionic liquid to obtain an efficient adsorbent for the removal of Pd(II) from water. The optimum pH was found to be 3.5–4.0, while the maximum adsorption capacity was 187.61 mg/g.

9.1.4.6 Miscellaneous

Total suspended solids (TSS) are described as solids found in water which can be trapped using a filter (Ismail et al. 2019). They are also defined as those solids which cannot pass via a 0.45 micron filter. Actually, TSS encompasses all the non-dissolved solid species and not just the dissolved solids that do not sink under the force of gravity. The standard test for the determination of TSS bears a value with the units of mg/L (Woodard and Curran 2006a). According to Simpson (2017) TSS is one of the most common water quality parameters. It is synonymous to "turbidity"; its units are formazin turbidity units (FTU), which are also called as formazin nephelometric units (FNU) by ISO. According to Simpson et al. since TSS remain suspended in the solution for immeasurable amount of time there is no method to accurately determine the TSS of any sample.

The dye cast facilities employ "wet scrubbers" to remove adulterants like cyanide, phenols, ammonia, magnesium, sulfide, copper, zinc, iron, and TSS. All these species are generally found in scrubber blowdown from various dye cast facilities and they are treated as a waste. Methods like coagulation, filtration, ion exchange, precipitation, and sedimentation are utilized to treat scrubber blowdown after the process of recycling at an industrial level (Woodard and Curran 2006b). Treatment of wastewater and sewage is done by acquiring purification with respect to chemical oxygen demand (COD), biochemical oxygen demand (BOD), ammonia, total phosphorous, total nitrogen, and total suspended solids (TSS). Membrane bioreactors “MBR” are availed to cater all these water quality parameters and to purify wastewater, sewage etc. (Singh 2015).

Flocculation

Flocculation is defined as a procedure in which destabilized species are bound to one another via van der Waal forces or hydrogen bonding to create a larger clump “flocs” (Ratnayaka et al. 2009). Flocculation is a very versatile method and can be used for pre-concentration and pre-harvesting of microalgae which can be used for production of biodiesel, treatment of wastewater and drinking water, enhancement of cell culture clarification, impurity clearing, papermaking systems, and single-use manufacturing procedures (Chatsungnoen and Chisti 2019) (Gupta et al. 2019; Mehta and Vedantham 2018) (Bajpai 2018) (Ali Ayati et al. 2019).

A novel use of ionic liquids as flocculants has been reported by Isik et al. (2016). In this study the polymeric ionic liquid (PLMS)-based nanostructures have been employed as flocculants to remove TSS from water. In this study polyelectrolytes like poly(1-vinyl-3-ethylimidazolium bromide) [PViEtImBr], poly(allylamine hydrochloride) [PAA HCl], and poly(diallyldimethyl ammonium chloride) [PDADMAC] were used as precursors to form “amphiphilic poly(ionic) liquid nanoparticles” via the process of self-assembly. The self-assemblage of nanoparticles occurs via “partial anion-exchange reaction” using a salt called bis(trifluorome thanesulfonimide) lithium [LiTFSI] in aqueous media. Furthermore, in the experimental portion the scientists used various concentrations of LiTFSI salt to get an idea about the most optimum and efficient crop of poly-ionic nanoparticles that can precisely precipitate out the TSS from sample solution. Hence, after using many concentrations like 1.25%, 2.5%, 5%, and 10% LiTFSI, it was found that the 10% concentration gives the desired, macroscopic precipitation. In conclusion it was found that poly(ionic) liquid nanoparticles yield rapid flocculation of TSS, due to greater nanoparticle sizes along with higher amount of purified water comparable to flocculation efficacies of the pristine polyelectrolytes. And poly-ViEtImBr-63K_x $TFSI_y$ as flocculating agent depicted the best efficacy among all the tested amphiphilic poly(ionic liquid) nanoparticles. And upon comparison of PViEtImBr-63 K_{90} $TFSI_{10}$ with the parent polyelectrolyte “PViEtImBr-63 K” used to clear impure water sample within 48 h, the system achieved the final state in which the pure water concentration was higher as well as a compact and dense sediment layer was seen settled at the bottom of the vial, therefore proving the efficacy and precision of poly(ionic liquid) nanoparticles in the flocculation-based removal of colloidal particles from water.

9.1.4.7 Recovery of Ionic Liquids and Future Prospectus

At present, the water purification is an indispensable necessity which may turn out to be more crucial within future couple of decades. The majority of traditional water purification techniques offer limited efficiency, and cause secondary environmental or waste management problem. The application of ionic liquid may provide a sustainable and green water purification choice. Ionic liquid-mediated solvent extraction falls among easily scalable choices but it may cause another water quality concern particularly for water-miscible ionic liquids. The supported ionic liquid membranes (SILMs) have been extensively investigated but stability and cost of such membranes have challenged its commercialization. ILs are widely used as solvent systems due to their exceptional ability to dissolve a broad range of compounds (Dahi et al. 2014; Ghandi 2014). Poly-ILs have an amazing application as they are used to prepare thermo-responsive polyelectrolytes (Kohno et al. 2015). As of recently, they have been extensively used in separation technology, chromatography, extraction, as well as membrane processes (Han and Armstrong 2007). As discussed above pervaporation mediated with ILs also has the potential to be the commercial choice for water purification.

9.2 Conclusion

Ionic liquids are revolutionizing the entire world; these offer multitude uses in almost every sphere of life. The most beneficial use however is the use of ionic liquids in the domain of water purification. Tunable structure of ionic liquids makes them a flexible option to remove any kind of adulterant from water just by changing the cationic or anionic specie of the ILs. Either fabricated ionic liquids can be used directly for water purification or ionic liquids may be embedded in membranes for convenient and efficient water purification. In my opinion ionic liquids must be coupled with a membrane to improve the water purification efficacy. A valuable use of ionic liquids is that they can be employed for the removal of toxic heavy metals from water. Moreover, ILs may also be utilized to eliminate total suspended solids, volatile organic compounds, and heavy metals.

Acknowledgements I, Syeda Mariam Hasany, like to acknowledge the great deal of support and motivation offered by my family members, especially my mother Mrs. Samreen Shoaib, for providing me the moral support and assistance in carrying out this project.

References

Akbari F, Almutairi FD, Alavianmehr MM (2019) Solubility of gases in ionic liquids using PHTC equation of state. J Mol Liq 276:553–561. https://doi.org/10.1016/j.molliq.2018.11.151

Ali Ayati SR, Tanhaei B, Sillanpää M (2019) Ionic liquid-modified composites for the adsorptive removal of emerging water contaminants: a review. J Mol Liq:71–83. https://doi.org/10.1016/j.molliq.2018.11.016

Amde M, Liu J-F, Pang L (2015) Environmental application, fate, effects, and concerns of ionic liquids: a review. Environ Sci Technol 49(21):12611–12627. https://doi.org/10.1021/acs.est.5b03123

Bajpai P (2018) Chapter 10 - papermaking chemistry. In: Bajpai P (ed) Biermann's handbook of pulp and paper, 3rd edn. Elsevier, Amsterdam, pp 207–236

Baker SN, Baker GA, Bright FV (2002) Temperature-dependent microscopic solvent properties of 'dry' and 'wet' 1-butyl-3-methylimidazolium hexafluorophosphate: correlation with (30) and Kamlet–Taft polarity scales. Green Chem 4(2):165–169. https://doi.org/10.1039/B111285F

Becker D (2016) 44 - evolution of pigments and dyes. In: Becker D (ed) Color trends and selection for product design. William Andrew Publishing, Burlington, pp 213–222

Böddeker KW, Gatfield IL, Jähnig J, Schorm C (1997) Pervaporation at the vapor pressure limit: vanillin. J Membr Sci 137(1):155–158. https://doi.org/10.1016/S0376-7388(97)00187-7

Brinda Lakshmi A, Balasubramanian A, Venkatesan S (2013) Extraction of phenol and chlorophenols using ionic liquid [Bmim]+[BF4]− dissolved in Tributyl phosphate. Clean (Weinh) 41(4):349–355. https://doi.org/10.1002/clen.201100632

Chatsungnoen T, Chisti Y (2019) Chapter 11 - flocculation and electroflocculation for algal biomass recovery. In: Pandey A, Chang J-S, Soccol CR, Lee D-J, Chisti Y (eds) Biofuels from algae, 2nd edn. Elsevier, Amsterdam, pp 257–286

Chaudhary S, Kaur Y, Umar A, Chaudhary GR (2016) Ionic liquid and surfactant functionalized ZnO nanoadsorbent for recyclable proficient adsorption of toxic dyes from waste water. J Mol Liq 224:1294–1304. https://doi.org/10.1016/j.molliq.2016.10.116

Chen X, Li F, Asumana C, Yu G (2013) Extraction of soluble dyes from aqueous solutions with quaternary ammonium-based ionic liquids. Sep Purif Technol 106:105–109. https://doi.org/10.1016/j.seppur.2013.01.002

Dahi A, Fatyeyeva K, Langevin D, Chappey C, Rogalsky SP, Tarasyuk OP et al (2014) Supported ionic liquid membranes for water and volatile organic compounds separation: sorption and permeation properties. J Membr Sci 458:164–178. https://doi.org/10.1016/j.memsci.2014.01.031

Dahi A, Fatyeyeva K, Langevin D, Chappey C, Poncin-Epaillard F, Marais S (2017) Effect of cold plasma surface treatment on the properties of supported ionic liquid membranes. Sep Purif Technol 187:127–136. https://doi.org/10.1016/j.seppur.2017.05.055

Dai Z, Noble RD, Gin DL, Zhang X, Deng L (2016) Combination of ionic liquids with membrane technology: a new approach for CO2 separation. J Membr Sci 497:1–20. https://doi.org/10.1016/j.memsci.2015.08.060

Egorov VM, Smirnova SV, Pletnev IV (2008) Highly efficient extraction of phenols and aromatic amines into novel ionic liquids incorporating quaternary ammonium cation. Sep Purif Technol 63(3):710–715. https://doi.org/10.1016/j.seppur.2008.06.024

Elhamifar D, Shojaeipoor F, Roosta M (2016) Self-assembled ionic-liquid based organosilica (SAILBO) as a novel and powerful adsorbent for removal of malachite green from aqueous solution. J Taiwan Inst Chem Eng 59:267–274. https://doi.org/10.1016/j.jtice.2015.07.036

Endres F, Zein El Abedin S (2006) Air and water stable ionic liquids in physical chemistry. Phys Chem Chem Phys 8(18):2101–2116. https://doi.org/10.1039/B600519P

Fan J, Fan Y, Pei Y, Wu K, Wang J, Fan M (2008) Solvent extraction of selected endocrine-disrupting phenols using ionic liquids. Sep Purif Technol 61(3):324–331. https://doi.org/10.1016/j.seppur.2007.11.005

Gao H, Wang Y, Zheng L (2013) Hydroxyl-functionalized ionic liquid-based cross-linked polymer as highly efficient adsorbent for anionic azo dyes removal. Chem Eng J 234:372–379. https://doi.org/10.1016/j.cej.2013.08.078

Ghandi K (2014) A review of ionic liquids, their limits and applications. Green Sust Chem 4:44–53. https://doi.org/10.4236/gsc.2014.41008

Gupta SK, Dhandayuthapani K, Ansari FA (2019) Chapter 19 - techno-economic perspectives of bioremediation of wastewater, dewatering, and biofuel production from microalgae: an overview. In: Pandey VC, Bauddh K (eds) Phytomanagement of polluted sites. Elsevier, Amsterdam, pp 471–499

Han X, Armstrong DW (2007) Ionic liquids in separations. Acc Chem Res 40(11):1079–1086. https://doi.org/10.1021/ar700044y

Heitmann S, Krings J, Kreis P, Lennert A, Pitner WR, Górak A, Schulte MM (2012) Recovery of n-butanol using ionic liquid-based pervaporation membranes. Sep Purif Technol 97:108–114. https://doi.org/10.1016/j.seppur.2011.12.033

Hernández-Fernández FJ, de los Ríos AP, Tomás-Alonso F, Gómez D, Víllora G (2009a) Improvement in the separation efficiency of transesterification reaction compounds by the use of supported ionic liquid membranes based on the dicyanamide anion. Desalination 244(1):122–129. https://doi.org/10.1016/j.desal.2008.04.041

Hernández-Fernández FJ, de los Ríos AP, Tomás-Alonso F, Palacios JM, Víllora G (2009b) Preparation of supported ionic liquid membranes: influence of the ionic liquid immobilization method on their operational stability. J Membr Sci 341(1):172–177. https://doi.org/10.1016/j.memsci.2009.06.003

Isik M, Fernandes AM, Vijayakrishna K, Paulis M, Mecerreyes D (2016) Preparation of poly(ionic liquid) nanoparticles and their novel application as flocculants for water purification. Polym Chem 7(8):1668–1674. https://doi.org/10.1039/C5PY02001H

Ismaiel AA, Aroua MK, Yusoff R (2013) Palm shell activated carbon impregnated with task-specific ionic-liquids as a novel adsorbent for the removal of mercury from contaminated water. Chem Eng J 225:306–314. https://doi.org/10.1016/j.cej.2013.03.082

Ismail AF, Khulbe KC, Matsuura T (2019) Chapter 8 - RO membrane fouling. In: Ismail AF, Khulbe KC, Matsuura T (eds) Reverse Osmosis. Elsevier, Amsterdam, pp 189–220

Izák P, Ruth W, Fei Z, Dyson PJ, Kragl U (2008) Selective removal of acetone and butan-1-ol from water with supported ionic liquid–polydimethylsiloxane membrane by pervaporation. Chem Eng J 139(2):318–321. https://doi.org/10.1016/j.cej.2007.08.001

Izák P, Friess K, Hynek V, Ruth W, Fei Z, Dyson JP, Kragl U (2009) Separation properties of supported ionic liquid–polydimethylsiloxane membrane in pervaporation process. Desalination 241(1):182–187. https://doi.org/10.1016/j.desal.2007.12.050

Jayachandra R, Lakshmipathy R, Reddy SR (2016) Hydrophobic d-galactose based ionic liquid for the sequestration of Pb2+ ions from aqueous solution. J Mol Liq 219:1172–1178. https://doi.org/10.1016/j.molliq.2015.11.060

Jiang Y, Li F, Ding G, Chen Y, Liu Y, Hong Y et al (2015) Synthesis of a novel ionic liquid modified copolymer hydrogel and its rapid removal of Cr (VI) from aqueous solution. J Colloid Interface Sci 455:125–133. https://doi.org/10.1016/j.jcis.2015.05.030

Jordan A, Gathergood N (2015) Biodegradation of ionic liquids – a critical review. Chem Soc Rev 44(22):8200–8237. https://doi.org/10.1039/C5CS00444F

Karousos DS, Labropoulos AI, Tzialla O, Papadokostaki K, Gjoka M, Stefanopoulos KL et al (2018) Effect of a cyclic heating process on the CO2/N2 separation performance and structure of a ceramic nanoporous membrane supporting the ionic liquid 1-methyl-3-octylimidazolium tricyanomethanide. Sep Purif Technol 200:11–22. https://doi.org/10.1016/j.seppur.2018.02.013

Kermanioryani M, Abdul Mutalib MI, Gonfa G, El-Harbawi M, Mazlan FA, Lethesh KC, Leveque J-M (2016) Using tunability of ionic liquids to remove methylene blue from aqueous solution. J Environ Chem Eng 4(2):2327–2332. https://doi.org/10.1016/j.jece.2016.04.008

Khachatryan KS, Smirnova SV, Torocheshnikova II, Shvedene NV, Formanovsky AA, Pletnev IV (2005) Solvent extraction and extraction–voltammetric determination of phenols using room temperature ionic liquid. Anal Bioanal Chem 381(2):464–470. https://doi.org/10.1007/s00216-004-2872-y

Kim SH, Son YA (2011) 18 - near-infrared dyes. In: Clark M (ed) Handbook of textile and industrial dyeing, vol 1. Woodhead Publishing, Sawston, pp 588–603

Kohno Y, Saita S, Men Y, Yuan J, Ohno H (2015) Thermoresponsive polyelectrolytes derived from ionic liquids. Polym Chem 6(12):2163–2178. https://doi.org/10.1039/C4PY01665C

Kumar ASK, Sharma S, Reddy RS, Barathi M, Rajesh N (2015) Comprehending the interaction between chitosan and ionic liquid for the adsorption of palladium. Int J Biol Macromol 72:633–639. https://doi.org/10.1016/j.ijbiomac.2014.09.002

Li D, Yao J, Sun H, Liu B, van Agtmaal S, Feng C (2018) Recycling of phenol from aqueous solutions by pervaporation with ZSM-5/PDMS/PVDF hollow fiber composite membrane. Appl Surf Sci 427:288–297. https://doi.org/10.1016/j.apsusc.2017.08.202

Liu X, Yao X, Wang Y, Zhang S (2019) Mesoscale structures and mechanisms in ionic liquids. Particuology. https://doi.org/10.1016/j.partic.2018.08.003

Lu J, Yan F, Texter J (2009) Advanced applications of ionic liquids in polymer science. Prog Polym Sci 34(5):431–448. https://doi.org/10.1016/j.progpolymsci.2008.12.001

Ma Y, Liu Y, Su H, Wang L, Zhang J (2018) Relationship between hydrogen bond and viscosity for a series of pyridinium ionic liquids: molecular dynamics and quantum chemistry. J Mol Liq 255:176–184. https://doi.org/10.1016/j.molliq.2018.01.121

Mai NL, Kim SH, Ha SH, Shin HS, Koo Y-M (2013) Selective recovery of acetone-butanol-ethanol from aqueous mixture by pervaporation using immobilized ionic liquid polydimethylsiloxane membrane. Korean J Chem Eng 30(9):1804–1809. https://doi.org/10.1007/s11814-013-0116-6

McFarlane J, Ridenour WB, Luo H, Hunt RD, DePaoli DW, Ren RX (2005) Room temperature ionic liquids for separating organics from produced water. Sep Sci Technol 40(6):1245–1265. https://doi.org/10.1081/SS-200052807

McGregor IR, JGW, Hamza AA, Nye ME (2001)

McKeen LW (2012) 4 - markets and applications for films, containers, and membranes. In: McKeen LW (ed) Permeability properties of plastics and elastomers, 3rd edn. William Andrew Publishing, Oxford, pp 59–75

Mehta KK, Vedantham G (2018) Chapter 39 - next-generation process design for monoclonal antibody purification. In: Jagschies G, Lindskog E, Łącki K, Galliher P (eds) Biopharmaceutical processing. Elsevier, Amsterdam, pp 793–811

Noble RD, Gin DL (2011) Perspective on ionic liquids and ionic liquid membranes. J Membr Sci 369(1):1–4. https://doi.org/10.1016/j.memsci.2010.11.075

Nulwala H, Mirjafari A, Zhou X (2018) Ionic liquids and poly(ionic liquid)s for 3D printing – a focused mini-review. Eur Polym J 108:390–398. https://doi.org/10.1016/j.eurpolymj.2018.09.023

Ratnayaka DD, Brandt MJ, Johnson KM (2009) Chapter 7 - storage, clarification and chemical treatment. In: Ratnayaka DD, Brandt MJ, Johnson KM (eds) Water supply, 6th edn. Butterworth-Heinemann, Boston, pp 267–314

Rovina K, Prabakaran PP, Siddiquee S, Shaarani SM (2016) Methods for the analysis of sunset yellow FCF (E110) in food and beverage products - a review. TrAC Trends Anal Chem 85:47–56. https://doi.org/10.1016/j.trac.2016.05.009

Schubert JK, Miekisch W (2013) Chapter 9 - breath analysis in critically ill patients—potential and limitations. In: Amann A, Smith D (eds) Volatile Biomarkers. Elsevier, Boston, pp 155–176

Shalu, Chaurasia SK, Singh RK, Chandra S (2013) Thermal stability, complexing behavior, and ionic transport of polymeric gel membranes based on polymer PVdF-HFP and ionic liquid, [BMIM][BF4]. J Phys Chem B 117(3):897–906. https://doi.org/10.1021/jp307694q

Sheldon RA (2012) Fundamentals of green chemistry: efficiency in reaction design. Chem Soc Rev 41(4):1437–1451. https://doi.org/10.1039/C1CS15219J

Simpson DA (2017) Chapter seven - water collection and disposal. In: Simpson DA (ed) Practical onshore gas field engineering. Gulf Professional Publishing, Houston, pp 461–511

Singh R (2015) Chapter 3 - hybrid membrane systems – applications and case studies. In: Singh R (ed) Membrane technology and engineering for water purification, 2nd edn. Butterworth-Heinemann, Oxford, pp 179–281

Stepnowski P (2007) Chapter 16 - sorption, lipophilicity and partitioning phenomena of ionic liquids in environmental systems. In: Letcher TM (ed) Thermodynamics, solubility and environmental issues. Elsevier, Amsterdam, pp 299–313

Talbi Z, Haddou B, Ghouas H, Kameche M, Derriche Z, Gourdon C (2014) Cationic dye removal from aqueous solutions using ionic liquid and nonionic surfactant-ionic liquid systems: a comparative study based upon experimental design. Chem Eng Commun 201(1):41–52. https://doi.org/10.1080/00986445.2012.759563

Thurston GD (2017) Outdoor air pollution: sources, atmospheric transport, and human health effects. In: Quah SR (ed) International encyclopedia of public health, 2nd edn. Academic Press, Oxford, pp 367–377

Tisserand R, Young R (2014) 6 - the respiratory system. In: Tisserand R, Young R (eds) Essential Oil Safety, 2nd edn. Churchill Livingstone, St. Louis, pp 99–110

Uragami T, Fukuyama E, Miyata T (2016a) Selective removal of dilute benzene from water by poly(methyl methacrylate)-graft-poly(dimethylsiloxane) membranes containing hydrophobic ionic liquid by pervaporation. J Membr Sci 510:131–140. https://doi.org/10.1016/j.memsci.2016.01.057

Uragami T, Matsuoka Y, Miyata T (2016b) Permeation and separation characteristics in removal of dilute volatile organic compounds from aqueous solutions through copolymer membranes consisted of poly(styrene) and poly(dimethylsiloxane) containing a hydrophobic ionic liquid by pervaporation. J Membr Sci 506:109–118. https://doi.org/10.1016/j.memsci.2016.01.031

Wakelyn PJ (2007) 14 - health and safety issues in cotton production and processing. In: Gordon S, Hsieh YL (eds) Cotton. Woodhead Publishing, Sawston, pp 460–483

Wang J, Luo J, Feng S, Li H, Wan Y, Zhang X (2016) Recent development of ionic liquid membranes. Green Energy & Environment 1(1):43–61. https://doi.org/10.1016/j.gee.2016.05.002

WHO (2019) Drinking-water: key facts. World Health Organization 14 June 2019, Accessed 26 July 2019

Woodard, Curran I (2006a) 5 - waste characterization. In: Woodard, Curran I (eds) Industrial waste treatment handbook, 2nd edn. Butterworth-Heinemann, Burlington, pp 83–126

Woodard, Curran I (2006b) 10 - wastes from industries (case studies). In: Woodard, Curran I (eds) Industrial waste treatment handbook, 2nd edn. Butterworth-Heinemann, Burlington, pp 409–496

Xing X, Chang P-H, Lv G, Jiang W-T, Jean J-S, Liao L, Li Z (2016) Ionic-liquid-crafted zeolite for the removal of anionic dye methyl orange. J Taiwan Inst Chem Eng 59:237–243. https://doi.org/10.1016/j.jtice.2015.07.026

Yadav M, Gupta R, Sharma RK (2019) Chapter 14 - green and sustainable pathways for wastewater purification. In: Ahuja S (ed) Advances in water purification techniques. Elsevier, Amsterdam, pp 355–383

Yan X, Anguille S, Bendahan M, Moulin P (2019) Ionic liquids combined with membrane separation processes: a review. Sep Purif Technol 222:230–253. https://doi.org/10.1016/j.seppur.2019.03.103

Zhang D, Deng Y, Chen J (2010) Enrichment of aromatic compounds using ionic liquid and ionic liquid-based aqueous biphasic systems. Sep Sci Technol 45(5):663–669. https://doi.org/10.1080/01496390903571119

Zhou H, Lv L, Liu G, Jin W, Xing W (2014) PDMS/PVDF composite pervaporation membrane for the separation of dimethyl carbonate from a methanol solution. J Membr Sci 471:47–55. https://doi.org/10.1016/j.memsci.2014.07.068

Zhou Q, Chen F, Wu W, Bu R, Li W, Yang F (2016) Reactive orange 5 removal from aqueous solution using hydroxyl ammonium ionic liquids/layered double hydroxides intercalation composites. Chem Eng J 285:198–206. https://doi.org/10.1016/j.cej.2015.10.004

Chapter 10
Electrical Double-Layer Structure and Property of Ionic Liquid-Electrode System for Electrochemical Applications

Guocai Tian

Contents

G. Tian (✉)
State Key Laboratory of Complex Nonferrous Metal Resources Clean Utilization, Faculty of Metallurgical and Energy Engineering, Kunming University of Science and Technolgy, Kunming, Yunnan Province, China
e-mail: tiangc@163.com; tiangc@iccas.ac.cn

Inamuddin, A. M. Asiri (eds.), *Nanotechnology-Based Industrial Applications of Ionic Liquids*, Nanotechnology in the Life Sciences,
https://doi.org/10.1007/978-3-030-44995-7_10

Abbreviations

[BMA]PF_6	Butyltrimethylammonium hexafluorophosphate
[BMA]TFSI	Butyltrimethylammonium bis(trifluoromethyl-sulfonyl)imide
[BMIM]BF_4	1-*n*-Butyl-3-methylimidazolium tetrafluoroborate
[BMIM]Br	1-*n*-Butyl-3-methylimidazolium bromide
[BMIM]CH_3COO	1-*n*-Butyl-3-methylimidazolium acetate
[BMIM]CH_3SO_4	1-n-Butyl-3-methylimidazolium methylsulfate
[BMIM]Cl	1-*n*-Butyl-3-methylimidazolium chloride
[BMIM]DCA	1-*n*-Butyl-3-methylimidazolium dicyandiamide
[BMIM]FAP	1-*n*-Butyl-3-methylimidazolium tris(pentafluoroethyl) trifluorophosphate
[BMIM]FSI	1-*n*-Butyl-3-methylimidazolium bis (fluorosulfonyl) imide
[BMIM]I	1-*n*-Butyl-3-methylimidazolium Iodide
[BMIM]$N(CN)_2$	1-*n*-Butyl-3-methylimidazolium dicyanamide
[BMIM]NO_3	1-Butyl 3-methylimidazolium nitrate
[BMIM]OTF	1-*n*-Butyl-3-methylimidazolium trifluoromethane-sulfonate
[BMIM]PF_6	1-*n*-Butyl-3-methylimidazolium hexafluorophosphate
[BMIM]TF_2N	1-*n*-Butyl-3-methylimidazolium bis(trifluoromethylsulfonyl) amide
[BMIM]TFSA	1-*n*-Butyl-3-methylimidazolium trifluoromethane-sulfonamide
[BMIM]TFSI	1-*n*-Butyl-3-methylimidazolium bis(trifluoromethylsulfonyl)imide
[BMP]TFSA	1-Butyl-1-methylpyrrolidinium trifluoromethane-sulfonamide
[$C_6(MIM)_2$]$(TF_2N)_2$	1-Hexyl-3-dimethylimidazolium di[bis(trifluoro-methyl)imide]
[C_6(MIM)TMA)]$(Tf_2N)_2$	1-(1-Trimethylammonium-yl-hexyl)-3-methylimidazolium di[bis(trifluoromethane-sulfonyl)-imide]
[C_6MIM]BF_4 or [HMIM]BF_4	1-Hexyl-3-methylimidazolium tetrafluoroborate
[$C_n(MIM)_2$]$(BF_4)_2$	1-Alkyl-3-dimethylimidazoliumdi[tetrafluoroborate]
[$C_n(MIM)_2$]$(TF_2N)_2$	1-Alkyl-3-dimethylimidazolium di[bis(trifluoro-methyl)imide]
[C_nMIM]FSI	1-Alkyl-3-methylimidazolium bis(fluorosulfonyl) imide
[C_nMIM]TFSI	1-Alkyl-3-methylimidazolium bis(trifluoromethyl-sulfonyl)imide
[DEME]PF_6	N,N-diethyl-N-methyl-N-(2-methoxyethyl)ammonium hexafluorophosphate

[DMIM]Cl	1,3-Dimethyl-2-imidazolidinone chloride
[DMIM]PF_6	1, 3-Dimethylimidazolium chloride hexafluorophosphate
[EMIM]$B(CN)_2$	1-Ethyl-3-methylimidazolium dicyanoborate
[EMIM]BF_4	1-Ethyl-3-methylimidazolium tetrafluoroborate
[EMIM]$C(CN)_2$	1-Ethyl-3-methylimidazolium dicyanomethanide
[EMIM]DCA	1-Ethyl-3-methylimidazolium dicyandiamide
[EMIM]FAP	1-Ethyl-3-methylimidazolium tris (pentafluoroethyl) trifluorophosphate
[EMIM]GLY	1-Ethyl-3-methylimidazolium aminoethanic acid
[EMIM]I	1-Ethyl-3-methylimidazolium Iodide
[EMIM]$N(CN)_2$	1-Ethyl-3-methylimidazolium dicyanamide
[EMIM]OTF	1-Ethyl-3-methylimidazolium trifluoromethanesulfonate
[EMIM]PF_6	1-Ethyl-3-methylimidazolium hexafluorophosphate
[EMIM]PHE	1-Ethyl-3-methylimidazolium phenylalanine
[EMIM]SER	1-Ethyl-3-methylimidazolium serine
[EMIM]TF_2N	1-Ethyl-3-methylimidazolium bis(trifluoromethylsulfonyl)amide
[EMIM]TFSA	1-Ethyl-3-methylimidazolium trifluoromethanesulfonamide
[EMIM]TFSI	1-Ethyl-3-methylimidazolium bis(trifluoromethylsulfonyl)imide
[EMIM]VAL	1-Ethyl-3-methylimidazolium valine
[EMMIM]TFSI	1-Ethyl-2,3-dimethylimidazolium bis(trifluoromethylsulfonyl)imide
[$EtNH_3$]NO_3	Ethylammonium nitrate
[Et_4N]BF_4	Tetraethylammonium tetrafluoroborate
[Li(G_3)]TFSI	Li(triethylene glycol dimethyl ether) bis(trifluoromethanesulfonyl)imide
[N_{2228}]TF_2N	Triethyloctylammonium bis(trifluoromethylsulfonyl) amide
[OMIM]OTF	1-Octyl-3-methylimidazolium trifluoromethanesulfonate
[OMIM]PF_6 or [C_8MIM]PF_6	1-Octyl-3-methylimidazolium hexafluorophosphate
[OMIM]TFSA	1-Octyl-3-methylimidazolium trifluoromethanesulfonamide
[$Pip_{1,4}$]$B(CN)_4$	N-methyl-N-butyl piperidinium tetracyanoborate
[PMIM]I	1-Pentyl-3-methylimidazolium iodide
[PMPy]TF_2N	3-Methyl-1-propylpyridinium bis(trifluoromethylsulfonyl) amide
[$Py_{1,4}$]FAP	1-Butyl-1-methyl-pyrrolidinium tris (pentafluoroethyl) trifluorophosphate

$[Py_{1,4}]TF_2N$	1-Butyl-1-methyl-pyrrolidinium bis(trifluoromethyl-sulfonyl)amide
$[Pyr]PF_6$	1-Methyl-1- butylpyrrolidinium hexafluorophosphate
$[Pyr_{12}O_1]DCA$	1-Methoxy-1-ethyl-1-methylpyrrolidinium dicyandiamide
$[Pyr_{13}]DCA$	1-Methyl-1-propylpyrrolidinium dicyandiamide
$[Pyr_{14}]B(CN)_4$	1-Methyl-1-butylpyrrolidinium tetracyanoborate
$[Pyr_{14}]BF_4$	1-Methyl-1-butylpyrrolidinium tetrafluoroborate
$[Pyr_{14}]DCA$	1-Methyl-1-butylpyrrolidinium dicyandiamide
$[Pyr_{14}]TF_2N$	1-Methyl-1-butylpyrrolidinium bis(trifluoromethyl-sulfonyl) imide
$[Pyr_{14}]TFSI$	1-Methyl-1-butylpyrrolidinium bis(trifluoromethyl-sulfonyl)imide
$[PyrH_4]TFSI$	N-butyl-pyrrolidinium bis(trifluoromethylsulfonyl) imide
$[TEA]BF_4$	Tetraethylammonium tetrafluoroborate

10.1 Introduction

The structure and properties of the electrode-solution interface are important in understanding many phenomena and processes in material, metallurgical, chemical, physical, and biological sciences. Usually, the reaction always takes place at the solution-electrode interface in the electrochemical process. Therefore, the structure and properties of the interface will directly affect the whole electrochemical process (Zhang 2006; Lockett et al. 2008; Li 2008; Zha 2002). On the one hand, the intensity of interfacial electric field has a great influence on the activation energy of electrochemical reaction. The interface electric field is caused by the double-layer structure of the electrode-solution interface. The electric field directly affects the charge transfer, and can effectively and continuously change the electrode reaction rate by modulating the electrode potential. Because the distance between charges with opposite symbols in the double layer is very small, a strong electric field can be generated up to about 10^9 V/m. Such a huge electric field will exert a great force on the charge, and the reaction speed of the electrode may change greatly; even some difficult chemical reactions can be carried out accordingly. On the other hand, the properties of electrolyte and the types and surface states of electrode materials used in the experiment have significant effects on the electrode reaction. Electrode-solution interface is the place where electrons exchange and is the "objective environment" for realizing interface reaction. Therefore, it is very important to study the structure, dynamics, and properties of the solution-electrode interface.

When electrodes contact with electrolyte solution, there are great differences in physicochemical properties between them. The species on the interface of electrode-solution are subject to both internal forces of solution and forces of electrodes, and eventually there is a certain distribution on the interface of electrodes, thus forming a stable boundary denoted as surface double-layer structure. The interfacial double-layer phenomenon is not only unique to the metal electrode-solution interface, but also prevalent at other two-phase interfaces. For aqueous solution systems, lots of research works on the interface structure of electrode-solutions have been performed and reported after Helmholtz put forward the concept of double layer in the nineteenth century (Parsons 1990; Helmholtz 1879; Gouy 1910; Chapman 1913; Stern 1924; Grahame 1946; Grahame 1954). The theoretical double-layer models have been gradually developed, perfected, and further revised by Gouy, Chapman, Stern, and other scholars, who provided good explanations for the experiment results obtained in that time (Gouy 1910; Chapman 1913; Stern 1924; Grahame 1946; Grahame 1954). With the rapid development of experimental technology, electrochemical research on the interface structure, properties, and effects at the molecular level has been gradually performed. The double-layer theory and electrochemical dynamics have been further developed by using electrochemical methods, as well as high time, spatial, and energy resolution technique, such as electrochemical transient technology, alternating current impedance method linear potential scanning method, rotating disk electrode system, high-energy resolution spectroscopy, and high-space-resolution scanning probe microscopy (Zhang 2006; Lockett et al. 2008; Zha 2002).

Ionic liquid is a new kind of green solvent and electrolyte which emerged in the 1990s (Welton 2002). It has many advantages such as nonvolatility, good thermal stability, wide electrochemical window, relatively high conductivity and ionic mobility, wide liquid range, and low toxicity (Li 2004; Zhang et al. 2018; Deng 2006). It has been successfully and widely used in catalysis, synthesis, electrochemical research such as metal electrodeposition, batteries, and supercapacitors (Li 2004; Abbott and McKenzie 2006; Deng 2006; Zhang et al. 2018; Tian et al. 2009; Tian et al. 2010; Tian 2012). The research on the structure dynamics and property of electric double layer at the ionic liquid-electrode interface has always been a hot topic in scientific research because of the successful application of ionic liquids in metal electrodeposition, batteries, fuel cell, and supercapacitors.

In the liquid state, ionic liquids are completely composed of pure ions, which is different from the traditional aqueous or organic solvents. Therefore, deeper insight into the interfacial structure and property of ionic liquid-metal electrode is highly desired because it is beyond the description of classical electrochemical theory for the dilute aqueous electrolyte-metal electrode system (Armand et al. 2009). In the past decades, more and more theoretical and experimental methods have been used in this field, and great progress has been made (He et al. 2018; Burt et al. 2014; Hayes et al. 2015; Fedorov and Kornyshev 2014). In this chapter, the recent research progress and perspectives in theoretical and experimental studies of the structure, dynamics, and property of double layer of the ionic liquid-electrode interface are reviewed. The brief introduction of ionic liquids and its properties, basic theoretical model of electrical double layer, and theoretical and experimental studies of the structure dynamic and property of double layer of the ionic liquid-electrode interface are reviewed and discussed. Further directions of research are also presented.

10.2 Brief Introduction of Ionic Liquids

10.2.1 Ionic Liquids

Ionic liquids, also known as room-temperature molten salts or room-temperature ionic liquids, are a kind of liquid salts composed of organic cations and organic or inorganic anions at room temperature (Welton 2002; Li 2004; Zhang et al. 2018; Deng 2006). It is evolved from high-temperature molten salts, while traditional molten salts generally have the characteristics of high viscosity, high melting point, and strong corrosiveness. The distinct difference between ionic liquids and traditional molten salts is that their melting point is low and they are liquid in a wide temperature range near room temperature.

The earliest discoveries of ionic liquids can be traced back to 1914. Walden reported the synthesis of ethylamine nitrate which was the first liquid organic salt at room temperature with a melting point of 12 °C (Walden 1914). However, because of its instability in the air and explosion proneness, it was not paid attention to at that time. Until 1948, Hurley and Weir's group mixed $AlCl_3$ with ethyl pyridine halide to produce a colorless transparent liquid (Hurley and Wier 1951a; Hurley and Wier 1951b). This accidental discovery pioneered the first generation of ionic liquids. The properties of aluminochlorate-type ionic liquids are controlled by the ratio of organic halogenated salts and $AlCl_3$, but the decomposition of $AlCl_3$ by superabsorbency greatly limits its application. To overcome the shortcomings of ionic liquids absorbing water, in 1992 Wilkes's group synthesized dialkyl-imidazole tetrafluoroboron ionic liquids which were stable in water and air (Wilkes and Zaworotko 1992). Since then, a large number of dialkyl-imidazole ionic liquids with various anions such as tartaric acid, acetic acid, $CF_3SO_3^-$, PF_6^-, and PO_4^{3-} have appeared which greatly expanded the application of ionic liquids. Based on the research work of Wilkes et al., it is gradually recognized that ionic liquids are not confined to chloroaluminate-type ionic liquids (Wilkes and Zaworotko 1992). Since then, the preparation and application of ionic liquids have entered an era of rapid development. The combination of different cations and anions and the introduction of chirality can produce different kinds of functional ionic liquids with different properties.

There are many kinds of ionic liquids because different combinations of anions and cations can produce different types of ionic liquids (Li 2004; Zhang et al. 2018; Deng 2006). For a specific performance or application, the task-specific ionic liquids can be formed by selecting specific anions and cations. Ionic liquids can be classified according to their composition or properties. The ionic liquids can be divided into alkyl imidazoles, alkyl pyridines, quaternary ammonium salts, quaternary phosphorus salts, pyrrole, and terbium salts according to their different cationic structures. Ionic liquids can also be mainly classified into two categories according to their different anionic structures: (1) Polynuclear metal anions such as

$Al_2Cl_7^-$, $Al_3Cl_{10}^-$, $Cu_2Cl_3^-$, and $Fe_2Cl_7^-$: These anions are made of corresponding acids, which are not stable in water and air. (2) Mononuclear nonmetallic anions such as HSO_4^-, CH_3COO^-, NO_3^-, NO_2^-, BF_4^-, PF_6^-, $N(CF_3SO_2)_2^-$, and $CH_3SO_3^-$, which are alkaline or neutral anions. Ionic liquids can be divided into hydrophilicity and hydrophobicity according to their properties such as solubility in water. According to the acidity and basicity of ionic liquids, ionic liquids can be divided into Lewis acidity, Brønsted acidity, Lewis alkalinity, Brønsted alkalinity, and neutral ionic liquids. Table 10.1 gives the various common cations and anions of ionic liquids.

Since ionic liquids have various unique properties, the application of ionic liquids has expanded from organic synthesis to energy, resources, environment, materials, and aerospace. The theoretical research of ionic liquids has also begun to go from the structure-activity relationship of basic physical properties to the exploration of scientific essence and basic theoretical issues, and has rapidly become one of the frontier areas of emerging interdisciplinary disciplines. The research and application of ionic liquids are increasing year by year, and their results have been published in authoritative journals including Science, Nature, Physical Review Letters, and The Journal of American Chemical Society and important interdisciplinary journals. Figure 10.1 gives the published literature statistics of ionic liquids in recent years, which have been collected from Web of Science with the keywords of "ionic liquids."

Table 10.1 Common cations and anions for forming ionic liquids (Li 2004; Zhang et al. 2018; Deng 2006)

Types	Structure or chemical formula
Cations	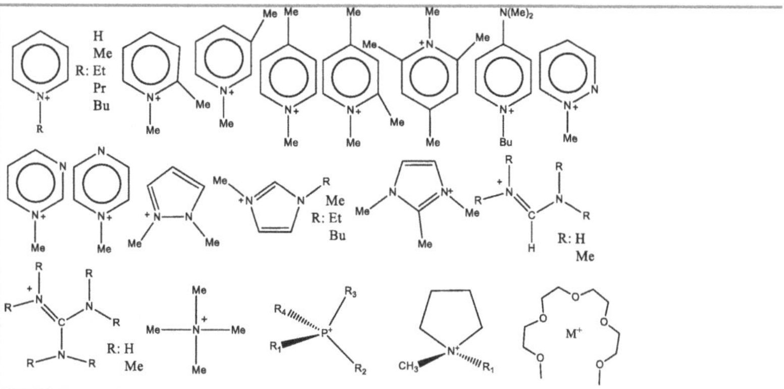
Anions	F^-, Cl^-, Br^-,I^-, $AlCl_4^-$, $Al_2Cl_7^-$, $Al_3Cl_{10}^-$, Cu_2Cl_3, $Fe_2Cl_7^-$, NO_3^-, NO_2^-, SO_4^{2-}, CH_3COO^-, $N(CF_3SO_2)_2^-$, $CH_3SO_3^-$, HSO_4^-, BF_4^-, PO_4^{3-}, $CF_3SO_3^-$, $C_3F_7COO^-$,$(C_2H_5SO_3)_2N^-$, $C_4F_9SO_3^-$, $C_2H_5BF_3^-$, $C_3H_7BF_3^-$, TF_2N^-,$C_4H_9BF_3^-$, $C_5H_{11}BF_3^-$, Lac^-, $CH_2CHBF_3^-$, FSI^-, $TFSI^-$, CH_3BF^-, $N(CN)_2^-$, $C_3F_7BF_3^-$, $C_4F_7BF_3^-$

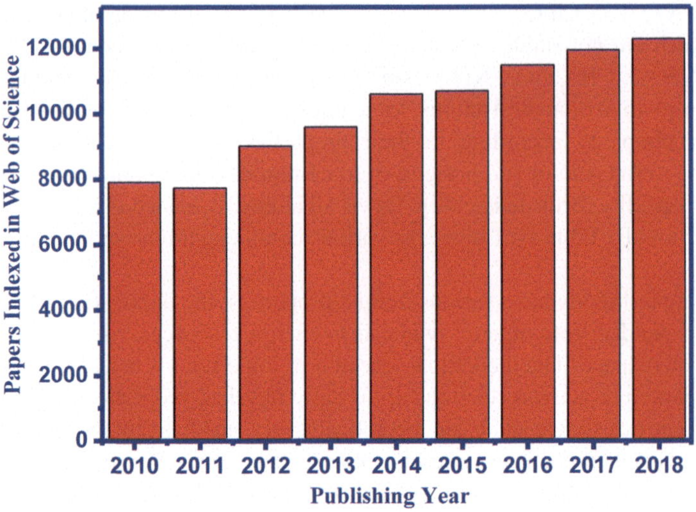

Fig. 10.1 Number of publications with "ionic liquids" or "ionic liquid" in subject from 2010 to 2018, data from Web of Science of Clarivate Analytics. The research and application of ionic liquids are increasing year by year

10.2.2 Physical and Chemical Property of Ionic Liquids

The physicochemical properties of ionic liquids often change with the structure of anions and cations in a wide range, such as density, melting point, thermal stability, viscosity, and hydrophilicity (Li 2004; Zhang et al. 2018; Deng 2006). Therefore, the study of the relationship between the physicochemical properties of ionic liquids and their microstructure information can help how to design and synthesize task-specific ionic liquids by choosing suitable anions and cations. In the following sections, the physical and chemical properties of ionic liquids will be briefly introduced.

10.2.2.1 Melting Point

Ionic liquids have a low melting point, usually at or near room temperature. The main factors affecting the melting point of ionic liquids are radius of ions, symmetry of ions, and volume of ions (Li 2004; Zhang et al. 2018; Deng 2006). The melting point depends mainly on the structure and type of cations and anions. For alkyl-imidazole cations and alkyl-pyridine ionic liquids, the charge is dispersed to the ring due to the formation of large π bonds, or the ion charge is shielded, which weakens the Coulomb interaction between ions. For the imidazole-type ionic liquids with same anion, with the increase of alkyl chain length, the cation size becomes larger and larger. Therefore the symmetry and the melting point of ionic liquids become lower. When the carbon number of side alkyl chain of ionic liquid is increased to a certain extent, the melting point will rise rapidly.

The effect of anion type on the melting point of ionic liquids is much less than that of cations. In general, the larger the volume of anions is, the weaker the coordination ability it has, and the lower the melting point it has. The melting point of ionic liquid formed by different anions and 1-ethyl-3-methylimidazolium cation $[Emim]^+$ decreases as the order of $Cl^- > PF_6^- > NO_2^- > NO_3^- > AlCl_4^- > BF_4^- > CF_3SO_3^- > CF_3COO^-$ (Hagiwara and Ito 2000). In addition, the anion radius of ionic liquids usually needs to be above 3.4–4 Å (Rooney and Seddon 2001).

10.2.2.2 Viscosity

Most ionic liquids have relatively high viscosities, ranging from 10 cP to 50 cP, and some even up to 500 cP. The viscosities of ionic liquids are higher than those of most organic solvents and water in 1–3 orders of magnitude (Mantz and Trulove 2002a). Because of the high viscosity and their adhesion to the reactor wall, it is difficult to deal with, which will also decrease the mass transfer and ion migration. Therefore, high viscosity is one of the disadvantageous factors restricting the large-scale application of ionic liquids (Gu and Brennecke 2002). The viscosity of ionic liquids is mainly controlled by hydrogen bond, van der Waals, and electrostatic interaction. The experimental results show that the types and structure of anions and cations of ionic liquids have an important impact on viscosity. For the ionic liquids formed with same anion, the longer the carbon chain length of the cation is, the stronger the van der Waals force between the anions and cations is, and the higher the viscosity of ionic liquids it has. For bis(trifluoromethylsulfonyl)imide $[(CF_3SO_2)_2N]^-$-type ionic liquid, the viscosity decreases in the following order: 1-octyl-3-methylimidazolium cation $[C_8MIM]^+$ > 1-hexyl-3-methylimidazolium cation $[C_6MIM]^+$ > 1-*n*-butyl-3-methylimidazolium cation $[BMIM]^+$ > 1-methyl-3-methylimidazolium cation $[MMIM]^+$ > 1-ethyl-3-methylimidazolium cation $[EMIM]^+$ (Tokuda et al. 2005). The viscosity of ionic liquids formed with same cation increases with the increase of alkylation in anion; its viscosity varies with different anions as bis(trifluoromethylsulfonyl) amide $TF_2N^- > BF_4^- > CF_3COO^- > CF_3SO_3^- > (C_2H_5SO_3)_2N^- > C_3F_7COO^- > CH_3CO_2^- > CH_3SO_3^- > C_4F_9SO_3^-$ (Mantz and Trulove 2002b).

It can be seen that the viscosity has small correlations with the anion volume, which is due to weak hydrogen bond formation between anions and cations. Temperature has a significant effect on the viscosity of ionic liquids. With the increase of temperature, the viscosity of ionic liquids decreases. In addition, the addition of mixed solvents, water content, and impurity ion content in ionic liquids have great influence on the viscosity of ionic liquids (Seddon et al. 2000). For example, the viscosity of dry 1-*n*-butyl-3-methylimidazolium hexafluorophosphate $[BMIM]PF_6$ is 375.9 mP.s, while the viscosity of water-saturated 1-*n*-butyl-3-methylimidazolium hexafluorophosphate $[BMIM]PF_6$ ionic liquids decreases to 84.8 mP.s when 20% molar fraction of water is added (Jacquemin et al. 2006).

10.2.2.3 Density

The density of matter is related not only to temperature and pressure, but also to relative molecular mass, intermolecular interaction force, and molecular structure. The density of most ionic liquids is greater than that of water, generally between 1.1 and 2.5 g/mL (Dzyuba and Bartsch 2002). The density of ionic liquids is related to the structure of anions and cations. The anions have more obvious influence on the density of ionic liquids than cations. Usually, the larger the anion volume is, the greater the density it has (Dzyuba and Bartsch 2002). For example, the weak coordination ions with larger volume, $CF_3CF_2CF_2SO_3^-$ and $(CF_3SO_2)_2N^-$, have relatively high density. On the contrary, the density of ionic liquids decreases with increasing cation size.

10.2.2.4 Conductivity and Electrochemical Window

Ionic liquids have excellent conductivity and wide electrochemical window, which are the basis of their electrochemical applications. Generally, ionic liquids have a conductivity of 10 $mS \cdot cm^{-1}$ at room temperature, which is almost same as that of nonaqueous solvents. The conductivity is related to the viscosity, molecular weight, density, and ionic size of ionic liquids. Viscosity is the primary factor affecting the conductivity of ionic liquids. The conductivity decreases with the increase of viscosity.

The size of the electrochemical window is mainly determined by the oxidation ability of anions and the reduction ability of cations. Wide electrochemical window of ionic liquids is one of the most remarkable advantage properties for their electrochemical application. The electrochemical windows of most ionic liquids are larger than about 4 V. So it is possible to carry out many research systems in ionic liquids to avoid the influence of hydrogen evolution in traditional aqueous solution. The order of electrochemical stability of ionic liquid cations was pyridine salt < pyrrole salt < imidazole salt < matte salt < quaternary ammonium salt (Trulove and Mantz 2002). The electrochemical window of tetravalent quaternary ammonium ionic liquids is very wide. It was reported that the electrochemical window of $[(C_3H_7)(CH_3)_3N](CF_3SO_2)_2N$ on glassy carbon electrode can reach 5.7 V (Matsumoto et al. 2000). At present, the widest electrochemical window reported in the literature is 7.1 V for [BMIM] PF_6 on tungsten electrode (Suarez et al. 1997).

The order of oxidation stability of ionic liquids formed with different anions was $Cl^- < F^- < Br^- < Al_2Cl_7^- < BF_4^- < PF_6^- < CF_3SO_3^-$, $(CF_3SO_2)_2N^-$, and $(C_2F_5SO_2)_2N^-$ (Trulove and Mantz 2002). Thus, if halogen ion impurities exist in ionic liquids, the anode limit of ionic liquids will be significantly reduced. In addition to halogen ions, the existence of water also has a great effect on the electrochemical window of ionic liquids. Because of the strong polarity of ionic liquids, it is easy to absorb water when it is exposed to air. For example, the electrochemical window of dried 1-*n*-butyl-3-methylimidazolium tetrafluoroborate $[BMIM]BF_4$ ionic liquids is

4.1 V, while that of water-containing 1-*n*-butyl-3-methylimidazolium tetrafluoroborate [BMIM]BF_4 is reduced to 1.95 V (Schröder et al. 2000).

Based on the above unique properties, ionic liquids were widely and successfully applied in electrodeposition, coasting, surface catalysis, and electrochemical fields such as supercapacitors, fuel cells, and dye-sensitized solar cells and other relative fields. In these applications, there inevitably exists an interface between ionic liquids and other charged solid surfaces or other materials that are transported into charge and discharge. This interface is very important for most reactions or mass transfer in these systems. Therefore a more detailed understanding of the structures and properties of the interface between ionic liquids and metal or electrode surfaces is curial in these applications, which will be beneficial for improving, optimizing, or designing new systems to achieve optimum performance.

10.3 The Theoretical Model of Electrical Double Layer

When electrodes contact with electrolyte solution, the species on the interface of electrode-solution are subject to both internal forces of solution and forces of electrodes because of the great differences in physicochemical properties of electrodes and solution. Therefore, a two-layer structure with opposite charge symbols will be formed spontaneously on the interface. The two-layer structure on the interface is called electrical double layer. In general, the thickness of electrical double layer on the interface is small, but there are electrical capacitance and electric field strength in electrical double layer. The electrical double layer formed at solution-metal electrode interface can be divided into ionic double layer, surface dipolar double layer, and adsorption double layer, which is given in Fig. 10.2.

The ion double layer consists of the excess charge on the surface of the electrode and the ions in the solution which is opposite to it. One layer is on the surface of the electrode and the other is in the solution close to the electrode. The dipole double layer consists of dipole molecules arranged in a directional direction on the surface of the electrode. The adsorbed double layer consists of ionic charges adsorbed on the surface of the electrode and another layer of ionic charges attracted by this layer of charges. Both dipole and adsorbed double layers exist in one phase.

There are two kinds of interfacial interactions at the interface of metal electrode-solution system: (1) the electrostatic long-range interaction caused by the residual charge in the two phases of the electrode and electrolyte solution and (2) the short-range interaction between the electrode and the particles in the electrolyte solution that occurs only within a few angstroms. In addition, there is also a thermal motion, which keeps charged particles in both phases in constant thermal motion. Due to the existence of residual charges such as electrons, ions, and charges on both sides of the electrode interface, the interface electrical double layer consists of compact layer and dispersion layer. The electrostatic effect makes the residual charges with opposite symbols to come close to each other and tend to be arranged close to the surface of the electrode. Therefore, it forms a compact double-layer structure,

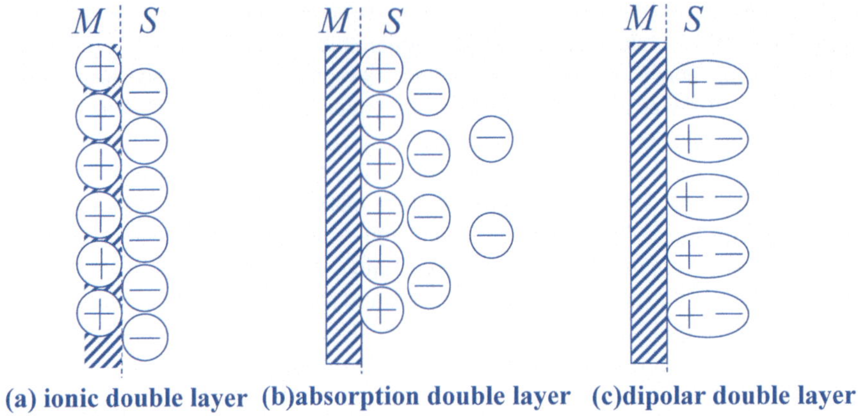

Fig. 10.2 Different models for double-layer structure at the aqueous solution-electrified metal interface. The symbols M and S at the top of the figure represent electrodes and electrolyte solution, respectively. The double layer formed at solution-metal electrode interface can be divided into three types: (**a**) Ion double layer: It consists of the excess charge on the surface of the electrode and the ions in the solution opposite to it. One layer is on the surface of the electrode and the other is in the solution close to the electrode. (**b**) Dipole bipolar layer: It is composed of dipole molecules arranged in a directional direction on the surface of the electrode. (**c**) Adsorption double layer: It composes of ionic charge adsorbed on the surface of the electrode and another layer of ionic charge attracted by this layer of charge [modified after Lockett et al. (2008)]

referred to as the compact layer. However, the charged particles tend to distribute uniformly in the thermal motion, so that the residual charge cannot be completely close to the surface of the electrode. It has a certain degree of dispersion and forms a dispersion layer. The properties of the compact layer depend on the structure of the interface layer, especially the degree to which the residual charge energy in the two phases is close to each other. The structure of the dispersion layer such as thickness and potential distribution is only related to temperature, electrolyte concentration including valence type, and residual charge density in the dispersion layer, but not to the individual characteristics of ions.

The theoretical model of electrical double layer had been firstly given by Helmholtz (1879). After that more models had been gradually developed, perfected, and further revised by Gouy, Chapman, Stern, and other scholars (Gouy 1910; Chapman 1913; Stern 1924; Grahame 1946; Grahame 1954; Bockris et al. 1963). In this section, we will discuss them briefly.

10.3.1 Helmholtz Double-Layer Model

Helmholtz model of double layer, also known as the flat-plate capacitor model or the compact electrical double-layer model, was proposed by Helmholtz in 1879 (Helmholtz 1879). The model regards the double layer of the electrode-solution on the interface as a flat capacitor. The charge in the electrode is located on the surface of the electrode. The charge in the solution is concentrated on a plane close to the electrode, forming a compact layer. Figure 10.3a shows the charge and potential distribution of Helmholtz model.

The potential distribution has a linear relation with the thickness of the double layer. The differential capacitance is independent of the potential and a certain value of the double layer. It is inversely proportional to the distance between ions in solution and the surface of the electrode. If it is assumed that the negative ion energy in solution is more near to the surface of electrode than the positive ion, the model can be used to explain the interfacial tension variation with the electrode potential and the plateau region in the differential capacitance curve. However, the model cannot explain the experimental facts that the interface capacitance varies with the total concentration of solution and the electrode potential, and the minimum differential capacitance in dilute solution at zero charge potential.

10.3.2 Gouy and Chapman Dispersed Double-Layer Model

In 1910 and 1913, Gouy and Chapman revised the Helmholtz model respectively, and independently proposed the dispersed double-layer model (Gouy 1910; Chapman 1913). They proposed a hypothesis about the ion distribution on the solution side. This hypothesis has the same starting point as Debye-Hückle's premise on strong electrolyte theory in 1923. The model considers that under the action of static electricity and thermal motion. The ions on solution side are not arranged tightly on the interface in the solution adjacent to the interface, but should be distributed in the liquid layer adjacent to the interface according to the distribution law of particles in the potential energy field. That is to say, the "dispersed layer" of charge is formed. The ion distribution on one side of the solution decreases as it extends to the interior of the solution. The law of dispersion follows Boltzmann distribution, and the existence of compact layer is completely neglected. Figure 10.3b shows the charge and potential distribution of Gouy-Chapman model. The model can well explain the appearance of minimum differential capacitance and the change of capacitance with electrode potential. However, the differential capacitance calculated by theory is much larger than that measured by experiment, and it cannot explain the appearance of "platform area" on the differential capacitance curve.

10.3.3 Gouy-Chapman-Stern Model

In 1924, Stern proposed an improved double-layer model, also known as Gouy-Chapman-Stern GCS dispersed double-layer model or adsorption double-layer model (Stern 1924). He developed on the basis of Gouy-Chapman's dispersed double-layer model and absorbed the reasonable part of Helmholtz's compact

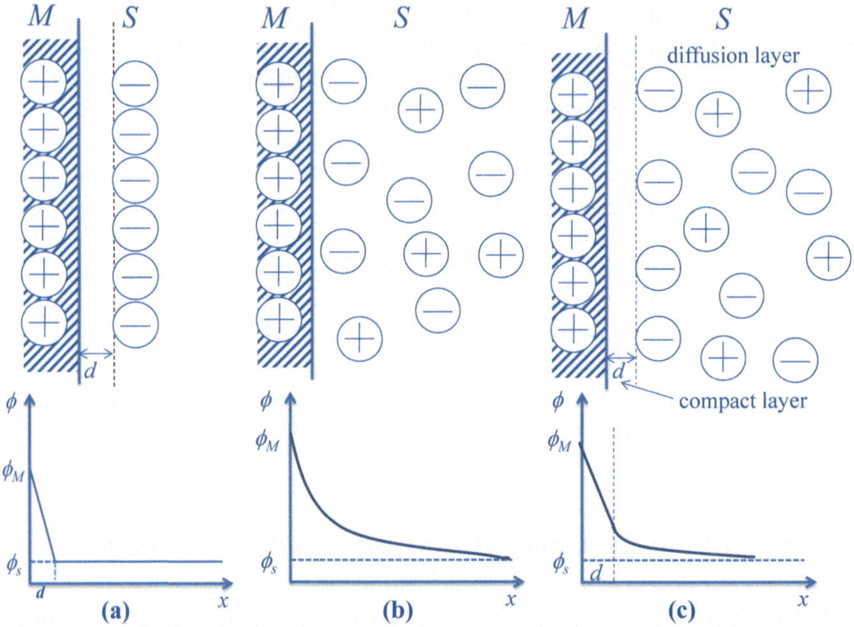

Fig. 10.3 Schematic illustration of the electrical double-layer models and the potential distribution in electrode-electrolyte interface are given on top and below, respectively. The symbols M and S at the top of the figure represent electrodes and electrolyte solution, respectively. The d is the distance of Helmholtz double layer. (**a**) is Helmholtz model. It regards the double layer as a flat capacitor. The charges on the electrode are located on the surface of the electrode. The charges in the solution are concentrated on a plane close to the electrode, forming a compact layer. The potential distribution of this model has a linear relation with the thickness *d* of the double layer. (**b**) is Gouy-Chapman model, which shows the charge and potential distribution of Gouy-Chapman model. The ions on solution side are not arranged tightly on the interface in the solution adjacent to the interface, but should be distributed in the liquid layer adjacent to the interface according to the distribution law of particles in the potential energy field. (**c**) is Stern model, which combines the reasonable parts of model (**a**) and model (**b**). It considers that the double layer is composed of compact layer and dispersive layer, and its potential can also be divided into compact layer potential and dispersive layer potential [modified after Lockett et al. (2008)]

double-layer model. It considers that the double layer is composed of compact layer and dispersive layer, and its potential can also be divided into compact layer potential and dispersive layer potential.

When the residual charge density on electrode surface and the concentration of electrolyte in the solution are high, the electrostatic interaction is dominant. Ions in the liquid tend to distribute tightly on the interface. The structure dynamics and property of the electrical double layer are basically compact. Its potential in this case is mainly composed of the close layer potential. When the residual charge density on electrode surface is small and the concentration of electrolyte in the solution is very thin, the thermal movement of the ions is dominant. The residual charges in the solution are impossible to be all concentrated on the interface, so that the charge distribution has a certain degree of "dispersion." The electrical double-layer structure is basically dispersed, and the potential of the electrical double layer is mainly from the dispersed layer. In this case, the electrical double layer consists of two parts such as the compact layer and the dispersed layer. This model has a detailed discussion of the dispersed layer. The description of the compact layer is very simple. The distribution of residual charge and potential in the dispersed layer is treated by the same mathematical method as Gouy-Chapman and the corresponding double-layer equation is deduced. Figure 10.2c shows the charge and potential distribution of Gouy-Chapman-Stern model.

Stern model can better reflect the real situation of the interface structure, but the electrical double-layer equation cannot be used for accurate calculation, and the description of the tight layer is too rough. When discussing the electrical double layer between metal and solution, only the interaction between the residual charge on metal and the residual charge of ion in solution is considered. Admittedly, this is the main reason for the electrical double layer between metal and solution. In addition, the interaction between polar molecules of solvents (e.g., water) and residual charges on metals and the characteristic adsorption of some negative ions on metals in solution will affect the structure of the interface electric double layer.

10.3.4 Modified Gouy-Chatman-Stem Model

In the late 1940s, Grahame first clearly pointed out that when species are adsorbed at the electrode-liquid interface, the compact layer had more fine structure (Grahame 1946; Grahame 1954).

As shown in Fig. 10.4, Grahame divided the electrolyte solution-metal interface as diffuse layer and inner or compact layer. The boundary is outer Helmholtz plane OHP, which is the plane closest to the center of the solvation ion on the metal surface. Grahame's modified Gouy-Chapman-Stern GSC model, also known as Grahame's double-layer model, is the basis of modern double-layer theory. However, Grahame does not consider the effect of solvent molecules on the structure and properties of double layer in this model. The role of solvent molecular layers has become one of the main topics in the double-layer theory since the 1960s.

10.3.5 Bockris-Devanathan-Muller BDM Model

In 1963, Bockris further revised the Gouty-Chapman-Stern GCS model considering the qualitative adsorption of water and ions (Bockris et al. 1963). To have a good description of the electrical double-layer structure, they believe that they must consider the characteristic adsorption of ions and the directional arrangement of water molecules in the double layer. The first water molecule layer on the electrode surface is all aligned, while the second layer of water molecules is partially aligned,

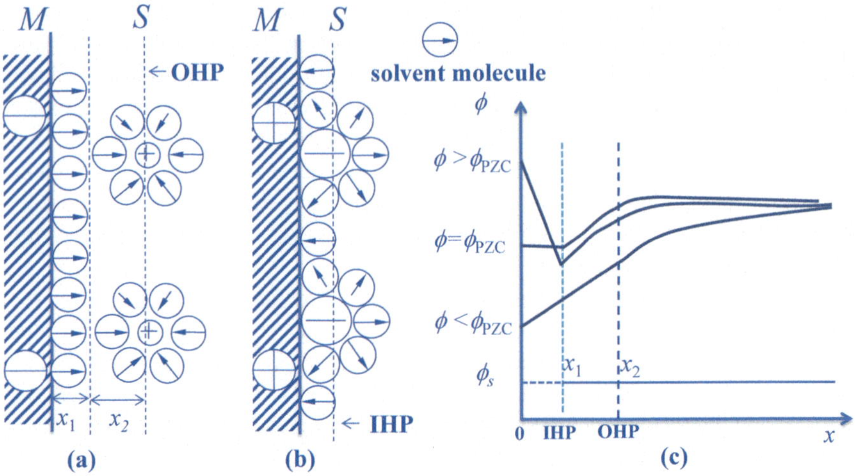

Fig. 10.4 Grahame modified Gouy-Chapman-Stern model of electrical double layer. The symbols M and S at the top of the figure represent electrodes and electrolyte solution, respectively. (**a**) Outer compact layer or outer Helmholtz plane OHP without characteristic adsorption. Its thickness is the distance from the surface of the electrode to the charge center of the hydrated cation. If x_1 is the thickness of the first water molecule layer and x_2 is the radius of a hydrated cation, $d = x_1 + x_2$. (**b**) Inner compact layer plane or inner Helmholtz plane IHP with characteristic adsorption. Grahame divided the electrolyte solution-metal interface as diffuse layer and compact or inner layer. The boundary is outer Helmholtz plane OHP, which is the plane closest to the center of the solvation ion on the metal surface. ϕ_{zpc} is the potential of zero charge [modified after Lockett et al. (2008)]

some of which are primary hydrated water of ions. In this model, the electrical double layer is considered to have three parts which consist of inner Helmholtz layer or inner compact layer, outer Helmholtz layer, or outer compact layer and dispersed layer.

The inner Helmholtz layer IHP is a compact layer composed of adsorbed water ions and characteristic adsorbed ions. The outer Helmholtz layer or outer compact layer OHP is a compact layer formed by hydrated positive ions, and the dispersion layer is between the outer Helmholtz layer or outer compact layer and the solution body. When the electric field between the compact layer and the surface of the electrode is strong, the compact layer contains a water molecule dipole layer, which is adsorbed to a certain extent on the electrode surface. The compact layer composed of this ionic charge is called the outer compact layer OHP, as shown in Fig. 10.5a. When the residual charge on electrode surface is positive, the hydration film of the anion hydrate forming the double layer is destroyed, and it can squeeze out the water molecules adsorbed on electrode surface and contact directly with the electrode surface. In this case, the distance between the center line of negative ions and the surface of the electrode in the compact layer is much smaller than that of positive ions, which can be called the inner compact layer IHP as shown in Fig. 10.5b.

Therefore, according to the different ion properties of the double layer, the compact layer can be divided into outer layer and inner layer. As mentioned above, it can

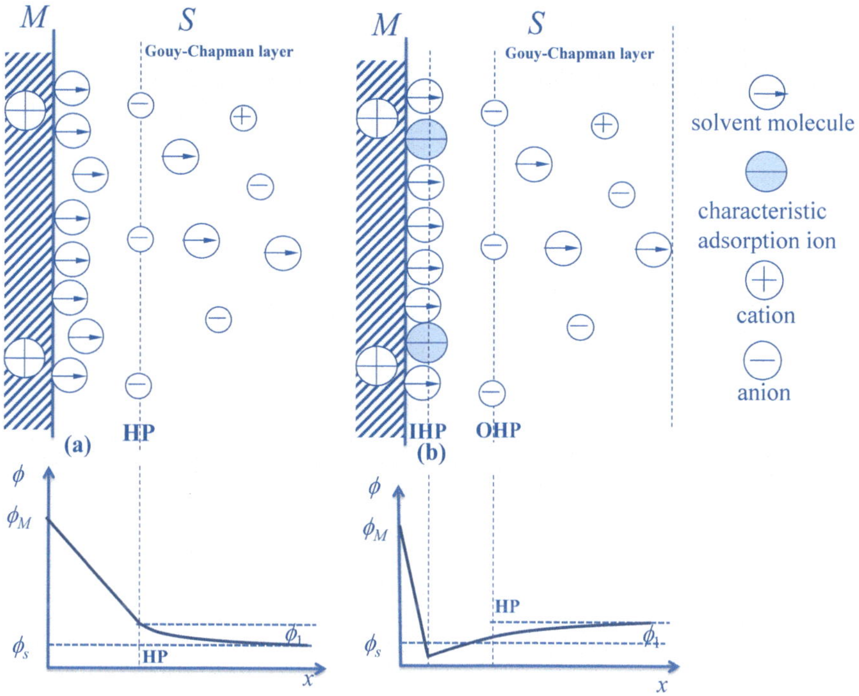

Fig. 10.5 Bockris-Devanathan-Muller BDM electrical double-layer model. The symbols M and S at the top of the figure represent electrodes and electrolyte solution, respectively. The charge distribution of the outer Helmholtz plane abbreviated as OHP is given in (**a**). The charge distribution of the inner Helmholtz plane abbreviated as IHP is given in (**b**). (**c**) is the distribution of electric potential. The outer Helmholtz plane OHP is also the plane where the diffuse layer begins and which refers to that of the nonspecifically adsorbed ions. The inner Helmholtz plane IHP refers to the distance of closest approach of specifically adsorbed ions [modified after Lockett et al. (2008)].

explain why the measured capacitance is larger when the surface of the electrode is charged positively than when the surface of the electrode is charged negatively. The research method of classical double-layer theory is mainly based on the hypothesis model to calculate the interface parameters and compare with the experimental results. The coincidence shows that the hypothesis model is valid.

The classical model does not consider the solvents and solute molecules, the particularity of ions, and the interaction between particles. It is considered that every point in Helmholtz plane is equal. However, in fact, there are different potential values at different points in this plane. If we consider the "mirror charge" caused by the ionic charge in Helmholtz plane as a point charge in the metal surface, the distribution of the metal surface charge is also uneven.

Since the 1970s, the theory of double layer has been further developed. For example, the idea of solution ions as charged hard spheres has been put forward. Solvents are treated as continuous medium or as point-dipole hard spheres. Based on fluid physics, classical models are modified by computer simulation technology. In double layer, solvents are treated as continuous medium or as point-dipole hard spheres. Some progress has been made in the study of the theory. In recent years, some advances have been made in the study of double layer by means of electrochemical scanning tunneling microscopy and other advanced physical and chemical means.

10.4 Experimental Study Progress of Electrical Double Layer in Ionic Liquids

In recent years, with the development of experimental technology, electrochemical methods, scanning tunneling microscopy (STM) imaging technology, atomic force microscopy (AFM) imaging technology, and spectroscopy technology (infrared spectroscopy, surface-enhanced Raman, sum frequency spectroscopy, etc.) have been used to reveal the structure, properties, and behavior of electrical double-layer structure of ionic liquid-electrode system, shape of capacitance curve, interface adsorption, and orientation (He et al. 2018; Burt et al. 2014; Hayes et al. 2015; Fedorov and Kornyshev 2014).

10.4.1 Electrochemical Measurement Technology

The earliest experimental study of differential capacitance curve of ionic liquids was performed in 1997. The differential capacitances of various imidazole ionic liquids at Hg electrodes were measured (Nanjundiah and McDevitt 1997). It was found that the potential of zero charge determined by the maximum surface tension on the capillary curve was not the minimum of differential capacitance, but the larger differential capacitance. They did not discuss this disagreement from the point of view of the interface double-layer structure, but thought that Hg electrode had begun to oxidize under the negative potential of zero charge, which led to the increase of capacitance, and concealed the minimum capacitance at the potential of zero charge.

Osaka team reported a series of differential capacitance curves of imidazole ionic liquids with different chain lengths on glassy carbon, mercury, and gold electrodes (Alam et al. 2007; Alam et al. 2008). It is found that the differential capacitance curve is parabolic, and the minimum value of differential capacitance of ionic liquids appears at the potential of zero charge. The characteristics of differential capacitance are closely related to the size of anions and cations and properties of electrodes in ionic liquids. But in non-imidazole ionic liquids, the differential capacitance curve measured on gold and platinum electrodes is bell shaped, and the maximum value of the differential capacitance curve appears at the potential of zero charge. Baldelli group used electrochemical impedance spectroscopy EIS to study the differential capacitance curves (Rivera-Rubero 2004; Baldelli 2008). It was found that the Helmholtz layer thickness of interface double layer of 1-*n*-butyl-3-methylimidazolium tetrafluoroborate [BMIM]BF_4-Pt system was about 5 Å.

10.4.2 Atomic Force Microscopy (AFM) Technology

Atomic force microscopy technology depends on the interaction force between the probe and the surface as a signal. It can not only achieve high-resolution imaging of the surface, but also study the distribution of charges in the dispersed layer by measuring the electrostatic force between the probe and the surface. Atkin's team first carried out the determination of force curves in ionic liquids. In 2007, they first detected the multilayer solvent layer structure formed by ionic liquids at the mica interface using AFM technology (Atkin and Warr 2007). The largest number of solvent layers (about nine layers) was measured in ethylammonium nitrate salt [$EtNH_3$]NO_3, which was consistent with the results obtained by Liu et al. (2006).

By comparing the measured results in different substrates, they believe that both surface charge and surface roughness of substrates have important effects on the formation of solvent layers in ionic liquids. Compared with Si and graphite substrates, mica has a smoother surface and a higher surface charge, so it is advantageous to form a stable multilayer solvent layer structure. In addition, the effects of temperature and anion and cation on the structure of interfacial solvent layer were also investigated with atomic force microscopy (AFM) (Wakeham et al. 2009; Atkin et al. 2011). It was shown that the number and strength of interfacial solvent layer decreased with the increase of temperature. The flexibility of alkyl chains of cations increased, and the formation of ions weakened. That is, the more carbon atoms on the side chains, the better the flexibility is. The hydrogen bond network structure will reduce the number of layers and increase the compressibility of the solvent layer at the interface. In addition, seven layers of ionic liquids on mica surface were founded in first time. The interfacial layers were related to the electrode potential. With the potential deviation away from the path potential, the number of layers of ionic liquids increased and the stability increased (Hayes et al. 2011).

10.4.3 Scanning Tunneling Microscopy (STM) Technology

Scanning tunneling microscopy (STM) technology is an important method for surface electrochemistry. It can study the structure and process of the electrode-solution interface at the level of atomic resolution by using the tunnel current which decreases exponentially with the tip-to-surface distance as a signal. In 2006, Pan and Freyland reported the first work with in situ scanning tunneling microscopy (STM) technology to study the 1-*n*-butyl-3-methylimidazolium hexafluorophosphate [BMIM]

PF_6-Au(111) interface (Pan and Freyland 2006). It was found that some phase transitions of PF_6^- adsorption occurred with the change of site. The Kolb group studied the interface of 1-*n*-butyl-3-methylimidazolium hexafluorophosphate [BMIM] PF_6-Au(111) (Gnahm et al. 2010) and 1-*n*-butyl-3-methylimidazolium hexafluorophosphate [BMIM]PF_6-Au(100) (Pajkossy and Kolb 2011) by combining scanning tunneling microscopy (STM), cyclic voltammetry, and impedance technique. Using in situ STM technique, they observed blurred surface imaging when the potential was positively controlled on both sides of the potential of zero charge. They believed that the ionic liquid adsorption was responsible for the blurring of the surface, while the surface clarity was significantly improved when the potential step was negatively shifted to the negative side of the potential of zero charge.

10.4.4 Second Harmonic Generation (SHG) and Sum Frequency Generation (SFG) Technology

Sum frequency generation (SFG) and second harmonic generation SHG technology have specific surface selection rules, that is, only molecules without central symmetry can give vibration signals. The general species have central symmetry and have no contribution to the signals of sum frequency generation and second harmonic generation. Therefore, sum frequency generation and second harmonic generation technology have unique advantages in identifying the orientation of adsorbed species.

Baldelli group studied the adsorption of a series of ionic liquids on the surface of polycrystalline Pt electrodes using sum frequency generation (Rivera-Rubero 2004; Baldelli 2008). The change of sum frequency generation intensity with potential in 1-*n*-butyl-3-methylimidazolium hexafluorophosphate [BMIM]PF_6 and 1-*n*-butyl-3-methylimidazolium tetrafluoroborate [BMIM]BF_4 systems shows that in the negative potential range, the cations of imidazole ring lie almost flat on the surface of the electrode with neutral and surface charges. With the potential moving forward gradually, the imidazole ring inclines along the normal direction of the surface, leaving space for anions such as BF_4^- and PF_6^- near the surface of the electrode.

They also studied various interfaces including 1-*n*-butyl-3-methylimidazolium dicyandiamide [BMIM]DCA-polycrystalline Pt (Aliaga and Baldelli 2006), titanium dioxide-[BMIM]DCA, and 1-*n*-butyl-3-methylimidazolium methyl sulfate [BMIM]CH_3SO_4-titanium dioxide and -quartz (Romero et al. 2006; Aliaga and Baldelli 2008; Baldelli 2008). The adsorption of anions and cations on the metal interface was observed on these three systems. It was found that the positive potential shift induced 1-*n*-butyl-3-methylimidazolium cation $[BMIM]^+$ to be adsorbed from lying down to being excluded, while dicyandiamide anion DCA^- strongly adsorbed on the positively charged surface. 1-*n*-Butyl-3-methylimidazolium cation $[BMIM]^+$ tended to be adsorbed parallel on the nonmetallic electrode titanium dioxide. It was found that dicyandiamide anion DCA^- had a characteristic adsorption on titanium dioxide, while a weak adsorption was found for methyl sulfate anion $CH_3SO_4^-$ on titanium dioxide.

The potential-induced adsorption or desorption process of anions and cations on the surface of ionic liquid-polycrystalline Pt electrodes were studied by sum frequency generation (SFG) measurement (Lockett et al. 2008; Drüschler et al. 2010; Gore et al. 2010), and observed that the hysteresis effects of ion desorption or reabsorption caused by the direction of potential change are related to the hysteresis of the shape change of differential capacitance curve caused by different potential scanning directions.

10.4.5 Surface-Enhanced Raman Spectroscopy (SERS)

Surface-enhanced Raman spectroscopy (SERS) can provide not only abundant spectral information for identifying the adsorbed species on the surface, but also orientation information of the adsorbed molecules on the surface. By using surface-enhanced Raman spectroscopy (SERS) technique, the adsorption behavior of solvent ions on the 1-*n*-butyl-3-methylimidazolium hexafluorophosphate [BMIM] PF_6-Ag interface was studied by Rubim team (Santos et al. 2006). Similar to the results of sum frequency generation (SFG) detection, they also observed the adsorption orientation of 1-*n*-butyl-3-methylimidazolium cation $[BMIM]^+$ on the rough Ag electrode gradually parallel to the surface with the negative potential shift.

They also used surface-enhanced Raman spectroscopy (SERS) to study the hydrophilic ionic liquids 1-*n*-butyl-3-methylimidazolium tetrafluoroborate [BMIM] BF_4 on rough copper electrodes (Brandão et al. 2009). The electrochemical window of copper electrode in 1-*n*-butyl-3-methylimidazolium tetrafluoroborate [BMIM] BF_4 ionic liquid is 2.5 V. When the potential is negative to 1.5 V, a large number of tetrafluoroborate BF_4^- anions accumulate on the surface of copper electrode, while the vibration peak wave number and intensity of imidazole ring of 1-*n*-butyl-3-methylimidazolium cation $[BMIM]^+$ change significantly. When the potential is negative to zero charge potential, 1-*n*-butyl-3-methylimidazolium cation $[BMlM]^+$ reacts with the electrode. 1-*n*-Butyl-3-methylimidazolium cation $[BMIM]^+$ is also observed to be reduced when the potential is adjusted to a negative value below 2.6 V.

In addition to the above experiments, in recent years, with the development of spectroscopy technology, many experimental studies have been carried out on the structure and behavior of the electrical double layer at the interface of electrode-ionic liquids. Here we summarize them in Table 10.2 for reference.

It can be seen from Table 10.2 that most experiments use carbon, mercury, gold, and other single crystals as electrodes. The phenomenological study of the double-layer structure is carried out by means of cyclic voltammetry or spectroscopy by investigating capacitance changes and observing interface images. Up to now, experimental studies have shown that for different ionic liquid-electrode systems, the capacitance of the double layer is related to the size of cations and anions of ionic liquids and the growth of alkyl chain. At the same time, capacitance lag at

Table 10.2 Experimental progress on the electrical double-layer structure of the ionic liquid-electrode system

Year	IL(s)	Electrode	Measurement method	References
2014	[EMIM]BF_4, [EMIM]I	PG	Cyclic voltammetry, electrochemical impedance spectroscopy	Siinor et al. (2014)
2014	[BMIM]PF_6, [OMIM]PF_6	Au(111)	Ec-atomic force microscopy	Zhang et al. (2014b)
2014	[$Py_{1,4]}$FAP, [EMIM] FAP	Au(111)	Atomic force microscopy, scanning tunneling microscopy, electrochemical impedance spectroscopy	Atkin et al. (2014)
2014	[BMIM]OTF	Pt	Electrochemical impedance spectroscopy, cyclic voltammetry	Yang and Zhou (2014)
2014	[Et_4N]BF_4	Activated carbon	Nuclear magnetic resonance	Griffin et al. (2015)
2014	[BMIM]PF_6	Glassy carbon	Electrochemical impedance spectroscopy, cyclic voltammetry	Zhang et al. (2014a)
2014	[EMIM] OTF, [BMIM] OTF, [OMIM] OTF	Au/Pt/glassy carbon	Electrochemical impedance spectroscopy, cyclic voltammetry	Yang (2014)
2014	[C_nMIM] TF_2N(n = 2,4,6), [C_4MIM]PF_6, [C_4MIM]BF_4	Au, Pt	Cyclic voltammetry, electrochemical impedance spectroscopy	Gomes et al. (2014)
2014	[BMIM]OTF	Au	Cyclic voltammetry, infrared spectra analysis	Xu et al. (2014)
2015	[Pyr_{13}]DCA, [Pyr_{14}] DCA, [$Pyr_{12}O_1$]DCA	Pt	Linear sweep voltammetry, Galvanostatic charge–discharge	Wolff et al. (2015)
2015	[BMIM]BF_4	Au	Surface-enhanced Raman scattering	Zhang (2015)
2015	[EMIM]BF_4	Activated carbon	Scanning electron microscopy, X-ray diffraction	Thürmer et al. (2015)
2015	[BMIM]BF_4	Au	Surface-enhanced Raman scattering, cyclic voltammetry	Zhang et al. (2015)
2015	[EMIM]TF_2N, [EMIM]FAP	Hg	Electrochemical impedance spectroscopy	Ivaništšev et al. (2015)
2015	[BMIM]TF_2N	Au	Cyclic voltammetry, electrochemical impedance spectroscopy, scanning tunneling microscopy	Müller et al. (2015)
2015	[BMIM]FAP	Hg, Au, Pt, glassy-carbon	Cyclic voltammetry, X-ray photoelectron spectroscopy	Costa et al. (2015)
2015	[BMP]TFSA	Au(111)	Scanning tunneling microscopy, cyclic voltammetry	Wen et al. (2015)

(continued)

Table 10.2 (continued)

Year	IL(s)	Electrode	Measurement method	References
2016	[Pyr_{13}]TFSI, [Pyr_{14}] TFSI, [Pyr_{15}]TFSI	Carbon	Cyclic voltammetry, X-ray diffraction	Cho et al. (2016)
2016	[EMIM]BF_4, [EMIM]Br	Bi(111)	Cyclic voltammetry, electrochemical impedance spectroscopy	Siimenson et al. (2016)
2016	[MPIM]FSI, [MPyr] FSI, [MeoMPyr]FSI	Glassy carbon	Linear sweep voltammetry, differential scanning calorimeter	Lee et al. (2016)
2016	[EMIM]TF_2N, [EMIM]FAP	Hg	Cyclic voltammetry, electrochemical impedance spectroscopy, in situ-attenuated total reflection infrared spectroscopy	Costa et al. (2016)
2016	[BMP]TFSA	Au(111)	Cyclic voltammetry, atomic force microscopy	Zhong et al. (2016)
2016	[EMIM]TFSI	Ag(111)	In situ scanning tunneling microscopy, electrochemical impedance spectroscopy, atomic force microscopy	Li et al. (2016b)
2016	[EMIM]TFSI	Graphene	Cyclic voltammetry, electrochemical impedance spectroscopy, X-ray photoelectron spectroscopy	Zhang et al. (2016)
2017	[BMIM]Cl, [BMIM] Br, [BMIM]I	Activated carbon	X-ray diffraction, nuclear magnetic resonance, cyclic voltammetry, electrochemical impedance spectroscopy	Liew et al. (2017)
2017	[EMIM] TFSA, [BMIM] TFSA, [OMIM]TFSA	Ag(111)	Atomic force microscopy, cyclic voltammetry	Chen (2017)
2017	[EMIM]BF_4	Activated carbon	Cyclic voltammetry, electrochemical impedance spectroscopy	Kumagai et al. (2017)
2017	[BMIM]BF_4, [BMIM]TFSA, [BMIM]TFSA	Au	Surface-enhanced infrared absorption spectroscopy, cyclic voltammetry	Motobayashi et al. (2017)
2017	[BMIM]BF_4	C(0001), Bi, Pb, Pt, and Au	Cyclic voltammetry, electrochemical impedance spectroscopy	Oll et al. (2017a)
2017	[PMIM]I, [EMIM] BF_4	Bi(111), Bi(001), Bi(011)	Cyclic voltammetry, electrochemical impedance spectroscopy	Oll et al. (2017b)

(continued)

Table 10.2 (continued)

Year	IL(s)	Electrode	Measurement method	References
2017	[EMIM]Cl	Graphene	X-ray diffraction, scanning electron microscope, transmission electron microscopy, X-ray photoelectron spectroscopy, atomic force microscopy	Lei et al. (2017)
2017	[BMIM]PF_6, [DEME]PF_6	Au	Electrochemical impedance spectroscopy, linear sweep voltammetry	Matsumoto et al. (2017)
2018	[EMIM] FAP	Polycrystalline gold	Cyclic voltammetry, electrochemical impedance spectroscopy	Lucio and Shaw (2018)
2018	[BMIM]OTF	Au	Cyclic voltammetry, electrochemical impedance spectroscopy	Zhao and Zhou (2018)
2017	[EMIM]BF_4, [EMIM]DCA, [EMIM]TF_2N, [BMIM]TF_2N	Polycrystalline gold	Cyclic voltammetry, electrochemical impedance spectroscopy	Jitvisate et al. (2017)
2018	[MAM] ClO_4(MAM:Li^+, Na^+, K^+, Rb^+, Cs^+)	Pt(111) and Au(111)	Cyclic voltammetry, electrochemical impedance spectroscopy	Garlyyev et al. (2018)
2018	[BMIM]PF_6, [BMIM]TF_2N, [BMIM]BF_4	Activated carbon	Three-dimensional full-spectrum automatic analysis	Liu (2018)
2018	[EMIM]BF_4	Carbon	Cyclic voltammetry, electrochemical impedance spectroscopy	Martins et al. (2018a)
2018	[TEA]BF_4	Activated carbon	Cyclic voltammetry, electrochemical impedance spectroscopy, TGA	Wu (2018)
2018	[BMIM]PF_6	Pt	Cyclic voltammetry, electrochemical impedance spectroscopy	Zhang et al. (2018)
2018	[BMIM]PF_6, [BMIM]FAP, [BMIM]TF_2N	Hg	Cyclic voltammetry, electrochemical impedance spectroscopy	Costa et al. (2018)
2018	[BMIM]PF_6, [BMIM]BF_4, [BMIM]TF_2N	Macroporous graphene	Scanning tunneling microscopy, Raman, X-ray photoelectron spectroscopy	Down and Banks (2018)
2018	[$Pip_{1,4}$]$B(CN)_4$, [$Pyr_{1,4}$]$B(CN)_4$, [$Pyr_{1,4}$]TF_2N	Activated carbon	Cyclic voltammetry, Galvanostatic charge–discharge, electrochemical impedance spectroscopy	Martins et al. (2018b)

(continued)

Table 10.2 (continued)

Year	IL(s)	Electrode	Measurement method	References
2018	$[Et_4N]BF_4$, $[Pyr_{14}]BF_4$, $[Pyr_{14}]$TFSI	Carbon	Cyclic voltammetry, energy-dispersive X-ray spectroscopy, scanning electron microscope, electrochemical impedance spectroscopy	Moreno-Fernández et al. (2018)
2018	$[TEA]BF_4$	Activated carbon	Cyclic voltammetry, electrochemical impedance spectroscopy, galvanostatic charge–discharge	Yang et al. (2018)
2019	$[Pyr_{14}]$TFSI	Graphene	Atomic force microscopy	Tsai et al. (2019)
2019	[EMIM]TFSI, [EMMIM]TFSI	Au(111)	Scanning tunneling microscopy, atomic force microscopy	Liu et al. (2019)
2019	$[PyrH_4]$TFSI	Activated carbon	Cyclic voltammetry, thermogravimetric analysis, energy-dispersive X-ray spectroscopy	Stettner et al. (2019)

In the abbreviations of the ionic liquids, EMIM or C_2MIM is 1-ethyl-3-methylimidazolium, *DMIM* 1,3-dimethyl-2-imidazolidinone, $Pip_{1,4}$ N-methyl-N-butyl piperidinium, *PMIM* 1-pentyl-3-methylimidazolium, *MPIM* 1-methyl-3-pentylimidazolium, $PyrH_4$ N-butyl-pyrrolidinium, $EtNH_3$ ethylammonium, *EMMIM* 1-ethyl-2,3-dimethylimidazolium, *DEME* N, N-diethyl-N-methyl-N-(2-methoxyethyl) ammonium, *MeoMPyr* N-methyl-N-ethoxymethyl pyrrolidinium, *TEA* tetraethylammonium, *BMIM or* C_4MIM 1-*n*-butyl-3-methylimidazolium, *OMIM or* C_8MIM 1-methyl-3-octylimidazolium, C_nMIM 1-alkyl-3-methylimidazolium, $C_n(MIM)_2$ 1-alkyl-3-dimethylimidazolium, $C_6(MIM)_2$ 1-hexyl-3-dimethylimidazolium, *Pyr* 1-butyl-1-methyl-pyrrolidinium, N_{2228} triethyloctylammonium, Et_4N tetraethylammonium, $C_6(MIM)TMA$ 1-(1-trimethylammonium-yl-hexyl)-3-methylimidazolium, *BMA* butyltrimethylammonium, *BMP* 1-butyl-1-methylpyrrolidinium, Pyr_{13} 1-methyl-1-propylpyrrolidinium, *SER* serine, Pyr_{14} 1-methyl-1-butylpyrrolidinium, BF_4 tetrafluoroborate, PF_6 hexafluorophosphate, *VAL* valine, *TFSI* bis-((trifluoromethyl)sulfonyl)imide, *FSI* bis(fluorosulfonyl)imide, TF_2N bis(trifluoromethylsulfonyl)amide, CH_3COO acetate, *SCN* thiocyanate, $B(CN)_4$ dicyanoborate, *OTF* trifluoromethanesulfonate, *GLY* aminoethanic acid, *PHE* phenylalanine, *DCA* dicyandiamide, *TFSA* trifluoromethanesulfonamide, *FAP* tris(pentafluoroethyl) trifluorophosphate

electrode-ionic liquid interface is observed with the increase of voltage. However, the causes of hysteresis, the mechanism of interaction between electrodes and ionic liquids, and the distribution and orientation of ions on the surface of electrodes cannot be known by experiments. There are many kinds of ionic liquids. It is undoubtedly a huge work to find the best combination of ionic liquid-electrodes through experiments to make capacitors or super-batteries with excellent performance. Systematically explore the structure, dynamic behavior, and properties of the electrical double layer at ionic liquid-electrode system at the molecular level that are very import to help solve these problems. Computer simulation is a useful method to study the structure and properties in an atomic or molecular scale. It can guide experiments or verify some theoretical assumptions to reduce the blindness of experiments.

10.5 Theoretical Study Progress of Electrical Double Layer in Ionic Liquids

Ionic liquids are high-concentrated electrolytes that are controlled by complex interactions of hydrogen bond, van der Waals, Coulomb, and dipole interactions. The classical Debye-Hückel theory and the traditional double-layer model will no longer be valid for ionic liquids. Again, the sizes of the anions and cations of ionic liquids are larger than traditional ions, so a finite size of ions in ionic liquids must be taken into account for developing new theory or model for ionic liquids (Kornyshev 2007). In recent years, both theoretical and computational methods were applied to give a deeper insight into the interfacial structure, behavior, and properties of ionic liquid-metal electrode.

Recently, the electrical double-layer model was developed according to the lattice-gas model for ionic liquids (Kornyshev 2007). Various analytical expressions of the double-layer capacitance for ionic liquids were also given. However, the intramolecular charge distribution of ionic liquids, asymmetry of cations and anions, and surficial adsorption of molecules were not considered in developing the double-layer model for ionic liquids (Kornyshev 2007). Again, the ion correlation effect in high-concentrated electrolytes was also not considered in the mean field approach (Kornyshev 2007; Keblinski et al. 2000).

Classical density functional theoryand computer simulation methods such as molecular dynamics (MD) and Monte Carlo (MC) are most used theoretical techniques. Both of these are used to reveal the structure, properties, and behavior of the electrical double layer at electrode-ionic liquid interface at the molecular level. Computer simulation is a useful method to systematically explore the structure, properties, and behavior of the electrical double layer of electrode-ionic liquid system at the molecular level. It can study the distribution and movement of atom-level particles at the solid-liquid interface under complex conditions which are difficult to achieve by experimental methods. We will give a brief introduction about the advance in these fields.

10.5.1 Classical Density Functional Theory

Based on statistical mechanics, the classical density functional theory (CDFT) method has natural advantages in predicting the microscopic structure and properties of inhomogeneous fluids, such as surface interfacial effect, adsorption, and dissolution. The classical density functional theory has much higher computational efficiency than traditional molecular simulation such as molecular dynamics simulation and Monte Carlo method, on the premise of guaranteeing the same computational accuracy. It can reflect the influence of molecular structure and interaction more precisely. The classical density functional theory can systematically study the effects of pore geometry, pore size, surface functional groups, solvent type, concentration, electrolyte ion composition, valence, and size or capacity.

Wu et al. studied the interfacial structure and behavior of ionic liquids close to the neutral or charged surfaces by using the classical density functional theory (Wu et al. 2011). Electrostatic correlations, molecular size, formation of long range, topology, and alternating structures of anions and cations at charged interfaces were examined for electric double layers of ionic liquid. The classical density functional theory was used to study the interfacial structure and dynamic properties of electrode-ionic liquid system over a range of electrode charges and two electrolyte concentrations (Henderson and Wu 2012). In this study, all of the spheres have been considered to have the same diameter for simplicity. It was found that differential capacitance curve has only a single hump. They found that the geometry of ionic species has an unexpected effect on the electrochemical properties of the electrical double layer in ionic liquids. They also found in other works that the double or single hump of differential capacitance curve is determined by the charge concentration of electrolyte (Henderson et al. 2013). With the increase of charge concentration, the differential capacitance curve can be changed from a double hump type to a single hump type continuously. Lamperski et al. used the density functional theory to study the effect of ion shapes on the differential capacitance and structure of the electrical double layer (Lamperski et al. 2013). It was found that the differential capacity curve evolves from a camel shape at low concentrations of electrolyte to a bell-like at higher concentrations.

The capacitance structure of electrical double layer of 1-ethyl-3-methylimidazolium tetrafluoroborate [EMIM]BF_4 and 1-ethyl-3-methylimidazolium bis(trifluoromethylsulfonyl)imide [EMIM]TFSI and their mixture was studied by classical density functional theory (Lian et al. 2016). It was shown that with the increase of the molar fraction of 1-ethyl-3-methylimidazolium tetrafluoroborate [EMIM]BF_4 in the mixture, the capacitance increases first and then decreases. The maximum capacitance of 1-ethyl-3-methylimidazolium tetrafluoroborate [EMIM]BF_4 and 1-ethyl-3-methylimidazolium bis(trifluoromethylsulfonyl)imide [EMIM]TFSI electrolytes is 15% and 45% higher than that of pure electrolytes, respectively.

10.5.2 Monte Carlo Method

Monte Carlo simulation is based on statistical mechanics, which is using important sampling to sample the phase space of subsystems. The properties and the performance of the system can be averaged in a large number of sampling steps, so that the ensemble average of equilibrium structural and thermodynamic performance of the system can be calculated. Monte Carlo method is limited to calculate the equilibrium properties because of its statistical mechanics basis.

Forsman used Monte Carlo method to study the interface between planar electrodes and ionic liquids with high concentration (Forsman et al. 2010). The differential capacitance curve was humped when the dispersion force was taken into account. If the dispersion force was not taken into account, there was no minimum capacitance in the differential capacitance curve at the potential of zero charge. In addition to the differential capacitance, the adsorption behavior of ions on the electrode surface is also one of the important properties of the interface double layer.

The effects of short-range correlations and ion size asymmetry on the electrical double layer of ionic liquids are studied by Monte Carlo simulation (Fedorov and Kornyshev 2008). The asymmetric bell-shape capacitance curve was obtained and agreed well with recent mean field theory (MFT) and experimental results, which can lead to an overscreening effect due to the short-range ionic correlations, which was found at moderate and small charging. When cation radius is larger than that of anions, the maximum capacitance position is moved positively from the potential of zero charge.

The structure and thermodynamics of an electrical double layer in ionic liquid with strong electrostatic interactions and the nonspherical shapes of particles were simulated by Monte Carlo method (Bhuiyan et al. 2012). A strong oscillation of ionic density profiles in this system was founded and agreed well with results from the classical density functional theory.

A model ionic liquid system was simulated by Monte Carlo method by Fedorov et al. (2010). A camel shape in differential capacitance curve was found for the electrical double layer if the ions have neutral tails in this system. The neutral counterparts of ions act as a potential hole, which can be substituted by charge groups through ion translation and rotation. It provides additional degrees of freedom for charge rearrangement by field induced in the electrical double layer, and thus producing a unique double-hump profile.

The free energy profiles of 1,3-dimethylimidazolium chloride [Dmim]Cl ionic liquid near a charged wall were studied by Monte Carlo method (Lynden-Bell et al. 2012). An excellent electrostatic screening was found at a distance of 1–2 nm. For an ionic liquid near the electrodes, a minimum can be found in the free energy profiles. The oscillation of the free energy curve of the non-charged solute shows that the non-electrostatic force plays a very important role in this system, which is a function of the distance from the wall.

10.5.3 Molecular Dynamics Simulation

Molecular dynamics simulation is common simulation method to study the interactions, structure, properties, and transport properties of many-body systems. The basic principle of molecular dynamics simulation is solving Newton's equation of motion of a many-body system under a given molecular mechanics force field over a simulated timestep. The structure and properties of the whole system can be calculated from the ensemble average of all particles in whole simulation times based on the ergodic hypothesis. Molecular dynamics simulation, as an important theoretical research method, has been widely used to study the adsorption of ionic liquids, and their distribution and interaction in the electrical double layer on the ionic liquid-electrode interface in recent years. The single wall or bilateral wall model with a vacuum slab is widely applied to reveal the structure, property, and behavior of interface of electrode-ionic liquids at the atomic and molecular level. Figure 10.6 shows the bilateral wall model with a vacuum slab that is widely used to study the electrical double layer on the interface of electrode-ionic liquids with molecular dynamics simulation.

Komyshev is a leader in the study of the theory of the interface double layer of electrode-ionic liquids with molecular dynamics simulation (Fedorov and Kornyshev 2008, 2009; Kornyshev 2007; Fedorov et al. 2010; Georgi et al. 2010). It was shown that the thickness of the double layer of ionic liquids is not the thickness of the single layer of ions, but the layers of anions and cations (Fedorov and Kornyshev 2009; Fedorov and Maxim 2008). When the electrode surface is in a non-polarized state, i.e., under the potential of zero charge, there are equal numbers of anions and cations in each layer of the solution side parallel to the surface. When the electrode is polarized, i.e., deviating from the potential of zero charge, the cation-rich or anion-rich planes are arranged alternately perpendicular to the surface direction, rather than in alternating arrangement of pure cations and anions.

In 2007, based on the mean field theory, the nature of Coulomb interaction of ions with high concentration in ionic liquids, the equal size assuming of anions and cations, and a differential capacitance equation of metal electrode-ionic liquid interface were proposed by introducing a lattice saturation parameter γ and without considering the characteristic adsorption of ion (Kornyshev 2007). It was found that the larger γ, the higher ion concentration and the more ion sites occupied in the system, and the more significant the lattice saturation effect. Using this equation, they predicted the shape of differential capacitance curves in ionic liquids. It was shown that the turning point of differential capacitance curves appeared with $\gamma = 3$. The differential capacitance curve presents characteristics of hump shape with $\gamma < 1/3$ and or bell shape with $\gamma > 1/3$.

By using molecular dynamic simulation, Komyshev found that the increase of ion concentration in the interfacial double layer is limited in high-concentration

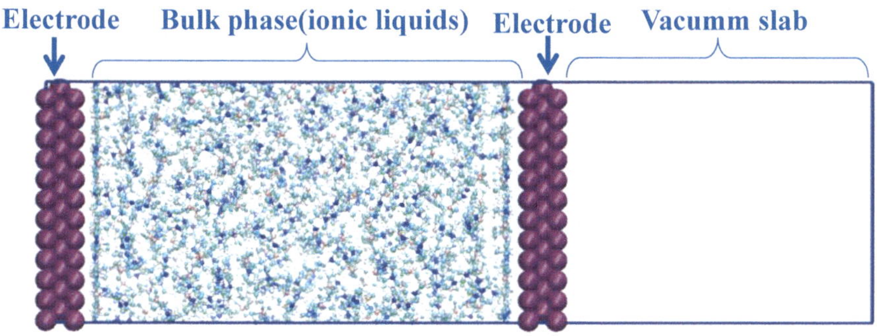

Fig. 10.6 The bilateral layer wall model used to simulate the electrical double layer in ionic liquids. The simulation box contains three parts: (1) the electrode used in experiment showed as two three-layer walls in this figure, (2) the bulk phase with ionic liquids, and (3) a vacuum slab

ionic systems (Fedorov et al. 2010). With the increase of potential deviation from the potential of zero charge, the position of the counter charge mass center is farther away from the electrode (lattice saturation effect). Therefore, the differential capacitance decreases gradually with the increase of polarization of the electrode, thus showing the hump or bell characteristics mentioned above. Because of the high concentration of ionic liquids, usually the γ is larger than 1/3, the bell curve is considered to be the characteristic of the differential capacitance curve at ionic liquid interface, and the peak potential can be correlated with the potential of zero charge.

The effect of side chain length of ionic liquids on the electrical double layer was studied with molecular dynamics simulation by Georgi et al. (2010). It was found that the differential capacitance curve will change from bell-shaped to hump-shaped, and the capacitance value will decrease with the increase of the neutral side chain length. They believed that the neutral side chains provided space for the reversal rearrangement of cations. Therefore, near potential of zero charge, the micro-polarization of the electrodes would change the configuration of cations and increase the density of anisotropic charges on the surface of the electrodes, but would not cause the significant compression of the double layer, so the interfacial capacitance would increase. When the polarization intensity increases, the interfacial capacitance would increase. Because of the steric resistance effect, the ion inversion will be blocked. At this time, the lattice saturation effect is prominent with the increase of the electrical double-layer thickness, which results in decrease in the capacitance value, thus forming a hump-shaped differential capacitance curve.

Oldham and Lauw deduced the differential capacitance curves of similar shapes based on Gouy-Chapman-Stern model and self-consistent mean field theory, respectively, which corresponded to the local minimum of differential capacitance in ionic liquids, a high-concentration ionic system (Oldham 2008; Lauw et al. 2009). Lauw's calculation results show that if the bulk dielectric constant of ionic liquids is used directly (Lauw et al. 2009), they obtained a bell-shaped differential capacitance curve. If the effective dielectric constant of the interface is considered, the hump-shaped differential capacitance curve is obtained, in which the potential of zero charge corresponds to the valley of the hump curve with local minimum of differential capacitance, while the hump-shaped differential capacitance curve appears such as micro-peak. The maximum value of capacitance appears at the position where the electrode potential deviates from the potential of zero charge about +0.8 V.

Lauw et al. thought that in water solution system, the hump-shaped differential capacitance curve was also theoretically obtained (Lauw et al. 2009). It was only because the maximum potential of the differential capacitance was outside the potential window of the water solution that the hump-shaped differential capacitance curve could not be observed in the experiment. In view of the prediction of differential capacitance curves in specific systems, some achievements have also been made by using molecular dynamics simulation.

Qiao et al. simulated the structure, property, and behavior of the double layer at 1-butyl-3-methylimidazolium nitrate [BMIM]NO_3-electrode interfaces, and deduced the corresponding differential capacitance curve (Feng et al. 2009). The thickness of about 1.1 nm for the double layer in ionic liquids was obtained. Borodin's group used the molecular dynamics simulation method to study the interfacial double-layer structure of 3-methyl-1-propylpyridinium bis(trifluoromethylsulfonyl) amide [PMPy]TF_2N-graphite electrode (Vatamanu et al. 2010). It was found that ionic liquids form a 2–3 nm multilayer structure on electrode surface.

Unlike the results of Qiao et al. (Feng et al. 2009), the Borodin group obtained the hump-shaped differential capacitance curve in 3-methyl-1-propylpyridinium bis(trifluoromethylsulfonyl) amide [PMPy]TF_2N, and the trough with local minimum of differential capacitance appeared at the potential of negative −1 V in the potential of zero charge (Vatamanu et al. 2010), while the two peaks with maximum of differential capacitance appeared at the potential of 1.5 V and −2.5 V relative to the potential of zero charge, respectively, as described earlier. The conclusion that the potential of zero charge corresponds to the valley of the hump-shaped differential capacitance curve is different.

Kislenko et al. used molecular dynamics simulation method to simulate the effect of interface electric field intensity on the adsorption of anions and cations on 1-butyl-3-methylimidazolium hexafluorophosphate [BMIM]PF_6-graphite interface ionic liquids (Kislenko et al. 2009). It was shown that the change of surface potential of graphite electrode would affect the distribution of anions and cations and adsorption orientation on the interface. The anions on the graphite surface with negative charge will not be adsorbed. The anions and cations can be simultaneously observed on the graphite electrode surface with the same positive charge. It can be concluded that the interaction between electrodes and ionic liquids at the interface, as well as the shape, structure, and size of the cations and anions, will significantly affect the properties of the double layer.

Molecular dynamics simulation, as an important theoretical research method, has been gradually applied to the study of electrode-ionic liquid interface adsorption in recent years (Feng et al. 2009; Kislenko et al. 2009; Vatamanu et al. 2010). Table 10.3 lists the advance and progress of theoretical research on the interface double-layer structure and property of ionic liquid-electrodes.

The main purpose of these works mentioned above is to investigate the adsorption of ionic liquids on electrodes and their distribution and interaction in the double layer. The published works show that theoretical calculation can provide a physical image of the interface double-layer structure of electrode-ionic liquids to a certain extent, which is very important for understanding the behavior and nature of electrical double layer on electrode-ionic liquid interface, for extending the electrochemical engineering application fields, and for improving application efficiencies of ionic liquids, such as metal electrodeposition, batteries, fuel cell, and supercapacitors.

Table 10.3 Progress on the theoretical studies of the electrical double-layer structure of the ionic liquid-electrode system

Year	IL(s)	Electrodes and shapes	Methods	References
2010	[BMIM]PF_6, [OMIM]PF_6	Graphite, massive on both sides	MD	Wang (2010)
2012	[DMIM]PF_6, [BMIM]PF_6	Graphite, bilateral three-layer plane	MD	Si (2012)
2014	[DMIM]Cl	Single-walled carbon nanotubes with radius of 6.78	MD	Huang (2014)
2014	[BMIM]BF_4	Spherical model plate electrode	MD	Ivaništšev et al. (2014)
2014	$[C_n(MIM)_2](BF_4)_2$, $[C_6(MIM)_2](TF_2N)_2$	Onion-like carbon	MD	Li et al. (2014)
2014	[BMIM]Li,K/BF_4	Bilateral monolayer graphite	MD	Méndez-Morales et al. (2014)
2014	[BMIM]PF_6	Graphite, bilateral three-layer plane	MD	Sha et al. (2014)
2014	[C_nMIM]TFSI, [C_nMIM]FSI	Planar and rough graphite electrodes	MD	Hu et al. (2014)
2014	[BMIM]PF_6, [BMIM]BF_4	Planar and rough graphite electrodes	MD	Bedrov et al. (2014)
2014	[Pyr]FSI	Graphite	MD	Vatamanu et al. (2014)
2014	[BMIM]PF_6	Graphite electrode	MD	Merlet et al. (2014)
2014	[EMIM]BF_4	Mica	MD	Turesson et al. (2014)
2015	[BMIM]PF_6	Ideal graphite electrodes, bilateral	MD	Jin (2015)
2015	[BMIM]PF_6	Graphene sheet	MD	Wang (2015)
2015	[BMIM]FSI, [BMIM]TFSI	Graphene bilayer	MD	Liu et al. (2015)
2015	[EMIM]TFSI	Graphene	MD	Feng et al. (2015)
2015	[EMIM]TF_2N	Graphene (bilateral three-layer plates)	MD	Black et al. (2015)
2015	[BMIM]PF_6	Planar graphite plate and atomic graphite plate	MD	Breitsprecher et al. (2015)
2015	[BMIM]PF_6	Bilateral graphene plates	MD	Jin et al. (2015)
2015	$[N_{2228}]NTf_2$	Bilateral graphene plates	MD	Sharma and Kashyap (2015)
2015	[BMIM]CH_3COO, [BMIM]SCN, [BMIM]$N(CN)_2$	Liquid-gas interface	MD	Mondal and Balasubramanian (2015)
2016	[EMIM]BF_4	Bilateral three-layer graphite	MD	Haskins and Lawson (2016)
2016	[BMIM]BF_4	Bilateral graphite plates	MD	Ivaništšev et al. (2016)
2016	[BMIM]PF_6	Nanocarbide electrodes	MD	Péan et al. (2016)

(continued)

Table 10.3 (continued)

Year	IL(s)	Electrodes and shapes	Methods	References
2016	$[Et_4N]BF_4$, $[C_2MIM]BF_4$, $[C_8MIM]BF_4$	Planar graphite electrodes and bilateral three-layer plane electrodes with different roughness	MD	Li et al. (2016a)
2016	$[EMIM]BF_4$, $[EMIM]TF_2N$, $[EMIM]PF_6$	Mica	MD	Black et al. (2016)
2016	$[C_n(MIM)_2](TF_2N)_2$, $[C_6(MIM)TMA](TF_2N)_2$	Bilateral three-layer graphite plates	MD	Li et al. (2016c)
2016	$[EMIM]BF_4$	Unilateral monolayer graphene	MD	Burt et al. (2016)
2017	$[EMIM]BF_4$, $[BMIM]BF_4$, $[HMIM]BF_4$	Bilateral graphene plates	MD	Jo et al. (2017)
2017	$[EMIM]TF_2N$, $[EMIM]BF_4$	Graphite	MD	Pal et al. (2017)
2018	$[EMIM]TF_2N$	Atomic smooth Pt electrode	MD	Chen et al. (2018)
2017	$[BMIM]PF_6$	Gold	MD	Vanin and Brodskaya (2017)
2017	$[EMIM]TF_2N$	Bilateral three-layer graphite	MD	Chen et al. (2017)
2017	$[EMIM]BF_4$	Multilayer transverse Graphene plate	MD	Lu et al. (2017)
2017	[EMIM]GLY, [EMIM]SER, [EMIM]VAL, [EMIM]PHE	Bilateral three-layer graphite plate	MD	Wu et al. (2017)
2017	[EMIM]DCA, [BMIM]DCA, [BMP]DCA	Bilateral graphene plates	MD	Begić et al. (2017)
2017	$[EMIM]TF_2N$	Unilateral grapheme plate	MD	Ma et al. (2017)
2018	$[EMIM]BF_4$	Au(111) plate	MD	Ruzanov et al. (2018)
2018	[Pyr]TFSI, [BMA]TFSI, $[Pyr]PF_6$, $[BMA]PF_6$	Bilateral graphene plates	MD	Chaban et al. (2018)
2018	$[Li(G_3)]TFSI$	Bilateral graphene plates	MD	Yu et al. (2018)
2018	$[BMIM]PF_6$	Graphite plate, three on each side	MD	Le Ma et al. (2018)
2018	$[EMIM]PF_6$	Unilateral graphene	MD	Razmkhah et al. (2018a)
2018	$[C_nMIM]TF_2N$ ($n = 2,4,6,10$), $[C_6MIM]BF_4$	Amorphous indium gallium zinc oxide transistor	MD	Zhao et al. (2018)
2018	$[EMIM],PF_6$	Graphene	MD	Razmkhah et al. (2018b)
2018	$[EMIM]BF_4$	Graphene	MD	Park et al. (2018)
2018	$[Pyr_{13}]TFSI$	Au(111) plate	MD	Wang et al. (2018)
2018	[EMIM]SCN, $[EMIM]N(CN)_2$, $[EMIM]C(CN)_2$, $[EMIM]B(CN)_2$	Bilateral graphene plates	MD	Jo et al. (2018)

(continued)

Table 10.3 (continued)

Year	IL(s)	Electrodes and shapes	Methods	References
2018	[BMIM]PF_6, [BMIM] BF_4	Amorphous silica sheet	MD	Neal et al. (2018)
2019	[Pyr_{14}]TFSI	HOPG/bilateral three-layer graphene sheets	MD	Liu et al. (2019)
2019	[BMIM]BF_4	Bilateral three-layer graphene sheets	MD	Stettner et al. (2019)

In the abbreviations of the ionic liquids, *EMIM* or *C_2MIM* 1-ethyl-3-methylimidazolium, *DMIM* 1,3-dimethyl-2-imidazolidinone, *BMIM* or *C_4MIM* 1-*n*-butyl-3-methylimidazolium, *OMIM* or *C_8MIM* 1-methyl-3-octylimidazolium, *C_nMIM* 1-alkyl-3-methylimidazolium, *$C_n(MIM)_2$* 1-alkyl-3-dimethylimidazolium, *$C_6(MIM)_2$* 1-hexyl-3-dimethylimidazolium, *Pyr* 1-butyl-1-methyl-pyrrolidinium, *N_{2228}* triethyloctylammonium.*Et_4N* tetraethylammonium, *C_6(MIM)TMA* 1-(1-trimethylammonium-yl-hexyl)-3-methylimidazolium, *BMA* butyltrimethylammonium, *BMP* 1-butyl-1-methylpyrrolidinium, *Pyr_{13}* 1-methyl-1-propylpyrrolidinium, *Li(G_3)* Li(triethylene glycol dimethyl ether), *Pyr_{14}* 1-methyl-1-butylpyrrolidinium, *BF_4* tetrafluoroborate, *PF_6* hexafluorophosphate, *TFSI* bis-((trifluoromethyl)sulfonyl)imide, *FSI* bis(fluorosulfonyl)imide, *TF_2N* bis(trifluoromethylsulfonyl)amide, *SCN* thiocyanate, *$N(CN)_2$* dicyanamide, *$C(CN)_2$* dicyanomethanide, *$B(CN)_2$* dicyanoborate, *GLY* aminoethanic acid, *CH_3COO* acetate, *SER* serine, *PHE* phenylalanine, *DCA* dicyandiamide, *VAL* valine, *TFSA* trifluoromethanesulfonamide, *OTF* trifluoromethanesulfonate, and *FAP* tris(pentafluoroethyl) trifluorophosphate, *MD* molecular dynamics simulation

10.6 Conclusions

Ionic liquids as a green solvent and electrolytes have been wildly and successfully applied in the electrodeposition of metals, catalysis, electrochemical applications such as batteries, fuel cells, supercapacitors, and solar cells. The obvious feature of these systems is that there is an interface between ionic liquids and other liquids, solids, or gases. In chemical and electrochemical reactions, most reactions occur at interface and the mass will transfer through the interface. The interface structure and dynamics are important for the reaction rate in these systems. However, ionic liquids are composed of pure ions in the liquid state, and are different from the traditional aqueous or organic solvents. Most theoretical models and knowledge are not suitable for description of the behavior of ionic liquids. Therefore, deeper insights into the interfacial structure and properties of ionic liquids are highly desired to understand the performance of ionic liquids in various applications.

In recent years, considerable advance and progress have been made in experimental characterization of the interface between electrodes and ionic liquids, but the research work is still in its infancy. Firstly, most of the existing literatures use metal polycrystalline electrodes, rough electrodes, mercury electrodes, and even glassy carbon electrodes with uncertain structure as working electrodes. Therefore, the results from different groups are not comparable because of the differences of experimental systems, but also bring difficulties to theoretical processing and model building. Secondly, due to the limitation of the principle of the method, it is impossible to use STM, AFM, or electrochemical vibration spectroscopy alone to conduct a comprehensive study of double layer on the ionic liquid-electrode interface.

Especially at present, the study of electrochemical vibration spectroscopy is still limited to polycrystalline or rough electrodes with uncertain structure, which cannot be related to the information of orderly adsorption of species obtained by STM on single-crystal electrodes. Therefore, the limited results are not enough to accurately and deeply understand the structure and properties of the ionic liquid-electrode interface at the molecular level. Finally, most of the ionic liquids have certain hygroscopicity. Even if drying measures are taken before use, it is sometimes difficult to avoid the existence of very trace water in the experimental process, and the presence of water may affect the structure of the ionic liquid interface and the electrochemical processes occurring on it. Therefore, the structure analysis of water at interface in ionic liquids and its influence on the interfacial process need to be further studied.

The further study of the interface between electrodes and ionic liquids requires the use of well-defined single-crystal electrode surfaces. It is very important to broaden the means of research and to characterize the interface jointly by different techniques. STM tunneling spectroscopy can be used to detect the electronic structure information of the species perpendicular to the solution side near the electrode surface. Combining with theoretical calculation, it is possible to expand the understanding of the solution side of the ionic liquid interface.

The theoretical calculation and simulation results can provide us with a physical image of the structure and property of the double layer at ionic liquid-electrode interface to a certain extent, which is of great guiding significance. However, there are still some shortcomings. One of the important points is that the model used is too simple to consider the tight-layer structure of the strong interaction between ionic liquids and electrode surface and its effect on the differential capacitance curve. In fact, the differential capacitance curves of ionic liquids and electrodes are obviously different in different systems when the studied system is concretized and the interfacial properties of ionic liquids and electrodes are investigated at the atomic level. Therefore, as in the study of aqueous solution, the combination of theory and experiment is very important for accurately describing the structure of the double layer at the ionic liquid-electrode interface. The interaction between electrodes and ionic liquids at the interface as well as the shape, structure, and size of the cations and anions have a very important effect on the structure and properties of the double layer. Therefore, improvement or development of new methods or theory can be combined with more factors to give more accurate description of structure and properties of the double layer which is a further direction. We believe that the comprehensive and systematic study of the behavior and nature of electrical double layer on electrode-ionic liquid interface will benefit their applications especially in electrochemical engineering application fields such as metal electrodeposition, batteries, fuel cell, and supercapacitors.

Acknowledgments The author would like to thank the financial support from National Natural Science Foundation of China (No. 51774158, 51264021, 50904031).

References

Abbott AP, McKenzie KJ (2006) Application of ionic liquids to the electrodeposition of metals. Phys Chem Chem Phys 8:4265–4279. https://doi.org/10.1039/b607329h

Alam MT, Islam MM, Okajima T, Ohsaka T (2007) Measurements of differential capacitance in room temperature ionic liquid at mercury, glassy carbon and gold electrode interfaces. Electrochem Commun 9:2370–2374. https://doi.org/10.1016/j.elecom.2007.07.009

Alam MT, Islam MM et al (2008) Capacitance measurements in a series of room-temperature ionic liquids at glassy carbon and gold electrode interfaces. J Phys Chem C 112(42):16600–16608. https://doi.org/10.1021/jp804620m

Aliaga C, Baldelli S (2006) Sum frequency generation spectroscopy and double-layer capacitance studies of the 1-butyl-3-methylimidazolium dicyanamide-platinum interface. J Phys Chem B 110(37):18481–18491. https://doi.org/10.1021/jp063476z

Aliaga C, Baldelli S (2008) A sum frequency generation study of the room-temperature ionic liquid–titanium dioxide interface. J Phys Chem C 112 (8):3064–3072. https://doi.org/10.1021/jp709753r

Armand M, Endres F, MacFarlane DR et al (2009) Ionic-liquid materials for the electrochemical challenges of the future. Nature Mater 8:621–629. https://doi.org/10.1038/nmat2448

Atkin R, Warr GG (2007) Structure in confined room-temperature ionic liquids. J Phys Chem C 111(13):5162–5168. https://doi.org/10.1021/jp067420g

Atkin R, Borisenko N, Drüschler M, El Abedin SZ, Endres F, Hayes R, Huber B, Roling B (2011) An in situ STM/AFM and impedance spectroscopy study of the extremely pure 1-butyl-1-methylpyrrolidinium tris(pentafluoroethyl)trifluorophosphate/au(111) interface: potential dependent solvation layers and the herringbone reconstruction. Phys Chem Chem Phys 13(15):6849. https://doi.org/10.1039/c0cp02846k

Atkin R, Borisenko N, Drüschler M et al (2014) Structure and dynamics of the interfacial layer between ionic liquids and electrode materials. J Mol Liq 192:44–54. https://doi.org/10.1016/j.molliq.2013.08.006

Baldelli S (2008) Surface structure at the ionic liquid-electrified metal interface. Acc Chem Res 41(3):421–431. https://doi.org/10.1021/ar700185h

Bedrov D, Vatamanu J, Hu Z (2014) Ionic liquids at charged surfaces: insight from molecular simulations. J Non-Cryst Solids 407:339–348. https://doi.org/10.1016/j.electacta.2014.08.072

Begić S, Jónsson E, Chen F et al (2017) Molecular dynamics simulations of pyrrolidinium and imidazolium ionic liquids at graphene interfaces. Phys Chem Chem Phys 19(44):30010–30020. https://doi.org/10.1039/c7cp03389c

Bhuiyan LB, Lamperski S, Wu J et al (2012) Monte Carlo simulation for the double layer structure of an ionic liquid using a dimer model: a comparison with the density functional theory. J Phys Chem B 116(34):10364–10370

Black JM, Okatan MB, Feng G et al (2015) Topological defects in electric double layers of ionic liquids at carbon interfaces. Nano Energy 15:737–745. https://doi.org/10.1016/j.nanoen.2015.05.037

Black JM, Zhu M, Zhang P et al (2016) Fundamental aspects of electric double layer force-distance measurements at liquid-solid interfaces using atomic force microscopy. Sci Rep 6:32389. https://doi.org/10.1038/srep32389

Bockris JOM, Devanathan MAV, Mueller K (1963) On the structure of charged interfaces. Proc Roy Soc Lond Ser A 274:55–79

Brandão CRR, Costa LAF, Breyer HS et al (2009) Surface-enhanced Raman scattering (SERS) of a copper electrode in 1-n-butyl-3-methylimidazolium tetrafluoroborate ionic liquid. Electrochem Commun 11(9):1846–1848. https://doi.org/10.1016/j.elecom.2009.07.034

Breitsprecher K, Szuttor K, Holm C (2015) Electrode models for ionic liquid-based capacitors. J Phys Chem C 119(39):22445–22451. https://doi.org/10.1021/acs.jpcc.5b06046

Burt R, Birkett G, Zhao XS (2014) A review of molecular modelling of electric double layer capacitors. Phys Chem Chem Phys 16(14):6519–6538. https://doi.org/10.1039/c3cp55186e

Burt R, Birkett G, Salanne M et al (2016) Molecular dynamics simulations of the influence of drop size and surface potential on the contact angle of ionic-liquid droplets. J Phys Chem C 120(28):15244–15250. https://doi.org/10.1021/acs.jpcc.6b04696
Chaban VV, Andreeva NA, Fileti EE (2018) Graphene/ionic liquid ultracapacitors: does ionic size correlate with energy storage performance. New J Chem 42(22):18409–18417. https://doi.org/10.1039/c8nj04399j
Chapman DL (1913) A contribution to the theory of electrocapillarity. Phil Mag 25:475–481
Chen L (2017) Study on the force curve of single crystal electrode/ionic liquid interface by AFM and the effects of cationic alkyl side chains and ferrocene, doctoral dissertation. Xiamen University, Xiamen, pp 11–16. (Doctoral dissertation)
Chen M, Li S, Feng G (2017) The influence of anion shape on the electrical double layer microstructure and capacitance of ionic liquids-based supercapacitors by molecular simulations. Molecules 22(2):241. https://doi.org/10.3390/molecules22020241
Chen M, Goodwin ZAH, Feng G et al (2018) On the temperature dependence of the double layer capacitance of ionic liquids. J Electroanal Chem 819:347–358. https://doi.org/10.1016/j.jelechem.2017.11.005
Cho J, Shin WK, Kim DW et al (2016) Electrochemical characterization of electric double layer capacitors assembled with pyrrolidinium-based ionic liquid electrolytes. J Electrochem Sci Technol 7(3):199–205. https://doi.org/10.5229/jecst.2016.7.3.199
Costa R, Pereira CM, Silva AF (2015) Charge storage on ionic liquid electric double layer: the role of the electrode material. Electrochim Acta 167:421–428. https://doi.org/10.1016/j.electacta.2015.02.180
Costa R, Pereira CM, Silva AF (2016) Role of the anion on the interfacial structure of ionic liquids binary mixtures at mercury interfaces. Electrochim Acta 195:150–157. https://doi.org/10.1016/j.electacta.2016.01.232
Costa R, Voroshylova IV, Cordeiro MNDS et al (2018) Enhancement of differential double layer capacitance and charge accumulation by tuning the composition of ionic liquids mixtures. Electrochim Acta 261:214–220. https://doi.org/10.1016/j.electacta.2017.12.134
Deng YQ (2006) Ionic liquids: properties, preparation and application. China SINO-PEC press, Beijing, pp 20–45
Down MP, Banks CE (2018) Freestanding three-dimensional graphene macroporous supercapacitor. ACS Appl Energy Mater 1(2):891–899. https://doi.org/10.1021/acsaem.7b00338
Drüschler M, Huber B, Passerini S et al (2010) Hysteresis effects in the potential-dependent double layer capacitance of room temperature ionic liquids at a polycrystalline platinum inter-face. J Phys Chem C 114(8):3614–3617. https://doi.org/10.1021/jp911513k
Dzyuba SV, Bartsch RA (2002) Influence of structural variations in 1-alkyl-aralkyl-3-methylimidazolium hexafluorophosphates and bis(trifluoro-methyl-sulfonyl)imides on physical properties of the ionic liquids. Chem Phys Chem 3:161–166. https://doi.org/10.1002/1439-7641(20020215)3:2<161::aid-cphc161>3.0.co;2-3
Fedorov MV, Kornyshev AA (2008) Ionic liquid near a charged wall: structure and capacitance of electrical double layer. J Phys Chem B 112(38):11868–11872. https://doi.org/10.1021/jp803440q
Fedorov MV, Kornyshev AA (2009) Ionic liquid near a charged wall: structure and capacitance of electrical double layer. J Phys Chem B 113(13):4500–4500. https://doi.org/10.1021/jp900599h
Fedorov MV, Kornyshev AA (2014) Ionic liquids at electrified interfaces. Chem Rev 114(5):2978–3036. https://doi.org/10.1021/cr400374x
Fedorov MV, Maxim V, Kornyshev AA (2008) Towards understanding the structure and capacitance of electrical double layer in ionic liquids. Electrochim Acta 53(23):6835–6840. https://doi.org/10.1016/j.electacta.2008.02.065
Fedorov MV, Georgi N, Kornyshev AA (2010) Double layer in ionic liquids: the nature of the camel shape of capacitance. Electrochem Commun 12(2):296–299. https://doi.org/10.1016/j.elecom.2010.03.023

Feng G, Zhang JS, Qiao R (2009) Microstructure and capacitance of the electrical double layers at the interface of ionic liquids and planar electrodes. J Phys Chem C 113(11):4549–4559. https://doi.org/10.1021/jp809900w
Feng G, Li S, Zhao W et al (2015) Microstructure of room temperature ionic liquids at stepped graphite electrodes. AICHE J 61(9):3022–3028. https://doi.org/10.1002/aic.14927
Garlyyev B, Xue S, Watzele S et al (2018) Influence of the nature of the alkali metal cations on the electrical double-layer capacitance of model Pt (111) and Au (111) electrodes. J Phys Chem Lett 9(8):1927–1930. https://doi.org/10.1021/acs.jpclett.8b00610
Georgi N, Kornyshev AA, Fedorov MV (2010) The anatomy of the double layer and capacitance in ionic liquids with anisotropic ions: electrostriction vs. lattice saturation. J Electroanal Chem 649(1–2):261–267. https://doi.org/10.1016/j.jelechem.2010.07.004
Gnahm M, Pajkossy T, Kolb DM (2010) The interface between Au(111) and all ionic liquid. Electrochim Acta 55:6212–6217. https://doi.org/10.1016/j.electacta.2009.08.031
Gomes C, Costa R, Pereira CM et al (2014) The electrical double layer at the ionic liquid/Au and Pt electrode interface. RSC Adv 4(55):28914–28921. https://doi.org/10.1039/c4ra03977g
Gore TR, Bond T, Zhang W et al (2010) Hysteresis in the measurement of double-layer capacitance at the gold–ionic liquid interface. Electrochem Commun 12(10):1340–1343. https://doi.org/10.1016/j.elecom.2010.07.015
Gouy C (1910) Constitution of the electric charge at the surface of an electrolyte. J Phys 9:457–467
Grahame DC (1946) Properties of the electrical double layer at a mercury surface. II. The effect of frequency on the capacity and resistance of ideal polarized electrodes. J. Am Chem Soc 68:301–310. https://doi.org/10.1021/ja01206a045
Grahame DC (1954) Differential capacity of mercury in aqueous sodium fluoride solutions. I. Effect of concentration at 250. J Am Chem Soc 76:4819–4823. https://doi.org/10.1021/ja01648a014
Griffin JM, Forse AC, Wang H et al (2015) Ion counting in supercapacitor electrodes using NMR spectroscopy. Faraday Discuss 2015(176):49–68. https://doi.org/10.1039/c4fd00138a
Gu ZY, Brennecke JF (2002) Volume expansivities and isothermal compressibilities of imidazolium and pyridinium-based ionic liquids. J Chem Eng Data 47:339–345. https://doi.org/10.1021/je010242u
Hagiwara R, Ito Y (2000) Room temperature ionic liquids of alkylimidazolium cations and fluoroanions. J Fluor Chem 105:221–227. https://doi.org/10.1016/s0022-1139(99)00267-5
Haskins JB, Lawson JW (2016) Evaluation of molecular dynamics simulation methods for ionic liquid electric double layers. J Chem Phys 144(18):184707. https://doi.org/10.1063/1.4948938
Hayes R, Borisenko N, Tam MK et al (2011) Double layer structure of ionic liquids at the Au(111) electrode interface: an atomic force microscopy investigation. J Phys Chem C 115(14):6855–6863. https://doi.org/10.1021/jp200544b
Hayes R, Warr GG, Atkin R (2015) Structure and nanostructure in ionic liquids. Chem Rev 115(13):6357–6426. https://doi.org/10.1021/cr500411q
He Y, Yang Y, Nie S et al (2018) Electric-double-layer transistors for synaptic devices and neuromorphic systems. J Mater Chem C 6(20):5336–5352. https://doi.org/10.1039/c8tc00530c
Helmholtz HF (1879) Studies of electric boundary layers. Ann Phys Chem 7:337–382
Henderson D, Wu J (2012) Electrochemical properties of the double layer of an ionic liquid using a dimer model electrolyte and density functional theory. J Phys Chem B 116(8):2520–2525. https://doi.org/10.1021/jp212082k
Henderson D, Lamperski S, Bari Bhuiyan L et al (2013) The tail effect on the shape of an electrical double layer differential capacitance curve. J Chem Phys 138(14):144704. https://doi.org/10.1063/1.4799886
Hu Z, Vatamanu J, Borodin O et al (2014) A comparative study of alkylimidazolium room temperature ionic liquids with FSI and TFSI anions near charged electrodes. Electrochim Acta 145:40–52. https://doi.org/10.1016/j.electacta.2014.08.072
Huang S (2014) Structural properties of ionic liquids in confined nanoscale space. Doctoral dissertation. Nankai University, Tianjin, pp 2–18

Hurley FH, Wier TP (1951a) Electrodeposition of aluminum from nonaqueous solutions at room temperature. J Electrochem Soc 98:207–212. https://doi.org/10.1149/1.2778133
Hurley FH, Wier TP (1951b) Electrodeposition of metals from fused quaternary ammonium salts. J Electrochem Soc 98:203–206. https://doi.org/10.1149/1.2778132
Ivaništšev V, O'Connor S, Fedorov MV (2014) Poly (a) morphic portrait of the electrical double layer in ionic liquids. Electrochem Commun 48:61–64. https://doi.org/10.1016/j.elecom.2014.08.014
Ivaništšev V, Kirchner K, Kirchner T et al (2015) Restructuring of the electrical double layer in ionic liquids upon charging. J Phys Condens Matter 27(10):102101. https://doi.org/10.1088/0953-8984/27/10/102101
Ivaništšev V, Méndez-Morales T, Lynden-Bell RM et al (2016) Molecular origin of high free energy barriers for alkali metal ion transfer through ionic liquid–graphene electrode interfaces. Phys Chem Chem Phys 18(2):1302–1310. https://doi.org/10.1039/c5cp05973a
Jacquemin J, Husson P, Padua AAH, Maje V (2006) Density and viscosity of several pure and water-saturated ionic liquid. Green Chem 8:172–180. https://doi.org/10.1039/b513231b
Jin WY (2015) Molecular dynamics simulation of interaction between ionic liquids and interfaces, doctoral dissertation. Nankai University, Tianjin, pp 8–18
Jin W, Liu X, Han Y et al (2015) Effects of repulsive interaction on the electric double layer of an imidazolium-based ionic liquid by molecular dynamics simulation. Phys Chem Chem Phys 17(4):2628–2633. https://doi.org/10.1039/c4cp04853a
Jitvisate M, Seddon JRT (2017) Direct measurement of the differential capacitance of solvent-free and dilute ionic liquids. J Phys Chem Lett 9(1):126–131. https://doi.org/10.1021/acs.jpclett.7b02946
Jo S, Park SW, Shim Y et al (2017) Effects of alkyl chain length on interfacial structure and differential capacitance in graphene supercapacitors: a molecular dynamics simulation study. Electrochim Acta 247:634–645. https://doi.org/10.1016/j.electacta.2017.06.169
Jo S, Park SW, Noh C et al (2018) Computer simulation study of differential capacitance and charging mechanism in graphene supercapacitors: effects of cyano-group in ionic liquids. Electrochim Acta 284:577–586. https://doi.org/10.1016/j.electacta.2018.07.126
Keblinski P, Eggebrecht J, Wolf D, Phillpot SR (2000) Molecular dynamics study of screening in ionic fluids. J Chem Phys 113(1):282–291. https://doi.org/10.1063/1.481819
Kislenko SA, Samoylov IS, Amirov RH (2009) Molecular dynamics simulation of the electrochemical interface between a graphite surface and the ionic liquid [BMIM][PF_6]. Phys Chem Chem Phys 11(27):5584–5590. https://doi.org/10.1039/b823189c
Kornyshev AA (2007) Double-layer in ionic liquids: paradigm change? J Phys Chem B 111(20):5545–5557. https://doi.org/10.1021/jp067857o
Kumagai S, Hatomi M, Tashima D (2017) Electrochemical performance of microporous and mesoporous activated carbons in neat and diluted 1-ethyl-3-methylimidazolium tetrafluoroborate. J Power Sources 343:303–315. https://doi.org/10.1016/j.jpowsour.2017.01.064
Lamperski S, Kaja M, Bhuiyan LB et al (2013) Influence of anisotropic ion shape on structure and capacitance of an electric double layer: a Monte Carlo and density functional study. J Chem Phys 139(5):054703. https://doi.org/10.1063/1.4817325
Lauw Y, Horne MD, Rodopoulos T, Leermakers FA (2009) Room-temperature ionic liquids: excluded volume and ion polarizability effects in the electrical double-layer structure and capacitance. Phys Rev Lett 103(11):117801. https://doi.org/10.1103/physrevlett.103.117801
Le Ma J, Meng Q, Fan J (2018) Charge driven lateral structural evolution of ions in electric double layer capacitors strongly correlates with differential capacitance. Phys Chem Chem Phys 20(12):8054–8063. https://doi.org/10.1039/c7cp08075a
Lee JH, Ryu JB, Lee AS et al (2016) High-voltage ionic liquid electrolytes based on ether functionalized pyrrolidinium for electric double-layer capacitors. Electrochim Acta 222:1847–1852. https://doi.org/10.1016/j.electacta.2016.11.176
Lei H, Tu J, Yu Z et al (2017) Exfoliation mechanism of graphite cathode in ionic liquids. ACS Appl Mater Interfaces 9(42):36702–36707. https://doi.org/10.1021/acsami.7b03306

Li RX (2004) Green solvents: synthesis and application of ionic liquids. Chemical Industry Engineering Press, Beijing, pp 27–39
Li D (2008) Principles of electrochemistry, 3rd edn. Beijing University of Aeronautics and Astronautics Press, Beijing, pp 25–38
Li S, Van Aken KL, McDonough JK et al (2014) The electrical double layer of dicationic ionic liquids at onion-like carbon surface. J Phys Chem C 118(8):3901–3909. https://doi.org/10.1021/jp409888f
Li J, Vatamanu M, Borodin O et al (2016a) A comparative study of room temperature ionic liquids and their organic solvent mixtures near charged electrodes. J Phys Condens Matter 28(46):464002. https://doi.org/10.1088/0953-8984/28/46/464002
Li MG, Chen L, Zhong YX et al (2016b) The electrochemical interface of Ag(111) in 1-ethyl-3-methylimidazolium bis (trifluoromethylsulfonyl) imide ionic liquid - a combined in-situ scanning probe microscopy and impedance study. Electrochim Acta 197:282–289. https://doi.org/10.1016/j.electacta.2015.12.227
Li S, Zhu M, Feng G (2016c) The effects of dication symmetry on ionic liquid electrolytes in supercapacitors. J Phys Condens Matter 28(46):464005. https://doi.org/10.1088/0953-8984/28/46/464005
Lian C, Liu K, Van Aken KL et al (2016) Enhancing the capacitive performance of electric double-layer capacitors with ionic liquid mixtures. ACS Energy Lett 1(1):21–26. https://doi.org/10.1021/acsenergylett.6b00010
Liew CW, Arifin KH, Kawamura J et al (2017) Effect of halide anions in ionic liquid added poly (vinyl alcohol)-based ion conductors for electrical double layer capacitors. J Non-Cryst Solids 458:97–106. https://doi.org/10.1016/j.jnoncrysol.2016.12.022
Liu Z (2018) Research on electrical double-layer structure and mechanical properties of ionic liquids at solid-liquid interface. Master's thesis. Southeast University, Nanjing, pp 3–15
Liu Y, Zhang Y, Wu G et al (2006) Coexistence of liquid and solid phases of Bmim-PF6 ionic liquid on mica surfaces at room temperature. J Am Chem Soc 128(23):7456–7457. https://doi.org/10.1021/ja062685u
Liu X, Wang Y, Li S et al (2015) Effects of anion on the electric double layer of imidazolium-based ionic liquids on graphite electrode by molecular dynamics simulation. Electrochim Acta 184:164–170. https://doi.org/10.1016/j.electacta.2015.10.064
Liu S, Peng J, Chen L et al (2019) In-situ STM and AFM studies on electrochemical interfaces in imidazolium-based ionic liquids. Electrochim Acta 309:11–17. https://doi.org/10.1016/j.electacta.2019.04.066
Lockett V, Sedev R, Ralston J et al (2008) Differential capacitance of the electrical double layer in imidazolium-based ionic liquids: influence of potential, cation size, and temperature. J Phys Chem C 112(19):7486–7495. https://doi.org/10.1021/jp7100732
Lu P, Dai Q, Wu L et al (2017) Structure and capacitance of electrical double layers at the graphene–ionic liquid interface. Appl Sci 7(9):939. https://doi.org/10.3390/app7090939
Lucio AJ, Shaw SK (2018) Capacitive hysteresis at the 1-ethyl-3-methylimidazolium tris (pentafluoroethyl)-trifluorophosphate–polycrystalline gold interface. Anal Bioanal Chem 410(19):4575–4586. https://doi.org/10.1007/s00216-018-0962-5
Lynden-Bell RM, Frolov AI, Fedorov MV (2012) Electrode screening by ionic liquids. Phys Chem Chem Phys 14(8):2693–2701. https://doi.org/10.1039/c2cp23267g
Ma K, Wang X, Cui Y et al (2017) Understanding the graphene-based electric double layer from dielectric perspective: a density functional study. Chem Phys Lett 677:137–142. https://doi.org/10.1016/j.cplett.2017.04.006
Mantz RA, Trulove PC (2002a) In: Wsserscheid P, Welton T (eds) Ionic liquids in synthesis. Wiley-VCH, Weinheim, pp 143–149
Mantz RA, Trulove PC (2002b) In: Wasserscheid P, Welton T (eds) In ionic liquids in synthesis. Wiley-VCH, Weinheim, pp 120–124
Martins VL, Rennie AJR, Sanchez-Ramirez N et al (2018a) Improved performance of ionic liquid supercapacitors by using tetracyanoborate anions. ChemElectroChem 5(4):598–604. https://doi.org/10.1002/celc.201701164

Martins VL, Torresi RM, Rennie AJR (2018b) Design considerations for ionic liquid based electrochemical double layer capacitors. Electrochim Acta 270:453–460. https://doi.org/10.1016/j.electacta.2018.03.094

Matsumoto H, Yanagida M, Tanimoto K, Nomura M, Kitagawa YY (2000) Highly conductive room temperature molten salts based on small trimethyl-alkylammonium canons and bistrifluoromethylsulfonyl imide. Chem Lett 922-923. https://doi.org/10.1246/cl.2000.922

Matsumoto M, Shimizu S, Sotoike R et al (2017) Exceptionally high electric double layer capacitances of oligomeric ionic liquids. J Am Chem Soc 139(45):16072–16075. https://doi.org/10.1021/jacs.7b09156

Méndez-Morales T, Carrete J, Pérez-Rodríguez M et al (2014) Molecular dynamics simulations of the structure of the graphene–ionic liquid/alkali salt mixtures interface. Phys Chem Chem Phys 16(26):13271–13278. https://doi.org/10.1039/c4cp00918e

Merlet C, Limmer DT, Salanne M et al (2014) The electric double layer has a life of its own. J Phys Chem C 118(32):18291–18298. https://doi.org/10.1021/jp503224w

Mondal A, Balasubramanian S (2015) A refined all-atom potential for imidazolium-based room temperature ionic liquids: acetate, dicyanamide, and thiocyanate anions. J Phys Chem B 119(34):11041–11051. https://doi.org/10.1021/acs.jpcb.5b02272

Moreno-Fernández G, Schütter C, Rojo JM et al (2018) On the interaction of carbon electrodes and nonconventional electrolytes in high-voltage electrochemical capacitors. J Solid State Electrochem 22(3):717–725. https://doi.org/10.1007/s10008-017-3809-7

Motobayashi K, Nishi N, Inoue Y et al (2017) Potential-induced restructuring dynamics of ionic liquids on a gold electrode: steric effect of constituent ions studied by surface-enhanced infrared absorption spectroscopy. J Electroanal Chem 800:126–133. https://doi.org/10.1016/j.jelechem.2017.02.049

Müller C, Vesztergom S, Pajkossy T et al (2015) The interface between Au(100) and 1-butyl-3-methyl-imidazolium-bis (trifluoromethylsulfonyl) imide. J Electroanal Chem 737:218–225. https://doi.org/10.1016/j.jelechem.2014.06.010

Nanjundiah C, McDevitt SF (1997) Differential capacitance measurements in solvent-rice ionic liquids at Hg and C interfaces. J Electrochem Soc 144:3392–3397. https://doi.org/10.1149/1.1838024

Neal JN, Wesolowski DJ, Henderson D et al (2018) Electric double layer capacitance for ionic liquids in nanoporous electrodes: effects of pore size and ion composition. J Mol Liq 270:145–150. https://doi.org/10.1016/j.molliq.2017.10.128

Oldham KB (2008) A Gouy–Chapman–Stern model of the double layer at a (metal)/(ionic liquid) interface. J Electroanal Chem 613(2):131–138. https://doi.org/10.1016/j.jelechem.2007.10.017

Oll O, Romann T, Siimenson C et al (2017a) Influence of chemical composition of electrode material on the differential capacitance characteristics of the ionic liquidl electrode interface. Electrochem Commun 82:39–42. https://doi.org/10.1016/j.elecom.2017.07.015

Oll O, Siimenson C, Lust K et al (2017b) Specific adsorption from an ionic liquid: impedance study of iodide ion adsorption from a pure halide ionic liquid at bismuth single crystal planes. Electrochim Acta 247:910–919. https://doi.org/10.1016/j.electacta.2017.07.034

Pajkossy T, Kolb DM (2011) The interracial capacitance of Au(100) in an ionic liquid, l-butyl-3-methyl-imidazolium hexafluorophosphate. Electrochem Commun 13:284–286. https://doi.org/10.1016/j.elecom.2011.01.004

Pal T, Beck C, Lessnich D et al (2017) Effects of silica surfaces on the structure and dynamics of room-temperature ionic liquids: a molecular dynamics simulation study. J Phys Chem C 122(1):624–634. https://doi.org/10.1021/acs.jpcc.7b10567

Pan GB, Freyland W (2006) 2D phase transition of PF_6 adlayers at the electrified ionic liquid/Au (111) interface. Chem Phys Lett 427(1–3):96–100. https://doi.org/10.1016/j.cplett.2006.05.114

Park JM, Montes-Campos H, Cabeza O et al (2018) 3D structure of the electric double layer of ionic liquid–alcohol mixtures at the electrochemical interface. Phys Chem Chem Phys 20(48):30412–30427. https://doi.org/10.1039/c8cp05632c

Parsons R (1990) Electrical double layer: recent experimental and theoretical developments. Chem Rev 90:813–826. https://doi.org/10.1021/cr00103a008

Péan C, Rotenberg B, Simon P et al (2016) Multi-scale modelling of supercapacitors: from molecular simulations to a transmission line model. J Power Sources 326:680–685. https://doi.org/10.1016/j.jpowsour.2016.03.095

Razmkhah M, Mosavian MTH, Moosavi F (2018a) Structural analysis of an amino acid ionic liquid: bulk and electrical double layer. J Mol Liq 268:506–516. https://doi.org/10.1016/j.molliq.2018.07.047

Razmkhah M, Mosavian MTH, Moosavi F (2018b) What is the effect of polar and nonpolar side chain group on bulk and electrical double layer properties of amino acid ionic liquids. Electrochim Acta 285:393–404. https://doi.org/10.1016/j.electacta.2018.07.225

Rivera-Rubero S (2004) Surface spectroscopy of room-temperature ionic liquids on a platinum electrode: a sum frequency generation study. J Phys Chem B 108(39):15133–15133. https://doi.org/10.1021/jp048260g

Romero C, Baldelli S (2006) Sum frequency generation study of the room-temperature ionic liquids/quartz interface. J Phys Chem B 110(12):6213–6223. https://doi.org/10.1021/jp0565150

Rooney DW, Seddon KR (2001) In: Wypych G (ed) In handbook of solvents. Chem Tec Publishing, Toronto, pp 1459–1484

Ruzanov A, Lembinen M, Jakovits P et al (2018) On the thickness of the double layer in ionic liquids. Phys Chem Chem Phys 20(15):10275–10285. https://doi.org/10.1039/c7cp07939g

Santos VO, Alves MB, Carvalho MS et al (2006) Surface-enhanced Raman scattering at the silver electrode/ionic liquid ($BMIPF_6$) interface. J Phys Chem B 110(41):20379–20385. https://doi.org/10.1021/jp0643348

Schröder U, Wadhawan JD, Compton RG et al (2000) Water-induced accelerated ion diffusion: voltammetric studies in 1-methyl-3-[2, 6-(S)-dimethyloctane-2-yl] imidazolium tetrafluoroborate, 1-butyl-3-methylimidazolium tetrafluoroborate and hexafluorophosphate ionic liquids[J]. New J Chem 24(12):1009–1015. https://doi.org/10.1039/b007172m

Seddon KR, Stark A, Torres MJ (2000) Influence of chloride, water, and organic solvents on the physical properties of ionic liquids. Pure Appl Chem 72:2275–2287. https://doi.org/10.1351/pac200072122275

Sha M, Dou Q, Luo F et al (2014) Molecular insights into the electric double layers of ionic liquids on Au (100) electrodes. ACS Appl Mater Interfaces 6(15):12556–12565. https://doi.org/10.1021/am502413m

Sharma S, Kashyap HK (2015) Structure of quaternary ammonium ionic liquids at interfaces: effects of cation tail modification with isoelectronic groups. J Phys Chem C 119(42):23955–23967. https://doi.org/10.1021/acs.jpcc.5b06460

Si XJ (2012) Molecular dynamics simulation of super capacitance based on ionic liquids. Master's thesis. Naikai University, Tianjin, pp 13–19

Siimenson C, Lembinen M et al (2016) Electrochemical investigation of 1-ethyl-3-methylimidazolium bromide and tetrafluoroborate mixture at Bi(111) electrode interface. J Electrochem Soc 163(9):H723–H730. https://doi.org/10.1149/2.0111609jes

Siinor L, Poom J, Siimenson C et al (2014) Electrochemical characteristics pyrolytic graphite mixture of 1-ethyl-3-methylimidazolium tetrafluoroborate and 1-ethyl-3-methylimidazolium iodide interface. J Electroanal Chem 719:133–137. https://doi.org/10.1016/j.jelechem.2014.01.035

Stern OZ (1924) The theory of the electrolytic double-layer. Elektrochem 30:508–516

Stettner T, Gehrke S, Ray P et al (2019) Water in protic ionic liquids: properties and use of a novel class of electrolytes for energy storage devices. Chem Sus Chem. https://doi.org/10.1002/cssc.201901283

Suarez PAZ, Selbach VM, Dullius JEL, Einloft S, Piatnicki CMS, Azambuja DS, de Souza RF, Dupont J (1997) Enlarged electrochemical window in dialkyl-imidazolium cation based room-temperature air and water-stable molten salts. Electrochim Acta 42:2533–2535. https://doi.org/10.1016/s0013-4686(96)00444-6

Thürmer S, Kobayashi Y, Ohba T et al (2015) Pore-size dependent effects on structure and vibrations of 1-ethyl-3-methylimidazolium tetrafluoroborate in nanoporous carbon. Chem Phys Lett 636:129–133. https://doi.org/10.1016/j.cplett.2015.07.042

Tian G (2012) Application of ionic liquids in extraction and separation of metals. In: Mohammad A (ed) Green solvents II prop. Appl. Ion. Liq. Springer, Dordrecht, pp 119–153. https://doi.org/10.1007/978-94-007-2891-2_5

Tian G, Li J, Hua Y (2009) Application of ionic liquids in metallurgy of nonferrous metals. Chin J Process Eng 10:42–54. https://doi.org/10.3321/j.issn:1009-606X.2009.01.039

Tian G, Li J, Hua Y (2010) Application of ionic liquids in hydrometallurgy of nonferrous metals. Trans Nonferrous Met Soc China 20:513–520. https://doi.org/10.1016/S1003-6326(09)60171-0

Tokuda H, Hayamizu K, Ishii K, Susan MABH, Watanabe M (2005) Physicochemical properties and structures of room temperature ionic liquids. 2. Variation of alkyl chain length in imidazolium cation. J Phys Chem B 109:6103–6110. https://doi.org/10.1021/jp044626d

Trulove PC, Mantz RA (2002) In: Wasserscheid P, Welton T (eds) In ionic liquids in synthesis. Wiley-VCH, Weinheim, pp 16–38

Tsai WY, Come J, Zhao W et al (2019) Hysteretic order-disorder transitions of ionic liquid double layer structure on graphite. Nano Energy 60:886–893. https://doi.org/10.1016/j.nanoen.2019.04.022

Turesson M, Szparaga R, Ma K et al (2014) Classical density functional theory & simulations on a coarse-grained model of aromatic ionic liquids. Soft Matter 10(18):3229–3237. https://doi.org/10.1039/c3sm53169d

Vanin AA, Brodskaya EN (2017) Simulation study of influence of component polarizability on the properties of the electric double layer of an ionic micelle. Colloids Surf A Physicochem Eng Asp 522:58–65. https://doi.org/10.1016/j.colsurfa.2017.02.024

Vatamanu J, Borodin O, Smith GD (2010) Molecular insights into the potential and temperature dependences of the differential capacitance of a room-temperature ionic liquid at graphite electrodes. J Am Chem Soc 132(42):14825–14833. https://doi.org/10.1021/ja104273r

Vatamanu J, Xing L, Li W et al (2014) Influence of temperature on the capacitance of ionic liquid electrolytes on charged surfaces. Phys Chem Chem Phys 16(11):5174–5182. https://doi.org/10.1039/c3cp54705a

Wakeham D, Hayes R, Warr G G et al, (2009) Influence of temperature and molecular structure on ionic liquid solvation layers. J Phys Chem B 113(17):5961–15966. https://doi.org/10.1021/jp900815q

Walden P (1914) Molecular weights and electrical conductivity of several fused salts. Bulletin del Academie Imperiale des Sciences de St.-Petersbourg, pp 405–422

Wang S (2010) Molecular dynamics simulation of ionic liquids at graphite interface, Master's thesis. Nankai University, Tianjin, pp 12–16. https://doi.org/10.7666/d.y1813657

Wang YY (2015) Computer simulation of the interface structure of ionic liquids/graphene, Master's thesis. Nankai University, Tianjin, pp 3–18

Wang R, Bi S, Presser V et al (2018) Systematic comparison of force fields for molecular dynamic simulation of Au(111)/ionic liquid interfaces. Fluid Phase Equilib 2018(463):106–113. https://doi.org/10.1016/j.fluid.2018.01.024

Welton T (2002) Room-temperature ionic liquids. Solvents for synthesis and catalysis. Chem Rev 99:2071–2083. https://doi.org/10.1021/cr980032t

Wen R, Rahn B, Magnussen OM (2015) Potential-dependent layer structure and dynamics at the ionic liquid/Au(111) interface: a molecular-scale in situ video-STM study. Angew Chem Int Ed 54(20):6062–6066. https://doi.org/10.1002/anie.201501715

Wilkes JS, Zaworotko MJ (1992) Air and water stable 1-ethyl-3-methylimidazolium based ionic liquids. J Chem Soc Chem Commun 965-967. https://doi.org/10.1039/c39920000965

Wolff C, Jeong S, Paillard E et al (2015) High power, solvent-free electrochemical double layer capacitors based on pyrrolidinium dicyanamide ionic liquids. J Power Sources 293:65–70. https://doi.org/10.1016/j.jpowsour.2015.05.065

Wu BJ (2018) Application of organic electrolyte in double-layer capacitors, Master's thesis. Tianjin Industrial University, Tianjin, pp 24–28
Wu J, Jiang T, Jiang D et al (2011) A classical density functional theory for interfacial layering of ionic liquids. Soft Matter 7(23):11222–11231. https://doi.org/10.1039/c1sm06089a
Wu M, Li W, Li S et al (2017) Capacitive performance of amino acid ionic liquid electrolyte-based supercapacitors by molecular dynamics simulation. RSC Adv 7(46):28945–28950. https://doi.org/10.1039/c7ra00443e
Xu Y, Qi C, Ying R, Cheng D, Zhou W (2014) In-situ surface-enhanced infrared spectroscopy for the study of ionic liquid/electrode interface structure. J Fudan Univ Nat Sci 53(02):163–168
Yang J (2014) Differential capacitance method for studying the electrochemical interface double layer structure of electrode/ionic liquid-lithium salt mixture system, doctoral dissertation. Shanghai University, Shanghai, pp 10–15
Yang J, Zhou W (2014) Electrochemical differential capacity of the mixture of [bmim][otf] ionic liquid and lithium ion on gold electrode. J Fudan Univ Nat Sci 53(02):175–180
Yang I, Kwon D, Kim MS et al (2018) A comparative study of activated carbon aerogel and commercial activated carbons as electrode materials for organic electric double-layer capacitors. Carbon 132:503–511. https://doi.org/10.1016/j.carbon.2018.02.076
Yu Z, Fang C, Huang J et al (2018) Solvate ionic liquids at electrified interfaces. ACS Appl Mater Interfaces 10(38):32151–32161. https://doi.org/10.1021/acsami.8b10387
Zha QX (2002) Introduction to electrical process dynamics, 3rd edn. Science Press, Beijing, pp 31–45
Zhang MJ (2006) Principle and application of molten salt electrochemistry. Chemical Industry Press, Beijing, pp 24–30
Zhang N (2015) Electrochemical Sers study of electric field distribution at the interface of ionic liquids/electrodes, Master's thesis. Suzhou University, Suzhou, pp 2–18
Zhang SJ, Lv XM (2006) Ionic liquids from fundamental study to industrial application. Science Press, Beijing, pp 35–95
Zhang Q, Han Y, Wang Y et al (2014a) Comparing the differential capacitance of two ionic liquid electrolytes: effects of specific adsorption. Electrochem Commun 38:44–46. https://doi.org/10.1016/j.elecom.2013.10.027
Zhang X, Zhong B, Yan J, Mao B (2014b) In-situ AFM force curve investigations on layered structures of Au(111)-ionic liquid interfaces and temperature dependence. J Electrochem 20(4):295–301. https://doi.org/10.13208/j.electrochem.130892
Zhang N, Wang XR, Yuan YX et al (2015) Probing double layer structure at au/[BMIM]BF_4 interface by molecular length-dependent SERS Stark effect. J Electroanal Chem 751:137–143. https://doi.org/10.1016/j.jelechem.2015.05.041
Zhang L, Dong K, Chen S et al (2016) Novel ionic liquid based electrolyte for double layer capacitors with enhanced high potential stability. SCIENCE CHINA Chem 59(5):547–550. https://doi.org/10.1007/s11426-016-5588-x
Zhang Q, Liu X, Yin L et al (2018) Electrochemical impedance spectroscopy on the capacitance of ionic liquid–acetonitrile electrolytes. Electrochim Acta 270:352–362. https://doi.org/10.1016/j.electacta.2018.03.059
Zhao J, Zhou W (2018) Differential capacitance study on the double layer structure of ionic liquids and 1,2-dichloroethane. J Fudan Univ Nat Sci 57(04):509–516. https://doi.org/10.15943/j.cnki.fdxb-jns.2018.04.013
Zhao W, Bi S, Balke N et al (2018) Understanding electric double-layer gating based on ionic liquids: from nanoscale to macroscale. ACS Appl Mater Interfaces 10(49):43211–43218. https://doi.org/10.1021/acsami.8b15199
Zhong Y, Yan J, Li M et al (2016) The electric double layer in an ionic liquid incorporated with water molecules: atomic force microscopy force curve study. Chem Electro Chem 3(12):2221–2226. https://doi.org/10.1002/celc.201600177

Chapter 11
Role of Ionic Liquid-Based Multipurpose Gas Hydrate and Corrosion Inhibitors in Gas Transmission Pipeline

Ali Qasim, Bhajan Lal, Azmi Mohammad Shariff, and Mokhtar Che Ismail

Contents

Abbreviations

AAs	Anti-agglomerants
ADT	Average depression temperature
DES	Deep eutectic solvents
EIS	Electrochemical impedance spectroscopy
GHCIs	Gas hydrate and corrosion inhibitors
IE	Inhibition efficiency

A. Qasim · B. Lal (✉) · A. M. Shariff
Chemical Engineering Department, Universiti Teknologi PETRONAS, Seri Iskandar, Perak, Malaysia

CO2 Research Centre (CO2RES), Universiti Teknologi PETRONAS, Seri Iskandar, Perak, Malaysia
e-mail: bhajan.lal@utp.edu.my

M. C. Ismail
Department of Mechanical Engineering, Centre for Corrosion Research, Universiti Teknologi PETRONAS, Seri Iskandar, Perak, Malaysia

Inamuddin, A. M. Asiri (eds.), *Nanotechnology-Based Industrial Applications of Ionic Liquids*, Nanotechnology in the Life Sciences,
https://doi.org/10.1007/978-3-030-44995-7_11

IL	Ionic liquid
KHIs	Kinetic hydrate inhibitors
LDHIs	Low-dosage hydrate inhibitors
NGH	Natural gas hydrate
QAS	Quaternary ammonium salts
THIs	Thermodynamic hydrate inhibitors

11.1 Introduction

Gas hydrates, also known as clathrate hydrates, are formed in mixtures of water and nonpolar or slightly polar low-molecular-mass gases when subjected to low temperatures and relatively higher pressures (Huo et al. 2001; Joshi et al. 2013; Khan et al. 2016; Sloan and Koh 2008). The temperature ranges from 275 to 285 K while the pressure ranges are generally 2–10 MPa (Gupta et al. 2018; Nashed et al. 2018; Sloan et al. 2010a). Gas hydrates are nonstoichiometric solid inclusion compounds, which consist of polyhedral structures formed by hydrogen-bonded water molecules that are stabilized via encaged guest molecules through van der Waals forces. Methane (CH_4), ethane (C_2H_6), carbon dioxide (CO_2), propane (C_3H_8), hydrogen sulfide (H_2S), isobutane (i-C_4H_{10}), and n-butane (n-C_4H_{10}) are some of the examples of hydrate-forming gases (Khan et al. 2018d; Koh et al. 2011; Khan et al. 2018; Qasim et al. 2019; Qureshi et al. 2016). On the other hand, corrosion is the gradual degradation of material properties of metal caused by the chemical reaction with the environment (Burgazli et al. 2005; Heppner and Evitts 2008; Obanijesu 2012). The chemical reaction is actually an electrochemical reaction with its nearby atmosphere (Amin et al. 2012; Javadian et al. 2013; Srinivasan and Kane 2003; Umoren et al. 2013). In general, metal degradation takes place when it is in contact with moisture; water; bases such as sodium hydroxide or sodium bicarbonate; acids as hydrochloric acid, sulfuric acid, and nitric acid; salts including sodium chloride; liquid chemicals and pure gases such as carbon dioxide, methane, or formaldehyde; and ammonia- and sulfur-containing gases (El-Haddad 2014; Hamadi et al. 2018; Kennell and Evitts 2009; Tawfik 2016).

The pipeline industry annually invests millions of dollar on chemical inhibitors for smooth flowline. The plugging of the pipeline by natural gas hydrate (NGH) costs the gas industry around $1 million per day estimated at the shutting down of production (Obanijesu et al. 2010). In case of corrosion in oil and gas pipelines, the cost per annum is assessed to be $1.372 billion (Popoola et al. 2013). Flow assurance issue and its appropriate management are critical to the economical and successful operation (Bavoh et al. 2018a; Khan et al. 2017c; Patel and Russum 2009; Sloan et al. 2010b). Further flow assurance challenges become more dominant in deepwater thermodynamic conditions and are required to be addressed accordingly (Chen 2011; Mokhatab et al. 2007; Sloan 2005). In oil and gas pipeline, the gas hydrates and corrosion formation is a routine phenomenon under the suitable conditions such as acid gases, water, and impurities with thermodynamic condition (Burgazli et al. 2005; Obanijesu 2012). At each stage of the process of hydrate formation, the interaction takes place between the hydrate composition and the pipeline which eventually leads to reaction and initiates internal corrosion and vice versa. In addition to the pipe rupture from hydrate plug, the corrosion process provides an impetus to degrade the metal surface and deteriorate the integrity of the pipeline and thus pipeline leakage or full bore rupture. To tackle this problem, a variety of gas hydrates and corrosion inhibitors are separately used which cause further economic and ecological problems such as requirement of large storage space and regeneration and biodegradable problems besides adverse effect on each other's performances (Moore et al. 2009; Sheng et al. 2017).

Different approaches have been implemented to avoid aforementioned pipeline issues including dehydration, heating, and chemical and mechanical methods (Ke and Kelland 2016; Khan et al. 2018d). Apart from chemical method, all other methods either are incompetent or require a large quantity of chemical solvents. It raises the OpEx and CapEx and puts a harmful impact on the environment of operating gas and oil facilities (Dendy Sloan and Koh 2008). For this reason, chemical method is the only feasible choice for the pipeline industry. Thus, there is an urgent need to develop cost-effective and environmentally friendly chemicals which can address the two major flow assurance problems. Previous studies highlighted the performance of ionic liquids, deep eutectic solvents (DES), biomolecules, polymer, amino acids, and other organic and inorganic based gas hydrate and corrosion inhibitors but less stress has been given on ionic liquid-based multipurpose hydrate as well as corrosion inhibitor (GHCI) behavior or performance in open literature.

Currently under the field of "green chemistry" research is being done in order to develop techniques and processes that can cause reduction of toxic materials. It is an emerging field which deals with designing different methods and processes which are environmentally benign. The development of environment-friendly chemicals is also included in this category. Previously, researchers worked on natural plant extracts for this application due to their nontoxic and harmless nature. Along with it organic compounds were also employed (Verma et al. 2018a, 2017). Nevertheless, the process of extraction and purification of plants proved to be economically inviable and required large amounts of organic solvents and time. Therefore, scientists recognized the importance of synthesis processes for green inhibitors which involve cheap and nontoxic materials (Smiglak et al. 2014; Verma et al. 2018b). The emergence of utilization of ionic liquids is a significant step to address this particular issue. The choice among selection of ions including anions and cations is advantageous in case of ionic liquids. Designed chemicals can be developed with required properties. These properties of ionic liquids affirm their importance as designer chemicals especially in catalysis or their use as a solvent in various mixtures (Welton 1999; Gonfa et al. 2015; Khan et al. 2017a). As these chemicals are highly soluble in different organic and inorganic compounds the use of ionic liquids (ILs) in the field of chemistry and chemical engineering leads towards sustainable chemicals which are environment friendly (Yaqub et al. 2018).

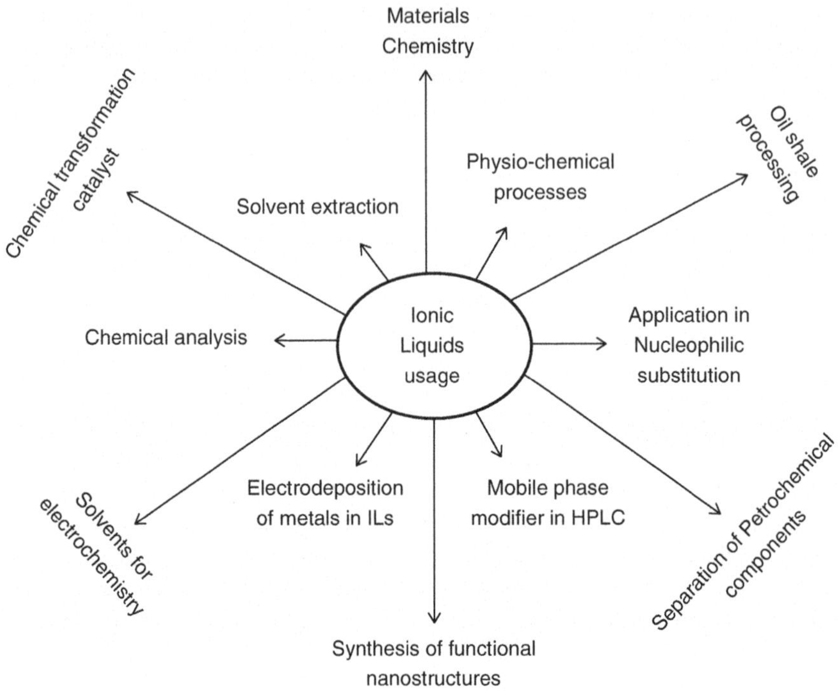

Fig. 11.1 Application of ionic liquids

These compounds are categorized as environment-friendly solvents that can be tuned or adjusted for specific tasks and applications (Verma et al. 2017; Welton 1999). ILs have exceptional structural and physiochemical properties that help in their tuneability (Khan et al. 2017c; Smiglak et al. 2014; Verma et al. 2017). The use of ILs has been employed in various scientific fields which are shown in Fig. 11.1.

Ionic liquids show some interesting properties regarding their behavior. These compounds have low melting point, i.e., less than 100 °C, low vapor pressure, low toxicity, and high polarity. These are thermally and chemically stable as well. These properties make these salts a suitable candidate to substitute traditional toxic corrosion inhibitors which also show high volatility (Welton 1999). It is found in literature that some of the synthesized ILs showed better corrosion inhibition behavior for mild steel in different electrolytic media (Verma et al. 2018a; Zhu et al. 2017). Although scientific research area is still in its early stage, sufficient data is available on the first-generation ILs for their potential application to act as dual-functional gas hydrate inhibitors (thermodynamic as well as kinetic) (Khan et al. 2018a, 2018c). In this book chapter, published data on ionic liquid chemistry for hydrate and corrosion inhibition is summarized. Also, hydrate and corrosion formation, occurrence, and mitigation methods for GHCls are discussed in such a way that important conclusion can be drawn regarding the development of this potentially interesting and important area of research. Moreover, challenge of application of ionic liquids in flow line is presented. Prospects of multipurpose GHCI for the gas industry are also discussed.

11.2 Gas Hydrate and Corrosion in Flow Pipeline

11.2.1 Gas Hydrate Formation and Inhibition

Three phases exist typically in flow pipeline when oil and gas are transported through it. These three phases are named as liquid hydrocarbon, an aqueous liquid, and gas. The flow model of hydrate formation is classified into four types. The first type among these four models is the oil-dominated system. In oil-dominated system all the water content is collectively present in oil phase as droplets and it is dominated by oil itself. The other system is a gas-dominated system in which the presence of aqueous liquid of liquid hydrocarbon is less. This kind of system does not usually cause hydrate blockage. It differs from oil-dominated systems in that it does not disperse the water in the liquid hydrocarbon phase. The third type is a condensate system in which the water dissolves in the condensate or suspends as droplets due to high shear. Among these systems the last one is called as high-water-cut system. This system contains high water content which may reach up to 70% in volume and it is immiscible in the oil phase. The temperature ranges from 275 to 285 K while the pressure ranges are generally 3–10 MPa (Koh et al. 2011; Sloan and Koh 2008). The gradual formation of gas hydrate is shown in Fig. 11.2.

Gas hydrate formation in the gas transmission line is prevented by several conventional methods like depressurization, heating, dehydration, and chemical which have their own pros and cons according to the encountered situation (Dendy Sloan and Koh 2008; Yaqub et al. 2018). Depressurization is generally considered as a remedial approach wherein pressure is reduced onto one side of the transmission line and used mostly when hydrate plug formation has already occurred. However, in long and high-pressure transmission pipelines, the approach of line depressurization is not practical (Drive and Carroll 2009). Furthermore, depressurization results in the velocity increase for the hydrate plug which is potentially destructive for gas pipeline bending and vents, hence rupturing the overall pipeline integrity. Thermal heating is also one of the conventional hydrate removal (remediate) techniques in which steam is applied along the gas pipeline to keep the temperature out of hydrate

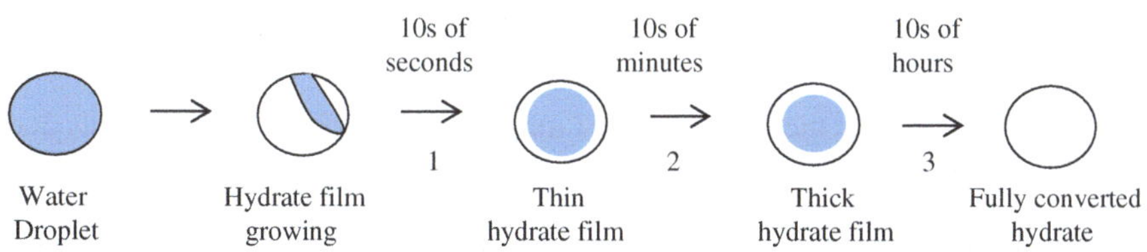

Fig. 11.2 Gradual formation of gas hydrate

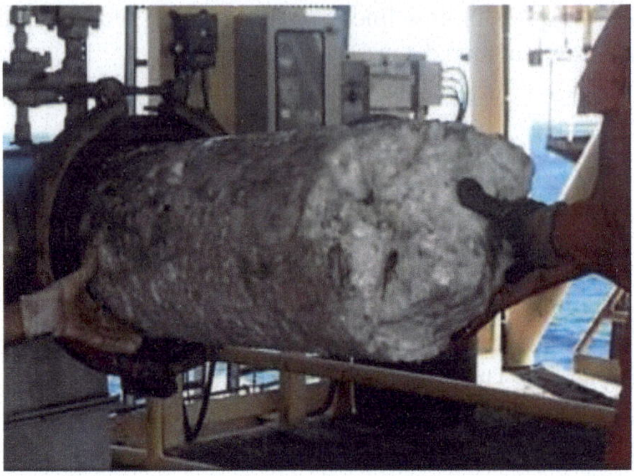

Fig. 11.3 Hydrate plug formed in subsea pipeline

formation temperature (Avula et al. 2014; Nešić 2007). However, it is one of the most expensive techniques which requires almost one million US$ per km of pipeline length for preventive measures (Gupta et al. 2012; Koh et al. 2011). Dehydration or water removal is considered as hydrate preventive method and it is a permanent solution as suggested in previous studies. If there is no water present in the stream, the problem of hydrate formation will cease to exist but due to human footprint constraints it is an economically inviable solution (Brustad et al. 2005; Fazlali et al. 2016; Yee Tak and Englezos 1996). This method is very effective for gas-processing plants or other downstream applications.

Hydrate mitigation can also be accomplished by inserting compounds which keep the operating temperatures and pressure out of hydrate formation zone and sustain hydrate-free region (Bavoh et al. 2018b; Koh et al. 2011; Patel and Russum 2009; Sloan 2005). These compounds either delay or change the phase behavior of the newly formed mixture (Del Villano et al. 2008; Partoon et al. 2013; Qasim et al. 2019). Among the methods mentioned above, chemical inhibition is the best preventive technique and widely employed in the gas pipelines. Chemical inhibitors are not intrusive and do not hinder pipeline flow harmfully, thus ensuring the smooth flow (Bavoh et al. 2018b; Patel and Russum 2009; Tabrez and Inam 2012).

Figure 11.3 shows the typical hydrate plug that caused the obstruction in subsea pipeline.

11.2.2 Corrosion Formation and Inhibition in Flow Pipeline

Corrosion is defined as the degradation of material or more specifically metals due to electrochemical reaction. The process of corrosion process can be defined in two electrochemical reactions. Formation of $FeCO_3$ compound which exists in solid form inside the pipeline is the main cause of corrosion (Tiu and Advincula 2015; Umoren et al. 2013).

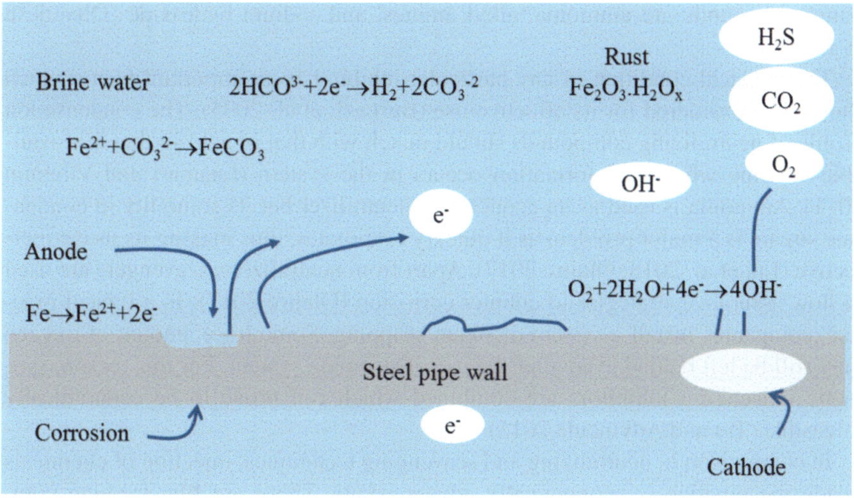

Fig. 11.4 Corrosion formation in pipeline

The pictorial representation of corrosion process for a pipeline can be seen in Fig. 11.4.

Fretting and erosion-corrosion are the two types of corrosion that usually occur inside the pipeline. Pipeline is more prone to both these types of corrosion when the fluid flow rate inside it is high. Fretting corrosion is witnessed when the load moves across the surface at a relatively high velocity (Srinivasan and Kane 2003; Xiao et al. 2008). When two rough surfaces come in contact with one another, mostly during transportation, the vibration caused in consequence results in fretting corrosion. Due to rubbing, the protecting layer on the surfaces is eliminated, and the metal is exposed to the corrosive activity of the surroundings. This problem becomes evident due to movement of fluid inside the oil and gas pipelines as it causes much vibration. The vibration is caused by the weight of fluid after coming in contact with inner surface of the pipeline (Srinivasan and Kane 2003). Erosion-corrosion occurs inside the pipeline that transports slurries and other particle-carrying liquids. Industry spends some heavy amounts every year to repair material damage from this type of corrosion (Rajahram et al. 2009; Wang and Wang 1998).

For oil and gas production processes, the appropriate treatment for corrosion-related problems is achieved by using a suitable corrosion inhibitor. The suitability of corrosion inhibitor is determined by its ability to dissolve into a wide variety of concentrations of salts and cope up with the harsh processing conditions. In this application, chemicals are characterized as neutralizers, scavengers, and absorbers or film formers. In refinery, hydrochloric, sulfuric, and sulfurous acids cause corrosion frequently. In process stream these acids exist in lower concentrations but they can be found in the condensate streams of distillation equipment or heat exchangers in mild or high concentrations. The use of neutralizers in the system reduces hydrogen ion concentration which in turn inhibits corrosion. The commonly used neutralizing compounds are ammonia, alkyl amines, and sodium hydroxide (Obanijesu et al. 2014).

The physical condition of any particular inhibitor is an important factor which should be considered for its effective use (Burgazli et al. 2005). The condensation profile of neutralizing compounds should match with that of acid so that it is available anytime when acid formation occurs in the system (Laamari and Villemin 2011). Ammonia is used as an economical neutralizer but its solubility in condensate stream is a major problem as it quickly evaporates, thus making its usage ineffective (Liu et al. 2018; Olajire 2017). Apart from neutralizers, scavengers are used in flow assurance strategies to counter corrosion (Olajire 2017). In a typical practice, companies install a system of steam stripping. Some trace amount of oxygen may still be left behind even after the use of stripping system. For this reason, synthetic scavenging inhibitors are combined which can prove to be economically infeasible (Tiu and Advincula 2015).

In comparison to neutralizing and scavenging techniques, injection of chemicals inside the pipeline is economically advantageous. These are film-forming compounds acting as corrosion inhibitors. These compounds are also known as barrier inhibitors. Their impact on the pipeline integrity is more encouraging as their effectiveness is independent of the interaction with corrosive agents inside it. Alternatively, these inhibitors form a protective layer on the metal surface through strong interactions. They adsorb electrostatically or in some cases establish chemisorption, thus diminishing the influence of corrosive substances significantly (Nešić 2007; Ullah et al. 2017; Umoren and Obot 2008).

Corrosion protection is that it is attained by covering the surface of the metal; that is, the inhibitor adsorbs onto the steel surface and affects one or more electrochemical reactions by slowing it down. The extent to which protection is achieved is directly proportional to the fraction of the steel surface covered by the inhibitor (Mendonca et al. 2017; Schweinsberg and Ashwortht 1988; Sherif et al. 2015). To explain this phenomenon mathematically, a relationship is established between the surface coverage and the inhibitor concentration in the solution. This is attained by the use of suitable adsorption isotherms such as Langmuir or Frumkin isotherm (Hegazy et al. 2014; Verma et al. 2018a).

11.2.3 Gas Hydrate Prevention via Chemical Injection

As discussed previously, chemical inhibition is a better preventive technique and widely employed in the gas pipelines. It does not affect the pipeline flow and proves to be nonintrusive. Chemical inhibitors are injected to avert hydrate materialization by reducing hydrate formation temperature below the pipeline operation temperature and also shifting the hydrate formation to higher pressure regions (Khan et al. 2018b; Koh et al. 2011; Sloan et al. 2010b). Chemical inhibitors are further classified into high-dosage hydrate inhibitors, i.e., thermodynamic hydrate inhibitors (THIs) and low-dosage hydrate inhibitors (LDHIs) (Nashed et al. 2018; Yaqub et al. 2018; Zerpa et al. 2011). THIs shift the hydrate formation equilibrium curve towards high-pressure and low-temperature regions, thus successfully keeping it outside of the domain of hydrate formation (Khan et al. 2018a; Qasim et al. 2018). The selection of thermodynamic inhibitors involves comparison of several features including cost of operation, safety, corrosion inhibition, physical properties, and ability of gas to dehydrate (Akhfash et al. 2017; Cha et al. 2013; Heidaryan et al. 2010). Among LDHIs, kinetic hydrate inhibitors (KHIs) prevent hydrate formation by delaying nucleation for a time in excess of the residence time of the free water phase in the flow line (Del Villano et al. 2008; Khan et al. 2018b; Moore et al. 2009). Hydrate formation or nucleation time is also known as induction time. Induction time is time elapsed from the equilibrium state to the point where first hydrate crystal materializes with stability. KHIs are used in low concentrations of 1–2 wt% (Kelland 2006; Khan et al. 2017b; Sa et al. 2013). KHI performance is dependent on time, unlike the chemicals that are used as thermodynamic inhibitors. KHIs prevent hydrate nucleation and do not allow hydrate growth at the crystal surface through adsorption (Kelland 2006). The other category of low-dosage hydrate inhibitors (LDHI) is anti-agglomerants (AAs) which are surface-active chemicals. These compounds allow hydrate crystals to form but disperse these crystals and keep them small which are eventually transported in normal pipeline operation without causing any disruption (Huo et al. 2001; Kelland et al. 2009). Anti-agglomerants (AAs) adsorb readily on the droplet surface which is hydrated and thus all the water content is converted to hydrate. The working principle of AAs is that the head group of chemical attaches to the hydrate, while the long hydrocarbon tail maintains the hydrate particle dispersed in the fluid phase (Erfani et al. 2013; Frostman 2000). Figure 11.5 summarizes the classification of gas hydrate inhibitors.

11.2.3.1 Ionic Liquids as Thermodynamic Hydrate Inhibitors (THIs)

For gas production systems, THIs are employed in hydrate inhibition applications in large amounts to avoid hydrate formation. For gas-water stream, these inhibitors are used in high quantity (Cha et al. 2015; Partoon et al. 2013).

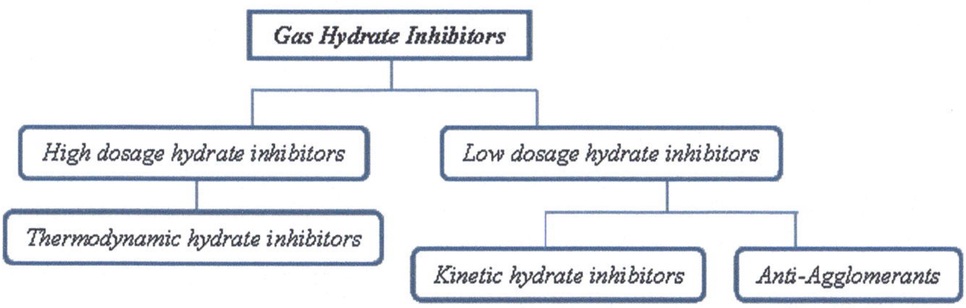

Fig. 11.5 Gas hydrate inhibitors

Different types of ionic liquids have been used as THIs. Phase boundaries of imidazolium-based ILs have been widely investigated in the hydrate inhibition application for carbon dioxide and methane gases (Pisarova et al. 2012; Yesudass et al. 2016). Among imidazolium-based ILs, the use of butyl methyl imidazolium in combination with tetrafluoroborate has been studied and reported in the literature (Sabil et al. 2015; Xiao and Adidharma 2009). The IL butyl methyl imidazolium tetrafluoroborate has been used both for carbon dioxide and methane hydrate inhibition applications. Similarly, ethyl methyl imidazolium bromide, ethyl methyl imidazolium ethylsulfate, ethyl methyl imidazolium chloride, and ethyl methyl imidazolium trifluoromethanesulfonate have been employed in THI applications (Sa et al. 2016; Tariq et al. 2014). These compounds have mostly been used for the inhibition of methane hydrates but in some cases also for carbon dioxide hydrate inhibition.

Among piperidinium-based compounds of ILs, ethyl methyl piperidinium bromide and ethyl methyl piperidinium tetrafluoroborate have been used as THIs in case of carbon dioxide hydrate inhibition (Sa et al. 2016). The preceding researches showed that ethyl methyl morpholinium bromide and ethyl methyl morpholinium tetrafluoroborate were also used for carbon dioxide hydrate inhibition (Cha et al. 2016; Kim and Kang 2011). Piperidinium-based compounds have not been studied as much as imidazolium-based ILs. Also the use of ammonium, morpholinium, and other groups among ILs as THIs has not been studied extensively and needs to be evaluated further as recommended by Yaqub et al. (Yaqub et al. 2018).

11.2.3.2 Ionic Liquids as Kinetic Hydrate Inhibitors (KHIs)

In preceding literature, the use of ILs as kinetic hydrate inhibitors has been reported. Nashed et al. worked on the performance of imidazolium-based ILs including (C_4C_1im) (CF_3SO_3), (C_4C_1im) (CH_3SO_4), and (OH-C_2C_1im) (Br) as kinetic hydrate inhibitor. Results showed that (OH C_2C_1im) (Br) increased induction time by 1.8 times than pure water at 7.1 MPa and 258.15 K (Nashed et al. 2017). Another study done by Saad and co-workers utilized four new ammonium-based ILs comprising tetramethyl ammonium hydroxide (TMAOH), tetraethyl ammonium hydroxide (TEAOH), tetrapropyl ammonium hydroxide (TPrAOH), and tetrabutyl ammonium hydroxide (TBAOH) for both THI and KHI behavior. The studies were performed using carbon dioxide-methane mixed-gas hydrates (Khan et al. 2019).

11.2.3.3 Ionic Liquids as Anti-agglomerants (AA)

Quaternary ammonium salts (QAS) and surfactants act as typical anti-agglomerants. These are longer chain compounds which are composed of hydrophilic and hydrophobic ends. In pipeline flow, hydrophilic end of the compound which is anion forms dipole-dipole interaction with the hydrogen ion of water molecule (Kelland 2006; Kuznetsova et al. 2010), while the hydrophobic end which is the cationic part of surfactant dissolves in the gas or oil phase. Shell Oil Company used surfactants which included two and three n-butyl, n-pentyl, and isopentyl groups. These compounds showed remarkably well anti-agglomeration behavior. Among QAS, tetrapentylammonium bromide (TPAB) was reported as hydrate inhibitor in the category of anti-agglomerants (Kelland 2006). Twin-tailed quaternary anti-agglomerants were synthesized such as tetraalkylammonium salts (Klomp et al. 1996). Among these salts, Dutch Shell team utilized dicocoyldibutylammonium bromide in field. Nevertheless, in terms of biodegradability, it did not perform well (Klomp et al. 1996). Huo et al. used some of the commercial surfactants, namely Sorbitan Monooleate (Span 80), Sorbitan Monooctadecanoate (Span 60), Sorbitan Monopalmitate (Span 40), and Sorbitan Monolaurate (Span 20) in their research. Results showed that surfactants kept the hydrate particles suspended at the conditions of 277 K and 8.20 MPa (Huo et al. 2001). Their performances were also compared with dodecyl-2-(2-caprolactamyl) ethanamide which is a synthesized anti-agglomerant. However, the synthesized chemical showed better dispersion at higher water cut condition, i.e., 0.75 wt% (Huo et al. 2001). Kelland et al. are the lead researchers of the oil and gas industry in developing cost-effective anti-agglomerant compounds. They further extended the research on AAs by utilizing polypropoxylates. It was found that polyamine polypropoxylates and other branched polypropoxylates dispersed gas hydrates in a hydrocarbon fluid in a well-agitated scenario (Kelland et al. 2009).

11.2.4 Ionic Liquids (ILs) as Corrosion Inhibitors

Figure 11.6 shows the classification of ILs that have been employed in gas hydrate and corrosion inhibition applications.

Ionic liquids, mostly imidazolium and ammonium based, have been used for corrosion study (Ma et al. 2016; Pisarova et al. 2012). In order to protect steel from corrosion, pyrrolidinium- and pyridazinium-based inhibitors have also been used for the application (Bousskri et al. 2015; El-Hajjaji et al. 2018). Corrosion prevention has been done in numerous electrolytic media including NaCl, H_2SO_4, and HCl (El-Shamy et al. 2015; Verma et al. 2017). Shamy et al. (El-Shamy et al. 2015) employed 1-butyl-1-methylpyrrolidinium trifluoromethylsulfonate,

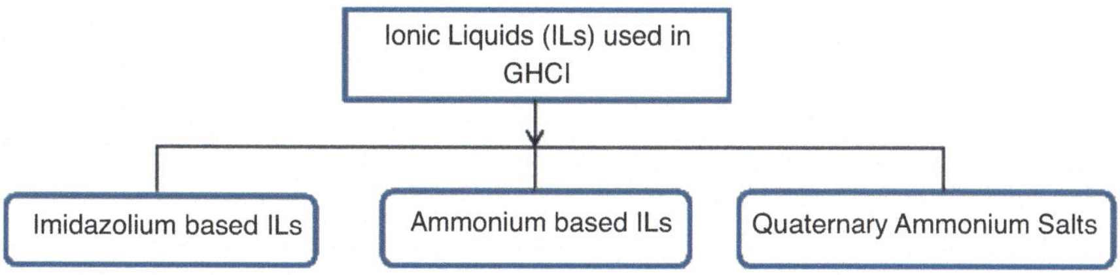

Fig. 11.6 Classification of IL use in GHCI

$[BMPyrr]^+[Otf]^-$ IL in electrolytic environment of 3.5% NaCl for mild steel protection. The chemicals showed mixed-type inhibition in this instance. Sasikumar et al. (2015) did Monte Carlo simulations on the 1-butyl-2,3-dimethylimidazolium tetrafluoroborate $[BDMIM]^+[BF_4]^-$ and 1-ethyl-3-methylimidazolium tetrafluoroborate $[EMIM]^+[BF_4]^-$ to probe their corrosion suppression performance on mild steel surface in 1 M HCl. Their findings revealed that both the studied ILs behaved as mixed-type inhibitors. Kannan et al. (2016) used 3-(4-chlorobenzoylmethyl)-1-methylbenzimidazolium bromide $[BMIMB]^+[Br]^-$ in corrosive media (1 M HCl) for mild carbon steel protection; the studied IL was also found to be a mixed-type inhibitor. Sami et al. (2016) studied the corrosion behavior of 1-butyl-3-methylimidazolium tetrachloroferrate $[BMIM]^+[FeCl_4]^-$ for mild steel protection in open and controlled environments using both weight loss and electrochemical methods. Ashassi et al. (2009) determined the corrosion-inhibiting performance of 1-butyl-3-methylimidazolium bromide $[BMIM]^+[Br]^-$ in 1 M HCl media for mild steel protection. They applied both weight loss and electrochemical methods and found that $[BMIM]^+[Br]^-$ acted as a mixed-type inhibitor. Yesudass and others (2016) examined five alkylimidazolium-based ionic liquids (ILs), namely 1-ethyl-3-methylimidazolium acetate $[EMIM]^+[Ac]^-$, 1-ethyl-3-methylimidazolium ethylsulfate $[EMIM]^+[EtSO_4]^-$, 1-butyl-3-methylimidazolium acetate $[BMIM] + [Ac]^-$, 1-butyl-3-methylimidazolium dicyanamide $[BMIM]^+[DCA]^-$, and 1-butyl-3-methylimidazolium thiocyanate $[BMIM]^+[SCN]^-$ using different approaches, i.e., electrochemical method, spectroscopy, and Monte Carlo simulation, to evaluate the corrosion inhibition performance of mild steel in acidic media (1 M HCl). For all the ILs involved in the study, mixed-type inhibition was observed by the researchers.

Yousefi et al. (2015) utilized six imidazolium-based ILs along with their mixtures with a surfactant of sodium dodecyl sulfate (SDS) which is anionic in nature using polarization and spectroscopy methods for thermodynamic hydrate inhibition. It was found that all the compounds presented mixed-type corrosion inhibition behavior. Due to highly acidic conditions of 2 M HCl, ions showed strong intermolecular interaction. Murulana et al. (2012) also employed imidazolium-based ILs as hydrate inhibitors and investigated four ILs as the corrosion inhibitors for mild steel protection in 1 M HCl solution. The ionic liquids selected for their performance included 1-butyl-3-methylimidazolium bis(trifluoromethyl-sulfonyl) imide $[BMIM]^+[NTf_2]^-$, 1-propyl-3-methylimidazolium bis(trifluoromethyl-sulfonyl) imide $[PMIM]^+[NTf_2]^-$, 1-propyl-2,3-methylimidazolium bis(trifluoromethyl-sulfonyl) imide $[PDMIM]^+[NTf_2]^-$, and 1-hexyl-3-methylimidazolium bis(trifluoromethyl-sulfonyl) imide $[HMIM]^+[NTf_2]^-$. Weight loss and electrochemical methods were used and the obtained results revealed that the ILs considered acted as mixed-type inhibitors. Among the studied systems, $[PDMIM]^+[NTf_2]^-$ showed the best inhibition efficiency perhaps due to the presence of propyl moiety in the imidazolium structure. Zarrouk et al. (2012) inspected two ILs, namely 3-(3-phenylpropyl)-1-propyl-1H-imidazol-3-ium bromide (PP_1B_1) and 3-(4-phenoxybutyl)-1-propyl-1Himidazol-3-ium bromide (PP_1B_4), for corrosion protection of carbon steel in 1 M HCl solution. Both ILs acted as mixed-type inhibitors in both weight loss and electrochemical methods.

Similarly, QASs were also examined for their corrosion inhibition performance. Like ionic liquids, these compounds are composed of ions; however, as the anions and cations have active forces of attraction, these ions are closely packed with each other and form solid (Li et al. 2012). The performances and results of cationic mixtures of cetyltrimethyl ammonium bromide (CTAB) and sodium dodecyl sulfate (SDS) to inhibit mild steel corrosion in an electrolytic media of 3.5% NaCl were inspected using EIS and potentiodynamic polarization scans. Results showed that both compounds worked as mixed-type inhibitors; however, CTAB (cationic surfactant) performed better than SDS (anionic surfactant). The enhanced corrosion inhibition of CTAB was found due to a strong interaction among the polar head groups of the metal and surfactant surface (Javadian et al. 2013). Moreover, Chong and others (2016) inspected the corrosion rates of organic salt comprising a protic imidazolinium cation and a 4-hydroxycinnamate anion in aqueous 0.01 M NaCl solution using electrochemical and weight-loss methods. The corrosion behaviors of 1-(4-sulfobutyl)-3-methylimidazolium hydrogen sulfate $[BsMIM]^+[HSO_4]^-$ and 1-(4-sulfobutyl)-3-methylimidazolium tetrafluoroborate $[BsMIM]^+[BF_4]^-$ were studied by Ma et al. (2016) by EIS and polarization method in 1 M H_2SO_4 solution. Hegazy et al. (2014) synthesized and applied the three new di-quaternary ammonium salts, namely N-3-2-isopropyl dimethylammonio acetoxy propyl-N,N-dimethyldodecane-1-aminium chloride bromide, N-3-2-2-hydroxyethyl dimethylammonio acetoxy propyl-N,N-dimethyldodecane-1-aminium chloride bromide, and N-3-2-phenyldiethylammonio acetoxy propyl-N,N-dimethyldodecane-1-aminium chloride bromide named as CI for steel pipeline (API X65) in 1 M HCl. The inhibition effect of the compounds was studied by all three methods which included electrochemical impedance spectroscopy, Tafel method, and weight-loss methods. The polarization curves showed that the QASs showed mixed-type inhibition characteristics. Also the process represented physical adsorption as Langmuir isotherm fitted well.

The key factors which influence the performance of ILs as corrosion suppressors are pH of the solution mixture, hydrodynamic conditions, chain length of inhibiting compound, and ability of the compound to adsorb onto the surface of metal creating a shielding layer. This layer outlines the extent of adsorption. Figure 11.7 demonstrates the factors that play a significant role in identifying the corrosion behavior of the system.

11.2.5 Ionic Liquids (ILs) as Gas Hydrate and Corrosion Inhibitors (GHCI)

As it is highlighted in the earlier sections gas hydrate and corrosion formation require appropriate measures independently. Corrosion and hydrate inhibitors generally have compatibility issues with each other. Gas hydrate inhibitors increase the corrosion rates whereas corrosion inhibitors provide the nucleation sites for extended hydrate formations especially in deepwater conditions. Currently, there is

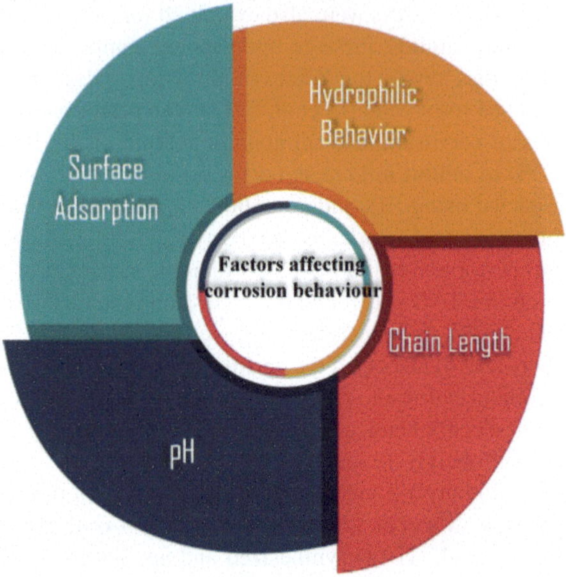

Fig. 11.7 Factors affecting corrosion behavior

a need for certain chemicals that can act as multipurpose gas hydrate and corrosion inhibitors (GHCIs). To address this issue in an effective manner, oil and gas industries are aiming for a single chemical that can be used in the flow pipeline instead of multiple chemical dosages. It will provide a cost-effective solution to this problem, thus resulting in potential economic savings. Table 11.1 shows the studied ILs as GHCIs for flow assurance problem.

According to Table 11.1, $[EMIM]^{+}[EtSO_4]^{-}$ shows the highest corrosion IE% among other ionic liquids and its performance as THI is also the best. It showed the highest value of average depression temperature (ADT) in THI applications (Yesudass et al. 2016).

The factors which affect significantly corrosion and gas hydrate inhibition are mentioned in Fig. 11.8. Among them, chain length has a major influence on GHCI behavior of a compound (Moore et al. 2009; Sheng et al. 2017). This is due to the requirement of shorter chain for gas hydrate mitigation whereas corrosion suppression requires relatively higher length of chain. Furthermore, the ability to make hydrogen bond and property regarding surface adsorption is mutually competitive and highly dependent on the chain length of the GHCIs. It is due to the reason that chemicals like ionic liquids form hydrogen bonding with water molecules and adsorb onto the surface of the pipeline, thus creating a shielding defense layer to mitigate hydrate and corrosion formations (Acidi et al. 2014; Yesudass et al. 2016). Therefore, the balanced or optimal chain length should be utilized depending upon the situation encountered in flow assurance applications. Charge density is another important factor which influences the hydrogen-bonding ability of a molecule. The ionic compounds with high charge density make strong hydrogen bonds. Charge density of electrolytic inhibitors such as ILs is dependent on their molecular structures (Zuriaga-Monroy et al. 2016).

Table 11.1 Ionic liquids as gas hydrate and corrosion inhibitor (GHCI)

Chemicals (ionic liquids)	Structure	Corrosion inhibitor specifications		Gas hydrate inhibitor specifications	Main findings			References
		Concentration	Electrolytic media		Corrosion (IE %)	THI (ADT) (K)	KHI (RIP)	
$[EMIM]^+[Cl]^-$	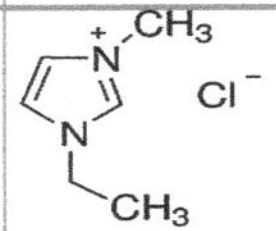	1, 5 mM	3.5 wt% NaCl	KHI + THI	70% (5 mM)	1.22	3.08	Bavoh et al. (2016); Sherif et al. (2015)
$[EMIM]^+[BF_4]^-$	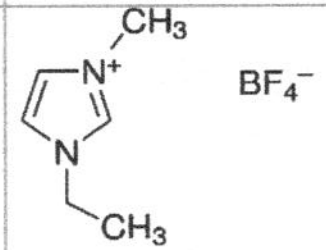	50–500 ppm	1 M HCl	KHI + THI	77.93% (500 ppm)	0.52	17	Sasikumar et al. (2015); Xiao and Adidharma (2009)
$[BMIM]^+[Br]^-$	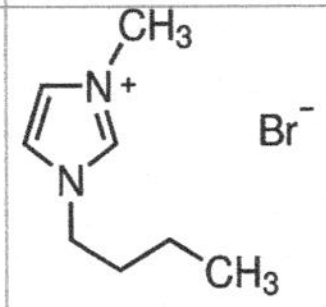	2–20 mM	1 M HCl	KHI + THI	94% (20 mM)	0.58	5.8	Ashassi-Sorkhabi and Es'haghi (2009); Xiao et al. (2010)
$[EMIM]^+[EtSO_4]^-$	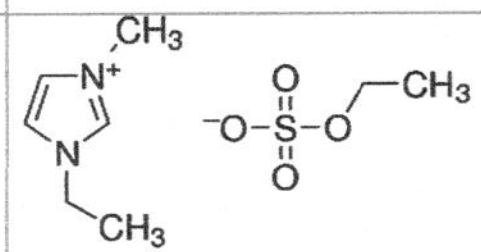	100, 300, 500 ppm	1 M HCl	KHI + THI	92.75% (500 ppm)	2.15	0.55	Talaghat (2014); Yesudass et al. (2016)
$[BMIM]^+[Cl]^-$	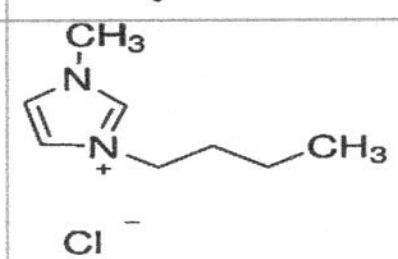	0.05–3 wt%	2 M HCl	KHI + THI	79% (1 wt%)	0.69	4.13	Xiao et al. (2010); Yousefi et al. (2015)

(continued)

Table 11.1 (continued)

Chemicals (ionic liquids)	Structure	Corrosion inhibitor specifications		Gas hydrate inhibitor specifications	Main findings			References
		Concentration	Electrolytic media		Corrosion (IE %)	THI (ADT) (K)	KHI (RIP)	
$[BMIM]^+[PF_6]^-$	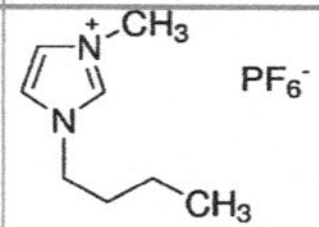	0.1–0.5 wt%	2 M HCl	THI	86% (0.1 wt%)	–	–	Peng et al. (2010); Yousefi et al. (2015)

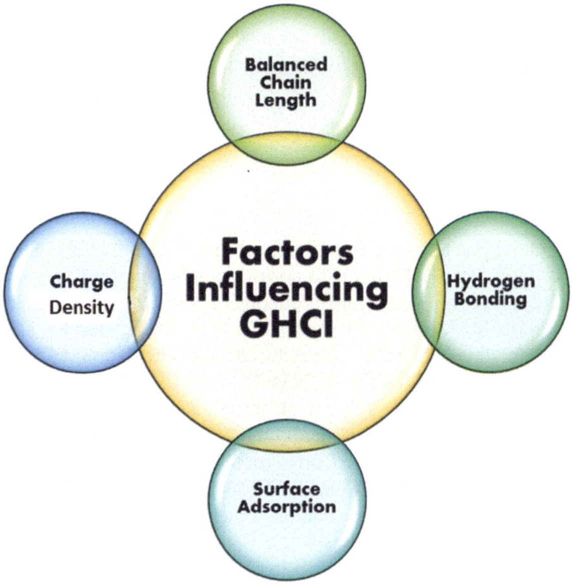

Fig. 11.8 Factors influencing GHCI behavior

11.3 Conclusion

In this book chapter, the role of ionic liquids as gas hydrate and corrosion inhibitors is examined for usage in offshore pipelines. It is found that among ILs, imidazolium-based, ammonium-based, and quaternary ammonium salts (QAS) have been utilized in this application. For hydrate inhibition study, ionic liquids were used in both THI and KHI applications. In case of researches involving corrosion inhibition, studies showed that these have been predominantly carried out involving electrolytic media with acidic conditions. The performance of 1-ethyl-3-methylimidazolium ethylsulfate $[EMIM]^+[EtSO_4]^-$ was found to be better in terms of hydrate and corrosion inhibition as compared to other ILs. The study revealed that optimum chain length, charge distribution, hydrogen-bonding ability, and surface adsorption are the main factors affecting the GHCI performance.

11.3.1 Future Prospects

In future, the work on the dual-functional compounds for hydrate and corrosion inhibition that show good biodegradability can be enhanced further. The properties of ionic liquids make these compounds a suitable candidate in this regard. Anionic or cationic groups from ILs can be co-tailored with some other charged moieties to develop a suitable compound in GHCI application. The advantage of using a single compound as GHCI can provide cost benefits in applications like equipment efficiency, operational savings in chemical deployment, and maintenance. In addition, QAS as GHCI can prove to be excellent dual-functional inhibitors for usage inside the pipeline.

Acknowledgements The authors would like to thank CO_2RES for providing necessary facilities. The authors gratefully also acknowledge the support of Chemical Engineering Department, Universiti Teknologi PETRONAS, Malaysia.

References

Acidi A, Hasib-ur-Rahman M, Larachi F, Abbaci A (2014) Ionic liquids [EMIM][BF_4], [EMIM][Otf] and [BMIM][Otf] as corrosion inhibitors for CO_2 capture applications. Korean J Chem Eng 31:1043–1048. https://doi.org/10.1007/s11814-014-0025-3

Akhfash M, Arjmandi M, Aman ZM, Boxall JA, May EF (2017) Gas hydrate thermodynamic inhibition with MDEA for reduced MEG circulation. J Chem Eng Data 62:2578–2583. https://doi.org/10.1021/acs.jced.7b00072

Amin MA, Hazzazi OA, Kandemirli F, Saracoglu M (2012) Inhibition performance and adsorptive behavior of three amino acids on cold-rolled steel in 1.0 M HCl-chemical, electrochemical, and morphological studies. Corrosion 68:688–698. https://doi.org/10.5006/0506

Ashassi-Sorkhabi H, Es'haghi M (2009) Corrosion inhibition of mild steel in acidic media by [BMIm]Br ionic liquid. Mater Chem Phys 114:267–271. https://doi.org/10.1016/j.matchemphys.2008.09.019

Avula VR, Gardas RL, Sangwai JS (2014) Modeling of methane hydrate inhibition in the presence of green solvent for offshore oil and gas pipeline. Proc Int Offshore Polar Eng Conf 3:49–54. https://doi.org/10.1016/j.chembiol.2007.07.009

Bavoh CB, Lal B, Keong LK, Jasamai MB, Idress MB (2016) Synergic kinetic inhibition effect of EMIM-cl + PVP on CO_2 hydrate formation. Procedia Eng 148:1232–1238. https://doi.org/10.1016/j.proeng.2016.06.474

Bavoh CB, Lal B, Khan MS, Osei H, Ayuob M (2018a) Inhibition effect of 1-ethyl-3-methylimidazolium chloride on methane hydrate equilibrium. J Phys Conf Ser:012060. https://doi.org/10.1088/1742-6596/1123/1/012060

Bavoh CB, Khan MS, Lal B, Bt Abdul Ghaniri NI, Sabil KM (2018b) New methane hydrate phase boundary data in the presence of aqueous amino acids. Fluid Phase Equilib 478:129–133. https://doi.org/10.1016/j.fluid.2018.09.011

Bousskri A, Anejjar A, Messali M, Salghi R, Benali O, Karzazi Y, Jodeh S, Zougagh M, Ebenso EE, Hammouti B (2015) Corrosion inhibition of carbon steel in aggressive acidic media with 1-(2-(4-chlorophenyl)-2-oxoethyl)pyridazinium bromide. J Mol Liq 211:1000–1008. https://doi.org/10.1016/j.molliq.2015.08.038

Brustad S, Løken KP, Waalmann JG (2005) Hydrate prevention using MEG instead of MeOH: impact of experience from major Norwegian developments on technology selection for injection and recovery of MEG, In: Offshore Technology Conference. p. 10

Burgazli CR, Navarrete RC, Mead SL (2005) New dual purpose chemistry for gas hydrate and corrosion inhibition. J Can Pet Technol 44:47–50. https://doi.org/10.2118/05-11-04

Cha M, Shin K, Kim J, Chang D, Seo Y, Lee H, Kang SP (2013) Thermodynamic and kinetic hydrate inhibition performance of aqueous ethylene glycol solutions for natural gas. Chem Eng Sci 99:184–190. https://doi.org/10.1016/j.ces.2013.05.060

Cha J-H, Ha C, Kang S-P, Kang JW, Kim K-S (2015) Thermodynamic inhibition of CO2 hydrate in the presence of morpholinium and piperidinium ionic liquids. Fluid Phase Equilib:2–6. https://doi.org/10.1016/j.fluid.2015.09.008

Cha JH, Ha C, Kang SP, Kang JW, Kim KS (2016) Thermodynamic inhibition of CO_2 hydrate in the presence of morpholinium and piperidinium ionic liquids. Fluid Phase Equilib 413:75–79. https://doi.org/10.1016/j.fluid.2015.09.008

Chen W (2011) Status and challenges of Chinese deepwater oil and gas development. Pet Sci 8:477–484. https://doi.org/10.1007/s12182-011-0171-8

Chong AL, Mardel JI, MacFarlane DR, Forsyth M, Somers AE (2016) Synergistic corrosion inhibition of mild steel in aqueous chloride solutions by an imidazolinium carboxylate salt. ACS Sustain Chem Eng 4:1746–1755. https://doi.org/10.1021/acssuschemeng.5b01725

Del Villano L, Kommedal R, Kelland MA (2008) Class of kinetic hydrate inhibitors with good biodegradability. Energy and Fuels 22:3143–3149. https://doi.org/10.1021/ef800161z

Dendy Sloan E, Koh CA (2008) Gas hydrates of natural gases, 3rd edn. CRC Press LLC, London

Drive C, Carroll J (2009) Natural gas hydrates: a guide for engineers. In: Industrial & engineering, Third edn. Elsevier, Amsterdam. https://doi.org/10.2118/23562-PA

El-Haddad MN (2014) Hydroxyethylcellulose used as an eco-friendly inhibitor for 1018 c-steel corrosion in 3.5% NaCl solution. Carbohydr Polym 112:595–602. https://doi.org/10.1016/j.carbpol.2014.06.032

El-Hajjaji F, Messali M, Aljuhani A, Aouad MR, Hammouti B, Belghiti ME, Chauhan DS, Quraishi MA (2018) Pyridazinium-based ionic liquids as novel and green corrosion inhibitors of carbon steel in acid medium: electrochemical and molecular dynamics simulation studies. J Mol Liq 249:997–1008. https://doi.org/10.1016/j.molliq.2017.11.111

El-Shamy AM, Zakaria K, Abbas MA, Zein El Abedin S (2015) Anti-bacterial and anti-corrosion effects of the ionic liquid 1-butyl-1-methylpyrrolidinium trifluoromethylsulfonate. J Mol Liq 211:363–369. https://doi.org/10.1016/j.molliq.2015.07.028

Erfani, A., Varaminian, F., Muhammadi, M., 2013. Gas hydrate formation inhibition using low dosage hydrate inhibitors, in: 2nd National Iranian Conference on gas hydrate (NICGH)

Fazlali A, Keshavarz-Moraveji M, Farjami M, Mohammadi AH (2016) Experimental study of the effect of kinetic inhibitor on methane hydrate formation. In: Taylor JC (ed) Advances in chemistry research. Nova Science Publishers, Inc., NY, pp 133–148

Frostman LM (2000) Anti-agglomerant hydrate inhibitors for prevention of hydrate plugs in Deepwater systems. SPE Annu Tech Conf Exhib:23–24. https://doi.org/10.2523/63122-MS

Gonfa G, Bustam MA, Sharif AM, Mohamad N, Ullah S (2015) Tuning ionic liquids for natural gas dehydration using COSMO-RS methodology. J Nat Gas Sci Eng 27:1141–1148. https://doi.org/10.1016/j.jngse.2015.09.062

Gupta, G., Limited, R.I., Singh, S.K., Limited, R.I., 2012. Hydrate inhibition _ optimization in deep water field 28–30

Gupta P, Chandrasekharan Nair V, Sangwai JS (2018) Phase equilibrium of methane hydrate in the presence of aqueous solutions of quaternary ammonium salts. J Chem Eng Data 63:2410–2419. https://doi.org/10.1021/acs.jced.7b00976

Hamadi L, Mansouri S, Oulmi K, Kareche A (2018) The use of amino acids as corrosion inhibitors for metals: a review. Egypt J Pet. https://doi.org/10.1016/j.ejpe.2018.04.004

Hegazy MA, Abdallah M, Awad MK, Rezk M (2014) Three novel di-quaternary ammonium salts as corrosion inhibitors for API X65 steel pipeline in acidic solution. Part I: experimental results. Corros Sci 81:54–64. https://doi.org/10.1016/j.corsci.2013.12.010

Heidaryan E, Salarabadi A, Moghadasi J, Dourbash A (2010) A new high performance gas hydrate inhibitor. J Nat Gas Chem 19:323–326. https://doi.org/10.1016/S1003-9953(09)60060-8

Heppner KL, Evitts RW (2008) Modelling of the effect of hydrogen ion reduction on the crevice corrosion of titanium. Environ Crack Mater 1:95–104. https://doi.org/10.1016/B978-008044635-6.50011-X

Huo Z, Freer E, Lamar M, Sannigrahi B, Knauss DM, Sloan ED (2001) Hydrate plug prevention by anti-agglomeration. Chem Eng Sci 56:4979–4991. https://doi.org/10.1016/S0009-2509(01)00188-9

Javadian S, Yousefi A, Neshati J (2013) Synergistic effect of mixed cationic and anionic surfactants on the corrosion inhibitor behavior of mild steel in 3.5% NaCl. Appl Surf Sci 285:674–681. https://doi.org/10.1016/j.apsusc.2013.08.109

Joshi SV, Grasso GA, Lafond PG, Rao I, Webb E, Zerpa LE, Sloan ED, Koh CA, Sum AK (2013) Experimental flowloop investigations of gas hydrate formation in high water cut systems. Chem Eng Sci 97:198–209. https://doi.org/10.1016/j.ces.2013.04.019

Kannan P, Karthikeyan J, Murugan P, Rao TS, Rajendran N (2016) Corrosion inhibition effect of novel methyl benzimidazolium ionic liquid for carbon steel in HCl medium. J Mol Liq 221:368–380. https://doi.org/10.1016/j.molliq.2016.04.130

Ke W, Kelland MA (2016) Kinetic hydrate inhibitor studies for gas hydrate systems: a review of experimental equipment and test methods. Energy Fuels 30:10015–10028. https://doi.org/10.1021/acs.energyfuels.6b02739

Kelland MA (2006) History of the development of low dosage hydrate inhibitors. Energy Fuels 20:825–847. https://doi.org/10.1021/ef050427x

Kelland MA, Svartås TM, Andersen LD (2009) Gas hydrate anti-agglomerant properties of polypropoxylates and some other demulsifiers. J Pet Sci Eng 64:1–10. https://doi.org/10.1016/j.petrol.2008.12.001

Kennell GF, Evitts RW (2009) Crevice corrosion cathodic reactions and crevice scaling laws. Electrochim Acta 54:4696–4703. https://doi.org/10.1016/j.electacta.2009.03.080

Khan MS, Lal B, Partoon B, Keong LK, Bustam AB, Mellon NB (2016) Experimental evaluation of a novel thermodynamic inhibitor for CH_4 and CO_2 hydrates. Procedia Eng 148:932–940. https://doi.org/10.1016/j.proeng.2016.06.433

Khan MS, Bavoh CB, Partoon B, Lal B, Bustam MA, Shariff AM (2017a) Thermodynamic effect of ammonium based ionic liquids on CO_2 hydrates phase boundary. J Mol Liq 238:533–539. https://doi.org/10.1016/j.molliq.2017.05.045

Khan MS, Lal B, Bavoh CB, Keong LK, Bustam A, Mellon NB (2017b) Influence of ammonium based compounds for gas hydrate mitigation: a short review. Indian J Sci Technol 10:1–6. https://doi.org/10.17485/ijst/2017/v10i5/99734

Khan MS, Partoon B, Bavoh CB, Lal B, Mellon NB (2017c) Influence of tetramethylammonium hydroxide on methane and carbon dioxide gas hydrate phase equilibrium conditions. Fluid Phase Equilib 440:1–8. https://doi.org/10.1016/j.fluid.2017.02.011

Khan MS, Yaqub S, Manner N, Karthwathi NA, Qasim A, Mellon NB, Lal B (2018) Experimental equipment validation for methane (CH_4) and carbon dioxide (CO_2) hydrates. IOP Conf Ser Mater Sci Eng 344:1–10. https://doi.org/10.1088/1757-899X/344/1/012025

Khan MS, Bavoh CB, Partoon B, Nashed O, Lal B, Mellon NB (2018a) Impacts of ammonium based ionic liquids alkyl chain on thermodynamic hydrate inhibition for carbon dioxide rich binary gas. J Mol Liq 261:283–290. https://doi.org/10.1016/j.molliq.2018.04.015

Khan MS, Cornelius BB, Lal B, Bustam MA (2018b) Kinetic assessment of tetramethyl ammonium hydroxide (ionic liquid) for carbon carbon dioxide, methane and binary mix gas hydrates. In: Recent Advances in Ionic Liquids, pp 159–179. https://doi.org/10.5772/intechopen.77262

Khan MS, Lal B, Keong LK, Sabil KM (2018c) Experimental evaluation and thermodynamic modelling of AILs alkyl chain elongation on methane rich gas hydrate system. Fluid Phase Equilib 473:300–309. https://doi.org/10.1016/j.fluid.2018.07.003

Khan MS, Lal B, Shariff AM, Mukhtar H (2018d) Ammonium hydroxide ILs as dual-functional gas hydrate inhibitors for binary mixed gas (carbon dioxide and methane) hydrates. J Mol Liq 274:33–44. https://doi.org/10.1016/j.molliq.2018.10.076

Khan MS, Lal B, Keong LK, Ahmed I (2019) Tetramethyl ammonium chloride as dual functional inhibitor for methane and carbon dioxide hydrates. Fuel 236:251–263. https://doi.org/10.1016/j.fuel.2018.09.001

Kim K, Kang SP (2011) Investigation of pyrrolidinium- and morpholinium-based ionic liquids into kinetic hydrate inhibitors on structure I methane hydrate, in: 7th International Conference on Gas Hydrates (ICGH) 2011. Edinburgh, Scotland, pp. 17–21

Klomp UC, Kruka VC, Reijnhart R (1996) International Patent Application WO 95/17579, 1995. There is no Corresp. Rec. this Ref

Koh CA, Sloan ED, Sum AK, Wu DT (2011) Fundamentals and applications of gas hydrates. Annu Rev Chem Biomol Eng 2:237–257. https://doi.org/10.1146/annurev-chembioeng-061010-114152

Kuznetsova T, Sapronova A, Kvamme B, Johannsen K, Haug J (2010) Impact of low-dosage inhibitors on clathrate hydrate stability. Macromol Symp 287:168–176. https://doi.org/10.1002/masy.201050124
Laamari R, Villemin D (2011) Corrosion inhibition of carbon steel in hydrochloric acid 0 . 5 M by hexa methylene diamine tetramethyl-phosphonic acid. Arab J Chem 4:271–277. https://doi.org/10.1016/j.arabjc.2010.06.046
Li XS, Zhan H, Xu CG, Zeng ZY, Lv QN, Yan KF (2012) Effects of tetrabutyl-(ammonium/phosphonium) salts on clathrate hydrate capture of CO_2 from simulated flue gas. Energy Fuels 26:2518–2527. https://doi.org/10.1021/ef3000399
Liu Z, Jackson TS, Ramachandran S (2018) Synergic Corrosion inhibitors. US2018/0201826A1
Ma Y, Han F, Li Z, Xia C (2016) Acidic-functionalized ionic liquid as corrosion inhibitor for 304 stainless steel in aqueous sulfuric acid. ACS Sustain Chem Eng 4:5046–5050. https://doi.org/10.1021/acssuschemeng.6b01492
Mendonca GLF, Costa SN, Freire VN, Casciano PNS, Correia AN, de Lima-Neto P (2017) Understanding the corrosion inhibition of carbon steel and copper in sulphuric acid medium by amino acids using electrochemical techniques allied to molecular modelling methods. Corros Sci 115:41–55. https://doi.org/10.1016/j.corsci.2016.11.012
Mokhatab S, Wilkens RJ, Leontaritis KJ (2007) A review of strategies for solving gas-hydrate problems in subsea pipelines. Energy sources, part a recover. Util Environ Eff 29:39–45. https://doi.org/10.1080/00908310933988
Moore JA, Ver Vers L, Conrad P (2009) Understanding kinetic hydrate inhibitor and corrosion inhibitor interactions, In: Offshore Technology Conference. Houston, Texas, pp. 1–20. https://doi.org/10.4043/OTC-19869-MS
Murulana LC, Singh AK, Shukla SK, Kabanda MM, Ebenso EE (2012) Experimental and quantum chemical studies of some bis(trifluoromethyl-sulfonyl) imide imidazolium-based ionic liquids as corrosion inhibitors for mild steel in hydrochloric acid solution. Ind Eng Chem Res 51:13282–13299. https://doi.org/10.1021/ie300977d
Nashed O, Sabil KM, Ismail L, Japper-Jaafar A, Lal B (2017) Mean induction time and isothermal kinetic analysis of methane hydrate formation in water and imidazolium based ionic liquid solutions. J Chem Thermodyn 1–8. https://doi.org/10.1016/j.jct.2017.09.015
Nashed O, Partoon B, Lal B, Sabil KM, Mohd A (2018) Review the impact of nanoparticles on the thermodynamics and kinetics of gas hydrate formation. J Nat Gas Sci Eng 55:452–465. https://doi.org/10.1016/j.jngse.2018.05.022
Nešić S (2007) Key issues related to modelling of internal corrosion of oil and gas pipelines - a review. Corros Sci 49:4308–4338. https://doi.org/10.1016/j.corsci.2007.06.006
Obanijesu EO (2012) Corrosion and hydrate formation in natural gas pipelines. Curtin University, Australia
Obanijesu EO, Pareek V, Tade MO (2010) Hydrate formation and its influence on natural gas pipeline internal corrosion rate. Environment 62:164–173. https://doi.org/10.2118/128544-MS
Obanijesu EO, Gubner R, Barifcani A, Pareek V, Tade MO (2014) The influence of corrosion inhibitors on hydrate formation temperature along the subsea natural gas pipelines. J Pet Sci Eng 120:239–252. https://doi.org/10.1016/j.petrol.2014.05.025
Olajire AA (2017) Corrosion inhibition of offshore oil and gas production facilities using organic compound inhibitors - a review. J Mol Liq 248:775–808. https://doi.org/10.1016/j.molliq.2017.10.097
Partoon B, Wong NMS, Sabil KM, Nasrifar K, Ahmad MR (2013) A study on thermodynamics effect of [EMIM]-cl and [OH-C_2MIM]-cl on methane hydrate equilibrium line. Fluid Phase Equilib 337:26–31
Patel ZD, Russum J (2009) Flow assurance: chemical inhibition of gas hydrates in deepwater production systems. Offshore Magzine 4. https://doi.org/10.4043/14010-MS
Peng X, Hu Y, Liu Y, Jin C, Lin H (2010) Separation of ionic liquids from dilute aqueous solutions using the method based on CO_2 hydrates. J Nat Gas Chem 19:81–85. https://doi.org/10.1016/S1003-9953(09)60027-X

Pisarova L, Gabler C, Dörr N, Pittenauer E, Allmaier G (2012) Thermo-oxidative stability and corrosion properties of ammonium based ionic liquids. Tribol Int 46:73–83. https://doi.org/10.1016/j.triboint.2011.03.014

Popoola LT, Grema AS, Gutti GKLB, Balogun SA (2013) Corrosion problems during oil and gas production and its mitigation. Int J Ind Chem 4:1–15

Qasim A, Shariff AM, Sufian S, Ayoub M, Lal B (2018) Phase identification of natural gas system with high CO_2 content through simulation approach using Peng-Robinson model IOP Conf. Ser Mater Sci Eng 458:1–5. https://doi.org/10.1088/1757-899X/458/1/012068

Qasim A, Khan MS, Lal B, Shariff AM (2019) Phase equilibrium measurement and modeling approach to quaternary ammonium salts with and without monoethylene glycol for carbon dioxide hydrates. J Mol Liq 282:106–114. https://doi.org/10.1016/j.molliq.2019.02.115

Qureshi MF, Atilhan M, Altamash T, Tariq M, Khraisheh M, Aparicio S, Tohidi B (2016) Gas hydrate prevention and flow assurance by using mixtures of ionic liquids and Synergent compounds: combined kinetics and thermodynamic approach. Energy Fuel 30:3541–3548. https://doi.org/10.1021/acs.energyfuels.5b03001

Rajahram SS, Harvey TJ, Wood RJK (2009) Erosion-corrosion resistance of engineering materials in various test conditions. Wear 267:244–254. https://doi.org/10.1016/j.wear.2009.01.052

Sa J-H, Kwak G-H, Lee BR, Park D-H, Han K, Lee K-H (2013) Hydrophobic amino acids as a new class of kinetic inhibitors for gas hydrate formation. Sci Rep 3:2428. https://doi.org/10.1038/srep02428

Sa J-H, Kwak G-H, Han K, Ahn D, Cho SJ, Lee JD, Lee K-H (2016) Inhibition of methane and natural gas hydrate formation by altering the structure of water with amino acids. Sci Rep 6:1–9. https://doi.org/10.1038/srep31582

Sabil KM, Nashed O, Lal B, Ismail L, Japper-Jaafar A, Japper A, Nashed O, Lal B, Ismail L, Japper-Jaafar A (2015) Experimental investigation on the dissociation conditions of methane hydrate in the presence of imidazolium-based ionic liquids. J Chem Thermodyn 84:7–13. https://doi.org/10.1016/j.jct.2014.12.017

Sasikumar Y, Adekunle AS, Olasunkanmi LO, Bahadur I, Baskar R, Kabanda MM, Obot IB, Ebenso EE (2015) Experimental, quantum chemical and Monte Carlo simulation studies on the corrosion inhibition of some alkyl imidazolium ionic liquids containing tetrafluoroborate anion on mild steel in acidic medium. J Mol Liq 211:105–118. https://doi.org/10.1016/j.molliq.2015.06.052

Schweinsberg DP, Ashwortht V (1988) The inhibition of the corrosion of pure iron in 0.5 M sulphuric acid by n-alkyl quaternary ammonium iodides. Corros Sci 28:539–545

Sheng Q, Da Silveira KC, Tian W, Fong C, Maeda N, Gubner R, Wood CD (2017) Simultaneous hydrate and corrosion inhibition with modified poly(vinyl caprolactam) polymers. Energy Fuels 31:6724–6731. https://doi.org/10.1021/acs.energyfuels.7b00525

Sherif ESM, Abdo HS, El Abedin SZ (2015) Corrosion inhibition of cast iron in Arabian gulf seawater by two different ionic liquids. Materials 8:3883–3895. https://doi.org/10.3390/ma8073883

Sloan ED (2005) A changing hydrate paradigm - from apprehension to avoidance to risk management. Fluid Phase Equilib 228–229:67–74. https://doi.org/10.1016/j.fluid.2004.08.009

Sloan ED, Koh CA (2008) Clathrate hydrates of natural gases, 3rd edn. CRC Press, Boca Raton

Sloan D, Koh C, Sum A (2010a) Natural gas hydrates in flow assurance. Gulf Professional Publishing, Atlantis

Sloan ED, Koh CA, Sum AK (2010b) Gas hydrate stability and sampling: the future as related to the phase diagram. Energies 3:1991–2000. https://doi.org/10.3390/en3121991

Smiglak M, Pringle JM, Lu X, Han L, Zhang S, Gao H, MacFarlane DR, Rogers RD (2014) Ionic liquids for energy, materials, and medicine. Chem Commun 50. https://doi.org/10.1039/c4cc02021a

Srinivasan S, Kane RD (2003) Critical issues in the application and evaluation of a corrosion prediction model for oil and gas systems. Corrosion 2003:1–18

Tabrez AR, Inam A (2012) Hydrocarbon potential in the Makran offshore area, In: Search and discovery article

Talaghat MR (2014) Experimental investigation of induction time for double gas hydrate formation in the simultaneous presence of the PVP and l-tyrosine as kinetic inhibitors in a mini flow loop apparatus. J Nat Gas Sci Eng 19:215–220. https://doi.org/10.1016/j.jngse.2014.05.010

Tariq M, Rooney D, Othman E, Aparicio S, Atilhan M, Khraisheh M (2014) Gas hydrate inhibition: a review of the role of ionic liquids. Ind Eng Chem Res 53:17855–17868. https://doi.org/10.1021/ie503559k

Tawfik SM (2016) Ionic liquids based gemini cationic surfactants as corrosion inhibitors for carbon steel in hydrochloric acid solution. J Mol Liq 216:624–635. https://doi.org/10.1016/j.molliq.2016.01.066

Tiu BDB, Advincula RC (2015) Polymeric corrosion inhibitors for the oil and gas industry: design principles and mechanism. React Funct Polym 95:25–45. https://doi.org/10.1016/j.reactfunctpolym.2015.08.006

Ullah S, Bustam MA, Shariff AM, Gonfa G, Izzat K (2016) Experimental and quantum study of corrosion of A36 mild steel towards 1-butyl-3-methylimidazolium tetrachloroferrate ionic liquid. Appl Surf Sci 365:76–83. https://doi.org/10.1016/j.apsusc.2015.12.232

Ullah S, Bustam MA, Shariff AM, Gonfa G, Ayoub M, Raihan M (2017) Experimental study of corrosion on A36 mild steel towards aqueous 2-amino-2-ethyl-1, 3-propanediol and diethanolamine. Int J Electrochem Sci 12:1642–1656. https://doi.org/10.20964/2017.02.36

Umoren SA, Obot IB (2008) Polyvinylpyrrolidone and polyacrylamide as corrosion inhibitors for mild steel in acidic medium. Surf Rev Lett 15:277–286. https://doi.org/10.1142/S0218625X08011366

Umoren SA, Banera MJ, Alonso-Garcia T, Gervasi CA, Mirífico MV (2013) Inhibition of mild steel corrosion in HCl solution using chitosan. Cellulose 20:2529–2545. https://doi.org/10.1007/s10570-013-0021-5

Verma C, Ebenso EE, Quraishi MA (2017) Ionic liquids as green and sustainable corrosion inhibitors for metals and alloys: an overview. J Mol Liq 233:403–414. https://doi.org/10.1016/j.molliq.2017.02.111

Verma C, Ebenso EE, Quraishi MA (2018a) Ionic liquids as green corrosion inhibitors for industrial metals and alloys. Green Chem:103–132

Verma C, Obot IB, Bahadur I, Sherif ESM, Ebenso EE (2018b) Choline based ionic liquids as sustainable corrosion inhibitors on mild steel surface in acidic medium: gravimetric, electrochemical, surface morphology, DFT and Monte Carlo simulation studies. Appl Surf Sci 457:134–149. https://doi.org/10.1016/j.apsusc.2018.06.035

Wang F, Wang LF (1998) Polydimethylsiloxane modification of segmented thermoplastic polyurethanes and Polyureas Polydimethylsiloxane modification of segmented thermoplastic polyurethanes and Polyureas. Virginia Polytechnic Institute and State University, Blacksburg

Welton T (1999) Room-temperature ionic liquids: solvents for synthesis and catalysis. Chem Rev 99:2071–2083

Xiao C, Adidharma H (2009) Dual function inhibitors for methane hydrate. Chem Eng Sci 64:1522–1527. https://doi.org/10.1016/j.ces.2008.12.031

Xiao K, Dong CF, Li XG, Wang FM (2008) Corrosion products and formation mechanism during initial stage of atmospheric corrosion of carbon steel. J Iron Steel Res Int 15:42–48. https://doi.org/10.1016/S1006-706X(08)60247-2

Xiao C, Wibisono N, Adidharma H (2010) Dialkylimidazolium halide ionic liquids as dual function inhibitors for methane hydrate. Chem Eng Sci 65:3080–3087. https://doi.org/10.1016/j.ces.2010.01.033

Yaqub S, Lal B, Partoon B, Mellon NB (2018) Investigation of the task oriented dual function inhibitors in gas hydrate inhibition: a review. Fluid Phase Equilib 477:40–57. https://doi.org/10.1016/j.fluid.2018.08.015

Yee Tak N, Englezos P (1996) Concentration of mechanical pulp mill effluents and NaCl solutions through propane hydrate formation. Ind Eng Chem Res 35:1894–1900. https://doi.org/10.1021/ie960001t

Yesudass S, Olasunkanmi LO, Bahadur I, Kabanda MM, Obot IB, Ebenso EE (2016) Experimental and theoretical studies on some selected ionic liquids with different cations/anions as corrosion inhibitors for mild steel in acidic medium. J Taiwan Inst Chem Eng 64:252–268. https://doi.org/10.1016/j.jtice.2016.04.006

Yousefi A, Javadian S, Dalir N, Kakemam J, Akbari J (2015) Imidazolium-based ionic liquids as modulators of corrosion inhibition of SDS on mild steel in hydrochloric acid solutions: experimental and theoretical studies. RSC Adv 5:11697–11713

Zarrouk A, Messali M, Aouad MR, Assouag M, Zarrok H, Salghi R, Hammouti B, Chetouani A (2012) Some new ionic liquids derivatives: synthesis, characterization and comparative study towards corrosion of C-steel in acidic media. J Chem Pharm Res 4:3427–3436. https://doi.org/10.1038/439399a

Zerpa LE, Salager J-L, Koh CA, Sloan ED, Sum AK (2011) Surface chemistry and gas hydrates in flow assurance. Ind Eng Chem Res 50:188–197. https://doi.org/10.1021/ie100873k

Zhu Y, Free ML, Woollam R, Durnie W (2017) A review of surfactants as corrosion inhibitors and associated modeling. Prog Mater Sci 90:159–223. https://doi.org/10.1016/j.pmatsci.2017.07.006

Zuriaga-Monroy C, Oviedo-Roa R, Montiel-Sánchez LE, Vega-Paz A, Marín-Cruz J, Martínez-Magadán JM (2016) Theoretical study of the aliphatic-chain length's electronic effect on the corrosion inhibition activity of methylimidazole-based ionic liquids. Ind Eng Chem Res 55:3506–3516. https://doi.org/10.1021/acs.iecr.5b03884

Chapter 12
Production of Biodiesel Using Ionic Liquids

Seán O'Connor, Suresh C. Pillai, Ehiaze Ehimen, and John Bartlett

Contents

Abbreviations

$AgBF_4$	Silver tetrafluoroborate
ASTM	American Society for Testing and Materials
BF4	Tetrafluoroborate

S. O'Connor (✉) · E. Ehimen · J. Bartlett
Department of Environmental Science, School of Science, Institute of Technology Sligo, Sligo, Ireland
e-mail: sean.oconnor2@mail.itsligo.ie; ehimen.ehiaze@itsligo.ie; bartlett.john@itsligo.ie

S. C. Pillai
Nanotechnology and Bio-Engineering Research Group, Department of Environmental Science, School of Science, Institute of Technology Sligo, Sligo, Ireland

Centre for Precision Engineering, Materials and Manufacturing Research (PEM), Institute of Technology Sligo, Sligo, Ireland
e-mail: pillai.suresh@itsligo.ie

Inamuddin, A. M. Asiri (eds.), *Nanotechnology-Based Industrial Applications of Ionic Liquids*, Nanotechnology in the Life Sciences,
https://doi.org/10.1007/978-3-030-44995-7_12

BMIm.CH_3SO_3	1-Butyl-3-methylimidazolium methanesulfonate
BMIm.DCA	1-Butyl-3-methylimidazolium dicyanamide
BMIm.MeSO4	1-Butyl-3-methylimidazolium methyl sulfate
BMIm.OH	1-Butyl-3-methylimidazolium hydroxide
BMIm.PF_6	1-Butyl-3-methylimidazolium hexafluorophosphate
BSP_y.CF_3SO_3	1-4-Sulfonic acid butylpyridinium trifluoromethanesulfonate
C	Conversion
CaO	Calcium oxide
ChOH	Choline hydroxide
ChOMe	Choline methoxide
DAIL	Dicationic Brønsted acidic ionic liquid
DES	Deep eutectic solvent
EMIm	1-Ethyl-3-methylimidazolium methyl sulfate
EMIm.PF_6	1-Ethyl-3-methylimidazolium
EMIm.TfO	1-Ethyl-3-methylimidazolium trifluoromethanesulfonate
FAME	Fatty acid methyl ester
FFAs	Free fatty acids
H_2SO_4	Sulfuric acid
HPyrBr	1-Hexylpyridinium bromide
IL	Ionic liquid
LGCPO	Low-grade crude palm oil
PF_4	Tetrafluorophosphite
PUFA	Polyunsaturated fatty acids
SBA-15	Mesoporous silica
uFAME	Unsaturated fatty acid methyl ester
wt	Weight
Y	Yield

12.1 Introduction

There is an immediate threat of climate change, where the average global temperature has increased by 1.1 °C since pre-industrial levels because of human activity (IPCC 2013, 2014, 2018). If current trends continue, the temperature will likely surpass 1.5 °C between 2030 and 2052, resulting in potentially disastrous consequences, including the mass die-off of coral reefs, worsening food shortages, more frequent wildfires, and an increase in extreme weather events (IPCC 2018). The release of harmful and hazardous exhaust gases from the combustion of petroleum-based fuels has contributed significantly to the advancement of climate change, representing 98% of carbon emissions released into the atmosphere (Aransiola et al. 2012). Therefore, the need to pursue and develop alternative, environmentally friendly sources of fuel is apparent (Farrell et al. 2006; Semelsberger et al. 2006).

One alternative energy source is biodiesel, which has a chemical composition similar to that of diesel, with the added benefit of being biodegradable, non-toxic, renewable, and environmentally friendly (Demirbas 2008). Biodiesel is an oil-based diesel-type fuel consisting of long-chain alkyl (ethyl, methyl, or propyl) esters typically obtained by the transesterification of lipids, either animal fats or vegetable oils, with short-chain alcohols to produce fatty acid monoesters (Van Gerpen 2005; Leung et al. 2010; Zhao and Baker 2013). This reaction is usually catalysed using acids or bases (homogenous or heterogeneous), or via enzymatic routes, i.e. using lipases. This conversion process is required to produce a fuel product which allows for its direct combustion in modern diesel engines, as the viscosity of the starting feedstock is often too high (Dupont et al. 2009). Various parameters can affect the process and subsequently influence the biodiesel yield, including the moisture content, free fatty acids (FFAs), loading time, reaction temperature, type of catalysts, and molar ratio of alcohol to oil (Parawira 2010; Highina et al. 2011). Biodiesel is considered carbon neutral; that is, the amount of carbon spent on the formation of biomass (processed into fuel) is equal to the amount of carbon returned to the atmosphere as a result of fuel combustion (Mathews 2008). This therefore allows for the reduction in the release of polluting emissions including hydrocarbon, particulate matter, and carbon monoxide while having a comparable fuel economy to diesel fuels (Gashaw and Tesita 2014).

Homogeneous catalysts are conventionally used for the preparation of biodiesel, despite having serious disadvantages, such as issues regarding its corrosivity, energy-intensive operations, product purification difficulties, and production of a high amount of wastewater (Antolin et al. 2002; Wu et al. 2007a; Liang et al. 2009; Stojković et al. 2014; Veljković et al. 2014; Khan et al. 2015). Therefore, because of the limitations associated with these solvents, the need for a viable alternative becomes a priority. One environmentally friendly alternative to conventional chemical transesterification is the enzymatic transesterification method, because it involves milder operating conditions, lower energy demand, reduced excess volumes, and reusability of enzymes, and permits a minor volume of water in substrates (Akoh et al. 2007; Troter et al. 2016). However, this method has several difficulties, such as leftover impurities, inactivation of lipase, and high cost of enzymes, which have greatly limited the potential of this approach (Troter et al. 2016).

In recent years, ionic liquids (ILs) have emerged as a promising alternative solvent group, extracting solvent, or catalyst in the production of biodiesel. ILs have generated significant attention as an adjustable (task-driven) and budding 'green' solvent, with the promise of being a low-cost alternative to conventional volatile organics. Additionally, a new generation of ILs, called deep eutectic solvents (DESs), have generated significant interest among researchers, as they have the major advantages of potentially being a more cost-effective and environmentally friendly approach. The properties of DESs can be finely tuned, similar to those of ILs, by selecting different cation and anion combinations. Many benefits can therefore be gained, such as high purity, ease of preparation, non-toxicity, biodegradability, requirement of mild reaction conditions, and insensitivity to water (Abbott et al.

2004). As a result, researchers have predicted DESs to progressively replace ILs in the biodiesel production processes in the near future (Troter et al. 2016).

The purpose of this chapter is to present an overview of the reported use of ILs and DESs in the production of biodiesel. This review focuses on the application of ILs as either a catalyst or co-solvent components, or for extracting solvents from numerous biodiesel feedstocks. In addition, the work critically examines and discusses the potential replacement of ILs with DESs in biodiesel synthesis. The work highlights progress in the field, problems and difficulties faced, and future research outlook.

12.2 Ionic Liquids

ILs are defined as a class of organic salts, comprised of organic cations and either organic or inorganic anions with a melting temperature of less than 100 °C. The process of IL synthesis consists of cation formation and anion exchange. The study of ILs dates back to 1914 by Walden (1914) who recorded the physical properties of ethyl ammonium nitrate, which was formed by reacting ethylamine with concentrated nitric acid. However, the discovery did not generate significant attention, as its explosive nature limited widespread application. In general, ILs continued to receive little attention until their use as a chemical synthesis in the 1990s, producing ILs with unique physical-chemical properties (Seddon 1997).

In recent years, ILs have gained significant interest as a promising green chemical solvent, where the abundance of structures available and their adjustable nature have allowed their application in various commercial segments. This second generation of ILs has garnered significant attention as their properties can be optimised for the reaction through the careful selection of the cation and anion (Vancov et al. 2012). Researchers have taken advantage of this characteristic by developing ILs finely tuned to meet the desired needs of specific applications. These needs have included specific melting points, viscosities, volatility, or both thermal and electrochemical stability (Wasserscheid and Welton 2002; MacFarlane et al. 2014). A variety of processes have used and applied the benefits of ILs, including the substituting of high-volatility solvents, biomass pretreatment, gas purification, reaction media, catalysis, cellulose dissolution, and elimination of heavy metal traces (Keskin et al. 2007; Tadesse and Luque 2011).

Today, a third generation of ILs is being developed to achieve specific desirable biological properties, such as enzyme stabilisation and activation as seen in Fig. 12.1 (Farra et al. 2015). Based on these developments, various enzymatic reactions have been examined using a variety of ILs (van Rantwijk et al. 2007; Moniruzzaman et al. 2010; Zhao 2010; Tang et al. 2012). Additionally, ILs have generated interest in different processes as a 'green solvent' because of their low vapour pressure, minimising the negative impact on humans and the environment (Passos et al. 2014). Even though previous studies have reported some ILs as having poor biodegradability and relative toxicity, recent studies have found that specific mixed-strain

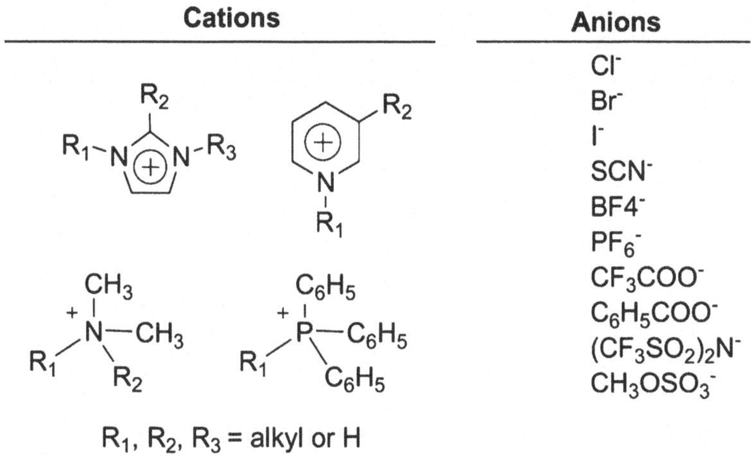

Fig. 12.1 Chemical structures of common ionic liquids. Reprinted with permission from Farra, A. et al. (2015) 'Green solvents in carbohydrate chemistry: From raw materials to fine chemicals', Chemical Reviews, 115(14), pp. 6811–6853. Copyright 2015 American Chemical Society

consortia have the potential to biodegrade (Troter et al. 2016). Owing to their environmentally friendly and tunable properties, ILs are considered a promising replacement for conventional solvents. Such studies, which have explored the use of ILs in biodiesel production, are shown in Table 12.1. The properties of ILs are particularly advantageous for the production of biodiesel as they can replace conventional solvents in the reaction and purification processes, by acting as co-solvents, catalysts, extracting solvents, or enzyme supports. Immediate benefits include a reduction in the volume of wastewater in comparison to conventional catalysts and fewer steps for product separation and purification (Yoo et al. 2017). However, many practical challenges must be overcome to realise wide-scale commercialisation, such as increased expenditures, high viscosities, toxicity, undesirable water tolerance, small solid loading, and complex recycling (Niedermeyer et al. 2013; Reddy 2015). The following sections discuss the IL application, recent progress, challenges, and future outlook in regard to the production of biodiesel.

12.3 Ionic Liquids as Catalysts in Biodiesel Synthesis

Several catalyst types are used in the biodiesel production process from vegetable oils or animal fats such as alkalines (Vicente et al. 2004; Dupont et al. 2009; Liang et al. 2009), acids (Zheng et al. 2006; Dupont et al. 2009; Liang et al. 2009), and enzymes (Iso et al. 2001; Gamba et al. 2008; Ranganathan et al. 2008). Notwithstanding their high adoption, serious disadvantages still exist with the use of such catalysts. For example, negatives associated with acid catalysis include the risk of corrosion, long retention times, and a large amount of alcohol required

Table 12.1 Biodiesel production using ionic liquids

Ionic liquid	Catalyst	Oil source	Reaction conditions	Biodiesel yield (%)	References
$[BMIm]PF_6$	Sn(3-hydroxy-2-methyl-4-pyrone)$_2$(H_2O)$_2$	Soybean	60 °C, 1 h	55	Abreu et al. (2005)
$[BMIm]InCl_4$	Sn(3-hydroxy-2-methyl-4-pyron)$_2$(H_2O)$_2$	Soybean	80 °C, 4 h	83	DaSilveira Neto et al. (2007)
$[Et_3NH]Cl$	$ALCl_3$	Soybean	70 °C, 3 h	98.5	Liang et al. (2009)
$[BMIm]NTf_2$	K_2CO_3	Soybean	70 °C, 1.5 h	98	Lapis et al. (2008)
$[BMIm]NTf_2$	H_2SO_4	Soybean	70 °C, 24 h	93	Lapis et al. (2008)
[EMIm]TfO	Lipase	Soybean	50 °C, 12 h	80	Ha et al. (2007)
$[BMIm]NTf_2$	Lipase	Soybean	25 °C, 30 h	96.3	Gamba et al. (2008)
$[BMIm]PF_6$	Lipase	Corn	40 °C, 25 h	69.7-86	Yang et al. (2010); Zhang et al. (2011)
$[BMIm]PF_6$	Lipase	Triolein	48-55 °C, 6 h	80	Ruzich and Bassi (2010)
$[C_{16}MIM]$ NTf_2	Lipase	Triolein	60 °C, 3 h	90.29	De Diego et al. (2011)
$[C_{18}MIM]$ NTf_2	Lipase	Triolein	60 °C, 6 h	96	Lozano et al. (2010)
1-(4-Sulfonic acid) butyl-pyridinium hydrogen sulphate	1-(4-Sulfonic acid) butyl-pyridinium hydrogen sulphate	Cotton seed	170 °C, 9 h	92	Wu et al. (2007a); Wu et al. (2007b)
$[BSPy]CF_3SO_3$	$[BSPy]CF_3SO_3$	Jatropha	100 °C, 5 h	92	Li et al. (2010)
$[BMIm]HSO_4$	$[BMIm]HSO_4$	Free fatty acid	120 °C, 2 h	91.2	Elsheikh et al. (2011)
$[NMP]CH_3SO_3$	$[NMP]CH_3SO_3$	Free fatty acid	70 °C, 8 h	93.6-95.3	Zhang et al. (2009)

(Liang et al. 2009). Alkalines are the preferred catalysts because of the lower costs and higher reaction rates achievable (Dupont et al. 2009; Atabani et al. 2012). However, to be used successfully, water or high levels of FFAs cannot be present in the reagent because of the possibility of saponification (Dupont et al. 2009). Subsequently, oils must undergo an often costly and time-consuming pretreatment.

Alternatively, enzymes can be used which are capable of overcoming the disadvantages of traditional catalyses, such as high energy demand and equipment corrosion (Andreani and Rocha 2012). However, many negatives are still associated with the process, including high costs, deactivation in the presence of alco-

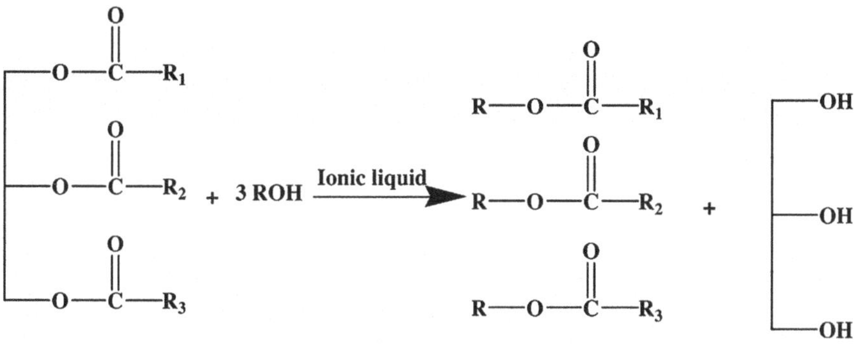

Fig. 12.2 Mechanism of biodiesel preparation via ionic liquid-catalysed process. Reprinted with permission from Liu, C,-Z. et al. (2015) 'An overview of the role of ionic liquids in biodiesel reaction', Journal of Industrial and Engineering Chemistry, 21, pp. 1-10. Copyright 2015 Applied Energy

hol, and potential for reduced enzyme immobilisation. ILs have shown promise as an alternative mechanism in the biodiesel preparation process as shown in Fig. 12.2.

12.3.1 Acidic Ionic Liquid Catalysts

Many acidic ILs have been explored by researchers as potential catalysts in biodiesel production such as Brønsted acidic ILs. These catalysts contain sulfonic acid groups in their cations and have been found to be effective in the transesterification of various oils or esterifying FFAs. Wu et al. (2007b) carried out one of the earliest studies, using ILs as catalysts in the transesterification of cottonseed oil with methanol. Three cations with different acidity strengths were applied to produce the ILs; these included N-methylimidazole and trimethylamine, in combination with the HSO_4 ion. Among the ILs evaluated, 1-(4-sulfonic acid) butyl pyridinium hydrogen sulphate achieved the highest yield of fatty acid methyl esters (FAME), reaching 92% under optimum conditions (oil/methanol/IL = 1:12:0.057 [molar ratio] at 170 °C for 5 h). In comparison, concentrated sulphuric acid (H_2SO_4) was used under the same reaction conditions. The FAME yield after 3 h of the IL catalyst was 81% with a H_2SO_4 concentration of 86%. It was concluded that the yields of IL catalysts were comparable to other studies. Another study performed the transesterification of jatropha oil using (BSP_y)(CF_3SO_3) resulting in FAME yields of up to 92% and consistent catalytic activity after seven successive cycles (Li et al. 2010). Similarly, Guo et al. (2011) examined a mixture of (BMIm)(CH_3SO_3) and $FeCl_3$ for the conversion of jatropha oil to biodiesel. In this mixture, metal ions acted as Lewis acidic sites resulting in a high FAME yield of 99.7% under optimal conditions.

Other ILs which have generated considerable attention as catalysts in biodiesel production processes are single- and multi-SO_3H functionalised Brønsted acidic ILs (Wu et al. 2007b; Han et al. 2009; Fan et al. 2012). Liang and Yang (2010) investigated the effectiveness of a multi-SO3H ILs, with high polarity and acidity, as a catalyst in the preparation of biodiesel through transesterification from rapeseed oil. The high polarity increases the ease of IL recovery, while the high acidity increases the amount of catalytic activity. Using this catalyst, a biodiesel yield of 98% was achieved after 7 h under optimised conditions. The catalyst showed good water-resistant ability, with the yield only decreasing by 3% after the water content had been enlarged to 2.0% from 0.3%.

Dicationic Brønsted acidic ionic liquids (DAILs) with propyl sulphonic acid groups on their cations have shown promise by demonstrating high catalytic activity and reusability in biodiesel synthesis (Zhou et al. 2011). Fang et al. (2011) examined the esterification of FFAs into biodiesel using low alcohols catalysed by both monocationic and dicationic ILs. The results found that higher conversion rates were obtained with the addition of the DAILs in the reaction (93–96%) in comparison to the monocationic IL (85–87%). Furthermore, high catalytic activity was demonstrated over six runs with minor losses.

12.3.2 *Basic Ionic Liquid Catalysts*

In addition to acidic ILs, basic ILs have also been applied to biodiesel synthesis such as imidazolium- and choline-based ILs, but not as regularly (Zhou et al. 2012; Luo et al. 2013; Ren et al. 2014). In most basic ILs, a substrate with a small FFA content is required to evade soap formation.

Studies involving basic ILs for the esterifications and transesterifications of animal fats and vegetable oils are limited in the academic literature, as shown in Table 12.2. One such study was by Liang et al. (2010) who explored the use of five basic binuclear functional ILs with an imidazolium structure as catalysts in the transesterification of cottonseed oil. The catalytic activity of the five ILs differed based upon the lengths of the carbon chain in the cations. Using a bis-(3-methyl-1-imidazolium)-ethylene dihydroxide catalyst concentration of 0.4%, the fraction and selectivity of the FAME in the substrate reached 98.5 and 99.9% under optimum conditions (oil/methanol = 1:12 (molar ratio) at 55 °C for 4 h), thus demonstrating the applicability of basic ILs as catalysts in the biodiesel synthesis of cottonseed oil. In addition, numerous imidazolium hydroxides have been invested for their use as recoverable catalysts in the transesterification of glycerol trioleate with methanol. Zhou et al. (2012) used 1-butyl-3-methylimidazolium hydroxide ([BMIm]OH) to achieve a methyl ester yield of 87% and the capability to be reused up to six times without a substantial reduction in yields.

Table 12.2 Application of basic ionic liquids as catalysts in transesterification

Feedstock	Type, volume of reactor, cm^3/type of agitator, agitation intensity, rpm	Alcohol	Alcohol:oil molar ratio, mol/mol	Catalyst/loading, wt% (to oil)	Temperature, °C	Optimal reaction conditions	Conversion (C) or yield (Y), %/time, h	Reference
Cottonseed oil	Three-neck flask, 100	Methanol 4:1–16:1	4:1–16:1	IMC_2OH/0.10–1.25 wt%	35–65	12:1; 0.4 wt%; 55 °C	98.5 (Y)/4	Vafaeezadeh and Hashemi (2014)
Glycerol trioleate		Methanol 3:1-9:1	3:1–9:1	(BMIm)(OH)	Room–120	9:1; 120 °C	87.2 (Y)/8	Zhou et al. (2012)
Rapeseed oil	Three-necked flask, 500/mechanical	Methanol 3:1–8:1	3:1–8:1	(BMIm)(Im)/ 1.0–10.0 wt%	30–70	6:1; 6.0 wt% 60 °C	~95.0 (Y)/l	Luo et al. (2013)
Soybean oil Sunflower oil Soybean oil	Flask/magnetic	Methanol 6:1-18:1	6:1–18:1	ChOH/2.0–10.0 wt% ChOMe/4.0 wt%	30–65	9:1; 4.0 wt% 60 °C	>95.0 (Y)/1 >95.0 (Y)/l 95.0 ± 1.7 (Y)/2.5	Fan et al. (2013)
Jatropha oil	Round-bottom flask/mechanical	Methanol 3:1–18:1	3:1–18:1	ChOH/1.0–5.0 wt%	30–70	9:1; 4.0 wt% 60 °C	88.3 ± 1.8 (Y)/2.5 76.9 ± 1.9 (Y)/2.5 95.0 ± 1.0 (Y)/4	Reddy et al. (2014)

12.4 Ionic Liquids as Solvents and Co-solvents

Conventionally, organic solvents (such as methanol) have been used with enzymes as catalysts in the biodiesel preparation process. However, many issues can arise from their use, such as poor miscibility with oils and fats, enzyme inactivation, and disruption of intra-protein hydrophobic interactions (Akoh et al. 2007). ILs are a promising alternative, where their beneficial properties such as hydrophobicity, hydrogen bond, and polarity provide enzyme stabilisation while avoiding environmental concerns (Zhao 2016). Moreover, because of the mild reaction conditions, ILs offer reusability while requiring a low energy demand.

Many groups have demonstrated the use of co-solvents for the chemical preparation of biodiesel (van Rantwijk et al. 2007; Moniruzzaman et al. 2010; Tang et al. 2012; Zhao and Baker 2013). For example, Ha et al. (2007) used immobilised *Candida antarctica* lipase as catalysts to test the applicability of 23 different ILs for the methanolysis of soybean oil. Hydrophilic IL (EMIm)(TfO) achieved the highest FAME yield of 80% after 12 h at 50 °C. Moreover, the yield was higher than the solvent-free alternative, and other widely used solvents. Another study by Sunitha et al. (2007) reported yields as high as 98% from FAME within 10 h of methanolysis of sunflower oil in hydrophobic 1-butyl-3-methylimidazolium (BMIm)(PF_6) and 1-ethyl-3-methylimidazolium (EMIm)(PF_6) with Novozym acting as the catalyst. An interesting observation is that hydrophilic BF4-based ILs under the same conditions returned low yields.

12.5 Ionic Liquids as Extraction Solvents in Biodiesel Synthesis

A promising application of ILs is as an extraction solvent in the various steps of the biodiesel production process. Although ILs hold attractive properties, such as thermal stability, synthetic flexibility, non-volatility, non-flammability, and recyclability, there are still considerable obstacles that must be overcome for widespread use in large-scale applications.

12.5.1 *Extraction of Lipids*

Lipid extraction from biomass is one of the essential steps among the various operations in the production of biodiesel, particularly for new promising oil feedstocks. One such feedstock is microalgae, which has gathered considerable attention in recent years, due to its attractive properties as a biofuel, including high lipid content, photosynthetic efficiency, production rate, and large abundance (Chisti 2007; Passos et al. 2014). Lipid extraction can be carried out after the microalgae biomass has been harvested and dewatered. Extraction may be achieved from the dried biomass powder (dry route) or directly from the wet concentrated biomass (wet route) (Xu et al. 2011). Solvents play a crucial role in both of these routes by facilitating cell disruption, increasing the efficiency of lipid extraction by improving the biomass mass transfer properties (Ehimen et al., 2012). The most common extraction techniques are Soxhlet extraction and Bligh and Dyer method with various solvents used to enhance extraction such as petroleum ether, n-hexane, ethanol mixture, and hexane–ethanol mixture (Bligh and Dyer 1959; Ahmad et al. 2011). However, there are major disadvantages to using these solvents, including volatility, toxicity, flammability, and non-miscibility with water (except ethanol) (Mata et al. 2010).

ILs have garnered much attention as a medium to enhance the efficiency of lipid extraction, primarily because of their ability to either auto-partition the lipids or improve access of co-solvents to the intracellular lipids. Other advantages ILs possess over conventional solvents include non-volatility, thermal stability, and synthetic flexibility. Subsequently, numerous ILs have been investigated, often using ILs and organic solvents such as methanol, acetic acid, chloroform, acetone, dimethyl sulfoxide, and isopropyl alcohol (Troter et al. 2016).

Young et al. (2010) investigated the capability of 1-ethyl-3-methylimidazolium methyl sulphate (EMIm) and methanol (mass ratio 1.2:1) to extract lipids from microalgae. Using a co-solvent enables the auto-partition of the lipids to a separate immiscible phase, therefore increasing the harvesting efficiency. The lipids extracted included microalgal biomass (Chlorella species and Dunaliella species), Kamani oil seed, Pongamia oil seed, and jatropha oil seed using an EMIm-methanol co-solvent and a number of polar covalent molecule types. The highest reported yield of lipids was obtained from jatropha oil seed at 49.9% with all other experimental yields within the ranges reported in the literature.

A similar study by Kim et al. (2012) trailed several IL-methanol co-solvents to extract lipids from *Chlorella vulgaris* microalgae to determine their effectiveness in comparison to the more commonly used Bligh and Dyer method. The IL co-solvents used include (BMIm)(MeSO4), (EMIm)($MeSO_4$), and (BMIm)(CF_3SO_3), extracting 11.84%, 11.88%, and 12.54%, respectively, achieving higher yields than Bligh and Dyer extraction method at 10.6%. This suggests that the extraction efficiency of the lipids was influenced by the anion structure, and the hydrophobicity or hydrophilicity of the ILs. Additionally, Choi et al. (2014) compared the influence of 12 ILs on the lipid extraction yield of *Chlorella vulgaris* microalgae. The results showed that ILs with organic solvents and IL mixtures generally displayed higher lipid yields because of the synergistic effects with different anions.

While ILs have proved promising for the effective extraction of lipids, more progress still needs to be made in order to achieve the economic and environmental standards for industrial-scale applications (Zhang et al. 2008; Harris et al. 2018). Although ILs are advantageous in many aspects, many difficulties still exist in their practical application such as the multiple steps required for synthesis, the use of potentially toxic or volatile solvents, and the environmental concerns related to their production. Much research is being conducted to overcome these issues: development of simpler and more efficient syntheses; improved methods for lipid extraction

(Yang et al. 2017); use of potentially cheaper cations to synthesise ILs such as pyridinium, ammonium, and phosphonium; and use of DESs with the potential of increasing efficiency and yields while reducing costs.

12.5.2 Extraction of Free Fatty Acids

Biodiesel derived from vegetable oil should have considerable potential but conversion is significantly reduced by the presence of FFAs. The conventional conversion process used is via transesterification with an alcohol. However, elevated levels of FFAs can greatly reduce the effectiveness of transesterification because of the reaction with the alkaline catalyst forming soap which subsequently reduces the ester yield. Conventional methods utilised for the removal of FFAs include caustic stripping, hydrolysis, and glycerolysis. Although these methods have proved to be effective, considerable drawbacks are associated with their use, such as the added complexity of an additional step and the harsh conditions required in operation (high temperatures and vacuum). ILs are one such option being explored by researchers to overcome these difficulties.

Manic et al. (2015) studied the use of poly(ethylene glycol) solvents with different molar masses and ILs for the extraction of FFA (linoleic acid) from soybean oil. The ILs investigated included 1-butyl-3-methylimidazolium dicyanamide (BMImDCA) and AMMOENG100, which are both miscible with the linoleic acid used. In the study AMMOENG100 was reported to having the highest values of linoleic acid distribution coefficient. Based on these findings, it can be concluded that ILs hold considerable potential in the biodiesel deacidification process.

A similar study by Grimes and Kewcharoenwong (2017) explored the use of a duel system with the purpose of using ILs (HPyrBr and Brønsted acid) to extract FFAs from waste vegetable oils and convert to biodiesel, thus allowing the ILs to be used for extraction and catalytic functionality without the uninhibited presence of the other. Of the 18 properties tested using ASTM standard methods, 14 of them met biodiesel fuel quality standards. Further research and innovation are required in this area to fully explore the potential role of ILs.

12.5.3 Extraction of Unsaturated Fatty Methyl Acid Esters

The removal of unsaturated fatty acid methyl esters (uFAMEs) is often necessary in biodiesel production to reduce the fuel's sensitivity to oxidation. In addition, this step is often necessary to meet the European standards for biodiesel such as EN 14214 and EN 14213, because they limit the total unsaturated biodiesel to 120 and 130 g iodine/100 g, respectively (European Committee for Standardization [DIN] 2003). Traditional methods employed for the removal of uFAMEs from biodiesel include urea inclusion complexation (Liu et al. 2006; Alavi Talab et al.

2010), low-temperature fractional crystallisation (Mudhaffar and Salimon 2010), molecular distillation (Jiang et al. 2006), supercritical fluid extraction (Rubio-Rodríguez et al. 2008, 2012), and lipase concentration (Shimada et al. 2001; Lee et al. 2011). Although effective, these methods have several inherent drawbacks which limit yields and application. ILs have garnered attention in a number of studies as an alternative mechanism to overcome such difficulties.

Several studies have explored the separation of uFAMEs in the biodiesel production process with ILs as extractants. Li et al. (2009a) extracted FAMEs in the form of methyl linoleate (18:2) and methyl linolenate (18:3) from soybean-derived biodiesel using novel π-complexing sorbents. The preparation of these π-complexing sorbents involved the covalent immobilisation of ILs onto silica followed by coating the silica-supported ILs with silver salts. The $AgBF_4$/SBA-15-IL-PF_4 sorbent produced exhibited a higher extraction capacity and greater reusability with eight sorption successive cycles. Building on this work, Li et al. (2009b) investigated the use of a IL-co-solvent mixture to improve upon the difficulties associated with using silver salts in the extraction process, including added cost, a larger viscosity, and an increased likelihood of losing activity. The study used novel π-complexing sorbents by covalently immobilising ILs onto mesoporous SBA-15 for separating polyunsaturated FAMEs, and prepared π-complexing sorbents by covalently immobilising ILs onto silica, followed by coating these silica-supported ILs with silver salts. Positive benefits were reported in the form of satisfactory extraction capacities and reusability. A similar study by Pham et al. (2009) used an adsorbent based on mesoporous silica for separating the polyunsaturated triacylglycerols, resulting in high selectivity. Additionally, Cheong et al. (2011) investigated the use of ILs comprised of aromatic rings for the extraction and enrichment of n-3 polyunsaturated fatty acids (PUFA) and ethyl esters through reversible ππ complexation. The use of ILs resulted in higher purity and increased extraction capabilities, strongly suggesting a correlation between the n-3 PUFA extractions and the aromatic/delocalised cation structure of the ILs.

12.6 Deep Eutectic Solvents: A New Generation of Ionic Liquids

The latest generation of ILs, named deep eutectic solvents (DESs), have garnered considerable interest in recent years, with predictions that they could potentially replace conventional ILs in biodiesel production. Figure 12.3 depicts the most common DESs used today. DESs are generally composed of a mixture of solid organic salt and a complexing agent that are liquid below temperatures of 100 °C (Zhao and Baker 2013). In such solvents, the complexing agent reacts with the anion which increases its effective size and reduces anion interaction with the cation, which results in a reduction in the mixture's freezing point (Zhao and Baker 2013). Abbott et al. (2003) was the first to demonstrate such solvents by reporting that a series of

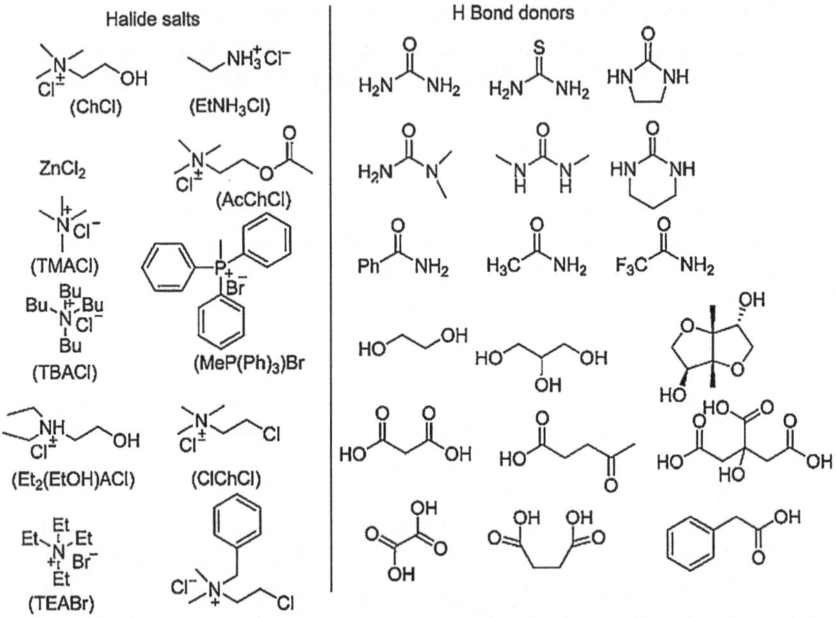

Fig. 12.3 Most common deep eutectic solvents. Reprinted with permission from Farra, A. et al. (2015) 'Green solvents in carbohydrate chemistry: From raw Materials to Fine Chemicals', Chemical Reviews, 115(14), pp. 6811–6853. Copyright 2015 American Chemical Society

DES from the mixture of a solid organic salt and a complexing agent can form a liquid at temperatures below 100 °C.

There has been much debate within the research community to whether DESs can be considered as ILs, mainly because of the molecular component in their structure (Zhao and Baker 2013). Much of the debate is centred around ILs being salts composed of one type of discrete anion and cation, whereas DESs are mixtures, potentially containing several anionic and/or cationic species, as well as non-ionic species. However, ILs and DESs do share many attractive solvent features, including their tunable nature. Considering the vast variety of possible combinations that can be used in the preparation of DESs the development of inexpensive, environmentally friendly solvents is therefore possible. Hence, it is now realistic to predict that DESs will play a significant role, and potentially replace ILs in biodiesel production.

12.6.1 Removal of Glycerol from Crude Biodiesel

One of the most crucial steps in the production of biodiesel is the purification process, which has proven to be one of the most process-intensive and economically costly steps. Before the fuel output can be classified as biodiesel, it must meet the specifications set by the international standards (i.e. EN 14214 and ASTM D6751) (European Committee for Standardization [DIN] 2003; ASTM D6751-19 2019). These standards outline the quality requirements for the fuel to be classified as alkyl ester-based biodiesel, contributing between 60 and 80% of the process costs (Atadashi et al. 2011). This involves the removal of unwanted impurities and by-products such as glycerol, which can have several consequences on the diesel engine such as fuel system clogging, injector fouling, engine corrosion, build-up in fuel tanks, valve deposits, and emission of harmful acrolein into the environment (Abbott et al. 2007; Berrios and Skelton 2008; Hayyan et al. 2010). The traditional methods used for the commercial removal of glycerol are through dry and water washing, and membrane extraction (Berrios and Skelton 2008; Santori et al. 2012). However, there are numerous difficulties associated with the use of these purification techniques in the industry, mainly because of the cost and complexity involved. DES has received significant attention as an alternative non-toxic, inexpensive, and environmentally benign extraction media.

Abbott et al. (2007) showed that DESs can be used for the extraction of glycerol from biodiesel using a mixture of quaternary ammonium salts and glycerol. In addition, Hayyan et al. (2010) demonstrated that for the purpose of removing glycerol both a 1:1 DES-to-biodiesel molar ratio and a 1:1 (salt:glycerine) DES molar composition were the most effective. Similarly, Shahbaz et al. (2010) showed effective glycerol extraction from palm oil-based biodiesel from two different DESs. The DESs used comprised of quaternary ammonium salt (choline chloride) with two different hydrogen-bond donors (ethylene glycol and 2,2,2-trifluoroacetamide). Shahbaz et al. (2011) improved this work, indicating that binary DESs synthesised using methyl triphenylphosphonium bromide and ethylene glycol can remove all free glycerol from the palm oil-derived biodiesel. Similar to Abbott et al. (2007), the study showed that the optimal molar ratio was 1:1 (DES:biodiesel) with no significant change after that point.

12.6.2 Deep Eutectic Solvents as Catalysts

Researchers have prepared various DESs for their use as co-solvents or catalyst for biodiesel preparation as presented in Table 12.3. Hayyan et al. (2013a) investigated the use of an ammonium-based DES as a novel catalyst in the transesterification of low-grade crude palm oil with a high FFA content. The DES utilised comprised of a salt (N-diethyl ethanol ammonium chloride) and a hydrogen bond donor (p-toluenesulfonic acid monohydrate). The results showed a reduction from 9.5% to less than 1% in the FFA content in the low-grade crude palm oil under optimum conditions. Hayyan et al. (2013b) also studied the use of a two-stage process using a phosphonium-based DES followed by an alkali treatment for the esterification of LGCPO with different dosages of DESs in the presence of alcohol; the FFA content was reduced to satisfactory levels for alkaline transesterification. The study demonstrated the efficient conversion of LGCPO to biodiesel using an alkali pretreatment.

Table 12.3 Application of deep eutectic solvents as catalysts in biodiesel production

Feedstock	Type, volume of reactor, cm^3/type of agitator, agitation intensity, rpm	Alcohol	Alcohol:oil molar ratio, mol/mol or ml.mg oil	DES/loading, wt% or ml/g oil	Other catalyst/ loading, wt% (to oil weight)	Temperature, °C	Optimal reaction conditions	Conversion (C) or yield (Y)/ time, h	Reference
Soybean oil	Round-bottom flask/100	Methanol	10:1–30:1	ChCI:$ZnCI_2$ (1:2)/10.0 wt%	–	50-90	16:1; 10.0 wt%; 70 °C	54.52 (C)/72	Long et al. (2010)
Palm oil	Three-neck flask	Methanol	10:1–15:1	ChCI:$ZnCI_2$ (1:2)/0.5-2.5 wt% ChCI:$FeCI_3$ (1:2)/0.5-2.5 wt% ChCI:$ZnCI_2$ (1:2)/2.5 wt% ChCl:$FeCl_2$ (1:2)/2.5 wt%	– – 95.0% H_2SO_4/20.0 wt%	65	15:1; 2.5 wt% DES; 65 °C	70.4 (Y)/4 67.4 (Y)/4 92.0 (Y)/4 89.5 (Y)/4	Isahak et al. (2011)
Palm oil, low grade (9.5% FFA)	Batch multi-unit reactor system/ mechanical, 100–500	Methanol	4:1–20:1	DEAC:PTSA (1:3)/0.25-3.5 wt%	–	40-80	1 step: 8:1; 0.75 wt% DES; 60 °C; 200 rpm	97.0 (Y)/0.5	Hayyan et al. (2013b)
Esterified oil (0.7% FFA)			10:1	–	KOH/1.0 wt%	60	II step: 10:1; 1.0 wt% KOH; 60 °C; 400 rpm	84.0 (Y)/1	

Palm oil, low grade (9.3% FFA)		Methanol	1:1–20:1	P-DES (1:3)/0.25–3.5 wt%	–.	40-80	I step: 10:1; 1.0 wt% 1.0 wt% DES; 60 °C; 300 rpm	96.0 (Y)/0.5	Hayyan et al. (2014)
Esterified oil (0.88% FFA)			10:1	–	KOH/1.0 wt%	60	II step: 10:1; 1.0 wt% KOH; 60 °C; 400 rpm	89.84 (Y)/1	
Palm oil, crude (9.0% FFA)	Reactor	Methanol	3:1–20:1	ChCI:PTSA (1:3)/0.25–3.50 wt%	–	40–70	I step: 10:1, 0.75 wt% DES, 60 °C	97.0 (Y)/0.5	Hayyan et al. (2014)
Esterified oil (<1% FFA)	Batch reactor with a reflux condenser, 1500/ stirrer, 400		10:1	–	KOH/1.0 wt%	60	II step: 10:1, 1.0 wt% KOH, 60 °C	92.0 (Y)/1	
Rapeseed oil	Round-bottom flask/vertical blender, 400	Methanol	14.28:1	ChCI:glycerol (1:2)/10.74 wt%	8.07 wt% CaO (commercial)	65	14.28:1; 10.74 wt% DES; 8.07 wt% CaO; 65 °C	91.9 (Y)/3	Huang et al. (2013)

(continued)

Table 12.3 (continued)

Feedstock	Type, volume of reactor, cm^3/type of agitator, agitation intensity, rpm	Alcohol	Alcohol:oil molar ratio, mol/mol or ml.mg oil	DES/loading, wt% or ml/g oil	Other catalyst/ loading, wt% (to oil weight)	Temperature, °C	Optimal reaction conditions	Conversion (C) or yield (Y)/ time, h	Reference
Miglyol 812	Micro-reaction vessel, 5/gentle	Methanol/20.0 vol%		ChOAc:glycerol (1:1.5)	Novozym 435/27.3 wt%	50	20.0 vol% Methanol; 1.0 mL mixture of DES and methanol; 27.3 wt% lipase; 1.0 vol% water; 50 °C	97.0 (C)/3	Zhao et al. (2011)
Soybean oil	Micro-reaction vessel, 5/gentle	Methanol/20.0–50.0 vol% (to IL)		ChCI:glycerol (1:2)	Novozym 435/20.0–60.0 wt%	50	7:3 (v/v) of DES and methanol, 40.0 wt% lipase, 0.2 vol% water	88.0 (C)/24	Zhao et al. (2013)
M. pinnata seed oil	Flask/shaker, 220	Methanol	3 mL/mg oil	ChOAc:glycerol (1:2)/7 mL DES/g oil	*P. expansum* lipase/1.0 wt% Novozym 435/1.0 wt% Lipozyme TLIM/1.0 wt%	50	3 mL Methanol/ mg oil; 7 mL DES/g oil; 1.0 wt% lipase; 50 °C	7.6 (C)/48 54.8 (C)/48 44.8 (C)/48	Huang et al. (2014)

Reprinted with permission from Troter, D. Z. et al. (2016) 'Application of ionic liquids and deep eutectic solvents in biodiesel production: A review', Renewable and Sustainable Energy Reviews, 66, pp. 473–500. Copyright 2016 Elsevier

In addition, the DES was recycled four times without a significant loss in its activity. For the transesterification of soybean oil, Long et al. (2010) used a Lewis acidic catalyst comprising of choline chloride and zinc chloride and methanol with a molar ratio of 1:2. The tests carried out showed a 55% conversion after 72 h under optimum conditions (10% catalyst, methanol/oil molar ratio 16:1, and 70 °C). A similar study by Isahak et al. (2011) used choline chloride:metal chloride and ILs as a catalyst in the transesterification of palm oil, resulting in ester yields of 70.4% and 67.4%. In addition, adding H2SO4 (95 vol%) further raised yields to 92.0%.

Some studies have explored combining DESs with other catalysts to improve the biodiesel production process. Huang et al. (2013) used DESs comprised of choline chloride and glycerol (1:2 molar ratio) as solvents in the CaO-catalysed transesterification of rapeseed oil. The solvent proved to be efficient in activating CaO through the removal of the inactive layers (calcium, carbonate, and calcium hydroxide) from the surface of the catalyst. The results showed a high ester yield of 91.9% in comparison to the low yield of 4.0% without the addition of the DES.

12.7 Summary and Future Perspectives

ILs, as 'green' tunable solvents, have significant potential in the biodiesel production process. From this review, it can be seen that ILs have already been shown to be promising as catalysts, co-solvents, and extracting solvents in biodiesel production. They offer substantial benefits including high product yields, negligible vapour pressure, high thermal stability, non-flammability, odourless, compatibility with facile post-synthesis separation, and enzyme catalysts. Furthermore, ILs can be designed for specific applications through the careful selection of the appropriate cation and anion combination.

Regardless of the recent progress made in the field, there are several shortcomings associated with the use of ILs in biodiesel production. One such shortcoming is the compromise often required between enzyme compatibility and solubilising power in the selection of an IL.

Another key obstacle to be overcome is the high costs associated with ILs, particularly for the achievement of industrial scale widespread use. However, when considering the overall advantages provided, this drawback can be justified in most cases. Alternatively, the costs associated with common ILs such as imidazolium and pyridiniums may be avoided by deriving its composition from low-cost alternatives. Moreover, the possibility of using inexpensive DESs comprised of choline salts allows for the development of new economic and environmentally friendly biodiesel production processes. An additional weakness is the high viscosity, typical in many ILs, which hinders mass transfer kinetics and leads to poor performance. Significant research has been conducted to reduce this issue by designing 'viscosity-reducing' anions, such as dicyanamide, tricyanomethanide, and tetracyanoborate.

From the studies presented in this work, it is clear that ILs and DESs have already made a significant contribution to improving the overall biodiesel production process for a range of feedstocks by increasing yields, aiding ease of preparation, and lowering costs. It is hoped that the continued research of such technology will greatly aid in the advancement of biodiesel as an alternative green fuel source.

Acknowledgements The authors are thankful for the funding received from the European Union's INTERREG VA Programme for the Renewable Engine project, managed by the Special EU Programmes Body (SEUPB), with match funding provided by the Department for the Economy and Department of Jobs, Enterprise and Innovation in Ireland.

References

Abbott AP et al (2003) Novel solvent properties of choline chloride/urea mixtures. Chem Commun 1:70–71. https://doi.org/10.1039/b210714g

Abbott AP et al (2004) Deep eutectic solvents formed between choline chloride and carboxylic acids: versatile alternatives to ionic liquids. American Chemical Society, Washington, D.C. https://doi.org/10.1021/JA048266J

Abbott AP et al (2007) Extraction of glycerol from biodiesel into a eutectic based ionic liquid. Green Chem 9(8):868. https://doi.org/10.1039/b702833d

Abreu FR et al (2005) New multi-phase catalytic systems based on tin compounds active for vegetable oil transesterification reaction. J Mol Catal A Chem 227(1–2):263–267. https://doi.org/10.1016/J.MOLCATA.2004.11.001

Ahmad AL et al (2011) Microalgae as a sustainable energy source for biodiesel production: a review. Renew Sust Energ Rev 15(1):584–593. https://ideas.repec.org/a/eee/rensus/v15y2011i1p584-593.html. Accessed 4 Aug 2019

Akoh CC et al (2007) Enzymatic approach to biodiesel production. J Agric Food Chem 55(22):8995–9005. https://doi.org/10.1021/jf071724y

Alavi Talab H et al (2010) Extraction and purification of omega-3 fatty acids concentrate from Hypophthalmichthys molitrix oil. Nutr Food Sci 40(2):193–203. https://doi.org/10.1108/00346651011029228

Andreani L, Rocha JD (2012) Use of ionic liquids in biodiesel production: a review. Braz J Chem Eng 29(1):1–13. https://doi.org/10.1590/S0104-66322012000100001

Antolin G et al (2002) Optimisation of biodiesel production by sunflower oil transesterification. Bioresour Technol 83(2):111–114. https://doi.org/10.1016/S0960-8524(01)00200-0

Aransiola EF et al (2012) Production of biodiesel from crude neem oil feedstock and its emissions from internal combustion engines. Afr J Biotechnol 11(22):6178–6186. https://doi.org/10.5897/AJB11.2301

ASTM D6751-19 (2019) Standard specification for biodiesel fuel blend stock (B100) for middle distillate fuels. ASTM International, West Conshohocken, PA. https://www.astm.org/Standards/D6751.htm. Accessed 4 Aug 2019

Atabani AE et al (2012) A comprehensive review on biodiesel as an alternative energy resource and its characteristics. Renew Sust Energ Rev 16(4):2070–2093. https://doi.org/10.1016/J.RSER.2012.01.003

Atadashi IM, Aroua MK, Aziz AA (2011) Biodiesel separation and purification: a review. Renew Energy 36(2):437–443. https://doi.org/10.1016/J.RENENE.2010.07.019

Berrios M, Skelton RL (2008) Comparison of purification methods for biodiesel. Chem Eng J 144(3):459–465. https://doi.org/10.1016/j.cej.2008.07.019

Bligh EG, Dyer WJ (1959) A rapid method of total lipid extraction and purification. Can J Biochem Physiol 37(8)

Cheong L-Z et al (2011) Extraction and enrichment of n-3 polyunsaturated fatty acids and ethyl esters through reversible π–π Complexation with aromatic rings containing ionic liquids. J Agric Food Chem 59(16):8961–8967. https://doi.org/10.1021/jf202043w

Chisti Y (2007) Biodiesel from microalgae. Biotechnol Adv 25:294–306. https://doi.org/10.1016/j.biotechadv.2007.02.001

Choi S-A et al (2014) Effects of ionic liquid mixtures on lipid extraction from Chlorella vulgaris. Renew Energy 65:169–174. https://doi.org/10.1016/j.renene.2013.08.015

DaSilveira Neto BA et al (2007) 1-n-Butyl-3-methylimidazolium tetrachloro-indate (BMI·InCl4) as a media for the synthesis of biodiesel from vegetable oils. J Catal 249(2):154–161. https://doi.org/10.1016/J.JCAT.2007.04.015

De Diego T et al (2011) An efficient activity ionic liquid-enzyme system for biodiesel production. Green Chem 13(2):444. https://doi.org/10.1039/c0gc00230e

Demirbas A (2008) Biodiesel. In: Biodiesel. Springer, London, pp 111–119. https://doi.org/10.1007/978-1-84628-995-8_4

Dupont J et al (2009) Catalytic production of biodiesel and diesel-like hydrocarbons from triglycerides. Energy Environ Sci 2(12):1258. https://doi.org/10.1039/b910806h

Elsheikh YA et al (2011) Brønsted imidazolium ionic liquids: synthesis and comparison of their catalytic activities as pre-catalyst for biodiesel production through two stage process. Energy Convers Manag 52(2):804–809. https://doi.org/10.1016/j.enconman.2010.08.005

European Committee for Standardization [DIN] (2003) European Standard EN 14214

Fan MM et al (2012) Effect of various functional groups on biodiesel synthesis from soybean oils by acidic ionic liquids. Chin Chem Lett 23(10):1107–1110. https://doi.org/10.1016/j.cclet.2012.07.009

Fan M et al (2013) Biodiesel production by transesterification catalyzed by an efficient choline ionic liquid catalyst. Appl Energy 108:333–339. https://doi.org/10.1016/J.APENERGY.2013.03.063

Fang D, Yang J, Jiao C (2011) Dicationic ionic liquids as environmentally benign catalysts for biodiesel synthesis. ACS Catal 1(1):42–47. https://doi.org/10.1021/cs100026q

Farra A et al (2015) Green solvents in carbohydrate chemistry: from raw materials to fine chemicals. Chem Rev 115(14):6811–6853. https://doi.org/10.1021/cr500719h

Farrell AE et al (2006) Ethanol can contribute to energy and environmental goals. Science 311(5760):506–508. https://doi.org/10.1126/science.1121416

Gamba M, Lapis AAMM, Dupont J (2008) Supported ionic liquid enzymatic catalysis for the production of biodiesel. Adv Synth Catal 350(1):160–164. https://doi.org/10.1002/adsc.200700303

Gashaw A, Tesita A (2014) Production of biodiesel from waste cooking oil and factors affecting its formation: a review. Int J Renew Sustain Energy, 3 (5). https://doi.org/10.11648/j.ijrse.20140305.12

Grimes SM, Kewcharoenwong P (2017) Dual-functionality ionic liquid mix for extraction and esterification of fatty acids as a step towards increasing the efficiency of conversion of waste cooking oils to biodiesel. J Chem Technol Biotechnol 92(8):2098–2105. https://doi.org/10.1002/jctb.5207

Guo F et al (2011) One-step production of biodiesel from Jatropha oil with high-acid value in ionic liquids. Bioresour Technol 102(11):6469–6472. http://www.ncbi.nlm.nih.gov/pubmed/23908993

Ha SH et al (2007) Lipase-catalyzed biodiesel production from soybean oil in ionic liquids. Enzym Microb Technol 41(4):480–483. https://doi.org/10.1016/j.enzmictec.2007.03.017

Han M et al (2009) Preparation of biodiesel from waste oils catalyzed by a Brønsted acidic ionic liquid. Bioresour Technol 100(7):2308–2310. https://doi.org/10.1016/j.biortech.2008.10.046

Harris J et al (2018) Advances in microalgal lipid extraction for biofuel production: a review. Biofuels Bioprod Bioref 12(6):1118–1135. https://doi.org/10.1002/bbb.1923

Hayyan M et al (2010) A novel technique for separating glycerine from palm oil-based biodiesel using ionic liquids. Fuel Process Technol 91(1):116–120. https://doi.org/10.1016/j.fuproc.2009.09.002

Hayyan A, Hashim MA et al (2013a) A novel ammonium based eutectic solvent for the treatment of free fatty acid and synthesis of biodiesel fuel. Ind Crop Prod 46:392–398. https://doi.org/10.1016/j.indcrop.2013.01.033

Hayyan A, Ali Hashim M et al (2013b) A novel phosphonium-based deep eutectic catalyst for biodiesel production from industrial low grade crude palm oil. Chem Eng Sci 92:81–88. https://doi.org/10.1016/J.CES.2012.12.024

Hayyan A et al (2014) A new processing route for cleaner production of biodiesel fuel using a choline chloride based deep eutectic solvent. J Clean Prod 65:246–251. https://doi.org/10.1016/J.JCLEPRO.2013.08.031

Highina BK, Bugaje IM, Umar B (2011) Biodiesel production from Jatropha caucus oil in a batch reactor using zinc oxide as catalyst. J Petrol Technol Alter Fuel. http://www.academicjournals.org/JPTAF

Huang W et al (2013) Activation of commercial CaO for biodiesel production from rapeseed oil using a novel deep eutectic solvent. Ind Eng Chem Res 52(34):11943–11947. https://doi.org/10.1021/ie401292w

Huang Z-L et al (2014) Deep eutectic solvents can be viable enzyme activators and stabilizers. J Chem Technol Biotechnol 89(12):1975–1981. https://doi.org/10.1002/jctb.4285

IPCC (2013) Climate change 2013: the physical science basis. In: Stocker TF et al (eds) Contribution of working group I to the fifth assessment report of the Intergovernmental Panel on Climate Change. Cambridge University Press, Cambridge

IPCC (2014) Climate change 2014 part a: global and sectoral aspects, climate change 2014: impacts, adaptation, and vulnerability. Part a: global and sectoral aspects. In: Field CB et al (eds) Contribution of working group II to the fifth assessment report of the Intergovernmental Panel on Climate Change. Cambridge University Press, New York. papers2://publication/uuid/B8BF5043-C873-4AFD-97F9-A630782E590D

IPCC (2018) Summary for policymakers. In: global warming of 1.5°C. In: Masson-Delmotte V et al (eds) An IPCC special report on the impacts of global warming of 1.5°C above pre-industrial levels and related global greenhouse gas emission pathways, in the context of strengthening the global response to. World Meteorological Organization, Geneva, p 32. https://www.ipcc.ch/site/assets/uploads/sites/2/2018/07/SR15_SPM_High_Res.pdf. Accessed 4 Dec 2018

Isahak WNRW et al (2011) Transesterification of palm oil by using ionic liquids as a new potential catalyst. Trend Appl Sci Res 6(9):1055–1062. https://doi.org/10.3923/tasr.2011.1055.1062

Iso M et al (2001) Production of biodiesel fuel from triglycerides and alcohol using immobilized lipase. J Mol Catal B Enzym 16(1):53–58. https://doi.org/10.1016/S1381-1177(01)00045-5

Jiang ST et al (2006) Molecular distillation for recovering tocopherol and fatty acid methyl esters from rapeseed oil deodoriser distillate. Biosyst Eng 93(4):383–391. https://doi.org/10.1016/J.BIOSYSTEMSENG.2006.01.008

Keskin S et al (2007) A review of ionic liquids towards supercritical fluid applications. J Supercrit Fluids 43(1):150–180. https://doi.org/10.1016/J.SUPFLU.2007.05.013

Khan Y et al (2015) Treatment of wastewater from biodiesel plants using microbiological reactor technology. Environ Sci Technol 12:297–306. https://doi.org/10.1007/s13762-014-0501-7

Kim Y-H et al (2012) Ionic liquid-mediated extraction of lipids from algal biomass. Bioresour Technol 109:312–315. https://doi.org/10.1016/j.biortech.2011.04.064

Lapis AAM et al (2008) Ionic liquid supported acid/base-catalyzed production of biodiesel. ChemSusChem 1(8–9):759–762. https://doi.org/10.1002/cssc.200800077

Lee JE et al (2011) Concentration of n-3 polyunsaturated fatty acids in cholesterol-reduced cod-liver oil by lipases. Int J Food Sci Technol 46(7):1462–1468. https://doi.org/10.1111/j.1365-2621.2011.02639.x

Leung DYC, Wu X, Leung MKH (2010) A review on biodiesel production using catalyzed transesterification. Appl Energy 87(4):1083–1095. https://doi.org/10.1016/J.APENERGY.2009.10.006

Li M, Pham PJ, Pittman CU et al (2009a) SBA-15-supported ionic liquid compounds containing silver salts: novel mesoporous π-complexing sorbents for separating polyunsaturated fatty acid methyl esters. Microporous Mesoporous Mater 117(1–2):436–443. https://doi.org/10.1016/J.MICROMESO.2008.07.017

Li M, Pham PJ, Wang T et al (2009b) Solid phase extraction and enrichment of essential fatty acid methyl esters from soy-derived biodiesel by novel π-complexing sorbents. Bioresour Technol 100(24):6385–6390. https://doi.org/10.1016/j.biortech.2009.07.054

Li K-X et al (2010) Application of pyridinium ionic liquid as a recyclable catalyst for acid-catalyzed transesterification of jatropha oil. Catal Lett 139(3–4):151–156. https://doi.org/10.1007/s10562-010-0409-x

Liang X, Yang J (2010) Synthesis of a novel multi-SO3H functionalized ionic liquid and its catalytic activities for biodiesel synthesis. Green Chem 12(2):201–204. https://doi.org/10.1039/B922966N

Liang X et al (2009) Highly efficient procedure for the synthesis of biodiesel from soybean oil using chloroaluminate ionic liquid as catalyst. Fuel 88(4):613–616. https://doi.org/10.1016/j.fuel.2008.09.024

Liang J et al (2010) Preparation of biodiesel by transesterification from cottonseed oil using the basic dication ionic liquids as catalysts. J Fuel Chem Technol 38(3):275–280. https://doi.org/10.1016/S1872-5813(10)60033-3

Liu S et al (2006) Concentration of docosahexaenoic acid (DHA) and eicosapentaenoic acid (EPA) of tuna oil by urea complexation: optimization of process parameters. J Food Eng 73(3):203–209. https://doi.org/10.1016/J.JFOODENG.2005.01.020

Long T et al (2010) Application of choline chloride·xZnCl2 ionic liquids for preparation of biodiesel. Chin J Chem Eng 18(2):322–327. https://doi.org/10.1016/S1004-9541(08)60359-6

Lozano P et al (2010) One-phase ionic liquid reaction medium for biocatalytic production of biodiesel. ChemSusChem 3(12):1359–1363. https://doi.org/10.1002/cssc.201000244

Luo H et al (2013) Biodiesel production using alkaline ionic liquid and adopted as lubricity additive for low-sulfur diesel fuel. Bioresour Technol 140:337–341. https://doi.org/10.1016/j.biortech.2012.11.112

MacFarlane DR et al (2014) Energy applications of ionic liquids. Energy Environ Sci 7(1):232–250. https://doi.org/10.1039/C3EE42099J

Manic MS, Najdanovic-Visak V, da Ponte MN (2015) Extraction of free fatty acids from soybean oil using ionic liquids or poly(ethylene glycol)s. AICHE J 57(5):1344–1355. https://doi.org/10.1002/aic.12349

Mata TM, Martins AA, Caetano NS (2010) Microalgae for biodiesel production and other applications: a review. Renew Sust Energ Rev 14(1):217–232. https://doi.org/10.1016/J.RSER.2009.07.020

Mathews JA (2008) Carbon-negative biofuels. Energy Policy 36(3):940–945. https://doi.org/10.1016/j.enpol.2007.11.029

Moniruzzaman M et al (2010) Recent advances of enzymatic reactions in ionic liquids. Biochem Eng J 48(3):295–314. https://doi.org/10.1016/j.bej.2009.10.002

Mudhaffar B, Salimon J (2010) Optimization of process variables using D-optimal design for separating linoleic acid in Jatropha curcas seed oil by urea complex fractionation. Biotechnology 9(3):362–367. https://doi.org/10.3923/biotech.2010.362.367

Niedermeyer H et al (2013) A step towards the a priori design of ionic liquids. Phys Chem Chem Phys 15(27):11566. https://doi.org/10.1039/c3cp50521a

Parawira W (2010) Biodiesel production from Jatropha curcas: a review. Sci Res Essays 5(14):1796–1808. https://www.researchgate.net/publication/228372818_Biodiesel_production_from_Jatropha_curcas_A_review

Passos H, Freire MG, Coutinho JAP (2014) Ionic liquid solutions as extractive solvents for value-added compounds from biomass. Green Chem 16(12):4786–4815. https://doi.org/10.1039/C4GC00236A

Pham PJ et al (2009) Selective extraction of polyunsaturated triacylglycerols using a novel ionic liquid precursor immobilized on a mesoporous complexing adsorbent. Biotechnol Prog 25(5):1419–1426. https://doi.org/10.1002/btpr.219

Ranganathan SV, Narasimhan SL, Muthukumar K (2008) An overview of enzymatic production of biodiesel. Bioresour Technol 99(10):3975–3981. https://doi.org/10.1016/J.BIORTECH.2007.04.060

van Rantwijk F et al (2007) Biocatalysis in ionic liquids. Chem Rev 107(6):2757–2785. https://doi.org/10.1021/CR050946X

Reddy P (2015) A critical review of ionic liquids for the pretreatment of lignocellulosic biomass. S Afr J Sci 111. https://doi.org/10.17159/sajs.2015/20150083

Reddy ER et al (2014) One-pot synthesis of biodiesel from high fatty acid Jatropha curcas oil using bio-based basic ionic liquid as a catalyst. Curr Sci:1394–1400. https://doi.org/10.2307/24102486

Ren Q et al (2014) Preparation of biodiesel from soybean catalyzed by basic ionic liquids [Hnmm] OH. Materials 7:8012–8023. https://doi.org/10.3390/ma7128012

Rubio-Rodríguez N et al (2008) Supercritical fluid extraction of the omega-3 rich oil contained in hake (Merluccius capensis–Merluccius paradoxus) by-products: study of the influence of process parameters on the extraction yield and oil quality. J Supercrit Fluids 47(2):215–226. https://doi.org/10.1016/j.supflu.2008.07.007

Rubio-Rodríguez N et al (2012) Supercritical fluid extraction of fish oil from fish by-products: a comparison with other extraction methods. J Food Eng 109(2):238–248. https://doi.org/10.1016/j.jfoodeng.2011.10.011

Ruzich NI, Bassi AS (2010) Investigation of enzymatic biodiesel production using ionic liquid as a co-solvent. Can J Chem Eng 88(2):n/a. https://doi.org/10.1002/cjce.20263

Santori G et al (2012) A review analyzing the industrial biodiesel production practice starting from vegetable oil refining. Appl Energy 92:109–132. https://doi.org/10.1016/j.apenergy.2011.10.031

Seddon KR (1997) Ionic liquids for clean technology. J Chem Technol Biotechnol 68(4):351–356. https://doi.org/10.1002/(SICI)1097-4660(199704)68:4<351::AID-JCTB613>3.0.CO;2-4

Semelsberger TA, Borup RL, Greene HL (2006) Dimethyl ether (DME) as an alternative fuel. J Power Sources 156(2):497–511. https://doi.org/10.1016/J.JPOWSOUR.2005.05.082

Shahbaz K et al (2010) Using deep eutectic solvents for the removal of glycerol from palm oil-based biodiesel. J Appl Sci 10(24):3349–3354. https://doi.org/10.3923/jas.2010.3349.3354

Shahbaz K et al (2011) Using deep eutectic solvents based on methyl triphenyl phosphunium bromide for the removal of glycerol from palm-oil-based biodiesel. Energy Fuel 25(6):2671–2678. https://doi.org/10.1021/ef2004943

Shimada Y, Sugihara A, Tominaga Y (2001) Enzymatic purification of polyunsaturated fatty acids. J Biosci Bioeng 91(6):529–538. https://doi.org/10.1016/S1389-1723(01)80169-9

Stojković IJ et al (2014) Purification technologies for crude biodiesel obtained by alkali-catalyzed transesterification. Renew Sust Energ Rev 32:1–15. https://doi.org/10.1016/J.RSER.2014.01.005

Sunitha S et al (2007) Ionic liquids as a reaction medium for lipase-catalyzed methanolysis of sunflower oil. Biotechnol Lett 29(12):1881–1885. https://doi.org/10.1007/s10529-007-9471-x

Tadesse H, Luque R (2011) Advances on biomass pretreatment using ionic liquids: an overview. Energy Environ Sci 4(10):3913. https://doi.org/10.1039/c0ee00667j

Tang S, Baker GA, Zhao H (2012) Ether- and alcohol-functionalized task-specific ionic liquids: attractive properties and applications. Chem Soc Rev 41(10):4030–4066. https://doi.org/10.1039/c2cs15362a

Troter DZ et al (2016) Application of ionic liquids and deep eutectic solvents in biodiesel production: a review. Renew Sust Energ Rev 61:473–500. https://doi.org/10.1016/j.rser.2016.04.011

Vafaeezadeh M, Hashemi MM (2014) Efficient fatty acid esterification using silica supported Brønsted acidic ionic liquid catalyst: experimental study and DFT modeling. Chem Eng J 250:35–41. https://doi.org/10.1016/J.CEJ.2014.04.001

Van Gerpen J (2005) Biodiesel processing and production. Fuel Process Technol 86(10):1097–1107. https://www.sciencedirect.com/science/article/pii/S0378382004001924. Accessed 31 July 2019

Vancov T et al (2012) Use of ionic liquids in converting lignocellulosic material to biofuels. Renew Energy 45:1–6. https://doi.org/10.1016/j.renene.2012.02.033

Veljković VB, Stamenković OS, Tasić MB (2014) The wastewater treatment in the biodiesel production with alkali-catalyzed transesterification. Renew Sust Energ Rev 32:40–60. https://

econpapers.repec.org/article/eeerensus/v_3a32_3ay_3a2014_3ai_3ac_3ap_3a40-60.htm. Accessed 31 July 2019

Vicente G, Martinez M, Aracil J (2004) Integrated biodiesel production: a comparison of different homogeneous catalysts systems. Bioresour Technol 92(3):297–305. https://doi.org/10.1016/j.biortech.2003.08.014

Walden P (1914) Molecular weights and electrical conductivity of several fused salts. Bull Acad Imp Sci Saint 405:42

Wasserscheid P, Welton T (eds) (2002) Ionic liquids in synthesis. ChemSusChem. https://doi.org/10.1002/cssc.200800126

Wu Q et al (2007a) Transesterification of cottonseed oil catalyzed by Brønsted acidic ionic liquids. Am Chem Soc. https://doi.org/10.1021/IE070678O

Wu Q et al (2007b) Transesterification of cottonseed oil to biodiesel catalyzed by highly active ionic liquids. Ind Eng Chem Res 46:7955–7960

Xu L et al (2011) Assessment of a dry and a wet route for the production of biofuels from microalgae: energy balance analysis. Bioresour Technol 102(8):5113–5122. https://doi.org/10.1016/j.biortech.2011.01.066

Yang Z et al (2010) Both hydrolytic and transesterification activities of Penicillium expansum lipase are significantly enhanced in ionic liquid [BMIm][PF6]. J Mol Catal B Enzym 63(1–2):23–30. https://doi.org/10.1016/J.MOLCATB.2009.11.014

Yang H et al (2017) New algal lipid extraction procedure using an amphiphilic amine solvent and ionic liquid. Biomass Bioenergy 100:108–115. https://doi.org/10.1016/j.biombioe.2017.03.018

Yoo CG, Pu Y, Ragauskas AJ (2017) Ionic liquids: promising green solvents for lignocellulosic biomass utilization. Curr Opin Green Sustain Chem 5:5–11. https://doi.org/10.1016/j.cogsc.2017.03.003

Young G et al (2010) Lipid extraction from biomass using co-solvent mixtures of ionic liquids and polar covalent molecules. Sep Purif Technol 72(1):118–121. https://doi.org/10.1016/j.seppur.2010.01.009

Zhang Y, Bakshi BR, Demessie ES (2008) Life cycle assessment of an ionic liquid versus molecular solvents and their applications. Environ Sci Technol 42(5):1724–1730. https://doi.org/10.1021/es0713983

Zhang L et al (2009) A Brønsted acidic ionic liquid as an efficient and environmentally benign catalyst for biodiesel synthesis from free fatty acids and alcohols. Bioresour Technol 100(19):4368–4373. https://doi.org/10.1016/j.biortech.2009.04.012

Zhang K-P et al (2011) Penicillium expansum lipase-catalyzed production of biodiesel in ionic liquids. Bioresour Technol 102(3):2767–2772. https://doi.org/10.1016/J.BIORTECH.2010.11.057

Zhao H (2010) Methods for stabilizing and activating enzymes in ionic liquids: a review. J Chem Technol Biotechnol 85(7):891–907. https://doi.org/10.1002/jctb.2375

Zhao H (2016) Lipase activation and stability enhancement in ionic liquids. Ionic Liquids in Lipid Processing and Analysis:99–152. https://doi.org/10.1016/B978-1-63067-047-4.00004-0

Zhao H, Baker GA (2013) Ionic liquids and deep eutectic solvents for biodiesel synthesis: a review. J Chem Technol Biotechnol 88(1):3–12. https://doi.org/10.1002/jctb.3935

Zhao H, Baker GA, Holmes S (2011) New eutectic ionic liquids for lipase activation and enzymatic preparation of biodiesel. Org Biomol Chem 9(6):1908. https://doi.org/10.1039/c0ob01011a

Zhao H, Zhang C, Crittle TD (2013) Choline-based deep eutectic solvents for enzymatic preparation of biodiesel from soybean oil. J Mol Catal B Enzym 85–86:243–247. https://doi.org/10.1016/j.molcatb.2012.09.003

Zheng S et al (2006) Acid-catalyzed production of biodiesel from waste frying oil. Biomass Bioenergy 30(3):267–272. https://doi.org/10.1016/J.BIOMBIOE.2005.10.004

Zhou JH et al (2011) Preparation of biodiesel from Tung oil catalyzed by sulfonic-functional Brŏnsted acidic ionic liquids. Adv Mater Res 314–316:1459–1462. https://doi.org/10.4028/www.scientific.net/AMR.314-316.1459

Zhou S et al (2012) Biodiesel preparation from transesterification of glycerol trioleate catalyzed by basic ionic liquids. Chin Chem Lett 23(4):379–382. https://linkinghub.elsevier.com/retrieve/pii/S1001841712000356. Accessed 3 Aug 2019

Chapter 13
Green Synthesis of Nanoparticles and Their Application for Sustainable Environment

Ardhendu Sekhar Giri and Sankar Chakma

Contents

A. S. Giri · S. Chakma (✉)
Department of Chemical Engineering, Indian Institute of Science Education and Research, Bhopal, M.P., India
e-mail: schakma@iiserb.ac.in

Inamuddin, A. M. Asiri (eds.), *Nanotechnology-Based Industrial Applications of Ionic Liquids*, Nanotechnology in the Life Sciences,
https://doi.org/10.1007/978-3-030-44995-7_13

13.1 Introduction

The nanomaterials have unique physicochemical, optical, and magnetic properties that are governed by the shape, size, and distribution of the nanoparticles (NPs). Due to these unique properties, nanotechnology has been used as one of the advanced and popular research areas (Daniel et al. 2004; Kumar et al. 2003). NPs have significantly a large surface area-to-volume ratio due to small size, and it considerably changes their properties (e.g., mechanical properties, catalytic activity, biological, electrical, and thermal conductivity) as compared to their bulk form (Perez et al. 2005). NPs are also used as catalysts for chemical reactions, sensors, pharmaceutical products, imaging for medical diagnostic, and medical treatment protocols. Some of the popular metallic nanoparticles such as gold (Au), silver (Ag), platinum (Pt), and palladium (Pd) are being comprehensively used in making the different electronic devices, pharmaceuticals, and grease paints. Gold NPs (AuNPs) have also been used in manufacturing pharmaceuticals but also in biomedical applications (Sperling et al. 2008; Puvanakrishnan et al. 2012). Silver nanoparticles (AgNPs) are being used to promote the faster spiral healing as they have anti-inflammatory and antibacterial properties for which the AgNPs are integrated into commercialized wound dressing kits, preparation of pharmaceutical, and medical transplant coverings (Pollini et al. 2008; Asha Rani et al. 2009). Platinum NPs (PtNPs), either in pure or alloy form, have been widely used in biomedical applications (Hrapovic et al. 2004). The palladium NPs (PdNPs) are used in catalysis, electro-catalysis, and antibacterial applications (Gopidas et al. 2003). Besides, the non-noble metallic nanoparticles like Fe, Co, Zn, and Se are also being utilized in medical applications, formulations of cosmetic, and antimicrobial applications (Njagi et al. 2011; Lee et al. 2011; Brayner et al. 2006).

With the increasing demand of different metallic and nonmetallic nanoparticles, several physicochemical techniques have been introduced to synthesize NPs with diverse shapes, sizes, and compositions. Both physical and chemical techniques are used to synthesize and stabilize the NPs. Among the different physical techniques, laser ablation (Mafune et al. 2001), lithography, and high-energy irradiation have been used widely throughout the world (Zhang et al. 2008; Treguer et al. 1998), while chemical techniques use chemical reduction or photochemical reduction (Chen et al. 2007; Eustis et al. 2005; Starowiicz et al. 2006). The interaction between the metal ions of the precursors is associated with several parameters like temperature, concentration, and kinetics of the process and the reducing agent, including the adsorption kinetics that helps to stabilize the NPs (Wang et al. 2005).

The toxicity is one of the problems for NPs which arises from the application of hazardous chemical reagents used for preventing the undesirable colloid agglomeration. Besides, some of the NPs are also observed to be toxic caused by the different compositions of metals, shape and sizes, and surface chemistry. This causes toxic NPs unusable in biomedical and clinical applications. However, these influencing parameters can be controlled by using biological mediated techniques, which are environment-friendly and green chemistry-based approaches (Ahmad et al.

2003; Gericke et al. 2006). One of such compounds is chayote, sometimes called *Sechium edule*, which looks like pears with coarse covering and an average length of 10–20 cm (Rao et al. 2017). Several polyphenolic compounds like phenylalanine and tyrosine along with natural antioxidants and amino acids are also the major compositions present in chayote (Sykora et al. 2010). The fruit is rich in flavonoids having 35 mg flavonoids per 10 g of dried chayote fruit and sugar approximately (Rao et al. 2017). The extracts of this fruit showed highly reducing properties and it converts potassium ferricyanide to potassium ferrocyanide (Torres-Chavolla et al. 2010).

In the synthesis of NPs, the metal ions from their corresponding salt precursors are reduced, which results in a color change in the reaction solution during the synthesis of NPs from both the chemical and biological processes. The color change is the first qualitative indication for the formation of NPs. The colloidal nanoparticles in the solution show a Tyndall effect which is used to perceive the existence of the nanoparticles in a solution (Poinern et al. 2013). High-speed centrifugation (~12,000 rpm) is generally used to separate the NPs from the colloid using different advanced techniques. Table 13.1 presents the different techniques for NP synthesis.

13.2 Biological Synthesis of Nanoparticles

The synthesis of NPs from different plants using biological methods is harmless, cheap, and environment friendly (Makarov et al. 2014). Both plants and microorganisms can absorb and build up the inorganic metallic ions from their surrounding environment. The ability of a biological entity in transforming the inorganic metal ions into their respective metal nanoparticles is a comparatively new and largely unexplored technique for NP synthesis (Baker et al. 2013). The ability of microorganisms for the accumulation of metallic ions surrounded in the environments has been exploited in various biotechnological applications like bioleaching and bioremediation processes (Stephen et al. 1999). The interaction capability of the microorganisms with the surrounding environment for oxidation-reduction mechanisms promotes the biochemical conversions. Several studies have shown that both extra- and intracellular synthesis of NPs can also assist them in promoting the redox phenomenon (Mandal et al. 2006). However, the nucleation rate, redox potential, and successive growth kinetics of NPs and their interaction with the metabolic process of microorganisms are yet to be explored (Lengke et al. 2006; Duran et al. 2005). In addition to that, the plants or their extracts have advantages over the other biological based eco-friendly systems like fungi and bacteria, which are needed to take care of the culture preparation or separation techniques. Conversely, due to relatively short production times, plant-based biosynthesis of NPs would be safe and have a negligible cultivation cost than the other living systems (Mittal et al. 2013). Furthermore, plant extract-based biosynthesis addresses a new process that can be easily climbed up to produce NPs in bulk.

Table 13.1 Conventional methods of NP fabrication (Rao 2018)

Fabrication methods	Advantages	Disadvantages
Chemical reduction method	• It was producing AgNPs almost instantly. • Very small particles are formed.	• This method is expensive as the process needs many chemicals as reducing and stabilizing agent.
Chemical vapor deposition	• Produces highly dense and pure materials. • High deposition rate. • Does not require high vacuum. • Low deposition temperature.	• Film deposited at elevated temperature. • Difficult to deposit multicomponent materials. • Use of more sophisticated reactor. • Toxic and corrosive gas generation.
Coprecipitation method	• Low reaction temperature needed. • Impurity level also remains low.	• Needs further grinding and milling to obtain desired NPs.
Bio-mediated synthesis	• Simple, eco-friendly, and energy-efficient synthesis. • Low-cost, non-hazard, and no additional chemicals needed other than metal precursors. • Does not require high temperature and pressure. • Almost free from contamination.	• Prolonged reaction time with some bio-extracts.
Flame spray pyrolysis	• Uniform particle distribution. • High-purity products.	• Raw materials are expensive and highly corrosive. • Controlling size, size distribution, and agglomeration of particles.
Laser ablation methods	• To produce AgNPs without any surface-active agents. • Laser ablation technique in liquid media is simple and efficient and can produce many NPs.	• It has a limitation of NP size control.
Sonochemical	• Very simple, efficient, and additive-free process. • Uniform particle size. • More crystalline in nature.	• More energy consumption.
Sol-gel method	• It is also a low-temperature technique. • Impurity level also remains low.	• Expensive. • Time-consuming process.
Solvothermal method	• This method is useful for synthesizing isotropic nanoparticles. • Particle size is highly controllable.	• Recovery of the product is difficult from the suspension without agglomeration.
Photoreduction	• Simple technique, and fast reaction; within 1–2 h the reaction is expected to complete.	• Expensive process as UV radiation is employed.
Wetness impregnation	• Low processing temperature from 100 to 800 °C. • Easy-to-produce nano-sized particles.	• Some precursor materials are costly. • Collection without aggregation is difficult. • Stoichiometric control can be difficult.

As mentioned above, from many biological substances like bacteria, algae, actinomycetes, plants, fungus, viruses, and yeast the NPs can be synthesized. However, different biological units have enormous skills for the synthesis of metallic NPs or metallic oxide NPs with different degrees of biochemical processing. Enzyme activity can significantly increase under the influence of nutrients, light source, pH of the solution, temperature, mixing speed, and buffer strength (Mukherjee et al. 2001). During the synthesis of NPs, the molecules found in the plant extracts act as both reducing and stabilizing agents (Narayanan and Sakthivel 2008; Sathishkumar et al. 2009); though the biological molecules are chemically complex they are eco-friendly. Different NPs like Co, Pt, and PtCo NPs can be obtained using *S. edule* fruit extract (which contains ascorbic acid) through a bio-mediated process (Rao and Golder 2016). The formation of face-centered cubic (FCC) crystals has been found due to metal ion reduction, and NP destabilization showed intermolecular attractions through conjugation (Tahir et al. 2015).

13.3 Plant Extract-Based Metal Nanoparticle Synthesis

Plants have a high potential biologically to accumulate and reduce the metallic ions (Kulkarni et al. 2014). Due to this property, plants have been considered a more environment-friendly way for biologically synthesizing metallic NPs that are used in detoxification reactions (Khan et al. 2013). Several chemicals like polyphenols, bioactive alkaloids, phenolic acids, proteins, and different other natural products obtained from plant extract play an important role in reducing and stabilizing the metallic ions (Castro et al. 2011). The selection of nanoparticle size and shapes depends on the main contributing factors like composition and concentration of these active biomolecules originated from different plants and their successive interaction between the metal ions in aqueous medium (Li et al. 2011). The synthesis of NPs occurred at room temperature from reducing metal salts via different plant extracts. After mixing the metal salt solution with the plant extract, the process starts, and the biochemical reduction reaction of the salts begins instantly. With the formation of the NPs the appearance of the reaction mixture color changes. Malik et al. (2014) proposed that during the synthesis of NPs, metal ions are oxidized in mono- or divalent oxidation states from their zerovalent states. After that, the smaller neighboring particles merge to form thermodynamically more stable larger NPs under reduction of metal ions that takes place biologically. Akhtar et al. (2013) suggested that nanoparticles are combined to form a variety of morphologies like triangles, spheres, cubes, pentagons, and hexagons. The quality of the synthesized NPs, including size and morphology, is influenced significantly by the properties or composition of the plants' extract. Also, the concentration of metal salt, extract, time of reaction, solution pH, and reaction temperature play a significant role in NP synthesis (Dwivedi and Gopal 2010). Figure 13.1 shows the bioinspired synthesis of Ag/CoNP-Gr using plant extracts.

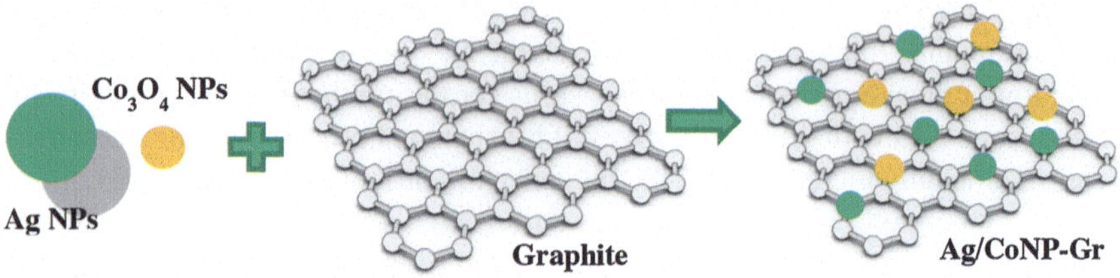

Fig. 13.1 Bio-inspired synthesis of the Ag/CoNP-Gr using plant extracts

13.4 Factors Affecting the Metal Nanoparticle Synthesis

A few controlling factors are involved during the biological synthesis of metallic stabilized nanoparticles. These factors that influence the NP synthesis are solution pH, reactant concentration, time of reaction, and temperature, which are discussed briefly in the subsequent sections.

13.4.1 Influence of pH

During the formation of NPs, the pH plays a significant role during the reaction for the formation of NPs in solution phase (Gardea-Torresdey et al. 1999). Nanoparticles with different shapes and sizes were produced due to the pH variation in the reaction medium. It was observed that at lower pH of the reaction solution, larger particles are produced compared to the particles produced at higher solution pH (Dubey et al. 2010). Armendariz et al. (2004) have suggested that *Avena sativa* biomass was used to synthesize the rod-shaped Au nanoparticles with the particle size in the range of 25–85 nm when synthesized at pH 2. However, the particle sizes are relatively smaller (5–20 nm) when they are synthesized at pH 4. The study reported that more available functional groups are available at pH 4 within the extract to participate in the nucleation process. However, only a few functional groups were available at lower pH value of 2 and led to the particle aggregation and AuNP formation.

In another similar study, *Cinnamon zeylanicum* bark extract has been used to synthesize AgNPs. Also, it has been observed that the number of synthesized particles could be increased with increasing the concentrations of the bark extract. In addition to that at higher pH (>5), the shape of the NPs tended to become spherical (Sathishkumar et al. 2009). On the other hand, at higher pH of the solution the synthesis of palladium (Pd) NPs using *Cinnamon zeylanicum* bark extract showed a slight increment in particle size. The particle size was in the range of 15–20 nm at below 5 of the solution, and at higher pH (>pH 5) the particle size when synthesized was in the range of 20–25 nm (Sathishkumar et al. 2009).

13.4.2 Effect of Concentration

The concentration of the biomolecules significantly influences the formation of metallic NPs. It has been revealed that under the varying concentration of *Cinnamomum camphora* (camphor) leaf extract, the shape of the AuNPs and AgNPs can be controlled (Huang et al. 2007). For example, when the extract concentration of the precursor chloroauric acid was increased, the shape of the resulting NPs was changed from triangular to spherical shape.

13.4.3 Influence of Reaction Time

Ahmad et al. (2012) reported that spherical AgNP synthesis using *Ananas comosus* (pineapple) has occurred within 2 min with a rapid color change. $Ag(NO)_3$ was used and rapidly reduced to Ag^0, and the nanoparticles appeared within 2 min. The produced nanoparticles were spherical in size with a diameter of 12 nm. Dwivedi and Gopal (2010) performed a similar study to produce Ag and Au nanoparticles using different leaf extracts. The formation of nanoparticles started within 15 min of reaction and continued till the next 2 h of reaction. However, the production of NPs was very few beyond the reaction time of 2 h. Prathna et al. (2011) suggested that the reaction time was increased while combining *Azadirachta indica* leaf extract and $Ag(NO)_3$ and also the size of the particles was increased. However, the range was between 10 and 35 nm when the reaction was conducted in 4 h.

13.4.4 Effect of Reaction Temperature

Temperature is an essential factor for the determination of both the size and shape in NP synthesis (Song et al. 2009). For example, when the AgNPs were synthesized using *Citrus sinensis* shell extract at a reaction temperature of 25 °C, the average particle size was around 35 nm. However, when the temperature of the reaction was 60 °C, the average particle size was decreased to 12 nm (Kaviya et al. 2011). Similarly, Song et al. (2009) took *Diospyros kaki* (persimmon) leaf extract for synthesizing AgNPs, and the temperature of the reaction was in between 25 °C and 95 °C. In their study, the particle size was decreased from 110 nm (synthesized at 25 °C) to 40 nm (synthesized at 95 °C). Therefore, by increasing the reaction temperature, the rate of reaction and particle formation appeared to be faster. Though the average particle size was decreased the rate of particle conversion was steadily increased with increasing reaction temperature.

13.5 Green Routes for NP Preparation

13.5.1 AgNPs and Ag-Doped NPs

Plant-based synthesis of NPs is highly favorable due to cost-effectiveness, eco-friendly nature, and their applications in a variety of fields. AgNPs were synthesized from a methanolic extract of *A. marmelos* fruit extract in a greener route. $AgNO_3$ of 1 mM aqueous solution was prepared for the synthesis of AgNPs. Co-precipitation technique using the fruit extract of *A. marmelos*, a reducing and plugging agent, was employed to synthesize the AgNPs from $AgNO_3$ solution (Devi et al. 2019). When the formation of NPs starts the color of the reaction solution changes as shown in Fig. 13.2. The bio-reduction of Ag^+ with the wavelength range from 350 to 680 nm in aqueous solution was detected using a UV spectrophotometer. To discover the mechanistic pathways of the reduction of Ag metal ions from its corresponding salt, the extract of the fungus was dialyzed for 42 h at 4 °C against distilled water. Then 200 μL of 20 mM nicotinamide adenine dinucleotide phosphate (NADPH) was mixed with 10 mM $AgNO_3$ solution (Boulch et al. 2001).

Moreover, the doped NPs are also becoming popular as they have multifunctional properties. Recently, Nigussie et al. (2018) have reported the Ag-doped TiO_2 synthesized using $TiCl_4$ combined with ethanol under continuous stirring. Then $AgNO_3$ was mixed gently with 0.5 mL deionized water (DI) to form gelation precipitation of AgNPs and dried at 100 °C for 24 h. The amorphous TiO_2 transformed into a crystalline structure using 460 °C for 5 h. Ag-doped ZnO nanopowder was also synthesized with a similar path using $AgNO_3$ which was added to the zinc solution containing NaOH solution, and then $Ag(OH)_2$ was formed as a precipitate. The Ag-doped ZnO NPs were obtained, and the powder was then calcined in an atmospheric air for 7 h at 460 °C (Nigussie et al. 2018).

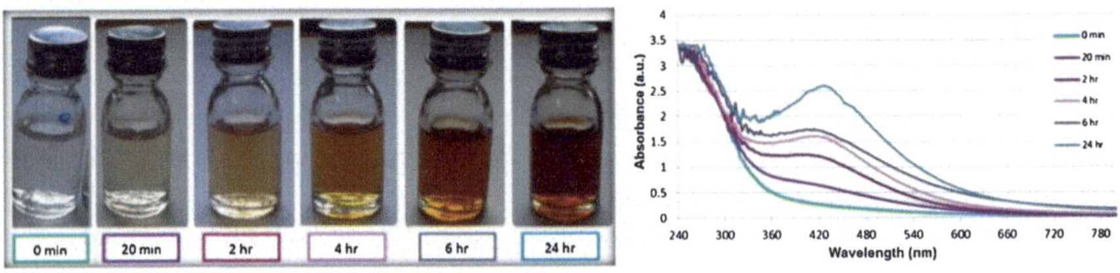

(a) (b)

Fig. 13.2 Change in solution color (**a**) and change of absorbance (**b**) in UV-vis spectra with the formation of AgNPs (Erjaee et al. 2017)

13.5.2 Zerovalent Iron NPs and Fe-Doped NPs

In various fields of medical science, the iron-based nanomaterials have shown a wide application in treating ecological pollution (Tripathi and Chunga 2019). Zerovalent iron nanoparticles (FeNPs) have been used in the field of biodegradation for the removal of different heavy metals like Hg, Ni, Cd, Pb, and Cr. The Fe^0 NPs can also be synthesized using the microbial biomasses or plant extracts. Mehrotra et al. (2017) showed that *yeast* extract has been used to synthesize the FeNPs. The solution of *yeast* extract was prepared by dissolving 2.0 g of *yeast* powder in 25 mL of deionized water followed by continuous boiling for 15 min (Tripathi and Chunga 2019). Ferric chloride ($FeCl_3$, 1 mM) solution was used as a source of Fe^{3+} ions for the reduction of Fe^{3+} ions to Fe^0 by dissolving about 750 μL of *yeast* extract solution in it. With the continuous development of the NPs, the color of the solution was transformed rapidly from light yellow to brown, indicating the formation of FeNPs.

13.5.3 AuNPs and Au-Doped NPs

Gold nanoparticles (AuNPs) attracted researchers due to their optical properties. Its colloidal solutions can be obtained by the reduction of Au(III) nanoparticles (Sun et al. 2017). The newly formed AuNPs in red color have surface plasmon resonance absorption (Sun et al. 2017). In addition, ligand-exchange reactions with functionalized molecules play an important role in the synthesis of optical materials and sensors from AuNPs (Zhang et al. 2012), catalysts (Deparis et al. 2009), etc. Recently, NPs have been prepared from different tea extracts. In green and black tea extracts both reductant and surfactant are present; they have been used in the "green" research of metal NPs (Cha et al. 2000; Zhou et al. 2014).

The NPs can be synthesized using green tea extract by using green tea leaves with 100 g to 750 mL of water and allowing this mixture in the refrigerator for 24 h at ~0 °C. About 500 mL of water was added to the tea extracts after filtration under identical conditions. An aqueous solution of 10 mM $HAuCl_4$ is then added to a solution of 900 μL tea extract at ambient temperature, having a metal concentration of 1 mM of AuNP solution. Das et al. (2011) have synthesized spherical shaped AuNPs using *Nyctanthes arbortristis* (night jasmine) flower extract ranging from 7 to 55 nm. The diverse shapes of decahedral, triangular, and spherical have been shown in the synthesized particles (Narayanan and Sakthivel 2008). Au/TiO_2 NPs can be produced from AuNP solution by mixing with 40 mM of TiF_4 with a continuous stirring followed by isolated Au/TiO_2 NPs having core-shell-like structure using centrifugation at 6000 rpm for 8 min (Sun et al. 2017). Coupling between TiO_2 and NPs, which are made by noble metal, has been used as an effective approach to overcome these walls (Zhang et al. 2012; Jiang et al. 2014).

13.5.4 CuO- and Cu-Doped NPs

CuNPs and copper-copper oxide (Cu/CuO) have been produced from Magnolia leaf extract ranging from 45 to 100 nm (Lee et al. 2011). CuNPs spherical in shape having the potential of antibacterial activity against *E. coli* cells are produced by a green route from *Syzygium aromaticum* (clove) extracts with a mean particle size range of 40–55 nm (Subhankari and Nayak 2013). An average particle size of cuprous oxide NPs was 4.8 nm that has been formed from the *Sterculia urens* (karaya gum) extract which has the ability to synthesize highly steady spherical NPs (Padil and Cerník 2013). The particles have been found to be effective in antimicrobial activity tests against common pathogens like *E. coli* and *Staphylococcus aureus*. Das et al. (2013) have also observed that both antioxidant and antibacterial behavior was found in CuONPs.

13.5.5 Mixed Metal/Metal Oxide NPs

Several metal oxide NPs such as titanium dioxide (TiO_2), zinc oxide (ZnO), and palladium oxide (PdO) nanoparticles with an effective size range from 100 to 150 nm have been reported using green route (Roopan et al. 2012). An extract from *Psidium guajava* was used to produce TiO_2 nanoparticles that have both antibacterial and antioxidant properties, which are estimated against *Aeromonas hydrophila*, *Proteus mirabilis*, and *E. coli* (Santhoshkumar et al. 2014). The antibacterial and antioxidant properties of TiO_2 were found to be most effective against *E. coli* and have also been observed and found to be harmful to several bacterial strains (Heinlaan et al. 2008).

The ZnO nanostructure shows high electron/hole (e^-/ h^+) binding energy (60 meV) with a wide bandgap (3.37 eV) (Mitra et al. 2012). ZnO NPs have been used in various fields of applications like in optical devices (Yude et al. 2006), biosensors (Hwa and Subramani 2014), solar cells (Al-Kahlout 2015), and photocatalysis devices (Tripathi et al. 2014). Rao et al. (2018) have synthesized Ag-doped ZnO using the analyte ascorbic acid (294 mg per kg fruit) extracted from *Sechium edule* in aqueous solution. The bandgap of ZnO was decreased to 2.85 eV from 3.13 eV at the optimum Ag loading with 1.18% (w/w) under the control catalytic system as shown in Fig. 13.3, while the commercial analyte ascorbic acid can diminish the bandgap up to 2.91 eV (Rao et al. 2018).

13.6 Characterization Techniques of NPs

The different techniques such as UV-vis spectroscopy, transmission electron microscopy (TEM), dynamic light scattering, X-ray powder diffraction, energy-dispersive spectroscopy, diffuse reflectance spectroscopy, and zeta potential are used for

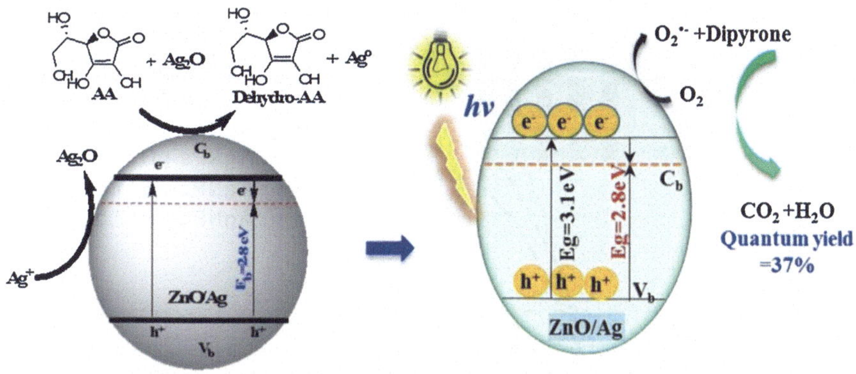

Fig. 13.3 Ag-doped ZnO supported by an ascorbic acid used as a potential bio-analyte rich for photocatalytic degradation of dipyrone drug (Rao et al. 2018)

characterizing the NPs. The detailed discussion is given below for the NP characterization.

13.6.1 UV-Vis Spectroscopy Method

The bandgap energy of any NPs can be determined by UV-visible absorption spectroscopy. For example, Co_3O_4 NPs were measured by UV-vis. The bands were found to be at 600–800 nm and 350–600 nm for their corresponding O^{2-}-to-Co^{3+} and the O^{2-}-to-Co^{2+} charge transfer transitions, respectively (Das et al. 2017). The corresponding bandgaps have been found to be 1.42 and 2.53 eV (Sharma et al. 2015). The absorption peaks of Co_3O_4 NPs are observed at 427 nm and 739 nm (Naveen and Selladurai 2015). Melissa et al. (2013) suggested that the photocatalytic-reactive oxygenated species (ROS) produced by metal oxide NPs have been detected, where TiO_2 or ZnO with the wide-bandgap semiconductors produced electron (e^{-1})/hole (h^+) pairs causing redox reaction under UV light irradiation.

Also, UV-visible absorption spectroscopy helps to identify the formation of different-sized NPs. In the case of variation in solution pH, the formation of AgNPs also changes, and accordingly the solution color also changes when synthesized with different pH (viz. pH 3–12.5). The variations of both color change of the reaction mixture and its optical absorbance at different pH are shown in Fig. 13.4. At pH 1, absorbance peak did not appear within the whole range of the wavelength because of no AgNP formation. However, with a gradual increase in pH the absorption peak visibility was increased. Rao and Golder (2016) observed a minor peak at a wavelength of 428 nm and pH 3 when AgNPs were synthesized. By increasing the peak intensity, the absorption was shifted to lower wavelengths up to 414 nm when synthesized at pH 12.5 as shown in Fig. 13.4. Smitha et al. (2008) suggested that the AgNP shows some characteristic peaks in the suspension obtained from shape- and

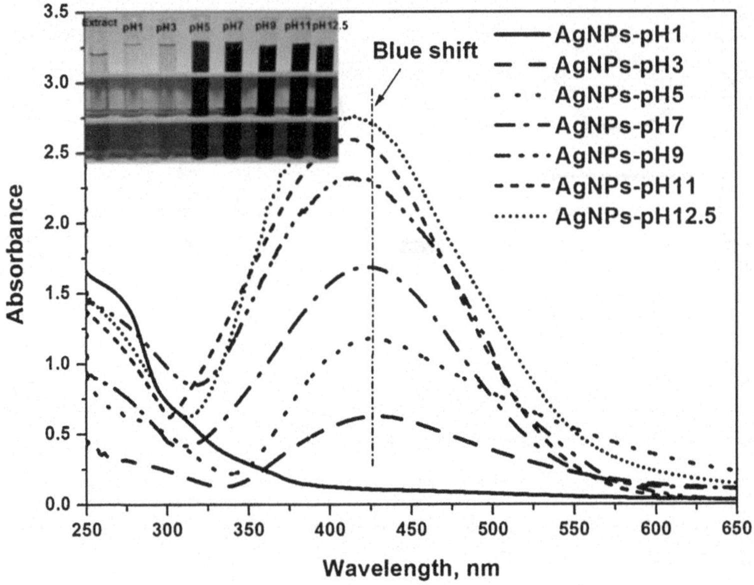

Fig. 13.4 Spectral absorbance of $AgNO_3$ and bio-extract mixture at different pH (Rao et al. 2016)

size-dependent surface plasmon resonance (SPR) effect. This reduction in the size and/or degree of anisotropy of the particles is indicated by the hypsochromic shift (Edison and Sethuraman 2013; Yilmaz et al. 2011). AgNPs show a very minute change in the color of the solution after 24 h of reaction time. The color was changed by increasing the pH from light red-brown to dark red-brown, as shown in Fig. 13.4 (inset). The conversion rate from Ag^+ to Ag^0 in AgNPs was faster with increasing pH at the initial period (Rao and Golder 2016). It has been found that about 88.5% of Ag^+ was converted to Ag^0 present in AgNP solution at pH 12.5 within 12 h. It was increased up to 96% when the reaction mixture was kept for 24 h. In another study, the UV-visible spectrum of the NiO-NPs has been dispersed in water existing at the wavelength of 319 nm with a strong band that may be caused by the presence of NiO-NPs. This is attributed to the electronic transition in $O_{(2p)}$ of the valence band to $Ni_{(3d)}$ of the conduction band in the NiO semiconductor (Barakat et al. 2013). Al-Sehemi et al. (2014) reported that the bandgap energy of the NiO-NPs was about 3.55 eV from the UV-vis spectroscopy experimentation.

13.6.2 Dynamic Light Scattering of NPs

The average particle size or size distribution has been found to be dependent significantly on the properties of the NPs, and this can be analyzed using the dynamic light scattering (DLS) analysis. A representative result of AgNPs is shown in Fig. 13.5.

From Fig. 13.5, it can also be observed that the particle size distribution is directly influenced by the solution pH. As the solution pH increases, the particle size becomes smaller. Rao and Golder (2016) have reported that the lowest and the highest particle size distribution was found to vary from 51 to 193.9 nm and 7.1 to 10.9 nm for AgNPs at pH 3 and pH 12.5, respectively. In another study by Rao and Golder (2019), it has been found that the hydrated layer of the organic molecules present on the surface of NPs is 60% greater than the hydrodynamic diameter of PtNPs, CoNPs, and PtCo (1:1) NPs. The change of the zeta potential was counted by the DLS analysis (Kuehner et al. 1997). The PtNPs showed the maximum mass loss due to smaller particle sizes (42.62 nm) with a high surface area surrounded by a more covering agent (Table 13.2).

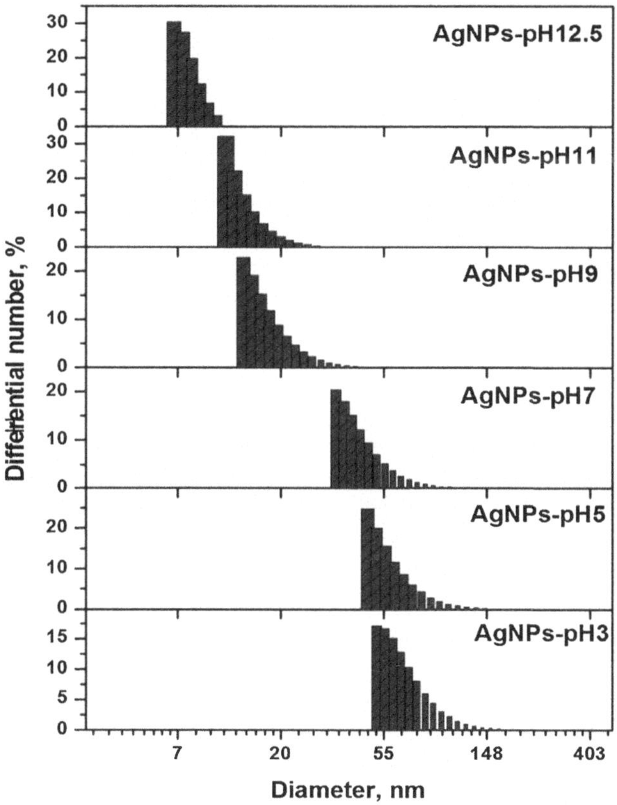

Fig. 13.5 Particle size distribution by dynamic light scattering with pH variation during AgNP synthesis (Rao et al. 2016)

Table 13.2 Characteristic parameters of AgNPs synthesized at different pH (Rao et al. 2016)

Sample	UV-vis absorption		Particle size calculated from XRD data		Particle size calculated from DLS data		Particle size calculated from TEM		Particle size calculated from FESEM	
	SPR peak (nm)	FWHM	d-Spacing (nm)	Crystallite size (nm)	Mean size	Size range	Mean	Size range	Mean	Size range
AgNPs at pH 3	428	168.7	0.23591	54.069	68.2	51–194	57.31	20–110	58.56	20–110
AgNPs at pH 5	427	146.9	0.2354	35.159	60.1	47–189				
AgNPs at pH 7	424	136.6	0.2354	34.479	44.9,	34–131				
AgNPs at pH 9	417	116.9	0.2354	22.247	31.2	14–58	25.94	5–50	25.37	5–50
AgNPs at pH 11	416	113.7	0.23543	22.344	14.5	12–44				
AgNPs at pH 12.5	414	108.9	0.23537	20.716	8.9	7–11				

13.6.3 TEM and FESEM Study

The surface morphology of the NPs can be measured by the SEM and TEM analysis. For example, Das et al. (2017) have used the SEM micrographs of Co_3O_4 to investigate the detailed surface morphology of Co_3O_4 NPs. They have also reported that Co_3O_4 NPs were mostly present in an irregular aggregate manner due to the presence of grain boundaries with a weak migration and a significant number of amorphous phases. The sizes of the Co_3O_4 NPs are normally in an average diameter of 20.8 nm and the size range varies from 5.8 to 38.1 nm.

The orientation of crystallites of Co_3O_4 at 500 °C for 12 h was found to be increased due to more distinct grain boundaries and the agglomeration of small grains shows the increase in average particle size to 28.12 nm. Rao and Golder (2016) have reported that the shape, size, and morphology of AgNPs were determined using TEM and FESEM analysis at pH range from 3 to 12.5, and the results are presented in Table 13.2. AgNP size was more substantial with a significant agglomeration at pH 3. However, at pH 9, the small particles of AgNPs were mostly separated, and the agglomeration was found to be reduced.

The interplanar spacing and lattice plane can also be measured from TEM analysis. A clear lattice boundary in a single AgNP with an interplanar d-spacing of 0.2354 and 0.236 nm was found by high-resolution TEM micrograph at pH 3 and pH 9, respectively (Rao and Golder 2019). The similar studies can also be

Table 13.3 Determination of particle sizes of PtCo mono- and bimetallic NPs (Rao et al. 2019)

NP sample	Calculated from FESEM analysis		Calculated from TEM analysis		Calculated from DLS analysis		Calculated from XRD analysis	
	Mean (nm)	Size range (nm)	Mean (nm)	Size range (nm)	Mean	Size range	d-Spacing (nm)	Crystallite size (nm)
Pt	28.76	10–70	–	–	43.62	10–122	0.2296	22.19
Co	47.55	40–110	–	–	69.00	21–143	0.2128	35.16
PtCo (1:1)	25.11	10–80	27.9	10–80	60.12	14–172	0.2248	32.08

performed to determine the d-spacing from XRD analysis as shown in Tables 13.2 and 13.3 (Rao and Golder 2016; Rao and Golder 2019). Sun et al. (2017) have shown that the TEM image of Au/TiO_2 shell-like NPs was made at 100 °C under an identical condition. Comparing with the shell-like NPs synthesized at 180 °C, the same NPs synthesized at 100 °C showed thicker shell width of 32 nm. The average diameter of 50 nm of the hollow structure of each AuNP appears as light gray thin shells that have also been verified by TEM image (Tu et al. 2015). Figure 13.6 shows the different morphological shapes of NPs using SEM analysis (Elechiguerra et al. 2006).

13.6.4 Influence of Zeta Potential (ζ)

The zeta potential basically shows the study of the stability of the NPs. Malika et al. (2016) have studied the destabilization tendency of NPs in aqueous media using zeta potential (ζ) analysis. Zhang et al. (2009) showed that the adsorption of biomolecules generally shows a negative surface charge due to the formation of AgNP-ascorbate layer. Zeta potential was inversely changed with pH from 3 to 12.5 and was found to be decreased from −3.8 to −25.8 mV between pH 3 and 11 (Rao and Golder 2016). It suggests that the higher pH of the reaction helps to achieve stable AgNPs (Fig. 13.7). Moreover, by increasing the pH the concentration of silver-ascorbate layer is increased which leads to the formation of nanocrystal surface (Oluwafemi et al. 2010). The change in ζ was measured in aqueous colloidal medium of PtNPs and CoNPs and was found to be negative for all metal NPs. These NPs were originated from different biomolecules such as ascorbate layer obtained from ascorbic acid acting as the formation of NP-ascorbate layer. The ζ was gradually reduced from −3.2 to −37.1 mV with increasing pH from 3 to 12 in the case of PtNPs. The ζ for CoNPs was decreased from −3.2 to −31.1 mV when pH reduced from pH 3 to pH 12. Similarly, the ζ for PtCo bimetallic NPs was changed from

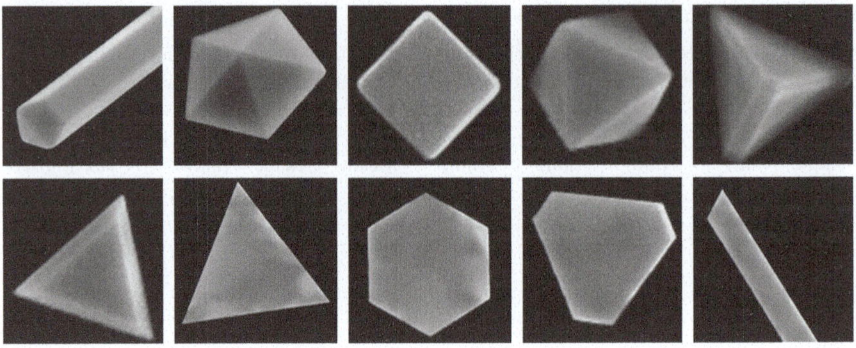

Fig. 13.6 Different morphologies of particles: nano-wire pentagonal shape, decahedral shape, cubic shape, octahedral shape, tetrahedral shape, truncated tetrahedral shape, and platelet shape (Elechiguerra et al. 2006)

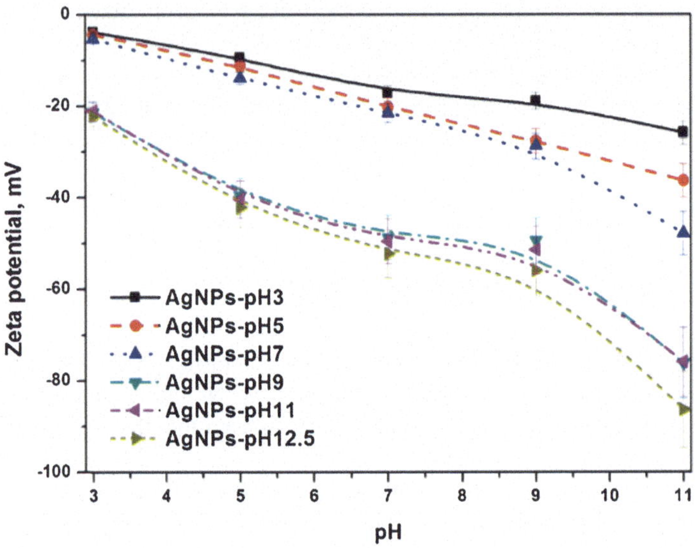

Fig. 13.7 Zeta potential of synthesized AgNPs with pH variation (Rao et al. 2016)

−5.1 mV at pH 3 to −46.2 mV at pH 12.5. The higher negative value of ζ shows the repulsion of the particles and helps to remain as an individual unit in a suspension. It is recommended that the bimetallic NPs like PtCoNPs act as an efficient catalyst compared to the monometallic NPs.

13.6.5 Structural Morphology

The crystalline structure of the NPs can be estimated using XRD data. The average particle size can also be determined using Scherrer's formula. Rao and Golder (2016) have synthesized AgNPs and calculated the crystallite size for the samples synthesized at different pH values (Fig. 13.8). Zuas et al. (2014) showed that the major peaks for an FCC crystal of AgNPs at 2θ value of 38, 46, 65, and 78° attributed to the lattice plane of (111), (200), (220), and (311), respectively. Rao and Golder (2016) reported a reduction in crystalline size due to the peak expansion from 54.1 nm to 21.7 nm for AgNPs at pH 3 and pH 12.5, respectively, during its formation. However, with pH variation (Table 13.2) the d-spacing did not change between the adjacent lattice planes. AgNPs were synthesized with a minor peak at 2θ = 32.2° at pH 3, indicating the existence of Ag_2O. Therefore, XRD analysis of AgNPs clearly showed the bi-crystalline structure for both samples synthesized at low as well as higher pH. AgNPs were synthesized from tea leaf extract and similar results have also been found (Sun et al. 2014). The average size of the CuNPs has been found to be 13.13 nm with a space group of pccn (56) and maximum peak intensity at 2θ value of 42.045 (Chaudhary et al. 2019). Chaudhary et al. (2019) also suggested the maximum intensity of the NiNPs of diffraction peak at 2θ = 12.50 with the size of 24.0 nm which indicates the presence of crystalline structure. The average grain size obtained from the XRD pattern of Au/TiO_2 NPs synthesized at 100 °C was calculated using the Scherrer's formula from the half-width of the anatase main (101) diffraction peak (Zhong et al. 2010). Sun et al. (2017) have

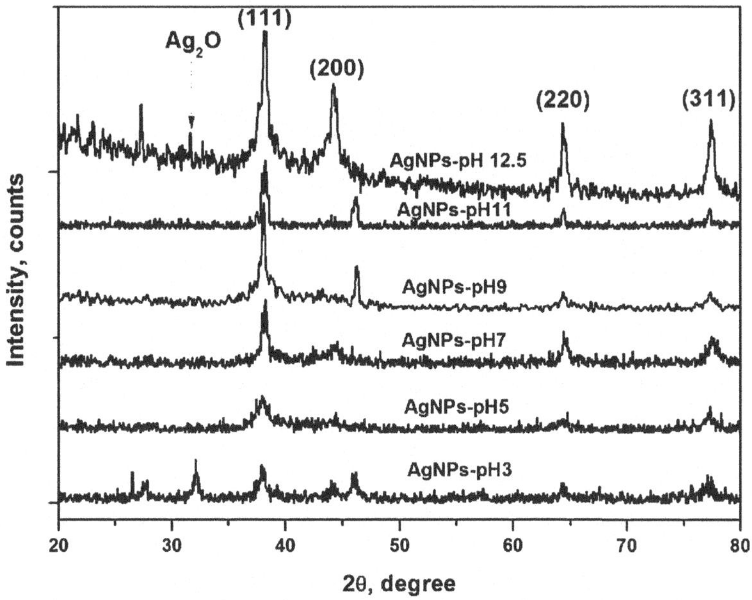

Fig. 13.8 XRD patterns of synthesized AgNPs (Rao et al. 2016)

estimated the Au/TiO_2 NP size as 14.2 nm and 12.6 nm for synthesizing at 180 °C and 100 °C, respectively. It is a sign of grain growth and the thinner shell appeared when synthesized at high temperature (Sun et al. 2017).

13.7 Mechanism of NP Formation Using Bio-extract

The synthesis of AgNPs using guava leaf extract has been described in this section as a representative mechanism of NP synthesis through bio-extract. The mechanism is as follows:

Chlorogenic acid derivative (dimethoxy cinnamoylquinic acid), D-glucose, and quercetin are the major constituents present in *P. guajava* leaves (Fierascu et al. 2017). AgNPs originated from Ag^+ which get reduced to Ag^0 forming silver nuclei, while the corresponding leaf components are oxidized to form CO_2 and H_2O as ultimate products along with other daughter ions as shown in Fig. 13.9. Glucose molecule with MW = 180 takes part in AgNP formation along with the other product P_4 (MW = 118.20). Then Ag^+ forms a complex (P_1) with MW 283.09 in the presence of deprotonated glucose molecule (β-D-glucopyranoside) under an alkaline medium (pH = 9.5) (Chakraborty et al. 2017) and oxidizes to acid (P_2, MW = 197.03). This is further converted to P_4 after decarboxylation reaction of P_2 followed by dihydroxylation of P_3 molecule (MW = 152.11). Hence, Ag^+ is coordinated with hydroxyl groups (-OH) (Baksi et al. 2015):

$$2Ag^+ + 2OH^- \rightleftarrows 2AgOH \leftrightarrow Ag_2O + H_2O \tag{13.1}$$

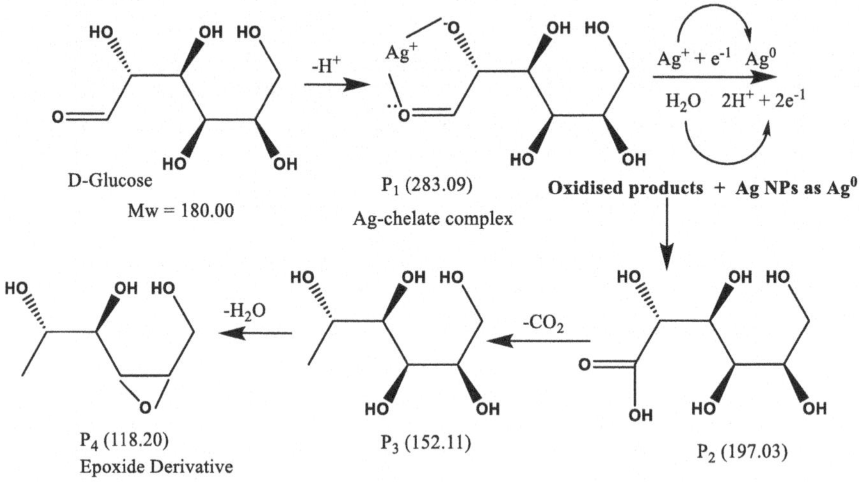

Fig. 13.9 Plausible mechanism of AgNP formation in the presence of D-glucose (present in guava leaf extract)

$$C_6H_8O_6 + Ag_2O \xrightarrow{OH^-} 2Ag^0 + C_6H_7O_6^{\bullet -} + H_2O \tag{13.2}$$

$$2C_6H_7O_6^{\bullet -} + Ag_2O + H^+ \rightarrow \text{Ascorbate anion} + \text{dehydro}\ \ AA + 2Ag^0 + H_2O \tag{13.3}$$

Bachmann et al. (2014) suggested that *S. edule* extract was used to produce dehydro-AA through a free radical intermediate originated from ascorbic acids (AA) at mild acidic and neutral pH in a reversible process (Eqs. 13.1–13.3). However, the higher pH shows an irreversible interconversion (Wechtersbach and Cigic 2007). Figure 13.8 shows a mechanistic route of interconversion of Ag^+ to Ag^0 for the formation of AgNPs. Glucose molecule as a sacrificial electron donor reduces Ag_2O (or Ag^+) and P_1 (m/z 283.09) followed by the formation of Ag-glucose chelate complex. After that, a fragment with P_2 (m/z 197.03) is formed on the cleavage of Ag-glucose complex, and simultaneously Ag^+ is reduced to Ag^0. The fragment P_3 with molar mass 152.11 g/mol is converted to an epoxide compound P_4 having m/z 118.20 through decarboxylation (-CO_2) reaction. This epoxide intermediate may be converted to oxalic acid, xylosone, and xylonic acid (Zhang et al. 2014).

13.8 Antimicrobial Activity Test of NPs

Among the various applications, NPs can also be used for the inhibition of bacteria. Saravanakumar et al. (2017) have employed various bio-inspired Ag-based NPs for their antibacterial activity including the NPs of Cu Pd (Surendra et al. 2016), Pt (Tahir et al. 2015), ZnO (Vijayakumar et al. 2018) (Lv et al. 2017), and Fe_2O_3 (Arokiyaraj et al. 2013). The antibacterial activity of biosynthesized AgNPs using *Ficus benghalensis* revealed that AgNPs with a concentration of 45 μg/mL showed the potential antibacterial activity with a minimum inhibitory concentration of 25 μg/mL (Tripathi and Chunga 2019; Saxena et al. 2012). The antibacterial activity of AgNPs has shown bacterium resistance to common antibiotics (Salomoni et al. 2017). Lv et al. (2017) have investigated the antibacterial activity of CuNPs and found that 100 μg/mL of CuNPs is required to have an effective antibacterial activity with the 10^5 CFU/mL *E. coli* suspension within 12 h (Lv et al. 2017). The biosynthesized PdNPs having the particle size of 27 nm with spherical shape exhibited a good antibacterial activity against *Staphylococcus aureus* and *E. coli* (Surendra et al. 2016). Tripathi et al. (2018) have investigated biogenic AgNPs to develop polyvinyl alcohol (PVA) used for food packaging materials and it also acts as an antibacterial biodegradable nanocomposite film (Tripathi et al. 2018). The bacterial growth inhibition experiment in the presence of AgNPs has been tested at pH 9 using two bacterial strains, i.e., *B. subtilis* and *E. coli*, and it has been observed that the rate of inhibition gradually increased with increasing particle concentration (Rao and Golder 2016). The potential toxicity may be altered by the different iron-based NPs due to their interactions with inorganic/organic contaminants. Iron-based

NPs (Fe_2O_3 NPs) due to their excellent adsorption properties may act as a carrier to transfer the target contaminants into cells (Hu et al. 2012). The toxicity of zerovalent iron NPs (Fe^0NPs) is a progressiveness of test organisms, function of their properties, and ecological conditions. They have been more considered in the field of water and wastewater remediation due to their adsorption-desorption efficiency. But at the same time, it is also becoming a source of NP contamination in the environment. NPs made by zerovalent iron reduce the concentrations of solvents nearly to zero within days, and at the same time, due to a significant change in oxygen levels, the pH level was also reduced (Fuhrer et al. 2000).

In additional information reported by Sadiq and Chandrasekaran (2010), TiO_2 nanoparticles with 20 nm diameter have been used for toxicity test to *E. coli, P. aeruginosa*, and *B. subtilis* through membrane damage under non-illuminated conditions. The production of reactive oxygen species (ROS) by the UV light irradiation and the existing rate of *E. coli* on ZnO, TiO_2, Al_2O_3, CeO_2, and Fe_2O_3NPs may have contributed to toxicity (Li et al. 2012). It has been observed that CuO microparticles have higher toxicity compared to CuONPs (Nutt et al. 2005). Likewise, the TiO_2NPs produced more DNA damage capacity compared to the ZnONPs (Wang and Zhang 1997). The toxicities of Ag, Ag, and Fe_3O_4NPs have been considered on plants and microorganisms (Elliott and Zhang 2001; Liu et al. 2005). Rao and Golder (2016) investigated the biological activity of both AgNPs and aqueous bio-extract and their results have been linked with the control media. An inactivation study of *B. subtilis* and *E. coli* bacteria has been chosen for the two antibiotics streptomycin and ampicillin, respectively. AgNPs exhibited the zone diameter of 22 mm with the lower inhibition of *B. subtilis,* while AgNPs showed to be most effective to suppress the growth of *E. coli* with 36 mm zone diameter of inhibition (Rao et al. 2017). This is explained by the extent of variation of permeability of AgNPs to the bacterial cell wall and the difference in the rate of growth inhibition for both gram-negative and gram-positive bacteria. A defensive peptidoglycan layer in outer membrane has been observed for the gram-negative bacteria, while gram-positive organisms do not have this (Li et al. 2009). Adegboyega et al. (2013) reported that AgNPs might cross the permeability barrier of the external membrane with free radical resulting in the seepage of cellular materials. Therefore, the disorder of membrane permeability could play a significant role in the inhibition of bacterial growth by AgNPs (Adegboyega et al. 2013). An antimicrobial activity of AgNPs for both types of bacterium growth is addressed in Fig. 13.10.

13.9 Overview

This chapter discussed the cost-effective and eco-friendly procedures for different NPs' synthesis by utilizing a natural resource, i.e., varieties of fruits and leaf extracts, which are excellent reducing agents, including the use of biological entities for the synthesis of metal and metal-doped nanoparticles discussed in detail. The potential of metal-metal oxide nanoparticles produced in different greener routes

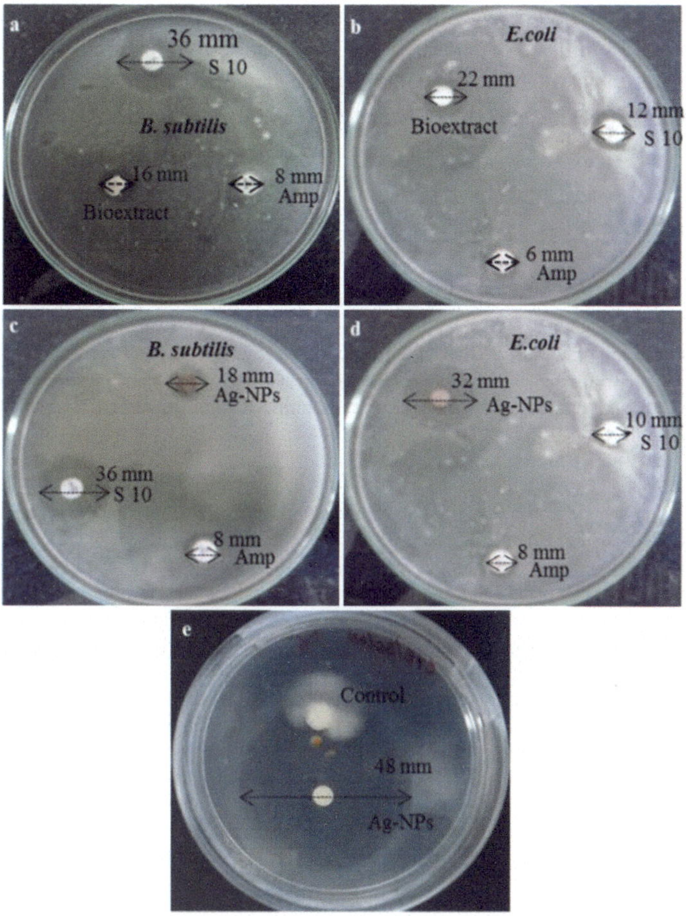

Fig. 13.10 Illustrative image showing AgNPs and their inhibiting growth of *B. subtilis* (**a**, **c**) and *E. coli* (**b**, **d**) in an antibiotic control media. (**e**) Antifungal activity of chayote extract against *Aspergillus thermomutatus* (Rao et al. 2017)

has been showing a significant interest in various areas like medicine, agriculture, and electronics. The biological things help to attain stable, cost-effective, nontoxic, and environment-friendly NPs through the green chemistry approach. The bio-inspired synthesis of NPs from various aqueous bio-extracts is strongly pH dependent, which may also play a role in determining the size, shape, stability, and purity of particles. At synthesized pH 3, 7, and 12.5, the corresponding diameters of NPs were found to be 68, 45, and 12.5, respectively. The higher pH facilitates the prevention of particle aggregation due to formation of a metal-ascorbate layer on NPs and the intense negative zeta potential. The NPs have a potential property in inhibiting the bacterium growth like *E. coli, B. subtilis,* and *P. aeruginosa*. In fact, NPs are synthesized through different bio-inspired methods for developing antibacterial biodegradable nanocomposite film for food packaging materials. The use of plant extracts for the synthesis of NPs is inexpensive and environmentally friendly, and can be scaled up easily. More interestingly, the plant extracts have the potential of producing NPs with the specific shape, size, or composition.

References

Adegboyega NF, Sharma VK, Siskova K, Zboril R, Sohn M, Schultz BJ, Banerjee S (2013) Interactions of aqueous Agl with fulvic acids: mechanisms of silver nanoparticle formation and investigation of stability. Environ Sci Technol 47:757–764

Ahmad A, Senapati S, Khan MI, Kumar R, Sastry M (2003) Extracellular biosynthesis of monodisperse gold nanoparticles by a novel extremophilic actinomycete, Thermomonospora sp. Langmuir 19(8):3550–3553

Ahmad N, Sharma S (2012) Green synthesis of silver nanoparticles using extracts of *Ananascomosus*. Gr Sustain Chem 2(4):141–147

Akhtar MS, Panwar J, Yun YS (2013) Biogenic synthesis of metallic nanoparticles by plant extracts. ACS Sustain Chem Eng 1(6):591–602

Al-Kahlout A (2015) Thermal treatment optimization of ZnO nanoparticles-photoelectrodes for high photovoltaic performance of dye-sensitized solar cells. J Assoc Arab Univ Basic Appl Sci 17:66–72

Al-Sehemi AG, Al-Shihri AS, Kalam A, Du G, Ahmad T (2014) Microwave synthesis, optical properties and surface area studies of NiO nanoparticles. J Mol Struct:56–61

Armendariz V, Herrera I, Peralta-Videa JR, Jose-Yacaman M, Troiani H, Santiago P, Gardea-Torresdey JL (2004) Size controlled gold nanopartilce formation by Avena sativa biomass: Use of plants in nanobiotechnology. J Nanopart Res 6(4):377–382

Arokiyaraj S, Saravanan M, Prakash NU, Arasu MV, Vijayakumar B, Vincent S (2013) Enhanced antibacterial activity of iron oxide magnetic nanoparticles treated with Argemone mexicana L. leaf extract: an in vitro study. Mater Res Bull 48:3323–3327

Asha Rani PV, Kah LG, Hande MP, Viliyaveettil S (2009) Cytotoxicity and Genotoxicity of silver nanoparticles in human cells. ACS Nano 3(2):279–290

Bachmann C, Probst B, Guttentag M, Alberto R (2014) Ascorbate as an electron relay between an irreversible electron donor and Ru(II) or re(I) photosensitizers. Chem Commun 50:6737–6739

Baker S, Harini BP, Rakshith D, Satish S (2013) Marine microbes: Invisible nanofactories. J Pharm Res 6(3):383–388

Baksi S, Zhenli l, Willie GH (2015) Natural nanoparticles: implications for environment and human. Health Environm Sci Technol 45:861–904

Barakat A, Al-Noaimi M, Suleiman M, Aldwayyan AS, Hammouti B, Hadda TB, Haddad SB, Boshaala A, Warad I (2013) One step synthesis of NiO nanoparticles via solid-state thermal decomposition at low-temperature of novel aqua(2,9-dimethyl-1,10 phenanthroline)$NiCl_2$ complex. Int J Mol Sci 14(12):23941–23954

Boulch F, Schouler MC, Donnadieu P, Chaix JM, Djurado E (2001) Domain size distribution of YTZP nanoparticles using XRD and HRTEM image. Anal Stereol 20:157–161

Brayner R, Ferrari-Iliou R, Brivois N, Djediat S, Benedetti M, Fiévet F (2006) Toxicological impact studies based on Escherichia coli bacteria in ultrafine ZnO nanoparticles colloidal medium. Nano Lett 6(4):866–870

Castro L, Blazquez ML, Munoz JA, Gonzalez F, Garcıa-Balboa C, Ballester A (2011) Biosynthesis of gold nanowires using sugar beet pulp. Process Biochem 46(5):1076–1082

Cha JN, Stucky GD, Morse DE, Deming TJ (2000) Biomimetic synthesis of ordered silica structures mediated by block copolypeptides. Nature 403:289–292

Chakraborty S, Rao CV, Das RK, Giri AS, Golder AK (2017) Bio-mediated silver nanoparticle: synthesis mechanism and microbial inactivation. Toxicol Environ Chem 99(3):434–447
Chaudhary J, Tailor G, Kumar D, Joshi A (2017) Synthesis and thermal properties of copper nanoparticles. Asian J Chem 29(7):1492–1494
Chaudharya J, Tailora G, Yadav BL, Michael O (2019) Synthesis and biological function of Nickel and Copper nanoparticles. Heliyon 5(6):e01878
Chen ZH, Jie JS, Luo LB, Wang H, Lee CS, Lee ST (2007) Applications of silicon nanowires functionalized with palladium nanoparticles in hydrogen sensors. Nanotechnol 18(34):345502
Daniel MC, Astruc D (2004) Gold nanoparticles: Assembly, supramolecular chemistry, quantum-size-related properties, and applications toward biology, catalysis, and nanotechnology. J Chem Rev 104:293–346
Das RK, Gogoi N, Bora U (2011) Green synthesis of gold nanoparticles using Nyctanthes arbor-tristis flower extract. Bioprocess Biosyst Eng 34:615–619
Das D, Nath BC, Phukon P, Dolui SK (2013) Synthesis and evaluation of antioxidant and antibacterial behavior of CuO nanoparticles. Colloids Surf B Biointerfaces 101:430–433
Das RK, Golder AK (2017) Co_3O_4 spinel nanoparticles decorated graphite electrode: Bio-mediated synthesis and electrochemical H_2O_2 sensing. Electrochimica Acta 251:415–426
Deparis O, Khuzayim N, Parker A, Vigneron JP (2009) Assessment of the antireflection property of moth wings by three-dimensional transfer-matrix optical simulations. Phys Rev E:79–87
Devi M, Devi S, Sharma V, Rana N, Bhatia RK, Bhatt AK (2019) Green synthesis of silver nanoparticles using methanolic fruit extract of Aegle marmelos and their antimicrobial potential against human. J Tradit Med Complement. https://doi.org/10.1016/j.jtcme.2019.04.007
Dubey SP, Lahtinen M, Sillanpaa M (2010) Tansy fruit mediated greener synthesis of silver and gold nanoparticles. Process Biochem 45(7):1065–1071
Duran N, Marcato PD, Alves OL, Souza GI, Esposito E (2005) Mechanistic aspects of biosynthesis of silver nanoparticles by several Fusarium oxysporum strains. J Nanobiotechnol 3:1–8
Dwivedi AD, Gopal K (2010) Biosynthesis of silver and gold nanoparticles using Chenopodium album leaf extract. Colloids Surf A 369(1–3):27–33
Edison JIETN, Sethuraman MG (2013) Electrocatalytic reduction of benzyl chloride by green synthesized silver nanoparticles using pod extract of Acacia nilotica. ACS Sustain Chem Eng 1:1326–1332
Elechiguerra JL, Reyes-Gasgab J, Yacaman MJ (2006) The role of twinning in shape evolution of anisotropic noble metal nanostructures. J Mater Chem 16:3906–3919
Elliott DW, Zhang W (2001) Field assessment of Nanoscale bimetallic particles for groundwater treatment. Environ Sci Technol 35:4922
Erjaee H, Rajaian H, Nazifi S (2017) Synthesis and characterization of novel silver nanoparticles using Chamaemelum nobile extract for antibacterial application. Adv Nat Sci Nanosci Nanotechnol 8:025004
Eustis S, Hsu HY, El-Sayed MA (2005) Gold nanoparticle formation from photochemical reduction of Au^{3+} by continuous excitation in colloidal solutions: A proposed molecular mechanism. J Phys Chem B 109(11):4811–4815
Fierascu I, Georgiev MI, Ortan A, Fierascu RC, Avramescu SM, Ionescu D, Sutan A, Brinzan A, Ditu LM (2017) Phyto-mediated metallic nano-architectures via Melissa officinalis L.: synthesis, characterization and biological properties. Sci Rep. https://doi.org/10.1038/s41598-017-12804-7
Fuhrer MS, Nygård J, Shih L, Forero M, Yoon YG, Mazzoni MSC, Choi HJ, Ihm J, Louie SG, Zettl A, McEuen PL (2000) Crossed nanotube junctions. Science 288:494
Gardea-Torresdey JL, Tiemann KJ, Gamez G, Dokken K, Tehuacamanero S, Jose-Yacaman M (1999) Gold nanoparticles obtained by bio-precipitation from gold (III) solutions. J Nanopart Res 1(3):397–404
Gericke M, Pinches A (2006) Biological synthesis of metal nanoparticles. Hydrometallurgy 83(1-4):132–140

Gopidas KR, Whitesell JK, Fox MA (2003) Synthesis, characterization and catalytic activity of a palladium nanoparticle cored dendrimier. Nano Lett 3:1757–1760

Heinlaan M, Ivask A, Blinova I, Dubourguier HC, Kakru A (2008) Toxicity of nanosized and bulk ZnO, CuO and TiO_2 to bacteria Vibrio fischeri and crustaceans Daphnia magna and Thamnocephalus platyurus. Chemosphere 71:1308–1316

Hrapovic S, Liu Y, Male KB, Luong JHT (2004) Electrochemical biosensing plate form using platinum nanoparticles and carbon nanotubes. Anal Chem 76:1083–1088

Huang J, Li Q, Sun D, Lu Y, Su Y, Yang X, Wang H, Wang Y, Shao W, He N (2007) Biosynthesis of silver and gold nanoparticles by novel sundried Cinnamomumcamphora leaf. Nanotechnol 18(10):105104:1–11

Hu XL, Wui S (2012) Synthesis of gold nanoparticles by solution plasma sputtering in various solvents. J Phys Conf Ser 417:012–030

Hwa KY, Subramani B (2014) Synthesis of zinc oxide nanoparticles on graphene-carbon nanotube hybrid for glucose biosensor applications. Biosens Bioelectron 62:127–133

Jiang RB, Li BX, Fang CH, Wang JF (2014) Metal/semiconductor hybrid nanostructures for plasmon-enhanced applications. Adv Mater 26:5274–5309

Kaviya S, Santhanalakshmi J, Viswanathan B, Muthumary J, Srinivasan K (2011) Biosynthesis of silver nanoparticles using citrus sinensis peel extract and its antibacterial activity. Spectrochim Acta A 79(3):594–598

Khan AA, Fox EK, Gorzny ML, Nikulina E, Brougham DF, Wege C, Bittner AM (2013) pH control of the electrostatic binding of gold and iron oxide nanoparticles to tobacco mosaic virus. Langmuir 29(7):2094–2098

Kuehner DE, Heyer C, Ramsch C, Fornefeld UM, Blanch HW, Prausnitz JM (1997) Interactions of lysozyme in concentrated electrolyte solutions from dynamic light-scattering measurements. Biophys J 73:3211–3224

Kulkarni N, Muddapur U (2014) Biosynthesis of metal nanoparticles: A review. J Nanotechnol 2014:1–8

Kumar A, Mandal S, Selvakannan PR, Parischa R, Mandale AB, Sastry M (2003) Investigation into the interaction between surface-bound alkylamines and gold nanoparticles. Langmuir 19:6277–6282

Lee HJ, Lee G, Jang NR, Yun JH, Song JY, Kim BS (2011) Biological synthesis of copper nanoparticles using plants extract. Nanotechnology 1:371–374

Lengke M, Ravel B, Fleet ME, Wanger G, Gordon RA, Southam G (2006) Mechanisms of gold bioaccumulation by filamentous cyanobacteria from gold (III) chloride complex. Environ Sci Technol 40(20):6304–6309

Li WR, Xie XB, Shi QS, Zeng HY, Yang O, Chen YB (2009) Antibacterial activity and mechanism of silver nanoparticles on Escherichia coli. Appl Microbiol Biotechnol 85:1115–1122

Li X, Xu H, Chen ZS, Chen G (2011) Biosynthesis of nanoparticles by microorganisms and their applications. J Nanomater 2011:1–16

Li Y, Zhang W, Niu J, Chen Y (2012) Mechanism of photogenerated reactive oxygen species and correlation with the antibacterial properties of engineered metal-oxide nanoparticles. ACS Nano 6:5164–5173

Liu Y, Choi H, Dionysiou D, Lowry GV (2005) Trichloroethene hydrodechlorination in water by highly disordered monometallic nanoiron. Chem Mater 17:5315

Lv Q, Zhang B, Xing X, Zhao Y, Cai R, Wang W, Gu Q (2017) Biosynthesis of copper nanoparticles using Shewanella loihica PV-4 with antibacterial activity: novel approach and mechanisms investigation. J Hazard Mater 347:141–149

Mafune F, Kohno J, Takeda Y, Kondow TJ (2001) Dissociation and aggregation of gold nanoparticles under laser irradiation. J Phys Chem B 105(38):9050–9056

Makarov VV, Love AJ, Sinitsyna OV, Makarova SS, Yaminsky IV, Taliansky ME, Kalinina NO (2014) Green nanotechnologies: Synthesis of metal nanoparticles using plants. Acta Naturae 6(1):35–44

Malik P, Shankar R, Malik V, Sharma N, Mukherjee TK (2014) Green chemistry based benign routes for nanoparticle synthesis. J Nanopart 2014:1–14

Malika M, Rao CV, Das RK, Giri AS, Golder AK (2016) Evaluation of bimetal doped TiO_2 in dye fragmentation and its comparison to mono-metal doped and bare catalyst. Appl Surf Sci 368:316–324

Mandal D, Bolander ME, Mukhopadhyay D, Sarkar G, Mukherjee P (2006) The use of microorganisms for the formation of metal nanoparticles and their application. Appl Microbiol Biotechnol 69(5):485–492

Mehrotra N, Tripathi RM, Zafar F, Singh MP (2017) Catalytic degradation of dichlorvos using biosynthesized zero valent iron nanoparticles. IEEE Transac Nano Biosci 16(4):280–286

Melissa A, Jones M, Gunsolus IL, Murphy CJ, Haynes CL (2013) Toxicity of engineered nanoparticles in the environment. Anal Chem 85:3036–3049

Mittal AK, Chisti Y, Banerjee UC (2013) Synthesis of metallic nanoparticles using plants. Biotechnol Adv 31:346–356

Mitra S, Chandra S, Laha D, Patra P, Debnath N, Pramanik A, Pramanik P, Goswami A (2012) Unique chemical grafting of carbon nanoparticle on fabricated ZnO nanorod: antibacterial and bioimaging property. Mater Res Bull 47:586–594

Mukherjee P, Ahmad A, Mandal D, Senapati S, Sainkar SR, Khan MI, Ramani R, Parischa R, Ajayakumar PV, Alam M (2001) Bioreduction of AuCl4 ions by the fungus, Verticillium sp. and surface trapping of the gold nanoparticles formed. Angew Chem Int Ed 40(19):3585–3588.

Narayanan KB, Sakthivel N (2008) Coriander leaf mediated biosynthesis of gold nanoparticles. Mater Lett 62:4588–4590

Naveen AN, Selladurai S (2015) Tailoring structural, optical and magnetic properties of spinel type cobalt oxide (Co_3O_4) by manganese doping. Physica B 457:251–262

Nigussie GY, Tesfamariam GM, Tegegne BM, Weldemichel YA, Gebreab TW, Gebrehiwot DG, Gebremichel GE (2018) Antibacterial activity of Ag-doped TiO_2 and Ag-doped ZnO nanoparticles. Int J Photoenergy. https://doi.org/10.1155/2018/5927485

Njagi EC, Huang H, Stafford L, Genuino H, Galindo HM, Collins JB (2011) Biosynthesis of iron and silver nanoparticles at room temperature using aqueous sorghum bran extracts. Langmuir 27:64–271

Nutt MO, Hughes JB, Wong MS (2005) Designing Pd-on-au bimetallic nanoparticle catalysts for Trichloroethene Hydrodechlorination. Environ Sci Technol 39:1346

Oluwafemi OS, Revaprasadu N, Adeyemi OO (2010) A facile green synthesis of ascorbic acid-capped ZnSe nanoparticles. Coll Surf B Biointer 79:126–130

Padil VVT, Cerník M (2013) Green synthesis of copper oxide nanoparticles using gum karaya as a biotemplate and their antibacterial application. Int J Nanomedicine 8:889–898

Perez J, Bax L, Escolano C (2005) Roadmap report on nanoparticles. Willems & van den Wildenberg, Barcelona, Spain

Poinern GEJ, Le X, Chapman P, Fawcett D (2013) Green biosynthesis of gold nanometre scale plate using the leaf extracts from an indigenous Australian plant Eucalyptus macrocarpa. Gold Bull 46(3):165–173

Pollini M, Paladini F, Sannino A, Picca RA, Sportelli MC, Nicola C, Nitti MA, Valentini M, Valentini A (2008) Individual, social, and environmental influences associated with HIV infection among injection drug users in Tijuana, Mexico. J Acquir Immune Defic Syndr 47(3):369–76

Prathna TC, Chandrasekaran N, Raichur AM, Mukherjee A (2011) Kinetic evolution studies of silver nanoparticles in a bio-based green synthesis process. Colloids Surf A 377(1–3):212–216

Puvanakrishnan P, Park J, Chatterjee D, Krishnan S, Tunnel JW (2012) In vivo tumor targeting of gold nanoparticles: Effect of particle type and dosing strategy. Int J Nanomed 7:1251–1258.

Rao CV (2018) Bio-inspired route of metal nanoparticles synthesis and photocatalysts doping. Ph.D. Dissertation. IIT Guwahati, Guwahati

Rao CV, Golder AK (2016) pH dependent size control, formation mechanism and antimicrobial functionality of bio-inspired AgNPs. RSC Adv 6(98):95483–95493

Rao CV, Golder AK (2016) pH dependent size control, formation mechanism and antimicrobial functionality of bio-inspired AgNPs. RSC Adv 6:95483–95493

Rao CV, Golder AK (2019) One pot green synthesis of Pt, Co and Pt@Co core-shell nanoparticles using Sechium edule. J Chem TechnolBiotechnol 94:911–918

Rao CV, Bag SS, Golder AK (2017) A biosynthesis route to nearly spherical AgNPs using chayote fruit extract. Environ Prog Sustain Energy 36(1):192–199

Rao CV, Chakraborty S, Golder AK (2018) Ag-doping on TiO_2 using plant-based glycosidic compounds for high photonic efficiency degradative oxidation under visible light. J Mol Liq 271:380–388

Roopan SM, Bharathi A, Prabhakarn A, Rahuman AA, Velayutham K, Rajakumar G, Padmaja RD, Lekshmi M, Madhumitha G (2012) Efficient phytosynthesis and structural characterization of rutile TiO2 nanoparticles using Annona squamosa peel extract. Spectrochim Acta A Mol Biomol Spectrosc 98:86–90

Sadiq IM, Chandrasekaran N (2010) Mukherjee, studies on effect of TiO_2 nanoparticles on growth and membrane permeability of Escherichia coli, Pseudomonas aeruginosa, and Bacillus subtilis. A Curr Nanosci 6:381–387

Salomoni R, Léo P, Montemor A, Rinaldi B, Rodrigues M (2017) Antibacterial effect of silver nanoparticles in Pseudomonas aeruginosa. Nanotechnol Sci Appl 10:115

Santhoshkumar T, Rahuman A, Jayaseelan C, Rajakumar G, Marimuthu S, Kirthi AV, Velayutham K, Thomas J, Venkatesan J, Kim SK (2014) Green synthesis of titanium dioxide nanoparticles using Psidium guajava extract and its antibacterial and antioxidant properties. Asian Pac J Trop Med 7:968–976

Saravanakumar A, Peng MM, Ganesh M, Jayaprakash J, Mohankumar M, Jang HT (2017) Low-cost and eco-friendly green synthesis of silver nanoparticles using Prunus japonica (Rosaceae) leaf extract and their antibacterial, antioxidant properties. Artif Cells Nanomed Biotechnol 45:1165–1171

Sathishkumar M, Sneha K, Yun YS (2009) Palladium nanocrystals synthesis using Curcuma longa tuber extract. Int J Mater Sci 4:11–17

Saxena A, Tripathi R, Zafar F, Singh P (2012) Green synthesis of silver nanoparticles using aqueous solution of Ficus benghalensis leaf extract and characterization of their antibacterial activity. Mater Lett 67:91–94

Sharma JK, Srivastava P, Singh G, Akhtar MS, Ameen S (2015) Green synthesis of Co_3O_4 nanoparticles and their applications in thermal decomposition of ammonium perchlorate and dye-sensitized solar cells. Mater Sci Eng B 193:181–188

Smitha SL, Nissamudeen KM, Philip D, Gopchandran KG (2008) Studies on surface plasmon resonance and photoluminescence of silver nanoparticles. Spectrochim Acta Part A Mol Biomol Spectrosc 71:186–190

Song JY, Jang HK, Kim BS (2009) Biological synthesis of gold nanoparticles using Magnolia kobus and Diopyros kaki leaf extracts. Process Biochem 44(10):1133–1138

Sperling RA, Gil PR, Zhang F, Zanella M, Parak WJ (2008) Biological applications of gold nanoparticles. Chem Soc Rev 37:1896–1908

Starowiicz M, Stypula B, Banas J (2006) Electrochemical synthesis of silver nanoparticles. Electrochem Commun 8:227–230

Stephen JR, Macnaughton SJ (1999) Developments in terrestrial bacterial remediation of metals. Curr Opin Biotechnol 10(3):230–233

Subhankari I, Nayak PL (2013) Synthesis of copper nanoparticles using Syzygium aromaticum (cloves) aqueous extract by using green chemistry. World J Nano Sci Technol 2:14–17

Sun Q, Cai X, Li J, Zheng M, Chen Z, Yu CP (2014) Green synthesis of silver nanoparticles using tea polyphenols and their catalytic activity in degradation of acid red B. Colloids Surfaces A Physicochem Eng Asp 444:226–231

Sun H, He Q, She P, Zeng S, Xu K, Li J, Liang S, Liu Z (2017) One-pot synthesis of Au/TiO_2 yolk-shell nanoparticles with enhanced photocatalytic activity under visible light. J Colloid Interf Sci 505:884–891

Surendra T, Roopan SM, Arasu MV, Al-Dhabi NA, Rayalu GM (2016) RSM optimized Moringa oleifera peel extract for green synthesis of M. oleifera capped palladium nanoparticles with antibacterial and hemolytic property. J Photochem Photobiol B Biol 162:550–557

Sykora D, Kasicka V, Miksik I, Rezanka P, Zaruba K, Matejka P, Kral V (2010) Applications of gold nanoparticles in separation sciences. J Sep Sci 33(3):372–387
Tahir K, Nazir S, Li B, Khan AU, Khan ZUH, Ahmad A, Khan QU (2015) Biodirected synthesis of palladium nanoparticles using Phoenix dactylifera leaves extract and their size dependent biomedical and catalytic applications. J Photochem Photobiol B Biol 153:261–266
Torres-Chavolla E, Ranasinghe RJ, Alocilja EC (2010) Characterization and functionalization of biogenic gold nanoparticles for biosensing enhancement. IEEE Trans Nanobiotechnol 9(5):533–538
Treguer M, Cointet C, Remita H, Khatouri J, Mostafavi M, Amblard J, Belloni JJ (1998) Dose rate effect on radiolytic synthesis of gold-silver bimetallic clusters in solution. J Phys Chem B 102(22):4310–4321
Tripathi RM, Chunga SJ (2019) Biogenic nanomaterials: synthesis, characterization, growth mechanism and biomedical applications. J Microbiol Methods 157:65–80
Tripathi R, Bhadwal AS, Gupta RK, Singh P, Shrivastav A, Shrivastav B (2014) ZnO nanoflowers: novel biogenic synthesis and enhanced photocatalytic activity. J Photochem Photobiol B Biol 141:288–295
Tripathi DK, Shivesh PA, Kumar SD, Nawal C, Dubey K (2018) Nanomaterials in Plants, Algae and Microorganisms: Concepts and Controversies 1:382–386
Tu WG, Zhou Y, Li HJ, Li P, Zou ZG (2015) Au/TiO_2 yolk-shell hollow spheres for plasmon-induced photocatalytic reduction of CO_2 to solar fuel via a local electromagnetic field. Nanoscale 7:14232–14236
Vijayakumar S, Krishnakumar P, Arulmozhi P, Mahadevan S, Parameaswari N (2018) Biosynthesis, characterization and antimicrobial activities of zinc oxide nanoparticles from leaf extract of Glycosmis pentaphylla (Retz.) DC. Microb Pathog 116:44–48
Wang CB, Zhang W (1997) Synthesizing Nanoscale Iron particles for rapid and complete dechlorination of TCE and PCBs. Environ Sci Technol 31:2154
Wang L, Chen X, Zhan J, Chai Y, Yang C, Xu L, Zhuang W, Jing B (2005) Synthesis of gold nano and microplates in hexagonal liquid crystals. J Phys Chem B 109:3189–3194
Wechtersbach L, Cigic B (2007) Reduction of dehydroascorbic acid at low pH. J. Biochem Biophys Methods 70:767–772
Yilmaz M, Turkdemir H, Kilic MA, Bayram E, Cicek A, Mete A, Ulug B (2011) Biosynthesis of silver nanoparticles using leaves of Stevia rebaudiana. Mater Chem Phys 130:1195–1202
Yude W, Shuo Z, Xinghui W, Qingju L (2006) Synthesis and optical properties of nano-ZnO particles/mesostructured SnO_2 composite. Mater Chem Phys 98:121–124
Zhang G, Wang DJ (2008) Fabrication of heterogeneous binary arrays of nanoparticles via colloidal lithography. J Am Chem Soc 130(17):5616–5617
Zhang F, Wu X, Chen Y, Lin H (2009) Application of silver nanoparticles to cotton fabric as an antibacterial textile finish. Fibers and Polymers 10(4):496–501
Zhang N, Liu SQ, Xu YJ (2012) Recent progress on metal core at semiconductor shell nanocomposites as a promising type of photocatalyst. Nanoscale 04:2227–2238
Zhang A, Liu M, Liu M, Xiao Y, Li Z, Chen J, Sun Y, Zhao J, Fang S, Jia D, Li JF (2014) Homogeneous Pd nanoparticles produced in direct reactions: green synthesis, formation mechanism and catalysis properties. Mater Chem A 2:1369–1374
Zhong J, Chen F, Zhang JL (2010) Carbon-deposited TiO_2: synthesis, characterization, and visible photocatalytic performance. J Phys Chem C 114:933–939
Zhou H, Ding L, Fan TX, Ding J, Zhang D, Guo QX (2014) Leaf-inspired hierarchical porous CdS/Au/N-TiO_2 heterostructures for visible light photocatalytic hydrogen evolution. Appl Catal B Environ 147:221–228
Zuas O, Hamim N, Sampora Y (2014) Bio-synthesis of silver nanoparticles using water extract of Myrmecodia pendan (Sarang Semut plant). Mater Lett 123:156–159

Chapter 14
Recent Advances in the Application of Greener Solvents for Extraction, Recovery and Dissolution of Precious Metals and Rare Earth Elements from Different Matrices

Philiswa N. Nomngongo, N. Raphael Biata, Masixole Sihlahla, Anele Mpupa, and Nomvano Mketo

Contents

P. N. Nomngongo (✉)
Department of Chemical Sciences, University of Johannesburg, Johannesburg, South Africa

DSI/Mintek Nanotechnology Innovation Centre, University of Johannesburg, Johannesburg, South Africa

DSI/NRF SARChI Chair: Nanotechnology for Water, University of Johannesburg, Johannesburg, South Africa
e-mail: pnnomngongo@uj.ac.za

N. R. Biata · M. Sihlahla · A. Mpupa
Department of Chemical Sciences, University of Johannesburg, Johannesburg, South Africa

N. Mketo
Department of Chemistry, College of Science and Engineering and Technology, Florida Science Campus, University of South Africa, Johannesburg, South Africa
e-mail: mketon@unisa.ac.za

Inamuddin, A. M. Asiri (eds.), *Nanotechnology-Based Industrial Applications of Ionic Liquids*, Nanotechnology in the Life Sciences,
https://doi.org/10.1007/978-3-030-44995-7_14

Abbreviations

Dca^-	Dicyanamide
DES	Deep eutectic solvents
ILs	Ionic liquids
IUPAC	International Union of Pure and Applied Chemistry
NdFeB	Neodymium magnet
REEs	Rare earth metals
SCN^-	Thiocyanate
Tf_2N^-	Bis(trifluoromethylsulphonyl)imide

14.1 Introduction

Recently, precious metals and rare earth metals (REEs) have been widely used in different industrial applications (Pavón et al. 2018; Perez et al. 2019; Serpe 2018; Wang et al. 2017b; Zhou et al. 2018). Their intensive application is due to the unique specific characteristics which include high chemical and thermal stability, malleability, ductility and non-corrosive nature (Pavón et al. 2018; Perez et al. 2019; Serpe 2018; Wang et al. 2017b; Zhou et al. 2018). The demand for these metals in various industrial applications is continuously rising globally (Geng et al. 2019; Su et al. 2018; Wang et al. 2017b; Wang et al. 2019; Xing and Lee 2019; Zhou et al. 2018). This can lead or is leading to challenges related to supply risks, economical significance and toxic influences. It is believed that waste from different industrial applications might contain recoverable amounts of precious metals and REEs. Therefore, there is a strong need to recover these elements from waste.

Various methodologies such as adsorption (Kumar et al. 2015; Roosen et al. 2016; Sharma and Rajesh 2016; Wei et al. 2016b), solvent extraction (Banda et al. 2019; Xie et al. 2014), chemical precipitation (Jorjani and Shahbazi 2016), ion exchange (Perez et al. 2019) and membrane separation (Kose Mutlu et al. 2018; Kubota et al. 2019) have been widely employed for the recovery of precious metals and REEs. Among the above-mentioned methods, solvent extraction has attracted a lot of attention; as a result it is one of the widely used extraction procedures (Banda et al. 2019; Entezari-Zarandi and Larachi 2019; Geng et al. 2019; Kubota et al. 2019; Perez et al. 2019; Swain and Mishra 2019; Xing and Lee 2019). The extensive use of solvent extraction is due to its remarkable properties such as high saturation capacity, selectivity and rapid extraction rate (Perez et al. 2019; Xie et al. 2014). Nonetheless, the key drawback of solvent extraction is the usage of large volumes of volatile organic solvents because they tend to cause secondary environmental pollution (Wang et al. 2017a). To overcome this drawback, a search for greener solvents has been reported in the literature (Makanyire et al. 2016; Wang et al. 2017a). Room-temperature ionic liquids have been considered as the alternative most promising greener solvents for recovery of REEs and precious metals (Makanyire et al. 2016; Wang et al. 2017a). This is due to their unique properties such as thermal and chemical stability, larger ionic conductivity, extensive electrochemical potential window, low flammability and insignificant vapor pressure (Boudesocque et al. 2019; Farzam and Feyzi 2019; Masilela and Ndlovu 2019; Van Roosendael et al. 2019; Wang et al. 2019). Ionic liquids (ILs) are mixtures made up of distinct cations and anions which are often in a liquid form at or below 100 °C (Singh et al. 2013). They display numerous attractive properties such as low vapour pressure, wide liquid range, adjustable viscosity, good extraction efficiency for several metal ions, selective dissolution, superior thermal stability and wide structural diversity (Martinis et al. 2014; Xu et al. 2015). Due to the aforementioned characteristics, ILs have been used in solvent extraction methods which results in high recoveries.

This chapter presents recent developments on the recovery of precious metals and REEs using ionic liquids. In this review the application of solvent extraction based on ILs and a brief introduction of newer division of ILs known as deep eutectic solvents as well as their application in solvent extraction/recovery of precious metals and REE process are covered.

14.2 Application of Ionic Liquids in Solvent Extraction of Precious Metals

Precious metals such as silver (Ag), gold (Au), rhodium (Rh), ruthenium (Ru), osmium (Os), iridium (Ir), palladium (Pd) and platinum (Pt) have excellent chemical and physical properties and are important for the production of advanced materials (Wang et al. 2019; Xing and Lee 2019). These materials are then used in different industries such as automobile, aviation, electronics and medical (Wang et al. 2019; Xing and Lee 2019). The demand for these metals has increased and natural resources are slowly decreasing. Therefore, there is a strong need to recover them from secondary resources such as industrial and mine wastes which in turn will address the issue of environmental pollution. Recently, researchers have focused on recovering precious metals from different secondary resources using ionic liquids. Boudesocque and co-workers reported the application of different ILs for extraction of Au(III), Pt(II) and Pt(IV) from aqueous effluents (Boudesocque et al. 2019). The ILs investigated are made up of cations such as tetrahexylammonium and tetraoctylammonium and anions such as bromide (Br^-), dicyanamide (Dca^-), thiocyanate (SCN^-) and bis(trifluoromethysulphonyl)imide (Tf_2N^-). The results obtained revealed that the ionic liquid systems used had a strong affinity towards the target precious metals except $Tf2N^-$-based ionic liquids which were specific to Au (Boudesocque et al. 2019). Farzam, and Feyzi, explored the use of 1-butyl-3-methylimidazolium hexafluorophosphate for recovery of gold from the aqueous hydrochloric acid medium (Farzam and Feyzi 2019). After system optimization, recovery greater than 99% was achieved. In a recent study by Rzelewska-Piekut and Regel-Rosocka, the recovery of precious metals such as palladium(II), rhodium(III), platinum(IV) and ruthenium(III) from synthetic mixtures derived from automobile catalytic converters was reported. The two-stage extraction, separation and recovery of precious metals, was achieved by the use of quaternary phosphonium salts (Rzelewska-Piekut and Regel-Rosocka 2019). Several researchers have proven that the unique properties of ion liquids are the best candidates for recovery of precious metals in different secondary resources (Cieszynska and Wieczorek 2018; Katsuta et al. 2016; Van Roosendael et al. 2019) (Table 14.1).

Table 14.1 Application of various ionic liquids for the recovery of precious metals from different secondary sources

Precious metals	Matrix	Ionic liquid type	Performance	Refs
Au	Metal mixtures	Aliquat-336	High-purity gold solutions 97–98.5	Wei et al. (2016b)
Pt and Pd	Acidic aqueous solution	1-Octyl-3-methylimdazolium bis(trifluoromethylsulphonyl)imide ([OMIM][NTf_2]) and trihexyltetradecylphosphonium bromide (Cyphos 102)	Extraction efficiency >99%	Papaiconomou et al. (2015)
Au	Acidic chloride media	Phosphonium-based ionic liquid (Cyphos IL 109)	Extraction efficiency of 99.4%	Nguyen et al. (2015)
Au, Pt, Pd	Aqueous solutions	Aliquat-336	High-purity metal solutions ranging from 95 to 99%	Wei et al. (2016a)
Au, Pd, Pt, Rh, and Ir	Synthetic HCl solution	Cyanex 272 and Aliquat 336	Recoveries up to 99.8% except for Pt (94%)	Xing and Lee (2019)
Ag and au	Chloride electronic waste leach solutions	[Bmim][Tf2N], [Bmim][PF6] and Cyphos 101	Extraction efficiency 95–97%	Masilela and Ndlovu (2019)
Pt	Chloride media	2,2-Diheptyl-1,1,3,3-tetramethylguanidinium bromide	Extraction efficiency and recovery were 99.5% and 99.0%	Wang et al. (2019)

14.3 Application of Solvent Extraction Based on Ionic Liquid Recovery of Rare Earth Elements

The International Union of Pure and Applied Chemistry (IUPAC) defines REEs as a group of 17 elements that have similar chemical and metallic properties (Wang et al. 2017a; Yang et al. 2015). These elements include lanthanides (La, Ce, Pr, Nd, Pm, Sm, Eu, Gd, Tb, Dy, Ho, Er, Tm, Y and Lu) as well as scandium and yttrium (Binnemans et al. 2013; Wang et al. 2017a; Yang et al. 2015). REEs have found their application in an ever-increasing variety of applications in modern technology such as permanent magnets, lamp phosphors, rechargeable batteries, catalysts, magnetic refrigeration, lasers and high-temperature superconductivity (Cardoso et al. 2019; Entezari-Zarandi and Larachi 2019; Swain and Mishra 2019; Xie et al. 2014). Due to the rapid industrial development, the amount of waste comprising REEs has significantly increased. Also, it has been reported that waste-containing REEs have a higher content of these elements as compared to natural resources such as ores (Liu et al. 2019). This then implies that these waste materials can become the secondary sources of REEs. In search of sustainable development of REE resources, their recovery from waste materials has gained significantly in recent years (Avdibegović et al. 2018; Cardoso et al. 2019; Devi and Sukla 2019; Entezari-Zarandi and Larachi 2019; Liu et al. 2019; Pavón et al. 2018; Su et al. 2018; Swain and Mishra 2019; Zhou et al. 2018). Among other extractants, different types of ionic liquids have been widely used for the recovery of REEs (Banda et al. 2019; Devi and Sukla 2019; Entezari-Zarandi and Larachi 2019; Makanyire et al. 2016; Su et al. 2018; Tian et al. 2010; Wang et al. 2017a) (Table 14.2).

For instance, Tunsu et al. performed the recovery of REEs from fluorescent lamp waste fractions using Cyanex 923 ionic liquid (Tunsu et al. 2016). The recovered REEs were converted to oxide after oxalic acid and thermal treatment and their purity was in the range of 99.96% REEs (94.61% yttrium and 5.09% europium) (Tunsu et al. 2016). Avdibegović and co-workers proposed the use of an innovative supported ionic liquid phase for separation and recovery of Sc, Y, Nd and Dy from bauxite residue (BR) (Avdibegović et al. 2018). Recovery of REEs from NdFeB magnets using carboxyl-functionalized ionic liquid with thermomorphic properties was proposed elsewhere (Dupont and Binnemans 2015). This developed system enabled the separation of Nd and Dy as well as other base elements such as Co from the iron. After suitable stripping and treatment process, high-purity REE oxides (99.9%) were recovered (Dupont and Binnemans 2015). The attractive feature of the proposed system was the regeneration of the ionic liquid which contributes towards the greener processes. Other applications of ionic liquid for recovery of REEs are summarised in Table 14.2.

Table 14.2 Application of various ionic liquids for the recovery of rare earth elements from different secondary sources

Precious metals	Matrix	Ionic liquid type	Performance	Refs
Pr, Nd, Gd and Dy	Simulated waste NdFeB feed solution	[Trihexyl(tetradecyl) phosphonium]2 [benzene-1,4-dioxydiacetate]	Recovery rate was up to 91% and the purity of the recovered REEs was 98.1 wt%	Su et al. (2018)
Y, Eu and Ce	Fluorescent lamp wastes	Primene 81R·Cyanex 572 and Primene 81R·D2EHPA	High purities >99.9 wt%	Pavón et al. (2018)
Sc, Y, Dy, Nd, Ce, La	Bauxite residue (red mud)	Betainium bis(trifluoromethylsulphonyl) imide (HbetTf2N)	70–85% REE recoveries	Davris et al. (2016)
Nd and Dy	Neodymium–iron–boron magnets (NdFeB)	Betainium bis(trifluoromethylsulphonyl) imide, [Hbet][Tf2N]	Recoveries of up to 99%; purity of >99.9 wt%,	Dupont and Binnemans (2015)
Sc, Y, Nd, Dy	Bauxite residue leachate	Betainium sulphonyl(trifluoromethanesulphonylimide) poly(styrene-*co*-divinylbenzene) [Hbet-STFSI-PS-DVB]	Recoveries ranged from 99 to 100%	Avdibegović et al. (2018)
Pr, Nd, Gd, Ho, Sm and Dy	Nd-Fe-B scrap	[Trihexyl(tetradecyl) phosphonium]2[4,4′-isopropylidenebis(phenoxyacetate)]	99.85% recovery rate	Zhou et al. (2018)
Sc, Y, La, Ce, Pr, Nd, Eu and Gd	Rödberg ore	Betainium bis(trifluoromethylsulphonyl) imide	65–100% recovery of light REE and 40–60% recovery of heavy REE	Davris et al. (2018)
Y	Chloride and nitrate solutions	[A336/Cy272] and Cyphos IL 104	Recoveries up to 82%	Devi and Sukla (2019)
Eu_2O_3 and Y_2O_3	Fluorescent lamp waste powder	Trihexyl(tetradecyl)phosphonium thiocyanate, [C101][SCN], and tricaprylmethylammonium thiocyanate, [A336][SCN]	Purity >98%	Banda et al. (2019)

14.4 Future Challenges in the Application of Ionic Liquid for Recovery of Precious Elements and Rare Earth Elements

Even though ionic liquids have been widely used in the solvent extraction of precious metals and REEs, some drawbacks include complex synthesis process, which is costly and difficult to purify (Xu et al. 2015). In addition, for ionic liquids to be appropriate for large-scale hydrometallurgy an extractant should have some of the following features: low cost, chemically stable, being environmentally friendly, being readily available in large quantities and having no restrictions (Jenkin et al. 2016). However, previous studies have shown that traditional ionic liquids do not meet these criteria (Jenkin et al. 2016). To overcome these challenges, Abbott and co-workers developed deep eutectic solvents as alternatives to traditional ionic liquids (Abbott et al. 2004). Deep eutectic solvents (DES) have similar physicochemical properties as ILs. However, they are considered different from ILs primarily because they do not totally consist of ionic species and they can be prepared from non-ionic species (Zhao et al. 2013). They vary from ionic liquids since they are classified to be ionic mixtures covering a mixture of organic halide salts (choline chloride) with hydrogen bond-donating compounds as Lewis acids (zinc chloride, carboxylic acids or urea, phenols, glycerol) (Abbott et al. 2007; Abbott et al. 2004; Kim et al. 2011; Singh et al. 2013). The capacity to form a hydrogen bond with the halide ion yields a eutectic combination (Gore et al. 2011). Deep eutectic solvents possess several attractive properties of ILs such as non-volatility and recyclability (Shahbaz et al. 2011). In addition, DES also has numerous advantages over usual ionic liquids such as (1) manageable preparation and storage; (2) no sensitivity towards moisture; (3) non-toxicity; (4) biodegradable nature; (5) cost-effectiveness due to cheap precursors; and (6) high possibility of industrial applicability (Abbott et al. 2007; Abbott et al. 2004; Shahbaz et al. 2012; Shahbaz et al. 2011; Shahbaz et al. 2010). Due to these outstanding advantages, DESs have been recently used in many research fields such as organic synthesis (Marset et al. 2019), dissolution media (Chen et al. 2019; Mamilla et al. 2019; Rodriguez Rodriguez et al. 2019), extraction processes (Bakkar and Neubert 2019; Entezari-Zarandi and Larachi 2019; Tran et al. 2019), electrochemistry (Vasilyev et al. 2019; Zhao et al. 2019) and material chemistry (Roda et al. 2019; Sánchez-Leija et al. 2019).

14.4.1 Application of Deep Eutectic Solvents for Recovery of Precious Elements and Rare Earth Elements

In a recent study by Entezari-Zarandi and Larachi (Entezari-Zarandi and Larachi 2019), DES was used as a greener solvent for dissolution of Y, La, Ce, Nd and Sm from carbonate salts, multicomponent mixtures and bastnäsite mineral. In their study, hydrogen donors such as urea, malonic acid and citric acid were evaluated in different proportions with choline chloride. The results obtained revealed that suitable DES could selectively dissolve higher-atomic-number rare earth elements at the expense of the lower ZREEs (Entezari-Zarandi and Larachi 2019). Riaño et al. (2017) reported the application of choline chloride-based DES alternative extractants to aqueous acid solutions for the recovery of Nd and Dy from NdFeB magnets. To separate other metals such as Fe, B and Co from Nd and Dy in the DEE, ionic liquid tricaprylmethylammonium thiocyanate (Aliquat 336 SCN, [A336][SCN]) diluted in toluene was employed. On the other hand, for the separation of Nd and

Dy, Cyanex 923 was used because it proved to be the most effective extractant. The purified Dy was recovered by elution using water while Nd present in DES was precipitated using oxalic acid (Riaño et al. 2017). Thepurity of the recovered oxides was 99.87% and 99.94% for Nd_2O_3 and Dy_2O_3, respectively.

14.5 Conclusions

Intending to address the escalating shortage of natural resources, recovery of rare earth elements and precious metals from secondary resources is crucial. Several methodologies have been used for this purpose and solvent extraction using greener solvents such as ionic liquid has been considered as one of the promising procedures. Ionic liquids have proven to be alternative greener solvents for the extraction separation and recovery of earth elements and precious metals. This is because of their unique properties such as tunable hydrophobicity and polarity, dissolution ability and non-volatility. However, the main challenge in using ionic liquids is the toxicity of some types of ionic liquids (e.g. imidazolium based) and production on a larger scale. These challenges have been solved by the introduction of DESs which are more environmentally friendly and biodegradable and can be produced in large quantities. Literature has proven that DES can serve as an alternative solvent for recovery of precious metals and REEs.

Acknowledgements This study was supported by the University of Johannesburg, South Africa (Department of Chemical Sciences), and National Research Foundation (grant nos. 99720; 91230), South Africa.

References

Abbott AP, Boothby D, Capper G, Davies DL, Rasheed RK (2004) Deep eutectic solvents formed between choline chloride and carboxylic acids: versatile alternatives to ionic liquids. J Am Chem Soc 126:9142–9147

Abbott AP, Barron JC, Ryder KS, Wilson D (2007) Eutectic-based ionic liquids with metal-containing anions and cations. Chem Eur J 13:6495–6501

Avdibegović D, Regadío M, Binnemans K (2018) Efficient separation of rare earths recovered by a supported ionic liquid from bauxite residue leachate. RSC Adv 8:11886–11893

Bakkar A, Neubert V (2019) Recycling of cupola furnace dust: extraction and electrodeposition of zinc in deep eutectic solvents. J Alloys Compd 771:424–432

Banda R, Forte F, Onghena B, Binnemans K (2019) Yttrium and europium separation by solvent extraction with undiluted thiocyanate ionic liquids. RSC Adv 9:4876–4883

Binnemans K, Jones PT, Blanpain B, Van Gerven T, Yang Y, Walton A, Buchert M (2013) Recycling of rare earths: a critical review. J Clean Prod 51:1–22. https://doi.org/10.1016/j.jclepro.2012.12.037

Boudesocque S, Mohamadou A, Conreux A, Marin B, Dupont L (2019) The recovery and selective extraction of gold and platinum by novel ionic liquids. Sep Purif Technol 210:824–834. https://doi.org/10.1016/j.seppur.2018.09.002

Cardoso CED, Almeida JC, Lopes CB, Trindade T, Vale C, Pereira E (2019) Recovery of rare earth elements by carbon-based. Nanomaterials 9:814
Chen Y-L, Zhang X, You T-T, Xu F (2019) Deep eutectic solvents (DESs) for cellulose dissolution: a mini-review. Cellulose 26:205–213
Cieszynska A, Wieczorek D (2018) Extraction and separation of palladium(II), platinum(IV), gold(III) and rhodium(III) using piperidine-based extractants. Hydrometallurgy 175:359–366. https://doi.org/10.1016/j.hydromet.2017.12.019
Davris P, Balomenos E, Panias D, Paspaliaris I (2016) Selective leaching of rare earth elements from bauxite residue (red mud), using a functionalized hydrophobic ionic liquid. Hydrometallurgy 164:125–135
Davris P, Marinos D, Balomenos E, Alexandri A, Gregou M, Panias D, Paspaliaris I (2018) Leaching of rare earth elements from 'Rödberg' ore of fen carbonatite complex deposit, using the ionic liquid HbetTf2N. Hydrometallurgy 175:20–27. https://doi.org/10.1016/j.hydromet.2017.10.031
Devi N, Sukla LB (2019) Studies on liquid-liquid extraction of yttrium and separation from other rare earth elements using bifunctional ionic liquids. Miner Process Extr Metall Rev 40:46–55
Dupont D, Binnemans K (2015) Recycling of rare earths from NdFeB magnets using a combined leaching/extraction system based on the acidity and thermomorphism of the ionic liquid [Hbet][Tf 2 N]. Green Chem 17:2150–2163
Entezari-Zarandi A, Larachi F (2019) Selective dissolution of rare-earth element carbonates in deep eutectic solvents. J Rare Earths 37:528–533. https://doi.org/10.1016/j.jre.2018.07.015
Farzam S, Feyzi F (2019) Response surface methodology applied to extraction optimization of gold(III) by combination of imidazolium-based ionic liquid and 1-octanol from hydrochloric acid. Sep Sci Technol:1–13. https://doi.org/10.1080/01496395.2019.1581217
Geng Y, Xiang Z, Lv C, Wang N, Wang Y, Yang Y (2019) Recovery of gold from hydrochloric medium by deep eutectic solvents based on quaternary ammonium salts. Hydrometallurgy
Gore S, Baskaran S, Koenig B (2011) Efficient synthesis of 3,4-dihydropyrimidin-2-ones in low melting tartaric acid–urea mixtures. Green Chem 13:1009–1013
Jenkin GRT et al (2016) The application of deep eutectic solvent ionic liquids for environmentally-friendly dissolution and recovery of precious metals. Miner Eng 87:18–24. https://doi.org/10.1016/j.mineng.2015.09.026
Jorjani E, Shahbazi M (2016) The production of rare earth elements group via tributyl phosphate extraction and precipitation stripping using oxalic acid. Arab J Chem 9:S1532–S1539. https://doi.org/10.1016/j.arabjc.2012.04.002
Katsuta S, Watanabe Y, Araki Y, Kudo Y (2016) Extraction of gold(III) from hydrochloric acid into various ionic liquids: relationship between extraction efficiency and aqueous solubility of ionic liquids. ACS Sustain Chem Eng 4:564–571. https://doi.org/10.1021/acssuschemeng.5b00976
Kim S-H, Yang S-T, Kim J, Ahn W-S (2011) Sonochemical synthesis of Cu3 (BTC) 2 in a deep eutectic mixture of choline chloride/dimethylurea. Bull Kor Chem Soc 32:2783–2786
Kose Mutlu B, Cantoni B, Turolla A, Antonelli M, Hsu-Kim H, Wiesner MR (2018) Application of nanofiltration for rare earth elements recovery from coal fly ash leachate: performance and cost evaluation. Chem Eng J 349:309–317. https://doi.org/10.1016/j.cej.2018.05.080
Kubota F, Kono R, Yoshida W, Sharaf M, Kolev SD, Goto M (2019) Recovery of gold ions from discarded mobile phone leachate by solvent extraction and polymer inclusion membrane (PIM) based separation using an amic acid extractant. Sep Purif Technol 214:156–161. https://doi.org/10.1016/j.seppur.2018.04.031
Kumar ASK, Sharma S, Reddy RS, Barathi M, Rajesh N (2015) Comprehending the interaction between chitosan and ionic liquid for the adsorption of palladium. Int J Biol Macromol 72:633–639. https://doi.org/10.1016/j.ijbiomac.2014.09.002
Liu B, Zhu N, Li Y, Wu P, Dang Z, Ke Y (2019) Efficient recovery of rare earth elements from discarded NdFeB magnets. Process Saf Environ Prot 124:317–325. https://doi.org/10.1016/j.psep.2019.01.026
Makanyire T, Sanchez-Segado S, Jha A (2016) Separation and recovery of critical metal ions using ionic liquids. Adv Manufact 4:33–46. https://doi.org/10.1007/s40436-015-0132-3

Mamilla JL, Novak U, Grilc M, Likozar B (2019) Natural deep eutectic solvents (DES) for fractionation of waste lignocellulosic biomass and its cascade conversion to value-added bio-based chemicals. Biomass Bioenergy 120:417–425

Marset X, Torregrosa-Crespo J, Martínez-Espinosa RM, Guillena G, Ramón DJ (2019) Multicomponent synthesis of sulfonamides from triarylbismuthines, nitro compounds and sodium metabisulfite in deep eutectic solvents green chemistry

Masilela M, Ndlovu S (2019) Extraction of Ag and au from chloride electronic waste leach solutions using ionic liquids. J Environ Chem Eng 7:102810. https://doi.org/10.1016/j.jece.2018.11.054

Nguyen VT, Lee J-c, Jeong J, Kim B-S, Gr C, Chagnes A (2015) Extraction of gold (III) from acidic chloride media using phosphonium-based ionic liquid as an anion exchanger. Ind Eng Chem Res 54:1350–1358

Papaiconomou N, Svecova L, Bonnaud C, Cathelin L, Billard I, Chainet E (2015) Possibilities and limitations in separating Pt (IV) from Pd (II) combining imidazolium and phosphonium ionic liquids. Dalton Trans 44:20131–20138

Pavón S, Fortuny A, Coll MT, Sastre AM (2018) Rare earths separation from fluorescent lamp wastes using ionic liquids as extractant agents. Waste Manag 82:241–248. https://doi.org/10.1016/j.wasman.2018.10.027

Perez JPH, Folens K, Leus K, Vanhaecke F, Van Der Voort P, Du Laing G (2019) Progress in hydrometallurgical technologies to recover critical raw materials and precious metals from low-concentrated streams. Resour Conserv Recycl 142:177–188. https://doi.org/10.1016/j.resconrec.2018.11.029

Riaño S et al (2017) Separation of rare earths and other valuable metals from deep-eutectic solvents: a new alternative for the recycling of used NdFeB magnets. RSC Adv 7:32100–32113

Roda A, Matias AA, Paiva A, Duarte ARC (2019) Polymer science and engineering using deep eutectic solvents. Polymers 11:912

Rodriguez Rodriguez N, Machiels L, Binnemans K (2019) p-Toluenesulfonic acid-based deep-eutectic solvents for solubilizing metal oxides. ACS Sustain Chem Eng 7:3940–3948

Roosen J, Van Roosendael S, Borra CR, Van Gerven T, Mullens S, Binnemans K (2016) Recovery of scandium from leachates of Greek bauxite residue by adsorption on functionalized chitosan–silica hybrid materials. Green Chem 18:2005–2013. https://doi.org/10.1039/c5gc02225h

Rzelewska-Piekut M, Regel-Rosocka M (2019) Separation of Pt(IV), Pd(II), Ru(III) and Rh(III) from model chloride solutions by liquid-liquid extraction with phosphonium ionic liquids. Sep Purif Technol 212:791–801. https://doi.org/10.1016/j.seppur.2018.11.091

Sánchez-Leija R et al (2019) Deep eutectic solvents as active media for the preparation of highly conducting 3D free-standing PANI xerogels and their derived N-doped and N, P-codoped porous carbons. Carbon 146:813–826

Serpe A (2018) 11 - green chemistry for precious metals recovery from WEEE** This chapter is dedicated to Prof. Paola Deplano on the occasion of her retirement. In: Vegliò F, Birloaga I (eds) Waste electrical and electronic equipment recycling. Woodhead Publishing, Sawston, pp 271–332. https://doi.org/10.1016/B978-0-08-102057-9.00011-1

Shahbaz K, Mjalli FS, Hashim M, Al Nashef IM (2010) Using deep eutectic solvents for the removal of glycerol from palm oil-based biodiesel. J Appl Sci 10:3349–3354

Shahbaz K, Mjalli F, Hashim M, AlNashef I (2011) Using deep eutectic solvents based on methyl triphenyl phosphunium bromide for the removal of glycerol from palm-oil-based biodiesel. Energy Fuel 25:2671–2678

Shahbaz K, Baroutian S, Mjalli F, Hashim M, AlNashef I (2012) Densities of ammonium and phosphonium based deep eutectic solvents: prediction using artificial intelligence and group contribution techniques. Thermochim Acta 527:59–66

Sharma S, Rajesh N (2016) Synergistic influence of graphene oxide and tetraoctylammonium bromide (frozen ionic liquid) for the enhanced adsorption and recovery of palladium from an industrial catalyst. J Environ Chem Eng 4:4287–4298. https://doi.org/10.1016/j.jece.2016.09.028

Singh BS, Lobo HR, Pinjari DV, Jarag KJ, Pandit AB, Shankarling GS (2013) Ultrasound and deep eutectic solvent (DES): a novel blend of techniques for rapid and energy efficient synthesis of oxazoles. Ultrason Sonochem 20:287–293

Su X, Guo X, Zhao Z, Dong Y, Wang Y, Li F, Sun X (2018) An efficient and sustainable [P6,6,6,14]2[BDOAC] ionic liquid based extraction–precipitation strategy for rare earth recovery. Chem Eng Res Des 136:786–794. https://doi.org/10.1016/j.cherd.2018.06.029

Swain N, Mishra S (2019) A review on the recovery and separation of rare earths and transition metals from secondary resources. J Clean Prod 220:884–898. https://doi.org/10.1016/j.jclepro.2019.02.094

Tian GC, Li J, Y-x H (2010) Application of ionic liquids in hydrometallurgy of nonferrous metals. Trans Nonferrous Metals Soc China 20:513–520. https://doi.org/10.1016/S1003-6326(09)60171-0

Tran MK, Rodrigues M-TF, Kato K, Babu G, Ajayan PM (2019) Deep eutectic solvents for cathode recycling of Li-ion batteries. Nat Energy 4:339

Tunsu C, Petranikova M, Ekberg C, Retegan T (2016) A hydrometallurgical process for the recovery of rare earth elements from fluorescent lamp waste fractions. Sep Purif Technol 161:172–186. https://doi.org/10.1016/j.seppur.2016.01.048

Van Roosendael S, Regadío M, Roosen J, Binnemans K (2019) Selective recovery of indium from iron-rich solutions using an Aliquat 336 iodide supported ionic liquid phase (SILP). Sep Purif Technol 212:843–853. https://doi.org/10.1016/j.seppur.2018.11.092

Vasilyev DV, Rudnev AV, Broekmann P, Dyson PJ (2019) A general and facile approach for the electrochemical reduction of carbon dioxide inspired by deep eutectic solvents. ChemSusChem 12:1635–1639

Wang K et al (2017a) Recovery of rare earth elements with ionic liquids. Green Chem 19:4469–4493. https://doi.org/10.1039/c7gc02141k

Wang M, Tan Q, Chiang JF, Li J (2017b) Recovery of rare and precious metals from urban mines—a review. Front Environ Sci Eng 11:1. https://doi.org/10.1007/s11783-017-0963-1

Wang N, Wang Q, Lu W, Ru M, Yang Y (2019) Extraction and stripping of platinum (IV) from acidic chloride media using guanidinium ionic liquid. J Mol Liq:111040. https://doi.org/10.1016/j.molliq.2019.111040

Wei W, Cho C-W, Kim S, Song M-H, Bediako JK, Yun Y-S (2016a) Selective recovery of au(III), Pt(IV), and Pd(II) from aqueous solutions by liquid–liquid extraction using ionic liquid Aliquat-336. J Mol Liq 216:18–24. https://doi.org/10.1016/j.molliq.2016.01.016

Wei W, Reddy DHK, Bediako JK, Yun Y-S (2016b) Aliquat-336-impregnated alginate capsule as a green sorbent for selective recovery of gold from metal mixtures. Chem Eng J 289:413–422. https://doi.org/10.1016/j.cej.2015.12.104

Xie F, Zhang TA, Dreisinger D, Doyle F (2014) A critical review on solvent extraction of rare earths from aqueous solutions. Miner Eng 56:10–28. https://doi.org/10.1016/j.mineng.2013.10.021

Xing WD, Lee MS (2019) A process for the separation of noble metals from HCl liquor containing gold (III), palladium (II), platinum (IV), rhodium (III), and iridium (IV) by solvent extraction. PRO 7:243

Xu K, Wang Y, Huang Y, Li N, Wen Q (2015) A green deep eutectic solvent-based aqueous two-phase system for protein extracting. Anal Chim Acta 864:9–20

Yang D, Ma P, Hou Z, Cheng Z, Li C, Lin J (2015) Current advances in lanthanide ion (Ln3+)-based upconversion nanomaterials for drug delivery. Chem Soc Rev 44:1416–1448. https://doi.org/10.1039/c4cs00155a

Zhao C, Ren J, Qu X (2013) G-quadruplexes form ultrastable parallel structures in deep eutectic solvent. Langmuir 29:1183–1191

Zhao H, Xu J, Sheng Q, Zheng J, Cao W, Yue T (2019) NiCo2O4 nanorods decorated MoS2 nanosheets synthesized from deep eutectic solvents and their application for electrochemical sensing of glucose in red wine and honey. J Electrochem Soc 166:H404–H411

Zhou H, Wang Y, Guo X, Dong Y, Su X, Sun X (2018) The recovery of rare earth by a novel extraction and precipitation strategy using functional ionic liquids. J Mol Liq 254:414–420. https://doi.org/10.1016/j.molliq.2018.01.078

Chapter 15
Applications of Ionic Liquids in Chemical Reactions

Venkata Durga Nageswar Yadavalli and Jayathirtha Rao Vaidya

Contents

V. D. N. Yadavalli (✉)
Former Chief Scientist, Medicinal Chemistry and Pharmacology Division, CSIR - Indian Institute of Chemical Technology, Hyderabad, India

J. R. Vaidya
Hetero Research Foundation, Plot No B80-81, Sanath Nagar - Balanagar, TSIE, Hyderabad, India

Inamuddin, A. M. Asiri (eds.), *Nanotechnology-Based Industrial Applications of Ionic Liquids*, Nanotechnology in the Life Sciences,
https://doi.org/10.1007/978-3-030-44995-7_15

15.1 Introduction

Chemical research has brought revolution in many areas particularly in medical and allied fields. Heterocyclic compounds have a significant role in both biological systems and our daily life. Being a major class of compounds, they also constitute prominently in the formation of natural products. With a wide range of applications these provided a greater platform for the rapid and multifold growth in chemical research contributing for the betterment of human life.

However, as a second side of the coin the ill effects of chemistry pronouncedly destroyed the serene qualities of our mother earth by enormously polluting soil, water, and air, with heavy loads of lethal contents. By the middle of the twentieth century realizing the bleak future scientists across the globe with the active support of government agencies conceptualized and propagated green chemistry—an environmentally benign chemical synthesis. It revolutionized the design, experimentation, manufacturing, and disposal processes and techniques in chemical research both at bench and scale-up levels. These sustainable chemical processes attempt to generate comparatively lesser hazardous materials and more eco-friendly sources of energy production.

In any chemical reaction the target products are produced when reactants are made to react in the presence of reagents if any under proper conditions such as solvent, temperature, and pressure. Laboratory as well as industrial procedures involve extraction, evaporation, and spillages of organic solvents, leading to environmental pollution. Due to the serious hazards caused by the use of a wide variety of volatile, nonvolatile, toxic, inflammable, nonpolar, as well as polar aprotic solvents, design and development of eco-friendly, sustainable reaction processes involving nonconventional alternate solvent media are increasingly being practiced by researchers across the world.

In view of their remarkable properties like high solubility of a wide range of organic and inorganic materials, thermal stability, nonflammability, nonvolatility, tunability, and recyclability ionic liquids have emerged as green solvents and are widely used as favorable alternate media for chemical processes. Broadly ionic liquids can be divided into two classes: protic ILs and aprotic ILs. Protic ILs are widely used as Brønsted acid or base in many acid-base-catalyzed organic reactions. Ionic liquids can also be generally categorized into four types based on their cationic species: (1) alkylammonium-, (2) dialkylimidazolium-, (3) phosphonium- and (4) *N*-alkylpyridinium-based ionic liquids. As ionic liquids are good microwave absorbents, they are also greatly applied as catalysts or media in many microwave-assisted reactions. Moreover, the discovery of task-specific ionic liquids has significantly contributed for the growth in the utility and importance of ionic liquids in various fields (Fig. 15.1).

With this short background this review emphasizes on the recent applications of ionic liquids in the synthesis of different heterocyclic systems by various enthusiastic scientists.

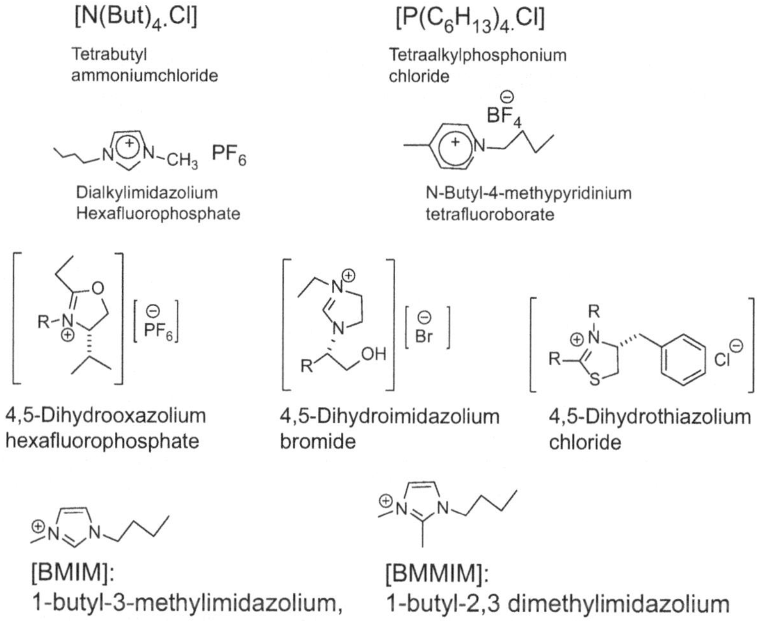

Fig. 15.1 Examples of certain ionic liquids

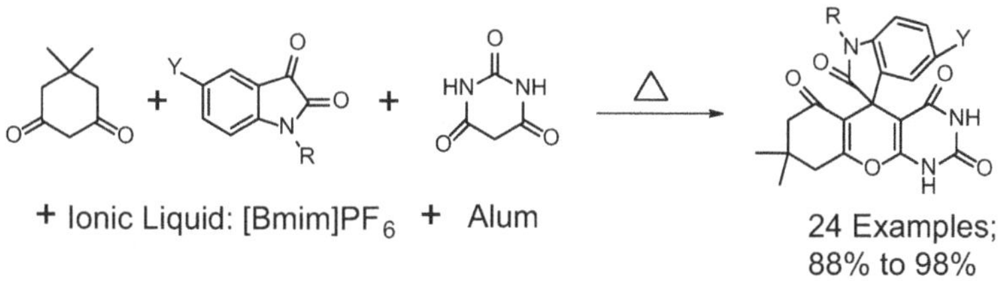

Fig. 15.2 Synthesis of spiro[chromeno[2,3-*d*]pyrimidine-5,3′-indoline]-tetraones

15.2 Examples of Different Heterocyclic Systems Synthesized by the Application of Ionic Liquids

15.2.1 Pyrroles, Indoles, and Fused Systems

Alum (catalyst) and ionic liquid [Bmim]PF_6medium were efficiently utilized by Mirhosseini Moghaddam and Mojtaba et al. to achieve one-pot three-component reaction leading to spiro[chromeno[2,3-*d*]pyrimidine-5,3′-indoline]-tetraones (Fig. 15.2) (Moghaddam et al. 2012). Synthesized spiro[chromeno[2,3-*d*]pyrimidine-5,3′-indoline]-tetraones were found to be important skeletal moieties for many pharmaceuticals. Dimedone, substituted isatin, barbituric acid, [Bmim]PF_6, and alum were involved in a reaction to provide the spiro-heterocyclic derivatives in excellent yields. A reasonably possible mechanism for the formation of complex spiro[chromeno[2,3-*d*]pyrimidine-5,3′-indoline]-tetraone structures was proposed.

15.2.2 Pyrazoles and Benzopyrazoles/Indazoles and Fused Systems

An efficient regioselective scale-up synthesis of pyrazoles by the reaction of hydrazines or hydrazides with various 1,3-diketones under mild conditions was reported by Shirin Safaei et al. employing a reusable multi-SO_3H Brønsted acidic room-temperature ionic liquid as catalyst in water medium (Fig. 15.3) (Safaei et al. 2012). Authors compared catalytic activity of different Brønsted and Lewis acidic ionic liquids for the above reaction under optimal conditions.

Ravi Balaskar et al. described the synthesis of dihydropyrano-pyrazole derivatives (Fig. 15.4) (Balaskar et al. 2010) using aromatic aldehydes, malononitrile, and 3-methyl-1-phenyl-2-pyrazoline-5-one in a one-pot three-component reaction at room temperature. Triethylammonium acetate ionic liquid was employed as medium as well as catalyst.

Narsidas Parmar and co-workers developed a highly efficient rapid one-pot eco-friendly protocol for a three-component Domino intermolecular Knoevenagel intermolecular hetero-Diels-Alder reaction providing indolyl and quinolyl pyrano[2,3-c]pyrazoles in high yields from pyrazalone, enolether, and heteroaldehyde in the presence of reusable ionic liquid—triethylammonium acetate (TEAA) under microwave conditions. During investigations, authors examined the role of different solvents and catalysts in the progress of the reaction (Fig. 15.5) (Parmar et al. 2013).

Several N-(phenylimino)indazole-1-carbothioamides were synthesized via an efficient one-pot three-component condensation of a wide range of aldehydes, dithizone, and dimedone, by Shirin Safaei et al., catalyzed by [bPy][$FeCl_4$] under

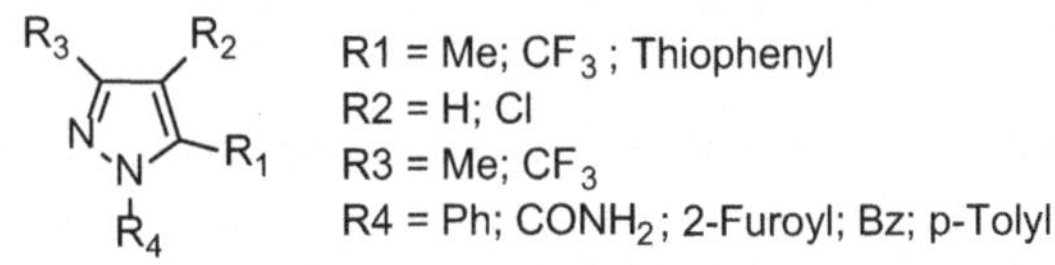

Fig. 15.3 Pyrazole derivatives

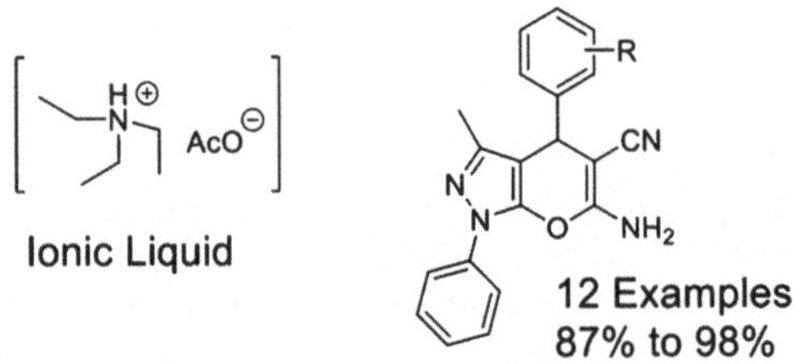

Fig. 15.4 Dihydropyrano-pyrazole derivatives

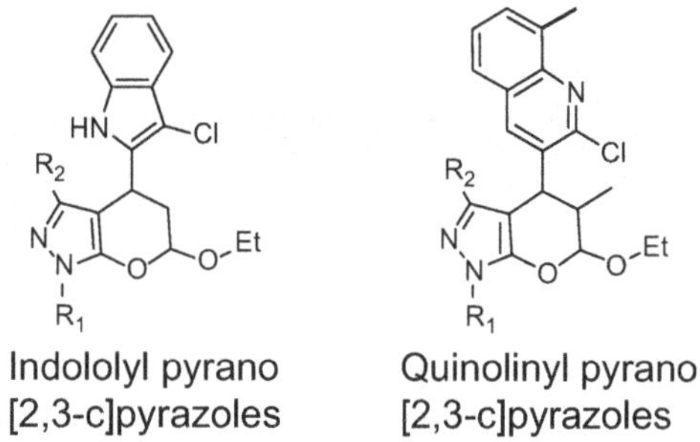

Fig. 15.5 Representative Scaffolds

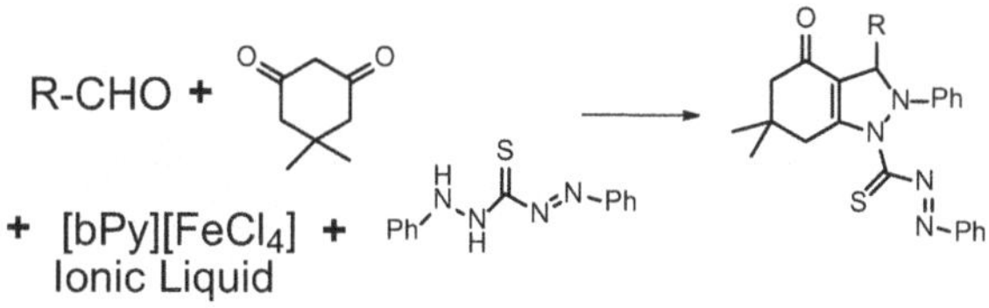

Fig. 15.6 Synthesis of N-(phenylimino) indazole-1-carbothioamides

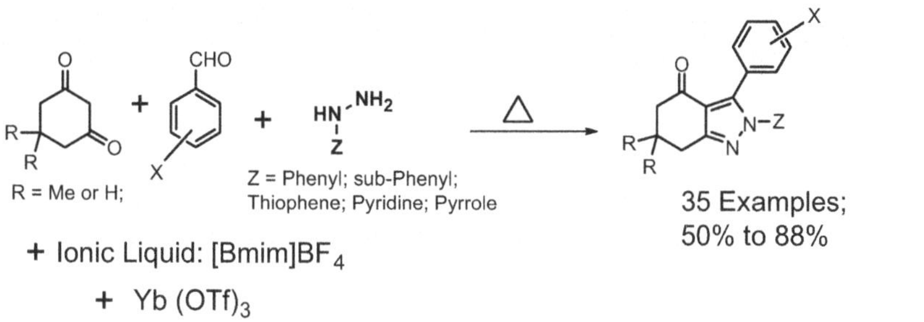

Fig. 15.7 Synthesis of tetrahydroindazolones

solvent-free conditions (Fig. 15.6) (Safaei et al. 2014). Authors examined various Lewis acid and Brønsted acid catalysts for their efficacy and it was observed that [byPy][$FeCl_4$] exhibited high catalytic activity.

Anil Kumar and co-workers developed a one-pot multicomponent reaction leading to tetrahydro-indazolones (Fig. 15.7) (Kameshwara Rao et al. 2012), with an objective to study the antiproliferative activity of synthesized compounds. Dimedone, diversely substituted benzaldehydes, and hydrazine derivatives were subjected to reaction in ionic liquid [Bmim]BF_4 as a medium and $Yb(OTf)_3$ as a catalyst to prepare tetrahydroindazolones. Single-crystal X-ray diffraction information was used to get solid-state structure of one of the compounds prepared.

A library of pyranopyrazoles was synthesized by Javad Ebrahimi et al. adopting a one-pot four-component protocol with ionic liquid as medium (Fig. 15.8) (Ebrahimi et al. 2012). Aldehydes, malononitrile, ethyl acetoacetate, and hydrazine

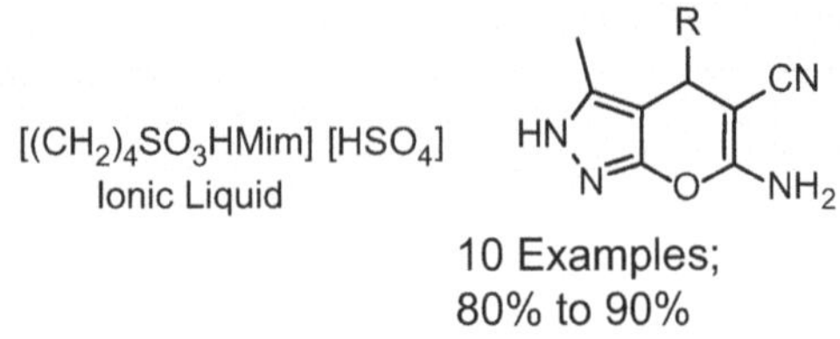

Fig. 15.8 Scaffolds of pyranopyrazoles

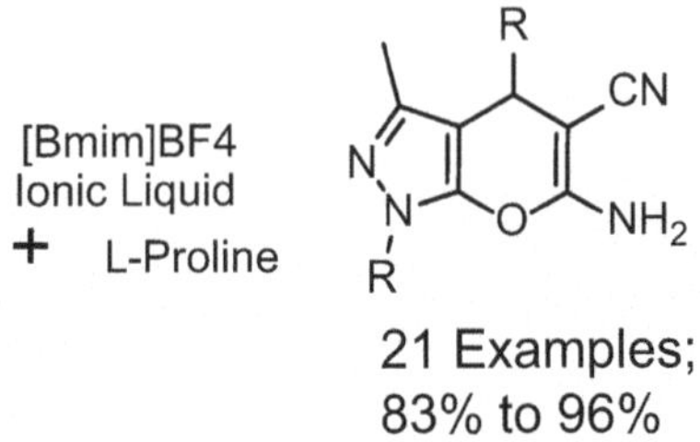

Fig. 15.9 Different pyrano pyrazole derivatives

were reacted in ionic liquid to make pyranopyrazole derivatives. In another report, several pyranopyrazole compounds were synthesized (Fig. 15.9) (Khurana et al. 2011) by Jitender Khurana et al. following a one-pot four-component approach with L-proline as a catalyst and ionic liquid [Bmim]BF_4 as medium.

Raghayeh Sharifi Aliabadi and Nosrat Mahmoodi disclosed green and efficient synthesis of diversified pyrano[2,3-c]pyrazoles, via a three-component cyclization reaction between pyrazolone, malononitrile, and various benzaldehydes employing a novel heterogenous imidazole-based ionic liquid catalyst, 1-butyl-3-methyl imidazolium hydroxide [Bmim][OH], in water under microwave conditions (Fig. 15.10) (Aliabadi and Mahmoodi 2016).

Maohua Ding and Hongyun Guo came up with efficient synthetic procedure for 2H-pyridazino[1,2-a]indazole-1,6,9(11H)-triones (Fig. 15.11) (Ding and Guo 2016). This three-component one-pot reaction mediated by ionic liquid [Hnmp]HSO_4 involves aryl aldehyde, dimedone, and maleic hydrazide as reactants. Experimental reaction conditions were improved by optimizing the yields.

15.2.3 Imidazoles and Benzimidazoles

Yatao Liang et al. disclosed Lewis acidic ionic liquid containing crown ether complex cation [18-C-6 K][$FeCl_4$]-catalyzed synthesis of 2-phenyl-1H-benzo[d]imidazoles and 1,2-disubstituted benzimidazole derivatives by the reaction of o-phenylenediamines and aromatic aldehydes at room temperature in ethanol (Fig. 15.12) (Liang et al. 2014).

Fig. 15.10 Synthesis of pyrano[2,3-c]pyrazole derivatives

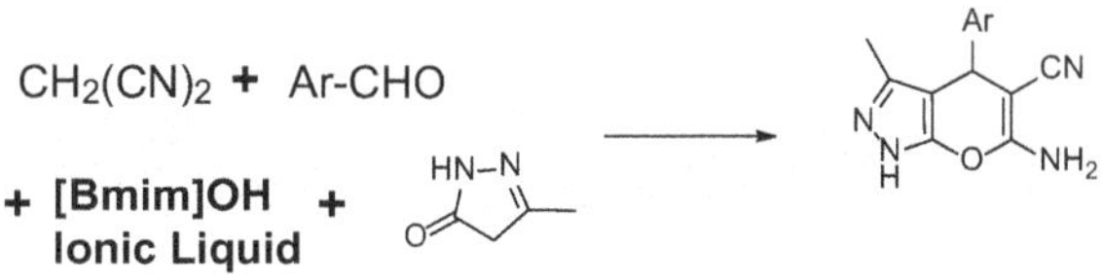

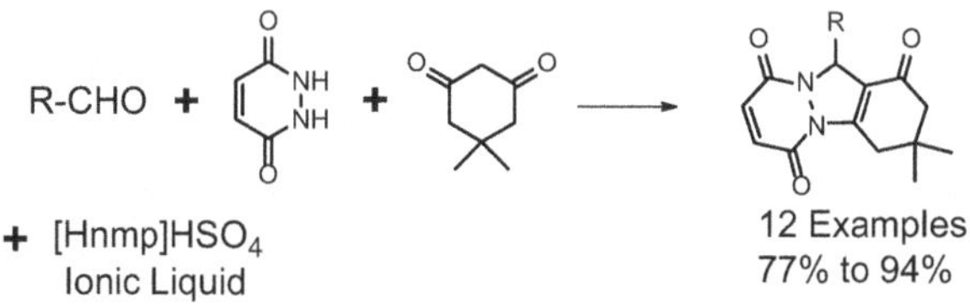

Fig. 15.11 Synthesis of 2H-pyridazino[1,2-a]indazole-1,6,9(11H)-triones

Fig. 15.12 Benzimidazole derivatives

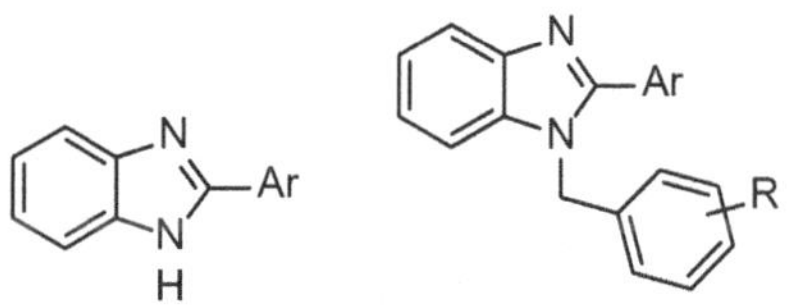

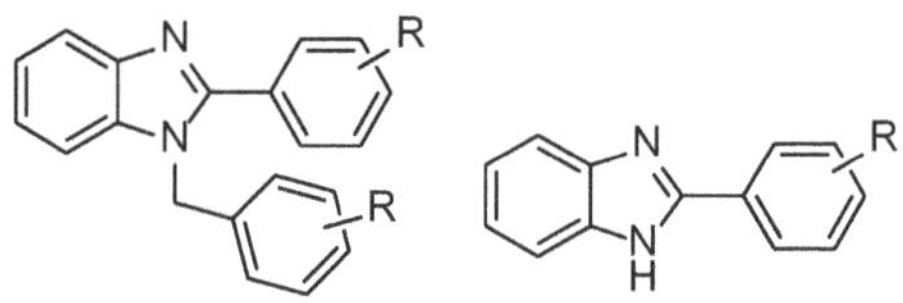

Fig. 15.13 Scaffolds of 2-substituted benzimidazoles and 2-aryl-1-arylmethyl-1H-1,3-benzimidazoles

A simple and eco-compatible methodology for a range of 2-substituted benzimidazoles and 2-aryl-1-arylmethyl-1H-1,3-benzimidazoles was disclosed by Swapan Majumdar et al. using grind stone chemistry with outstanding selectivity (Fig. 15.13) (Liang et al. 2014), in their interesting research communication. Authors achieved protic ionic liquid-tuned green synthesis by grinding a mixture of o-phenylenediamines, suitable aldehydes, and ionic liquid catalyst—imidazolium trifluoroacetate. It was successfully demonstrated that aromatic aldehydes without any substitution or with electron-releasing substituents formed 1,2-disubstituted benzimidazoles. Aromatic aldehydes bearing electron-withdrawing substituents or 2-alkoxy aldehydes provided 2-substituted benzimidazoles.

During the studies authors screened three protic ionic liquids, 1-butylimidazolium trifluoroacetate, 1-butyl-2-methylimidazolium trifluoroacetate, and 1-butylbenzimidazolium trifluoroacetate, for the catalytic efficiency towards highly selective annulations of aromatic aldehydes and o-phenylenediamines.

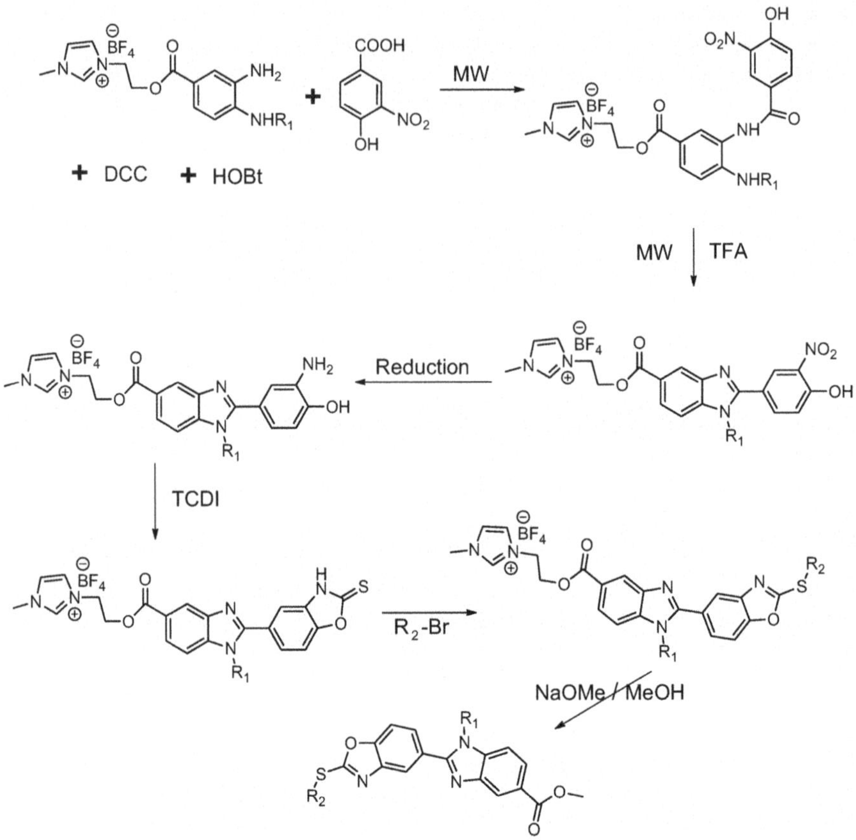

Fig. 15.14 Combinatorial library via ionic liquid

Benzoxazole-linked benzimidazole, tethered with ionic liquid moiety, was synthesized and it was utilized to generate a combinatorial library of chemicals having potential applications (Fig. 15.14) (Chanda et al. 2012). The diversity-oriented approach was adopted in making the ionic liquid-tethered heterocycles. 4-Hydroxy-3-nitrobenzoic acid was coupled with starting ionic liquid compound, and the coupled amide was cyclized to get imidazole. Then nitro group of imidazole derivative was reduced to amino group leading to hydroxyl–amino functionality, which was converted into oxazole moiety by allowing it to react with thiodiamide. Thus obtained thiono compound was reacted with bromo compound to make S-substituted derivative. Finally the ionic liquid moiety was detached by using sodium methoxide in methanol solvent to reach target molecule. Varying R_1 and R_2 and their combinations provided a combinatorial library of chemicals (Fig. 15.14). The role of ionic liquid moiety was found to be as a good microwave-absorbing residue and it facilitates the MW-induced chemical reactions.

15.2.4 Triazoles and Benzotriazoles

Jayant Patil and Dattaprasad Pore explored the synthesis of a library of novel 1,2,4-triazolidine-3-thiones in water medium by the reaction of thiosemicarbazides or 4-substituted semicarbazide derivatives with different aryl and heteroaryl aldehydes at ambient temperature, catalyzed by recyclable, efficient, and novel ionic liquid [C_{16}MPy]$AlCl_3$Br. The catalyst was derived by the reaction of $AlCl_3$ and [C_{16}MPy]Br, which was obtained from N-methyl pyrrolidine and 1-bromohexadecane. Authors investigated the efficiency of several catalysts on the formation of the title compounds (Fig. 15.15). (Patil and Pore 2014)

15.2.5 Tetrazoles

Majid Heravi and co-workers developed popular click reaction using ionic liquid [Bmim]N_3. Aldehyde, hydroxylamine, [Bmim]N_3, copper diacetate, and DMF were together mixed and used to conduct click reaction (Fig. 15.16) (Heravi et al. 2012). Aldehyde first reacted with hydroxylamine to form oxime, then it underwent dehydration to form nitrile, and this nitrile reacted with azide to give tetrazole. Copper diacetate acted as a catalyst and DMF was found to be the best solvent.

15.2.6 Furans and Benzofurans

Pharmaceutically important 2-aroylbenzofuran-3-ols were produced by Patil et al. in high yields from methyl salicylates and phenacyl bromides when reacted in the presence of ionic liquid [Bmim]OH (Fig. 15.17) (Patil et al. 2012).

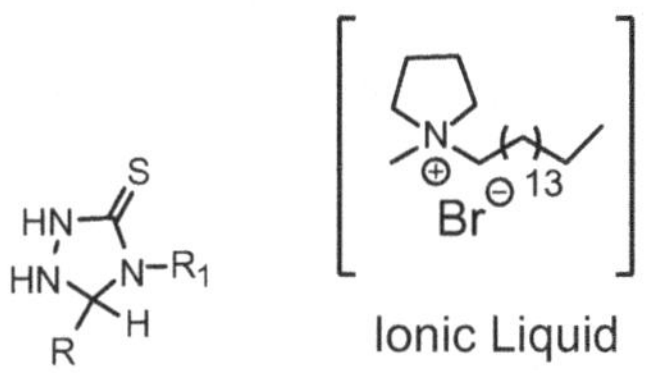

Fig. 15.15 Triazolidine-3-thiones

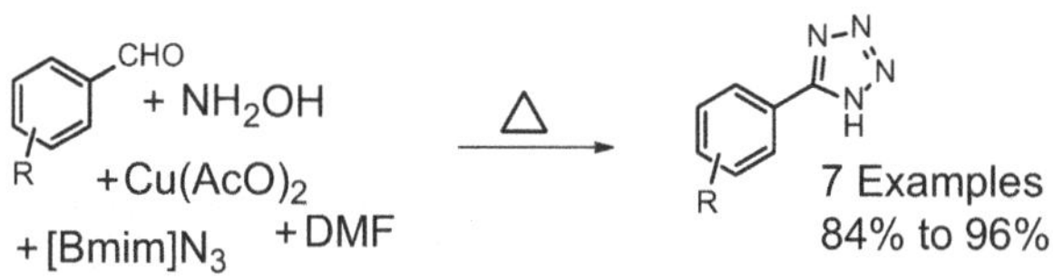

Fig. 15.16 Click reaction—tetrazoles

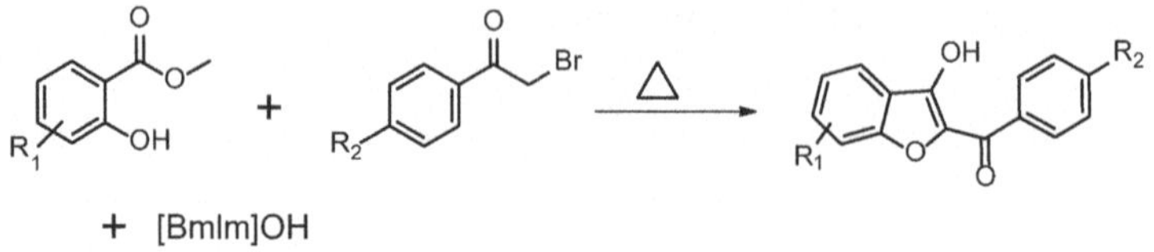

Fig. 15.17 Protocol for 2-aroylbenzofuran-3-ols

Fig. 15.18 Diazobenzoxazoles

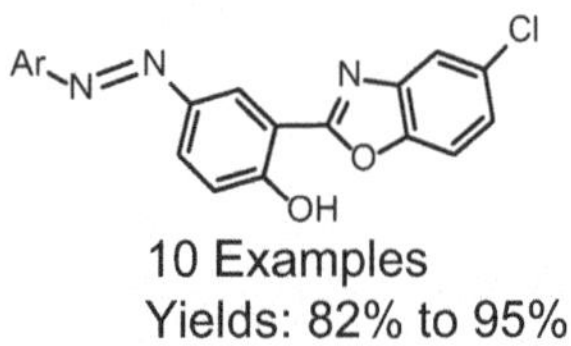

15.2.7 *Oxazoles, Isoxazoles, and Benzoxazoles*

Mohammad Nikpassand et al. employed an effective bis-ionic liquid [BDBDIm]Br for converting 2-amino-4-chlorophenol and arylazosalicylic acid into the corresponding oxazoles (Fig. 15.18) (Nikpassand et al. 2015) at room-temperature conditions. Bis-ionic liquid was recovered and reused without much reduction in activity.

Bhaskar Chakraborty and Esmita Chettri demonstrated synthesis of a new class of spiroisoxazolidine derivatives via ionic liquid. 1,3-Dipolar addition between *N*-benzyl-*C*-fluoro-substituted phenyl nitrones (polarophiles) and furan derivative provided spiroisoxazolidine compounds (Fig. 15.19) (Chakraborty and Chettri 2018). Regiochemistry observed in this transformation was predicted to be controlled by FMOs. The stereochemical outcome is noteworthy.

A series of new angular pyrazole and isoxazole heterocyclic derivatives were prepared by Tushar R. Suatiya et al. [Ref] using eco-friendly Domino synthetic methodology in 1,8-diazabicyclo[5.4.0] undec-7-ene-8-ium acetate ([DBU][Ac]), a recyclable ionic liquid, via Domino Knoevenagel-hetero-Diels-Alder approach [DKHDA]. Authors prepared various chromeno-fused pyrano[2,3-c]pyrazoles by the reaction of active methylene of pyrazolone with O-alkenylated acetophenones and chromeno-fused pyrano[2,3-c]isoxazoles by the reaction of O-alkenylated or alkynylated acetophenones with active methylene isoxazolones. Authors also examined different catalysts and solvent conditions while optimizing DKHDA reactions (Fig. 15.20) (Sutariya et al. 2015).

A series of 2-arylbenzoxazoles, 2-arylbenzimidazoles, and 2-arylbenzothiazoles were synthesized in a highly efficient and eco-friendly approach by Quang The Nguyen et al., catalyzed by recyclable phosphonium acidic ionic liquid, a homogeneous catalyst triphenyl(butyl-3-sulfonyl)phosphonium toluenesulfonate. The title

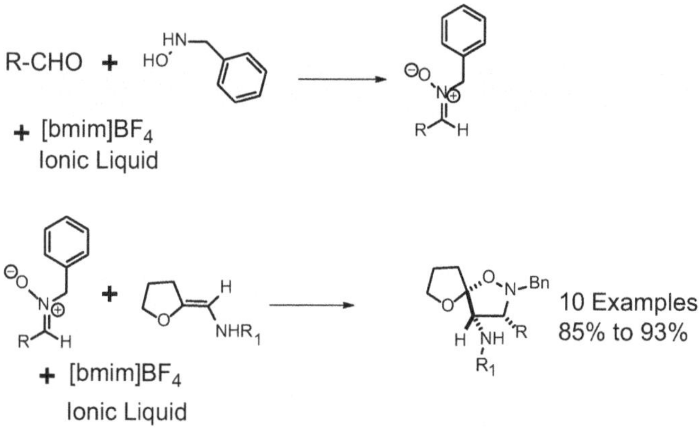

Fig. 15.19 Synthesis of spiroisoxazolidine derivatives

Fig. 15.20 Angularly fused isoxazole and pyrazole derivatives

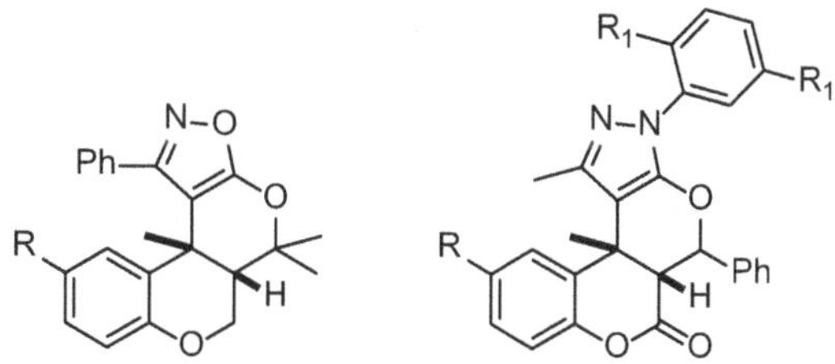

Fig. 15.21 Different benzoxazoles, benzimidazoles, benzothiazoles

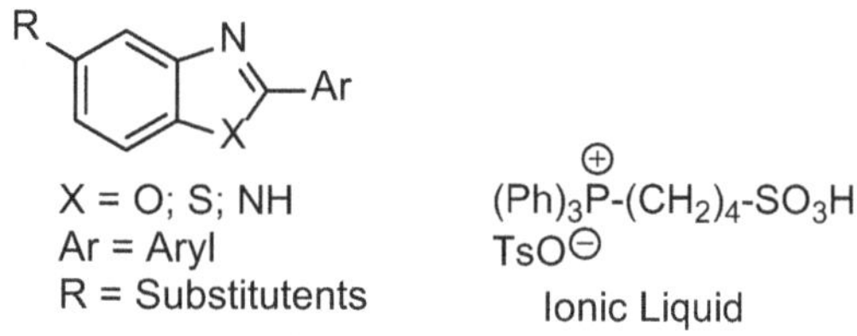

compounds were prepared by the condensation of various diversified aryl aldehydes with o-aminophenols, o-phenylenediamine, and o-aminothiophenols (Fig. 15.21) (Nguyen et al. 2018a).

An environmentally benign approach for the synthesis of 2-benzoxazolyl and 2-benzothiazolyl ketones by the condensation reaction of alkynylbromides with 2-amino(thio)phenols catalyzed by FeCl3.6H2O in [Bmim]BF4 was reported by Cui Liangyan et al. (Fig. 15.22) (Cui and Yan Fan 2012).

Kushal Lanjewar et al. generated a library of pyrimidinone-isoxazole hybrids mediated by ionic liquid [Bmim]PF_6. Suitable pyrimidinone derivative was treated with hydroxylamine hydrochloride, [Bmim]PF_6, KOH, and water were involved to produce pyrimidinone-isoxazole hybrids in very good yields (Fig. 15.23) (Lanjewar

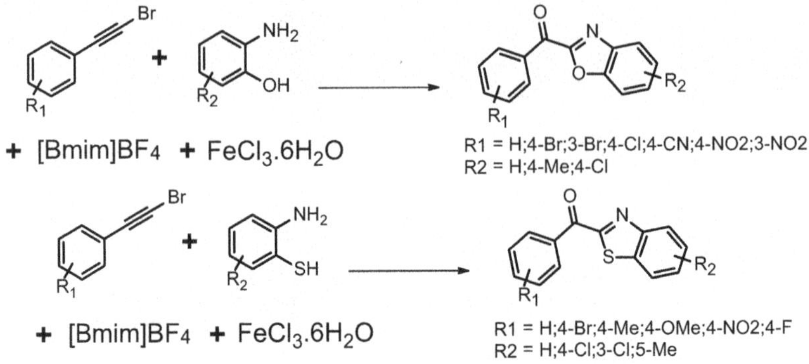

Fig. 15.22 Benzoxazole and benzothiazole derivatives

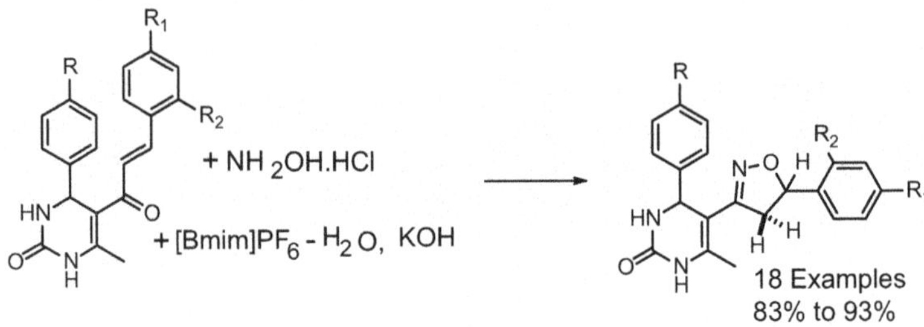

Fig. 15.23 Pyrimidinone-isoxazole hybrids

et al. 2011). Varying the substituents on the benzene residue produced a library of pyrimidinone-isoxazole hybrids.

Mei-Yan Wang et al. developed a simple cost-effective and environmentally benign protocol for the carboxylative cyclization of propargylic amines with CO2 at atmospheric pressure under solvent-free conditions, promoted by bifunctional halogen-free Cu(II) substituted polyoxometallate-based ionic liquid—$[(nC_7H_{15})_4N]_6[\alpha\text{-}SiW_{11}O_{39}Cu]$. Various 5-alkylidene oxazolidine-2-ones were produced in excellent yields (Fig. 15.24) (Wang et al. 2016).

15.2.8 Thiazoles and Benzothiazoles

Santosh Kumar et al. reported a novel multicomponent reaction of pyridine-2(1H)-one or thiazole/benzo[d]thiazol-2(3H)-ones with o-tosyl-hydroxylamine and acetophenones using copper triflate in $[Bmim]BF_4$ through C-H functionalization under microwave conditions.

The efficient protocol successfully provided a library of functionalized fused imidazo-heterocycles. The ionic liquid along with copper triflate was recycled

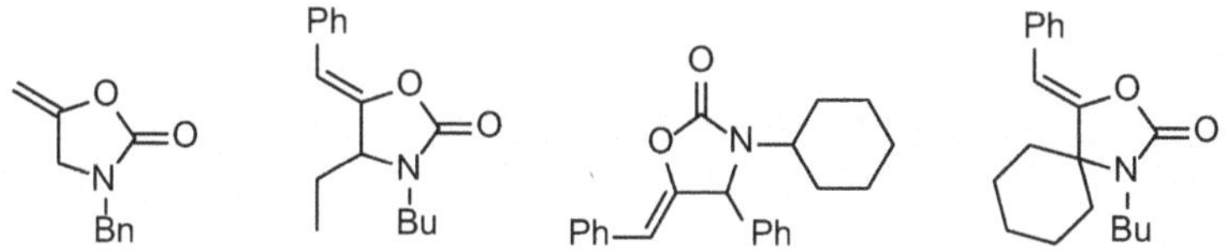

Fig. 15.24 Scaffolds of various 5-alkylidene oxazolidine-2-ones

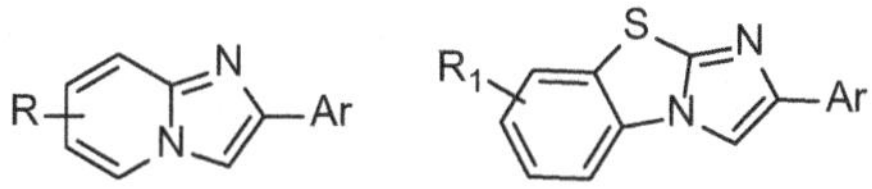

Fig. 15.25 2-Aryl imidazo[1,2-a]pyridines and thiazol/benzo[d]thiazole derivatives

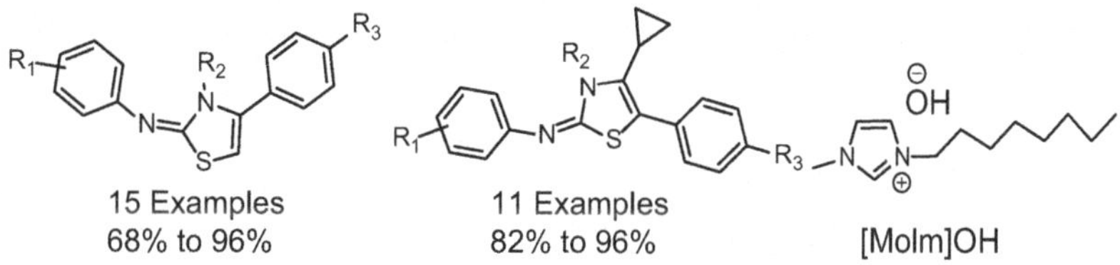

Fig. 15.26 Scaffolds of thiazole derivatives and ionic liquid

(Fig. 15.25) (Santosh Kumar et al. 2015). During the investigations authors examined different copper sources and ionic liquids for their efficacy. Possible mechanistic pathway was depicted.

Jafar Abbasi Shiran et al. informed the synthesis of thiazol-2-imine derivatives by a one-pot three-component reaction in the presence of ionic liquid acting as catalyst as well as reaction medium (Fig. 15.26) (Shiran et al. 2015). Aryl amines, alkylisothiocyanate, and various α-haloketones were mixed with ionic liquid [omim][OH] at room temperature to prepare the target products in a regioselective manner. X-ray structural analysis was reported for one of the target compounds.

15.2.9 Pyridines and Fused Analogues

An efficient ionic liquid-phase synthesis (IOLiPS) of 2-aminothiazoles and imidazo[1,2-a] pyridines was achieved by Sunita Choudhary et al. following a catch-and-release strategy, using various acetophenones with thiourea and 2-aminopyridines, respectively. The hypervalent iodine acted both as a reagent and as a soluble support (Fig. 15.27) (Choudhary et al. 2014).

Mohsen Abbasi, in a significant piece of research work, designed, synthesized, and characterized 1,3-disulfonic acid benzimidazolium chloride [DSbim]Cl as a

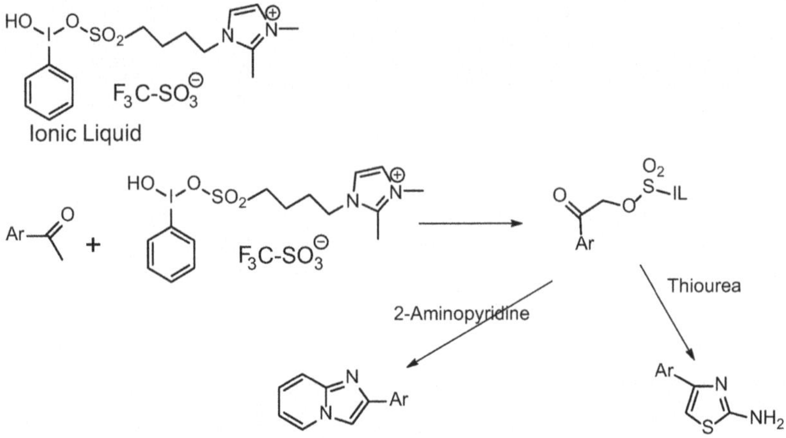

Fig. 15.27 2-Aminothiazoles and imidazo[1,2-a]pyridines

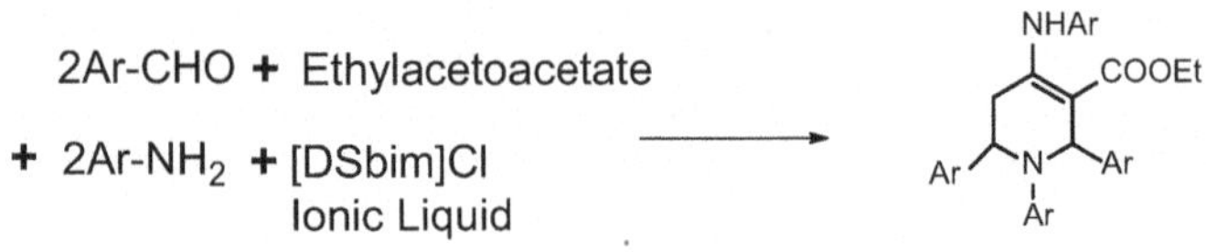

Fig. 15.28 Highly functionalized tetrahydropyridines

new ionic liquid. This was utilized as an efficient, recyclable, and homogenous catalyst in a one-pot multicomponent condensation reaction between aromatic aldehydes, anilines, and ethyl acetoacetate providing highly functionalized tetrahydropyridines in high yields under solvent-free conditions (Fig. 15.28) (Abbasi 2015). While conducting investigations, different solvents were evaluated and it was observed that solvent-free conditions were ideal. The scope was extended to include various diversely substituted aldehydes and amines.

In an interesting paper, Rajesh Kumar and co-authors presented an operationally simple and highly efficient microwave-assisted one-pot multicomponent synthetic strategy for diversely substituted symmetrical and unsymmetrical 1,4-dihydropyridines in higher yields. Aromatic aldehydes and heteroaromatic aldehydes, ammonium carbonate, methyl acetoacetate, dimedone, and cost-effective glycine nitrate as ionic liquid were the components (Figs. 15.29 and 15.30) (Kumar et al. 2014).

During the investigations authors examined several ammonium salts and different solvents and observed that $(NH_4)_2CO_3$ and ethanol provided better results. Authors monitored the progress of the reaction by Q-TOF, as well as direct-injection electrospray ionization mass spectrometry (EI-MS), and experimentally revealed that Hantzsch reaction predominantly follows the diketone pathway among four competing pathways.

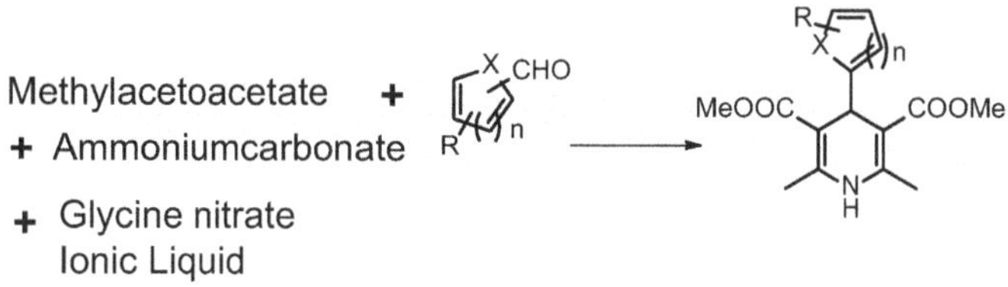

Fig. 15.29 Synthesis of completely substituted dihydropyridines

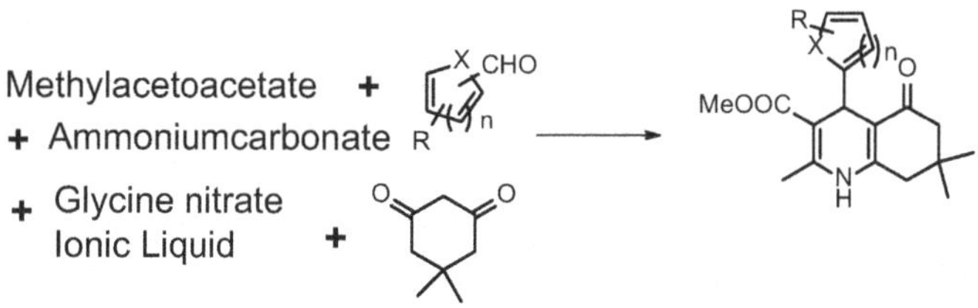

Fig. 15.30 Synthesis of fused dihydropyridine derivatives

Fig. 15.31 Fused pyrimido-pyrido-triones

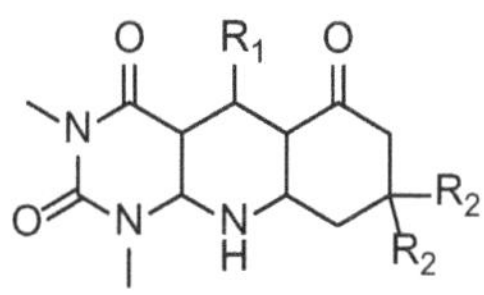

Kamal Mohammadi et al. demonstrated an eco-friendly and convenient protocol for the synthesis of pyrimido-[4,5-b]quinoline derivatives, via a one-pot multicomponent condensation reaction involving 6-amino 1,3-dimethyluracil, cyclic 1,3-diketones, and diversely substituted aldehydes in the presence of an efficient recyclable ionic liquid catalyst 1,3-disulfonic acid imidazolium hydrogen sulfate conducted in ethanol as medium (Fig. 15.31) (Mohammadi et al. 2015).

15.2.10 Pyrazines and Fused Analogues

1-Butyl-3-methyl imidazolium hydroxide ([BMIM]OH), an eco-friendly task-specific basic ionic liquid catalyst, was employed by Lingyu Jia et al., in an efficient one-pot synthesis of deoxy-fructosazine and fructosazine from D-glucosamine hydrochloride. The atom economical approach represented a simple method for the synthesis of hydroxy alkyl pyrazine derivatives from the self-condensation of GLcNH2.[BMIM]OH that acted as solvent and catalyst in the reaction (Fig. 15.32) (Jia et al. 2014).

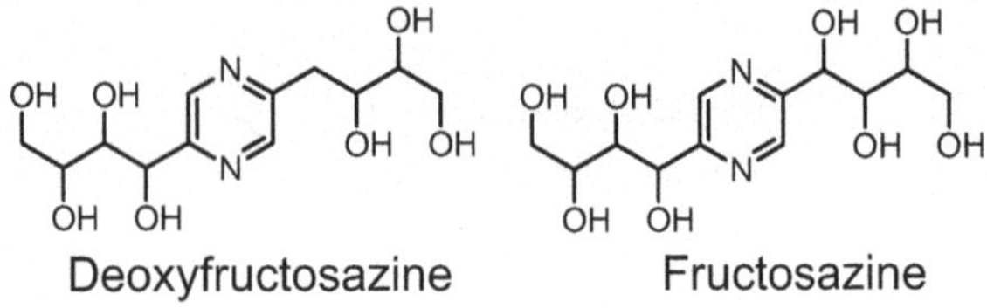

Fig. 15.32 Deoxyfructosazine and fructosazine

15.2.11 *Phthalazines*

In an interesting research communication, Jamal Davarpanah and Ali Reza Kiasat disclosed the synthesis of amorphous silica nanoparticles from low-cost rice husk ash (RH-SiO_2) modified by acidic ionic liquid, containing doubly charged diazabicyclo[2.2.2]octane chloride using sol-gel method. Authors established the structural properties of this heterogeneous Brønsted solid acid nanocomposite material and successfully employed it as a recyclable catalyst in the one-pot three-component synthesis of 1H-pyrazolo(1,2-b)phthalazine-5,10-diones. The target compounds were prepared by the cyclo-condensation of phthalhydrazide, substituted aromatic aldehydes, and malononitrile (Fig. 15.33) (Davarpanah and Kiasa 2015).

Dusic Simijonovic and co-workers developed an efficient one-pot multicomponent approach for the synthesis of pyrazolyl-phthalazinedione derivatives from acetylacetone, 2,3-dihydrophthalazine-1,4-dione, and diversely substituted aldehydes promoted by [HDEA][ClAc], a recyclable diethylammonium chloroacetate ionic liquid catalyst (Fig. 15.34) (Simijonović et al. 2018).

A simple and efficient one-pot four-component solvent-free protocol was developed by Hojat Veisi et al. for the synthesis of 2H-indazolo[2,1-b]phthalazine triones and also dihydro-1H-pyrano[2,3-c]pyrazol-6-ones catalyzed by mildly basic reusable ionic liquid N,N,N,N-tetramethyl guanidinium acetate [TMG][Ac] (Fig. 15.35) (Veisi et al. 2014). Preparation of 2H-indazolo[2,1-b]phthalazine triones was achieved by reacting phthalic anhydride, hydrazinium hydroxide, dimedone, and various aldehydes in the presence of [TMG][Ac].

15.2.12 *Quinazolines*

Ghazaleh Yassaghi et al. in their interesting research paper informed the preparation of a new heterogeneous acidic catalyst, by impregnating silica (Aerosil 300) with 1-(4-sulfonic acid)butyl pyridinium hydrogen sulfate [PyC_4SO_3H][HSO_4], an acidic ionic liquid (Fig. 15.36) (Yassaghi et al. 2012). Authors have characterized it using techniques such as FT-IR, TG/DTG, and XRD. The catalytic efficiency and recyclability of this material were examined for the synthesis of 2-aryl-2,3-dihydroquinazolin-4(H)-ones. Aromatic aldehydes are subjected to cyclo-condensation with 2-aminobenzamides under solvent-free conditions in short reaction times to obtain high yields of 2-aryl-2,3-dihydroquinazolin-4(H)-ones.

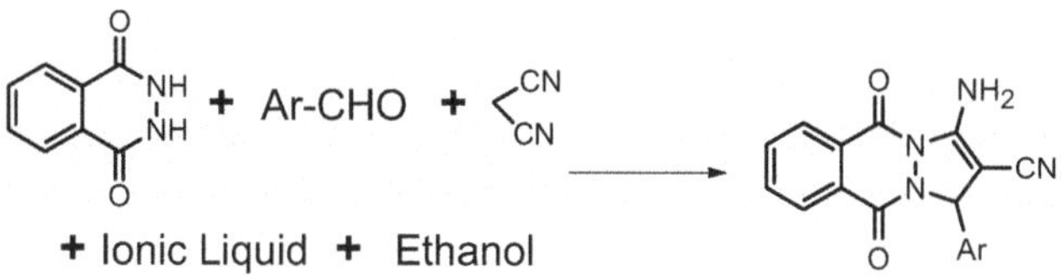

Fig. 15.33 Synthesis of 1H-pyrazolo(1,2-b)phthalazine-5,10-diones

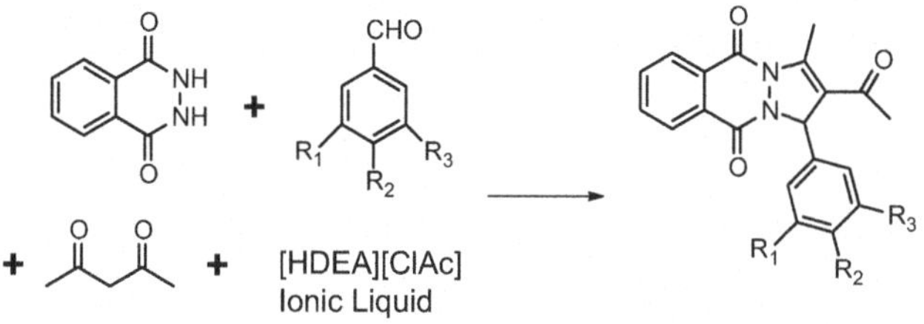

Fig. 15.34 Synthesis of pyrazolyl-phthalazinedione derivatives

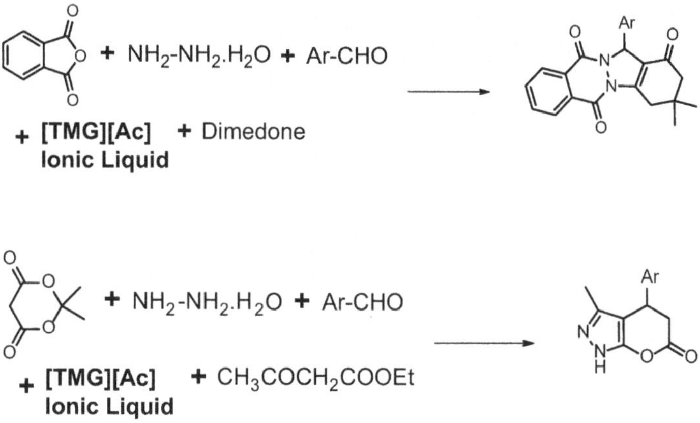

Fig. 15.35 Synthesis of 2H-indazolo[2, 1-b]phthalazine triones, and dihydro-1H-pyrano[2,3-c] pyrazol-6-ones

Fig. 15.36 Synthesis of quinazoline derivatives

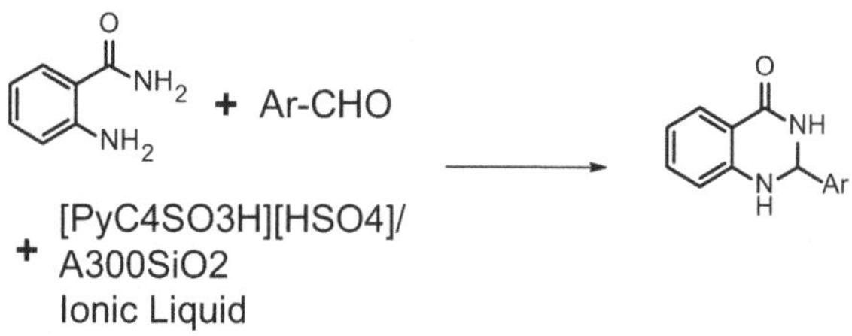

Maryam Hajjami and Bahman Tahmasbi prepared an efficient and magnetically reusable nanocatalyst by immobilizing glucosulfonic acid on Fe_3O_4 (GSA@MNPS) and applied for the synthesis of a library of polyhydroquinoline and 2,3-dihydroquinazolin-4(1H)one derivatives. Polyhydroquinolines were synthesized by reacting various benzaldehydes possessing electron-donating and electron-withdrawing groups on aromatic ring with dimedone, ethyl acetoacetate, and ammonium acetate in ethanol. 2,3-Dihydroquinolin-4-(1H)-ones were obtained by the reaction of aromatic aldehydes with anthranilamide in ethanol (Fig. 15.37) (Hajjami and Tahmasbi 2015).

Yue Wang et al. demonstrated that 2-(1H-benzo[d]imidazol-2-yl)anilines underwent reaction with 5-chloropentan-2-one and 6-chlorohexan-2-one in the presence of ionic liquid [BMIm]Br and also with iodine, acting as catalyst, to provide benzo[4,5]imidazo[1,2-c]pyrrolo[1,2-a]quinazoline and 6H-benzo[4,5]imidazo[1,2-c]pyrido[1,2-a]quinazoline derivatives (Fig. 15.38) (Wang et al. 2017). Interesting

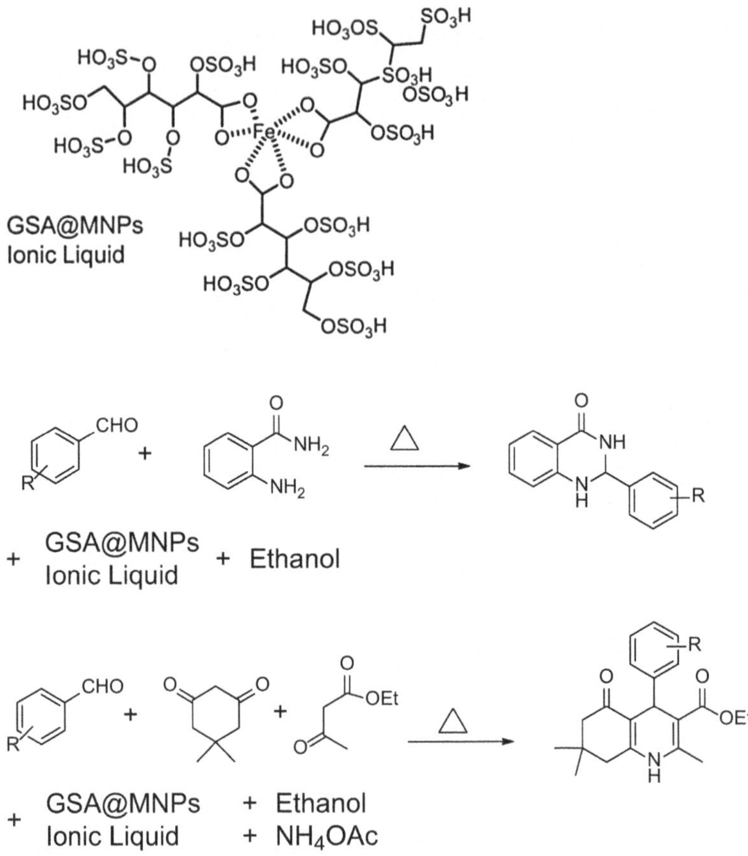

Fig. 15.37 GSA@MNPS-ionic liquid; synthesis of various polyhydroquinoline and 2,3-dihydroquinazolin-4(1H)one derivatives

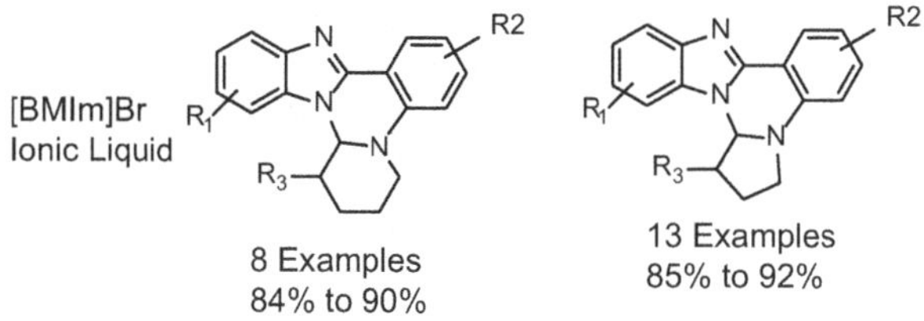

Fig. 15.38 Benzo[4,5]imidazo-quinazoline derivatives

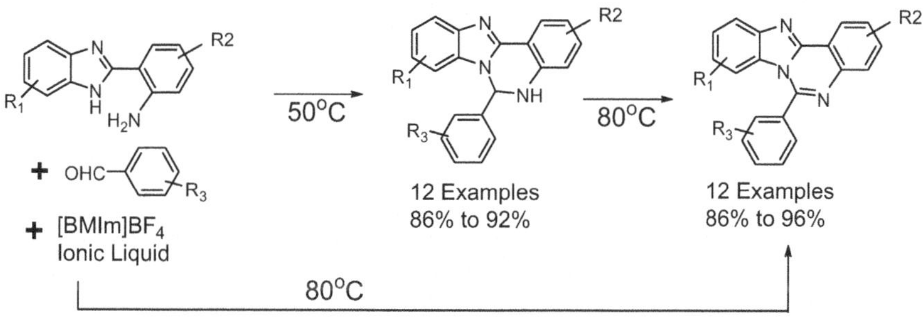

Fig. 15.39 Benzo[4,5]imidazo-quinazoline derivatives

aspect of this synthesis is that the target molecule is pentacyclic heterocycle and it contains benzimidazole and quinazoline moieties.

Dong-Sheng Chen et al. informed about the synthesis of benzo[4,5]imidazo-quinazoline derivatives(Fig. 15.39) (Chen et al. 2017) from 2-(1H-benzo[d]imidazol-2-yl)aniline and benzaldehydes in the presence of ionic liquid [BMIm]BF_4. In this interesting chemical transformation the intermediate dihydro product was isolated by adjusting reaction temperatures. Authors isolated dihydro products and further the same dihydro products were transformed to the final target molecules by adopting higher reaction temperatures. Crystal structure of dihydro compound was determined to confirm the structure unambiguously.

Wen-Wen Qiang et al. reported the formation of 5-arylimidazo[1,2-c]quinazoline derivatives from 2-(4,5-diphenyl-1H-imidazol-2-yl)aniline and substituted benzaldehydes under the influence of ionic liquid [BMIm]Br (Fig. 15.40) (Qiang et al. 2016). The dihydro intermediate was obtained by conducting the same reaction under the blanket of nitrogen. Crystal structure was deduced for the dihydro compound to prove the mechanism. Authors advocated that oxygen present during the reaction was found to be responsible for consecutive dehydrogenation reaction leading to the final target molecule.

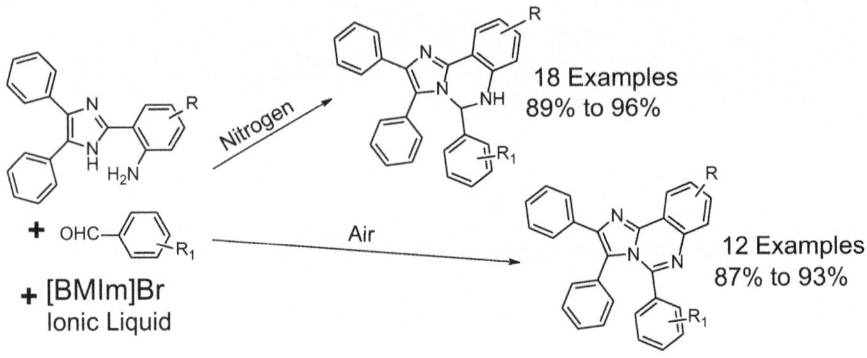

Fig. 15.40 Approach for 5-arylimidazo[1,2-c]quinazoline derivatives

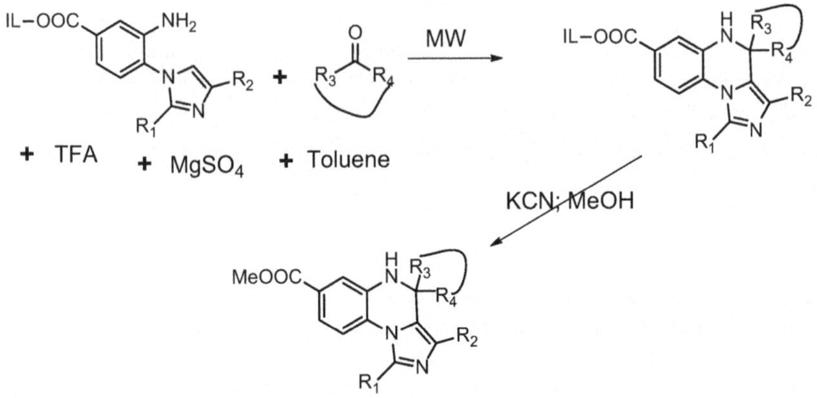

Fig. 15.41 Synthesis of imidazo [1,5-a]quinoxalines

15.2.13 *Quinoxalines*

Li-Hsun Chen et al. communicated rapid regioselective synthesis of imidazo[1,5-a] quinoxalines and methyl N-phenyl-benzimidates supported by 1-(2-hydroxyethyl)-3-methylimidazolium tetrafluoroborate ([hydemim][BF_4]), by unconventional Pictet-Spengler reaction. It was performed regioselectively at C-5 position of imidazole scaffold with different ketones and aliphatic aldehydes (Fig. 15.41) (Chen et al. 2016). However, with aromatic aldehydes authors reported the formation of unusual imidazo[1,5-a]quinoxaline ring opening.

15.2.14 Oxazines and Benzoxazines

Garima Khanna and co-authors presented a wide range of novel 3,4-dihydro-2H-naphtho[2,3-c][1,3]oxazine-5,10-dione derivatives, synthesized by a one-pot three-component condensation reaction between 2-hydroxy-1,4-naphthoquinone, formaldehyde, and various aromatic amines in the presence of a reusable ionic liquid [Bmim]BF4, under mild conditions (Fig. 15.42) (Khanna et al. 2015).

Shivaji Naidu and Sabbasini Rajasekhara Reddy reported a simple one-pot tandem reaction for synthesis of highly substituted pyrrolo/pyrido [2,1-b]benzo-(d)[1,3]oxazin-1-ones from 2-aminobenzyl alcohols and alkyne-containing carboxylic acids under ligand- and base-free conditions using copper acetate as a catalyst in the presence of green recyclable solvent medium—ionic liquid [Bmim]OTf at 100 °C (Fig. 15.43) (Naidu and Reddy 2016).

During the investigations authors studied the effect of different copper sources like Cu_2O, CuBr, CuI, and $Cu(OAc)_2$ and observed that $Cu(OAc)_2$ gave better results over other catalysts. DMF, [Bmim]Br, [Bmim]BF_4, and [Bmim]OTf were examined as solvents and finally [Bmim]OTf was reported to be yielding favorable results. Authors successfully recycled and reused both catalyst and solvent system.

Ebrahim Soleimani et al. synthesized benzo[b][1,4]oxazines in a one-pot three-component reaction involving *o*-aminophenol, aldehyde, isocyanide, and [Bmim]Br without any added catalyst (Fig. 15.44) (Soleimani et al. 2012). *o*-Aminophenol

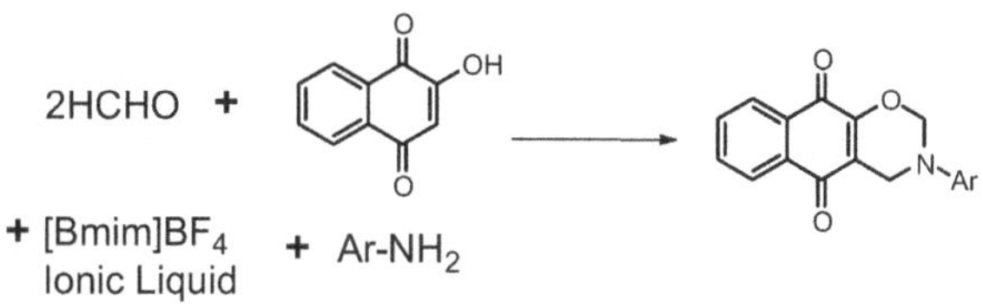

Fig. 15.42 Oxazine-fused naphthoquinones

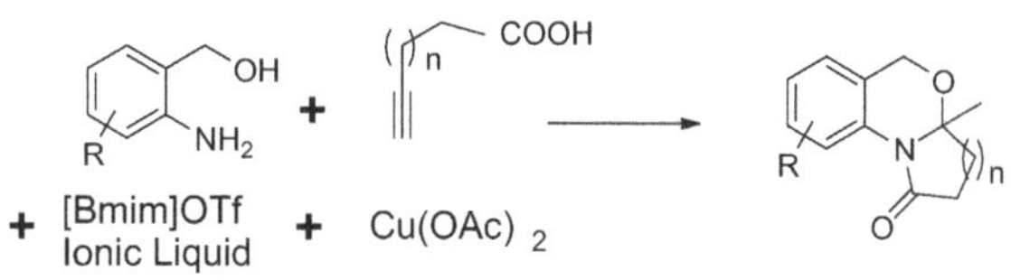

Fig. 15.43 Approach for highly substituted pyrrolo/pyrido [2,1-b]benzo-(d)[1,3]oxazin-1-ones

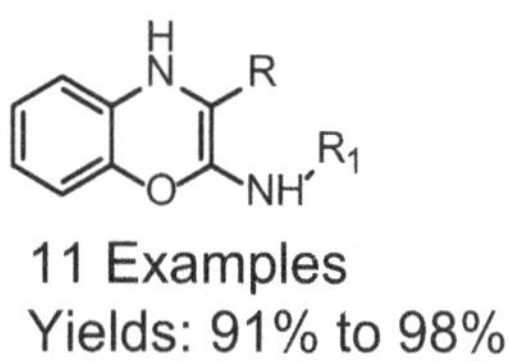

Fig. 15.44 Benzoxazine scaffolds

reacted with aldehyde to give Schiff's base which underwent reaction with isocyanide resulting in cyclization, followed by proton shift leading to the product benzoxazine.

15.2.15 Thiazines and Fused Derivatives

Kapil Arya and co-authors developed a reusable zeolite-supported Brønsted acid catalyst system ZSM-5[MIM]BF_4 and applied this efficiently in the synthesis of novel N-substituted spiro-heterocycles, which were prepared by reacting N-substituted isatins, 2-mercaptonicotinic acid, and a wide range of amines (Fig. 15.45) (Arya et al. 2012). It was reported that yields increased under ultrasonication compared to the conventional as well as microwave-irradiated conditions.

15.2.16 Pyrimidines and Fused Derivatives

Sumit Kumar Panja and Satyen Saha developed an environmentally benign, solvent-free high-yielding one-pot multicomponent strategy towards the synthesis of highly functionalized quinazoline derivatives from 2-aminobenzophenone, ammonium acetate, and various aromatic aldehydes catalyzed by recyclable magnetic ionic liquid—butyl methyl imidazolium tetrachloro ferrate—([Bmim][$FeCl_4$]) (Fig. 15.46) (Panja and Saha 2013). During the investigational studies authors screened different catalysts for their efficiency and observed that [Bmim][$FeCl_4$] exhibited excellent activity. Tentative reaction mechanism was also proposed.

Sunil Chavan and Mariam Degani developed a novel facile and efficient one-pot synthesis of different 6-aminouracils via in situ preparation of ureas and cyanoacetyl ureas by the action between various aliphatic as well as aromatic amines and cyanates in the presence of recyclable ionic liquid catalyst 1,1,3,3-tetramethyl guanidine acetate [TMG][Ac] (Fig. 15.47) (Chavan and Degani 2012).

Tuan-Jie Li et al. conducted a one-pot three-component reaction mediated by ionic liquid to make fused pyrimidine derivatives (Fig. 15.48) (Li et al. 2012). Arylaldehydes, trifluoromethyl-1,3-diketone or ethyl trifluoromethyl acetoacetate, and amino tetrazole with [Bmim]BF_4 ionic liquid were reacted together to make trifluoromethyl-pyrimidine derivatives. Arylaldehyde underwent reaction with active methylene compound to form diacylolefin intermediate (Fig. 15.48), which then reacted with aminotetrazole to produce fused pyrimidinone derivatives.

Dihydropyrimidinones were synthesized through Biginelli reaction using ionic liquid as medium under microwave irradiation. Hetero-poly-anion-based ionic liquid was employed (Fig. 15.49) (Renzhong et al. 2014). Aldehydes, urea or thiourea, and ethyl or methyl acetoacetate, or acetylacetone were employed along with ionic liquid in the reaction. Indeed, several bonds are formed in this chemical transformation leading to a variety of pyrimidinone derivatives.

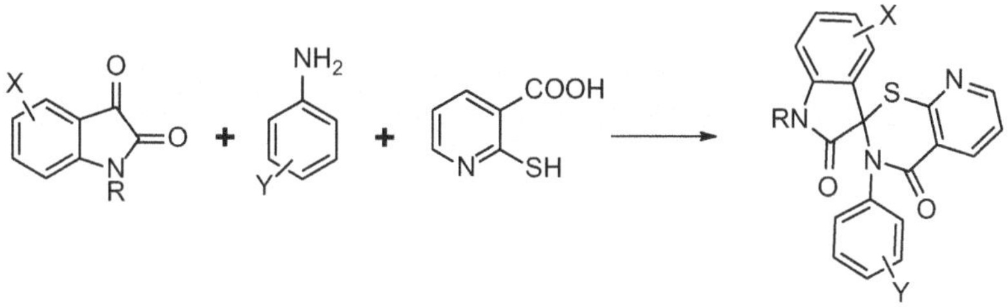

Fig. 15.45 Synthesis of novel N-substituted spiro-heterocyclic derivatives

Fig. 15.46 Synthesis of highly functionalized quinazoline derivatives

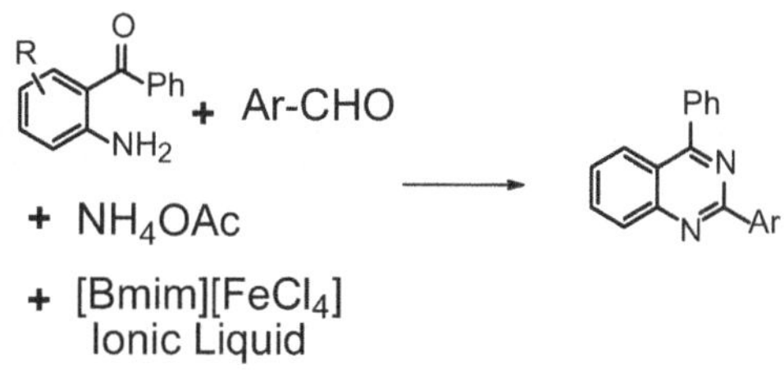

Fig. 15.47 Synthesis of different 6-aminouracils

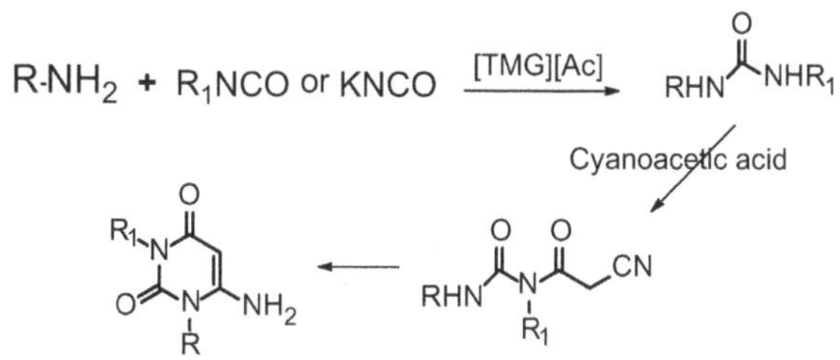

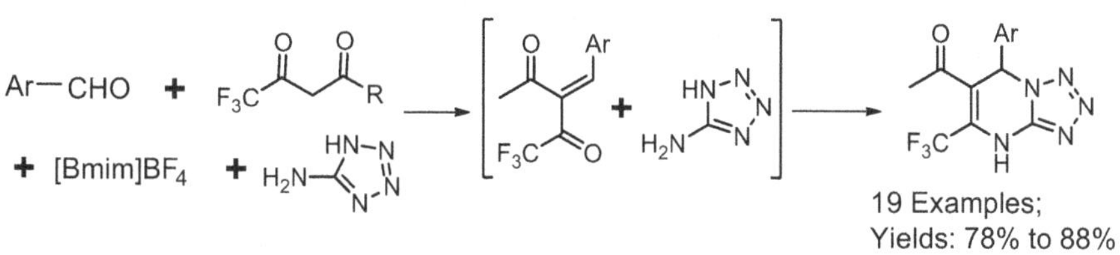

Fig. 15.48 Preparation of fused pyrimidine derivatives

Several bicyclic pyranopyrimidinone derivatives were synthesized by Hong-Yun Guo et al. (Fig. 15.50) (Guo et al. 2012) employing ionic liquid as medium as well as catalyst. Arylaldehydes, urea or thiourea, dihydropyran, and [Hnmp]HSO_4 were taken in one pot to conduct multicomponent reaction. Initially aldehyde reacted

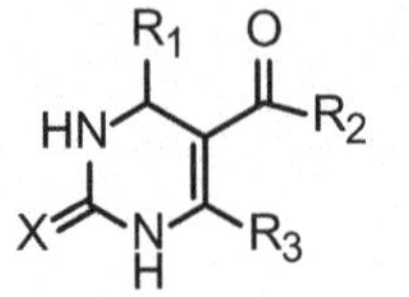

Fig. 15.49 Dihydropyrimidinones

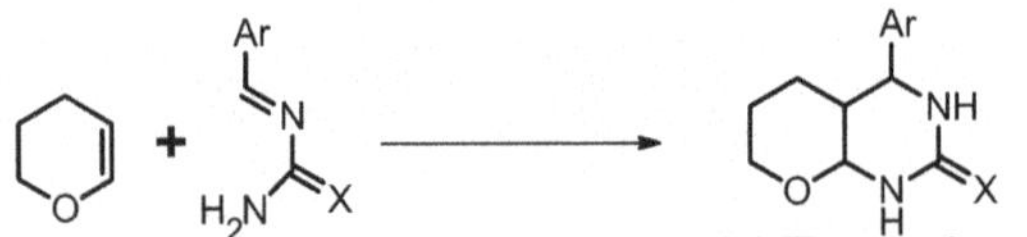

Fig. 15.50 Approach for pyrano-pyrimidinones

with urea to form imine and this in situ interacted with dihydropyran to give cyclized product (Fig. 15.50).

Abdol Hajipour et al. reported synthesis of pyrimidinone derivatives mediated by employing ionic liquid. Arylaldehyde, urea or thiourea, cyclic ketone, and [TEBSA]HSO_4 ionic liquid were taken together in one pot to conduct a three-component reaction to get fused pyrimidinones. It is rather a Biginelli-type reaction conducted with ionic liquid as medium (Fig. 15.51) (Hajipour et al. 2011).

An efficient protocol was developed by Sankari Kanakaraju et al. to make chromeno-pyrimidines involving a one-pot four-component reaction catalyzed by ionic liquid—[Bmim]BF_4 (Fig. 15.52) (Kanakaraju et al. 2012). α-Naphthol, malononitrile, and substituted benzaldehydes were taken in DMF along with a trace of triethylamine as well as ionic liquid and were heated, and then ammonium chloride was added in the same pot to prepare "chromeno-pyrimidines" in excellent yields. The third nitrogen fixed in the final product as an amino group was from the added ammonium chloride. Crystal structure for the final product was generated to confirm the structure unambiguously.

Janardhan Banothu et al. developed a highly efficient synthesis of chromeno-pyrimidinone derivatives by reacting 2-amino-5,6,7,8-tetrahydro-5-oxo-4-aryl-4H-chromene-3-carbonitriles, employing cost-effective, eco-friendly, reusable (4-sulfobutyl) tris (4-sulfophenyl)phosphonium hydrogen sulfate as Brønsted acidic ionic liquid catalyst (Fig. 15.53) (Banothu and Bavanthula 2012).

Fig. 15.51 Pyrimidinone derivatives

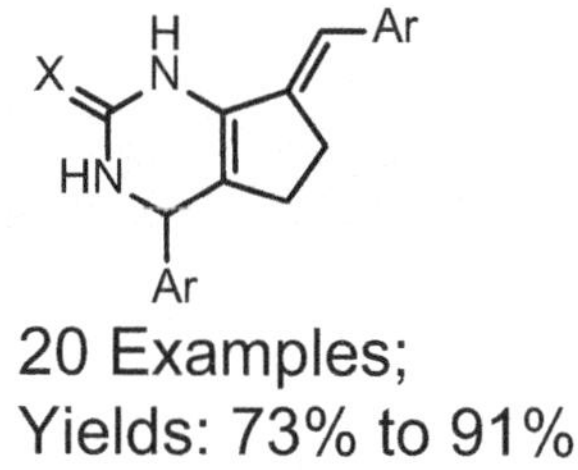

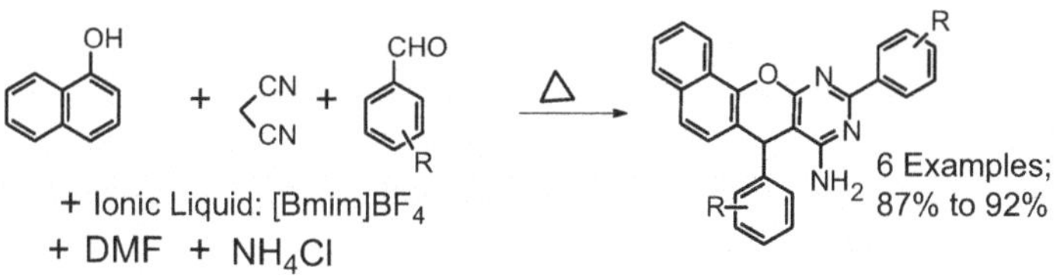

Fig. 15.52 Approach for chromeno-pyrimidines

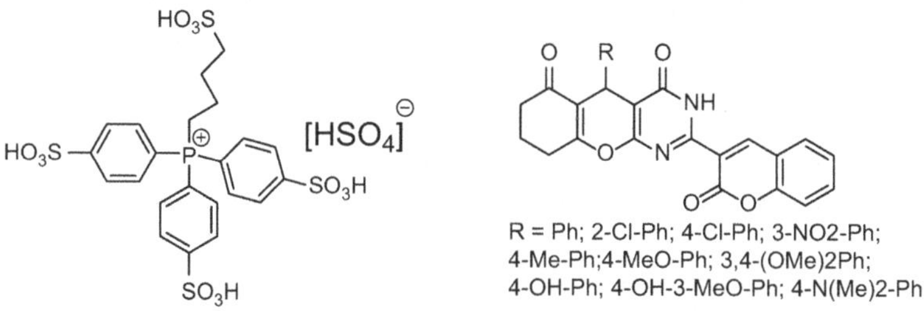

Fig. 15.53 (4-Sulfobutyl) tris (4-sulfophenyl)phosphonium hydrogen sulfate and Chromeno-pyrimidinone derivatives

15.2.17 Acridines

Nader Ghaffari Khaligh synthesized tetrahydrotriazoloacridines employing ionic liquid—[MSIm]HSO_4. Aldehydes, amino-benzotriazole, dimedone, and [MSIm]HSO_4 ionic liquid were microwaved to drive the three-component one-pot reaction to obtain tetrahydrotriazoloacridines (Fig. 15.54) (Khaligh 2017). A complex mechanism was proposed for the product formation.

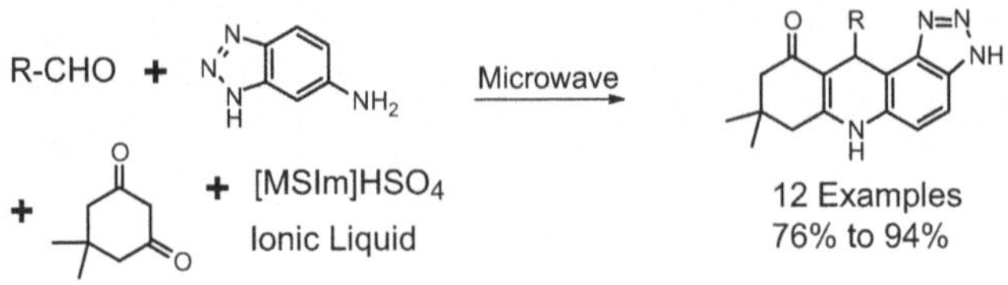

Fig. 15.54 Synthesis of various tetrahydrotriazoloacridines

15.2.18 Quinolines, Isoquinolines, and Fused Analogues

Tetrahydroquinoline derivatives were synthesized by Yu Wan et al. in a four-component one-pot reaction, employing arylaldehydes, malononitrile, cyclohexanone, and amine in ionic liquid as medium and catalyst (Fig. 15.55) (Yu et al. 2011). Three crystal structures were reported to strengthen structural aspects. Reaction conditions were also probed to define suitable experimental conditions. The interesting part of this report was that all the four reactants were added initially together to conduct the reaction.

Avik Kumar Bagdi and Alakananda Hajra developed a new green synthetic protocol for the regioselective synthesis of pyrano[3,2-c]quinoline-2-one derivatives by the tandem cyclization of 4-hydroxy-1-methyl-2-quinolone with chalcones and employing task-specific Brønsted acidic ionic liquid [BAIL] [1-methyl-3(4-sulfobutyl)imidazolium-4-methyl benzene sulfonate] as a reusable catalyst under solvent-free conditions (Fig. 15.56) (Bagdi and Hajra 2014). Chalcones having aryl and hetero-aryl substituents were used. The reaction was assumed to proceed through the Michael addition followed by annulation reaction. Scope of the reaction was extended to prepare a library of the title compounds. During the investigational studies, authors examined the efficacy of various ionic liquids and the catalyst loadings.

Amarsinh Deshmukh et al. executed synthesis of 3-substituted phenyl-2-(4-(tetrazolo[1,5-a]quinolin-4-yl methoxy)phenyl)thiazolidin-4-ones (Fig. 15.57) (Deshmukh et al. 2019), via a three-component one-pot reaction in the presence of a dicationic liquid. (4-(Tetrazolo[1,5-a]quinolin-4-ylmethoxy)benzaldehyde, substituted anilines, and mercaptoacetic acid were reacted in the presence of bis-ionic liquid without any catalyst to produce target molecules for the purpose of antitubercular activity. (4-(Tetrazolo[1,5-a]quinolin-4-ylmethoxy)benzaldehyde reacted with aniline to form imine followed by cyclization involving carboxylic group and amino function leading to thiazolidinone derivative.

Rolando Cannalire et al. reported ionic liquid [TBMA][MsO]-mediated elegant synthesis of ciprofloxacin. They have very successfully demonstrated three-step reaction in a single-pot reaction facilitated by [TBMA][MsO] as shown in Fig. 15.58 (Cannalire et al. 2018). Authors claim that the pharmaceutically relevant quinolone skeleton or ciprofloxacin synthesis is eco-friendly with shorter reaction times, complete conversions, easy workup procedure, and high yields without using conventional solvents.

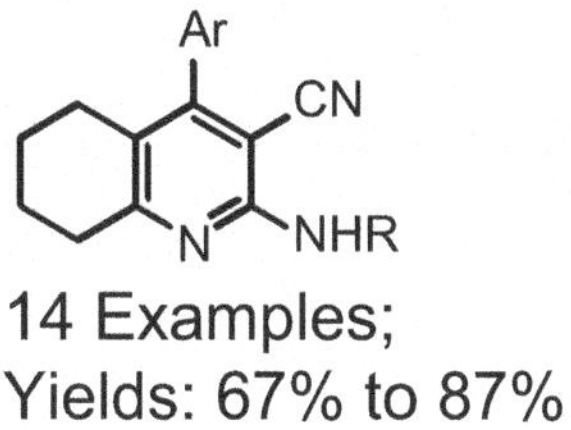

Fig. 15.55 Tetrahydroquinolines

Fig. 15.56 Approach for pyrano-quinolines

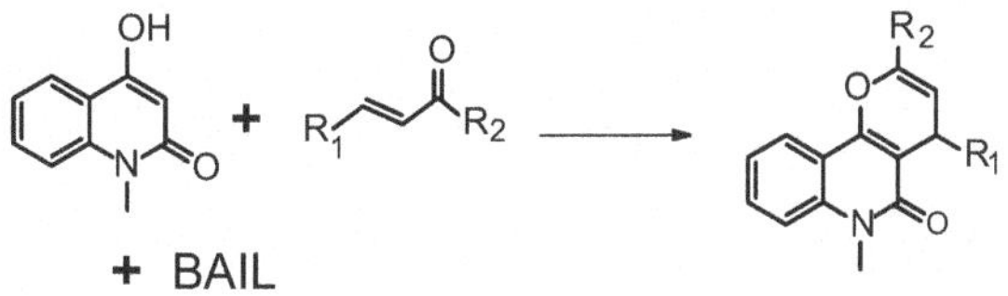

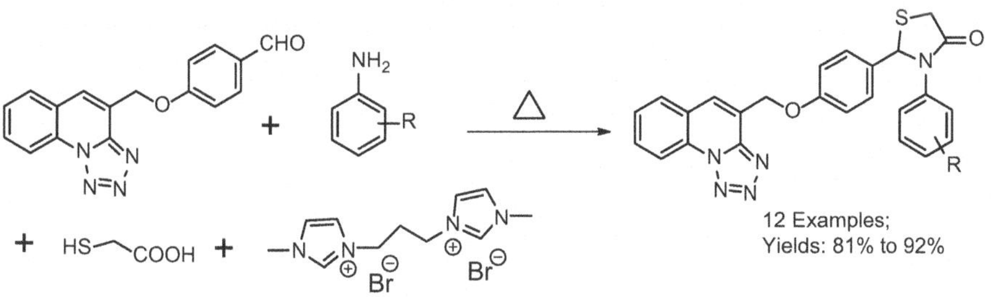

Fig. 15.57 Synthesis of tetrazoloquinoline derivatives

Bai Lin Li and co-workers developed a facile four-component one-pot synthesis for polyhydroquinolines (Fig. 15.59) (Li et al. 2014). Substituted benzaldehydes or aryl aldehydes, dimedone, ethyl acetoacetate, ammonium acetate, and a novel ionic liquid carrying two sulfonic acid groups are involved in reaction to provide high yields of polyhydroquinolines. The reaction was facilitated by ultrasound.

15.2.19 Pyrans, Chromans, and Fused Scaffolds

In a significant communication Nguyen and co-authors discussed the synthesis of a novel magnetic nanomaterial—immobilized, Lewis acidic ionic liquid by the covalent embedding of 3-(3-(trimethoxysilyl) propyl)-1H-imidazol-3-ium chlorozincate(II) ionic liquid onto the surface of Fe_3O_4 nanoparticles. Authors after characterizing the

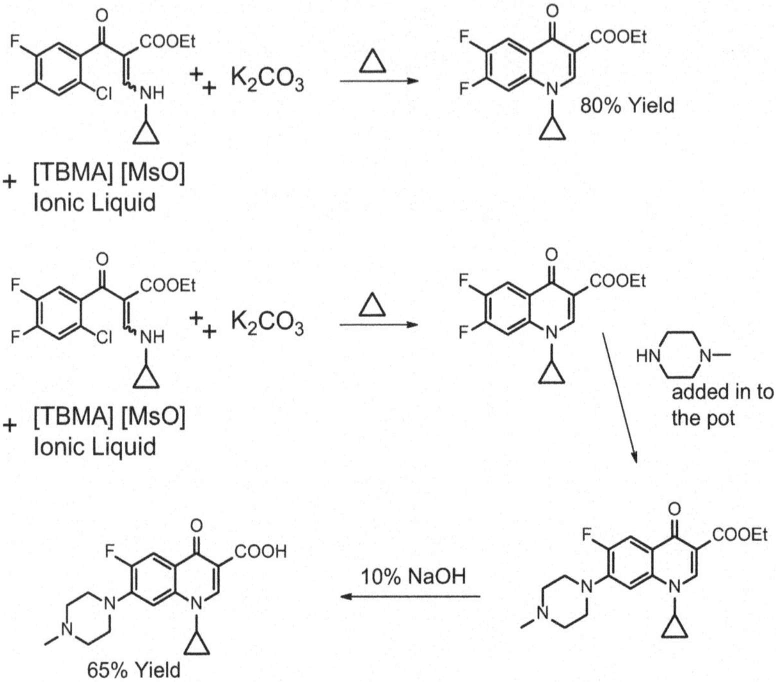

Fig. 15.58 Synthesis of ciprofloxacin

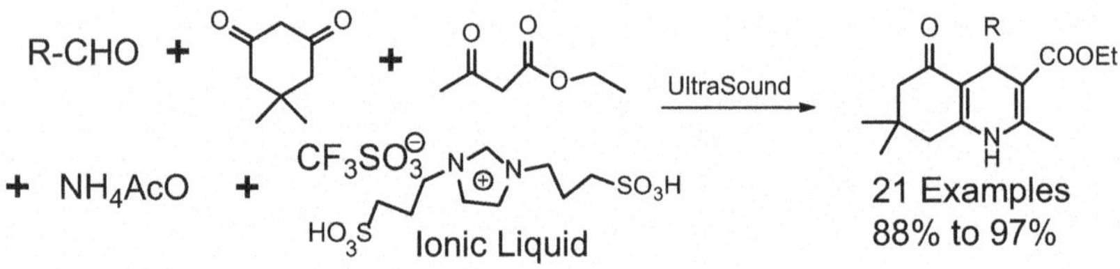

Fig. 15.59 Preparation of polyhydroquinoline derivatives

Fig. 15.60 Three-component reaction for fused tetracyclic pyran derivatives

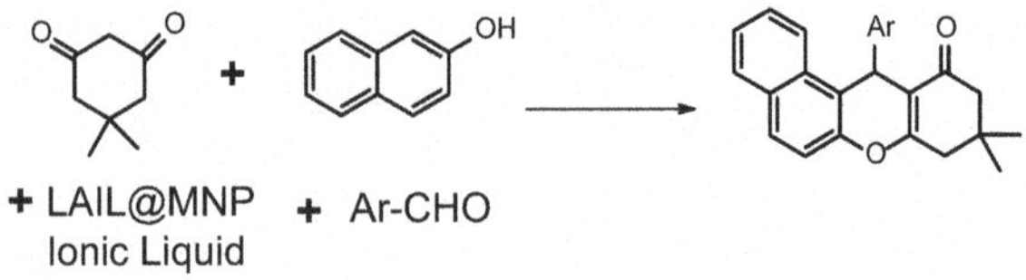

material successfully applied to the one-pot synthesis of benzoxanthenes and pyrroles, under solvent-free conditions (Fig. 15.60) (Nguyen et al. 2018b).

To prove the merit of the newly prepared magnetic nano-Fe3O4-immobilized Lewis acidic ionic liquid material (LAIL@MNP), authors screened and compared the efficiency of different catalysts in the above reaction. Synthesis of pyran-annulated heterocyclic derivatives was achieved from 2-naphthol, dimedone, and various benzaldehydes, under above mentioned reaction conditions.

Authors also extended the applicability of the new catalyst to the Paal-Knorr synthesis of 2,5-dimethyl-1-phenyl-1H-pyrroles from anilines and acetylacetone (Fig. 15.61). Plausible mechanism was explained by the authors for the reactions catalyzed by the new LAIL@MNP.

A multicomponent one-pot reaction involving 1,3-diketone (or imino analogue), substituted *ortho*-propargyloxy benzaldehyde, CuI, K_2CO_3, and [Bmim]NO_3 ionic liquid was reported by Saeed Balalaie et al. Substituted propargyloxy benzaldehyde reacted with 1,3-diketone (or imino analogue) to provide Knoevenagel condensation product as intermediate (Fig. 15.62) (Balalaie et al. 2012). The in situ-generated intermediate was in a suitable format with enone (diene equivalent) and acetylenic group (dienophile) for the possible hetero-Diels-Alder reaction, leading to fused pyran skeleton.

Jitender M. Khurana et al. reported an approach for cyclized benzochromene derivatives (Fig. 15.63) (Khurana et al. 2013) utilizing ionic liquids in a one-pot three-component reaction with aryl aldehydes, hydroxynaphthoquinone, malononitrile, or ethyl cyanoacetate and [Bmim]OH. The condensation product between aldehyde and active methylene group of malononitrile underwent reaction with hydroxynaphthoquinone to provide various benzochromene derivatives.

Wide range of 2-amino-4H-chromenes were synthesized by Arjun Kumbar et al. from readily available α- or β-naphthol, malononitrile, and aromatic aldehydes under microwave conditions catalyzed by mesitylene-tagged multicationic recyclable ionic liquid containing acetate in a one-pot multicomponent strategy (Fig. 15.64) (Kumbhar et al. 2016). Authors prepared a new class of mono-, bis-, and tris-imidazolium-based ILs containing 1,3,5-alkylidene 2,4,6-trimethylbenzene linkers and screened these ILS for the reaction to check the efficiency.

Cheng-Bin Li et al. developed an efficient synthesis of 2-amino-4*H*-chromenes using DABCO-based ionic liquid. Substituted salicylaldehydes, malononitrile, and substituted indole or pyrazalone or dimedone were reacted in the presence of [DABCO-H]HSO_4 in ethanol medium as a multicomponent one-pot reaction to provide 2-amino-4H-chromene and benzopyranone derivatives (Fig. 15.65) (Li et al. 2018).

Synthesis of a wide range of 3,4-dihydropyrano[c]chromene, 4,5-dihydropyrano [4,3-b]pyran, and tetrahydrobenzo[b] pyran scaffolds through eco-friendly Domino-Knoevenagel-Michael multicomponent reaction was demonstrated by Himani Sharma and Suman Srivastava (2018). Recyclable saccharin-based ionic liquid [Bmim]Sac was employed as catalyst (Fig. 15.66) (Sharma and Srivastava 2018). During the investigations, authors screened several ionic liquid catalysts for their efficacy for the above reactions and concluded that [Bmim]Sac in water medium at 80 °C was an ideal condition.

3,4-Dihydropyrano[c]chromenes were prepared from malononitrile, 4-hydroxycoumarin, and various aromatic aldehydes in the presence of [Bmim]Sac. Tetrahydrobenzo[b]pyrans were obtained by the reaction of malononitrile,

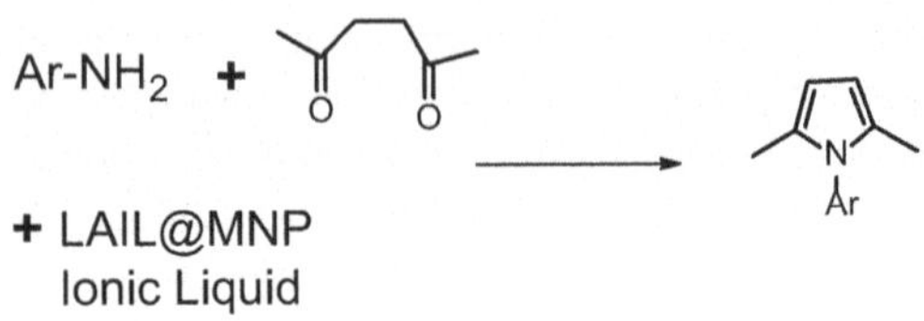

Fig. 15.61 Approach for pyrrole derivatives

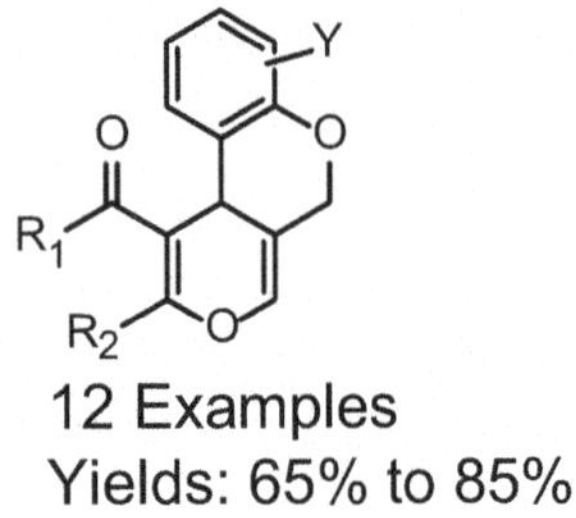

Fig. 15.62 Fused benzopyran skeleton derivatives from hetero-Diels-Alder reaction

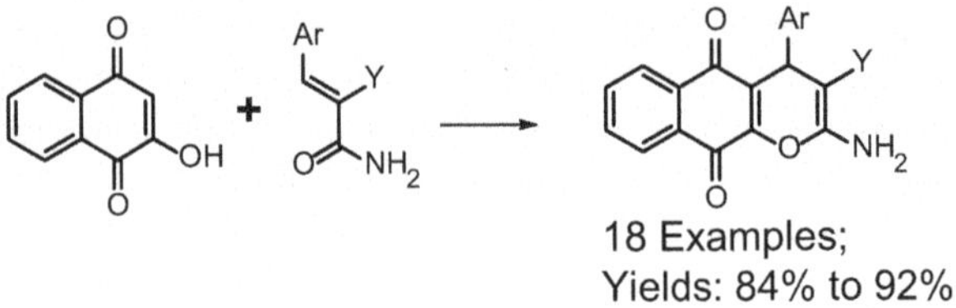

Fig. 15.63 Scheme for benzochromene derivatives

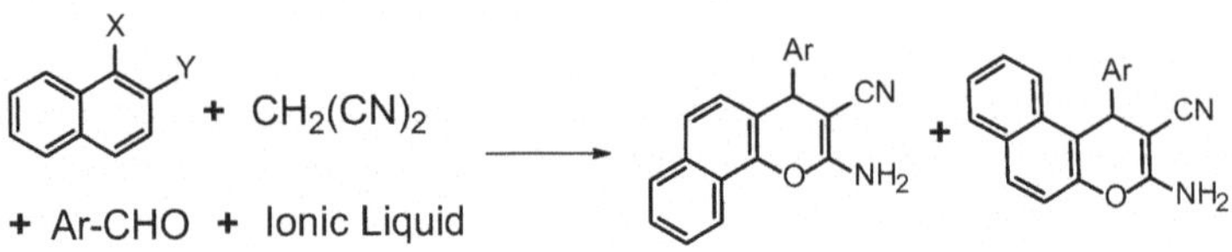

Fig. 15.64 Preparation of 2-amino-4H-chromenes

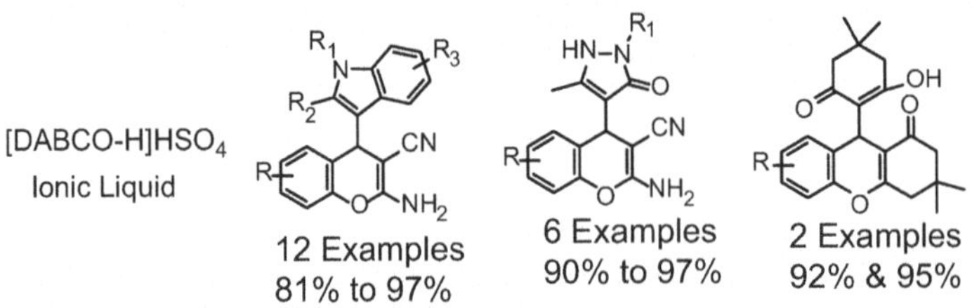

Fig. 15.65 2-Amino-4H-chromene and benzopyranone derivatives

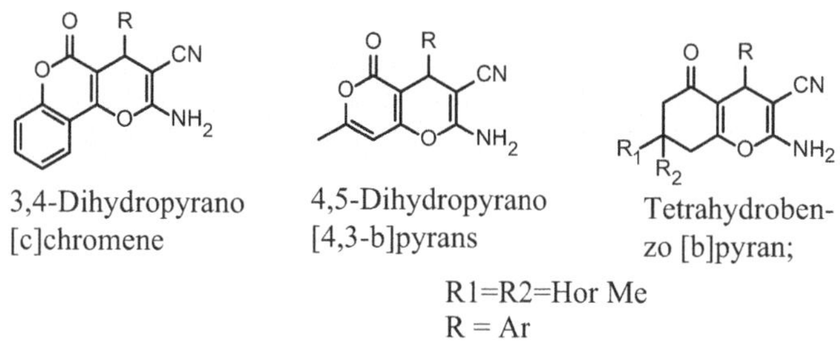

Fig. 15.66 Pyrano-chromenes and fused pyrano-chromenes

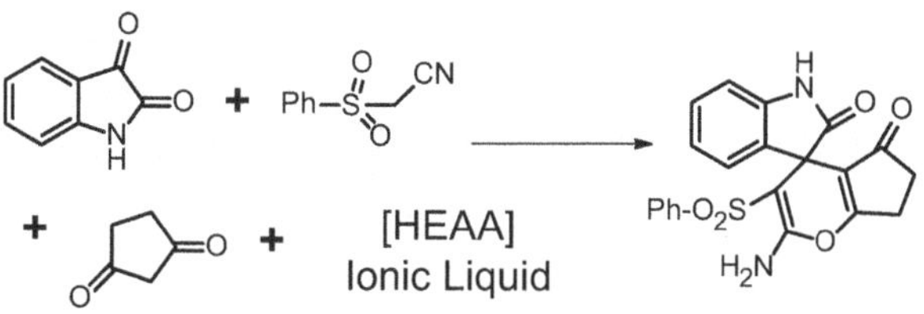

Fig. 15.67 Spiro derivatives from isatin

dimedone, or 1,3-cyclohexanedione with different aromatic aldehydes in the presence of [Bmim]Sac. Scope of the reaction was also extended to include 4-hydroxy-6-methyl-2-pyran as the cyclic 1,3-dicarbonyl compound to achieve the synthesis of 4,5-dihydropyrano[4,3-b]pyran derivatives.

Shuai-Shuai Jin et al. described the ionic liquid-catalyzed eco-friendly one-pot synthesis of a series of novel spiro-2-amino-3-phenylsulfonyl-4H-pyran derivatives via a three-component reaction of 1,3-dicarbonyl compounds and phenyl sulfonyl acetonitrile with isatins or acenaphthenequinone in ethanol under mild conditions. Authors have used basic ionic liquid 2-hydroxyethyl ammonium acetate, HEAA, [$NH_3CH_2CH_2OH$][CH_3COO], as a reusable catalyst (Fig. 15.67) (Jin et al. 2013).

15.2.20 Coumarins and Related Fused Scaffolds

Various indole-3-dihydrocoumarins were synthesized by the C-3 functionalization of indoles with dihydrocoumarin, as reported by Kumar et al., in the presence of reusable artificial saccharin-based functional ionic liquids (imidazolium saccharinate) ([C_4MIM]Sac). This new multicomponent methodology involves Knoevenagel coupling of salicylaldehyde and Meldrum's acid followed by a Michael-type reaction with indole followed by lactonization via decarboxylative elimination. Authors reported that among several imidazolium saccharinates examined [C_4MIM]Sac provided better results (Fig. 15.68) (Kumar et al. 2012).

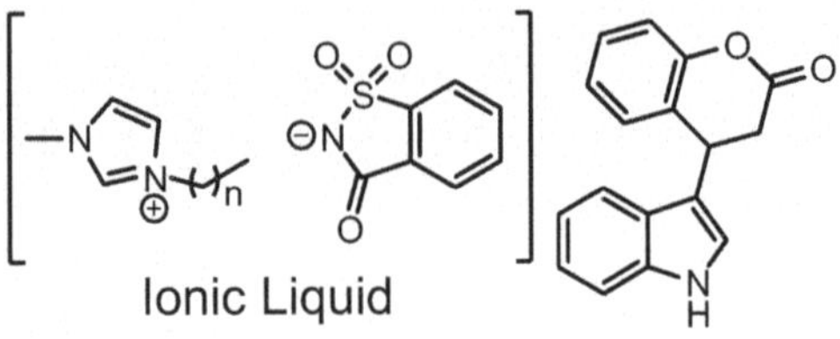

Fig. 15.68 Imidazolium saccharinate-induced chemistry

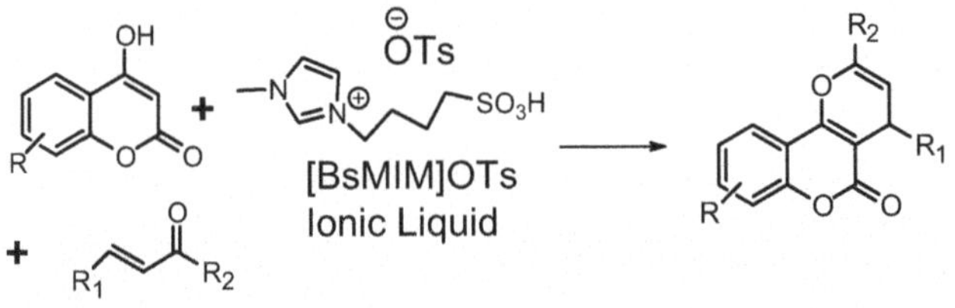

Fig. 15.69 Synthesis of pyrano-coumarin derivatives

Sachinta Mahto et al. described an effective regioselective synthesis of a wide range of pyrano(3,2-c)coumarin derivatives by the tandem cyclization of 4-hydroxy coumarin with various chalcones catalyzed by reusable Brønsted acidic ionic liquid—1-butane sulfonic acid-3-methylimidazolium tosylate—(BsMIM)OTs, under metal- and solvent-free conditions (Fig. 15.69) (Mahato et al. 2017). It was explained that the reaction proceeds through the Michael addition followed by the annulation reaction. During the study authors examined the effects of different other ionic liquids on the course of the reaction.

15.2.21 Thiazepines and Fused Analogues

An environmentally benign protocol for "on water"-mediated one-pot multicomponent synthesis of various spiro[pyrazolo(4,3-c)][1,5]benzothiazepines by the reaction of cyclic ketones or isatin with 3-methyl-1-phenyl-2-pyrazolin-5-one and substituted thiols, catalyzed by reusable ZSM-5-([BMIM]BF_4 catalyst system, was described by Kapil Arya and Biswajeet Prabhakar (Fig. 15.70) (Arya and Prabhakar 2013). During the course of investigations authors studied the effect of different ionic liquids confined into ZSM-5 zeolite as well as several solvent systems. The best results were reported in aqueous-mediated ZSM-5([BMIM)BF_4) catalytic system.

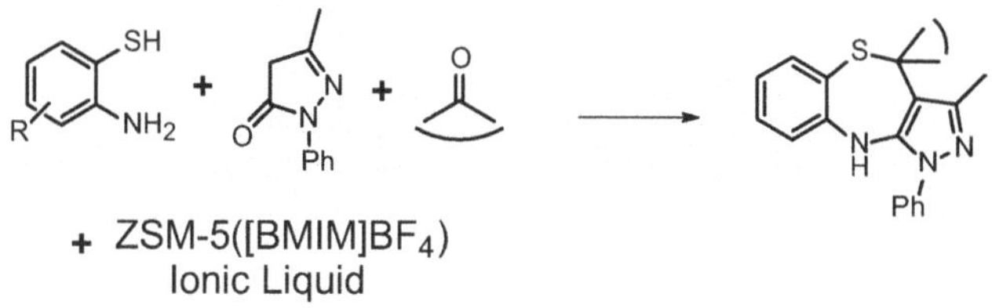

Fig. 15.70 Synthesis of various spiro[pyrazolo(4,3-c)][1,5]benzothiazepines

Fig. 15.71 Scaffolds of 1,8-dioxo-octahydroxanthenes

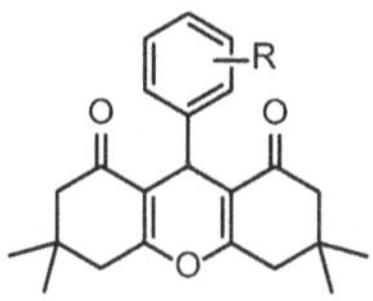

R = 4-NO_2; 2-Cl; 2-NO_2; 3-NO_2;3-Cl; 4-Cl; 4-OMe; 4-Me; 4-OH; 4-F; 2-Br; 3-Br; 2-OH; 2-Furyl; 2-Thienyl; 2-Pyridyl; 2,5-$(OMe)_2$; 3,4-$(OMe)_2$; 3,4,,5-$(OMe)_3$;

15.2.22 Xanthenes and Related Analogues

Abhishek Dadhania et al. have described an efficient eco-friendly procedure for the synthesis of 1,8-dioxo-octahydroxanthenes by the condensation of 5,5-dimethyl-1,3-cyclohexanedione with structurally diverse aldehydes employing carboxy-functionalized ionic liquid as a reusable catalyst ([MMIm]BF_4) under microwave conditions (Fig. 15.71) (Dadhania et al. 2017).

Somayyeh Sarvi Beigbaghlon et al. developed a new ionic liquid solid acid based on clay as an efficient, thermally stable, and reusable catalyst and applied for the synthesis of several xanthenes and coumarin derivatives under solvent-free conditions. Ammonium hydrogen sulfate propyl triethoxy silane was anchored on to Na+-montmorillonite by covalent bonds to provide required IL-AHS@MMT, which was fully characterized before evaluating in the organic transformations. Aryl-14H-dibenzo[a,j]xanthenes were prepared from 2-naphthol and different aromatic aldehydes catalyzed by IL-AHS@MMT. Different 12-aryl-8,9,10,12-tetrahydrobenzo[a]xanthenes-11-ones were prepared by reacting 2-naphthol, dimedone, and aromatic aldehydes in the presence of IL-AHS@MMT. Coumarin derivatives were also synthesized by heating a mixture of phenols and ethyl acetoacetate and IL-AHS@MMT (Fig. 15.72) (Beigbaghlou et al. 2016).

A simple eco-friendly solvent-free multicomponent reaction involving 2-naphthol, dimedone, and diversified aromatic aldehydes was described by Pranab Das and Jupitara Das and obtained a wide variety of xanthenes and benzoxanthene derivatives (Fig. 15.73) (Das and Das 2015). Authors prepared environmentally benign and recyclable secondary amine-based ionic liquids from easily available dialkylamines and concentrated sulfuric acid and applied them as catalysts in the above reaction to assess the efficiency. It was observed that [(n-propyl)2NH2][HSO4] was a better choice.

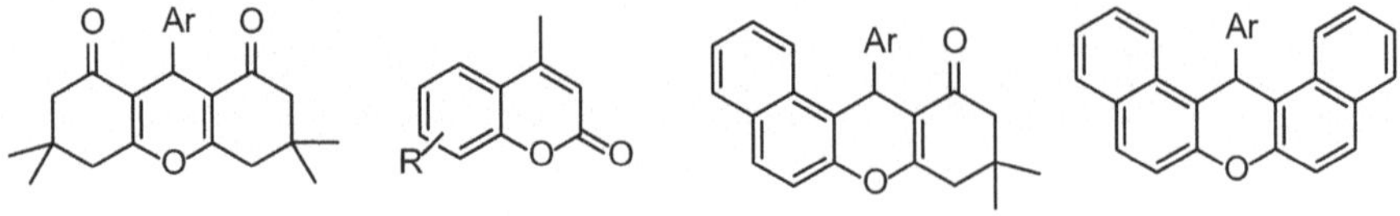

Fig. 15.72 Fused oxygen heterocycles

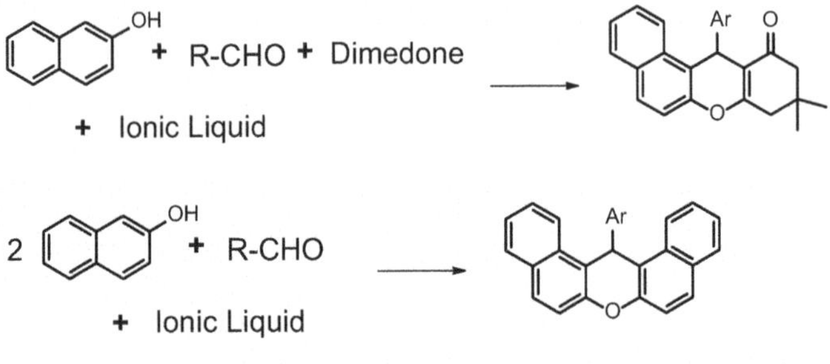

Fig. 15.73 Preparation of xanthenes and benzoxanthene derivatives

Fig. 15.74 Scaffolds of tetrahydrobenzoxanthines

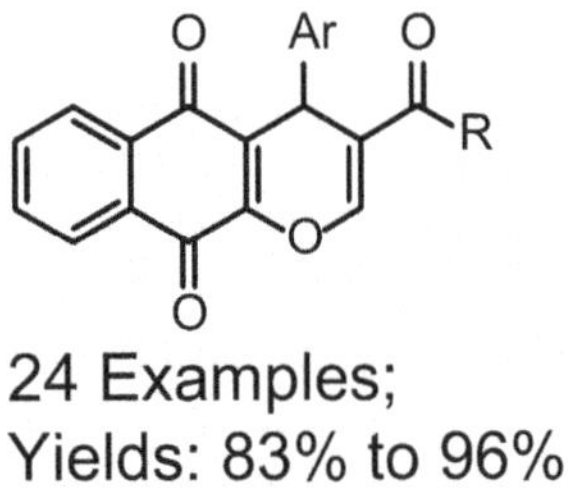

Tetracyclic tetrahydrobenzoxanthine derivatives (Fig. 15.74) (Khurana et al. 2012) were synthesized by Jitendar Khurana et al. using aryl aldehydes, hydroxynaphthoquinone, and 1,3-diones employing task-specific ionic liquid [Bmim]HSO_4 as the medium with catalytic amount of H_2SO_4. Aryl aldehyde underwent condensation with active methylene group of 1,3-dione compound to produce insitu an olefin intermediate, which then reacted with hydroxynaphthoquinone to afford cyclized product, benzoxanthine derivative.

Zhiwei Chen et al. reported the synthesis of a new unique ionic liquid DMDBSI carrying two sulfonic acid groups and chemically characterized. The DMDBSI ionic liquid can be used for alkylation, esterification, nitration, hydrolysis, and synthesis of heterocycles. It was employed for a one-pot multicomponent green synthesis of heterocycles, involving hydroxycoumarin, substituted benzaldehydes, and substituted cyclohexane-1,3-diones reacted in water. 10,11-Dihydrochromeno[4,3-b]chromene-6,8(7H,9H)-dione was found to be angular type indicating the selectivity observed in the reaction (Fig. 15.75) (Chen et al. 2011).

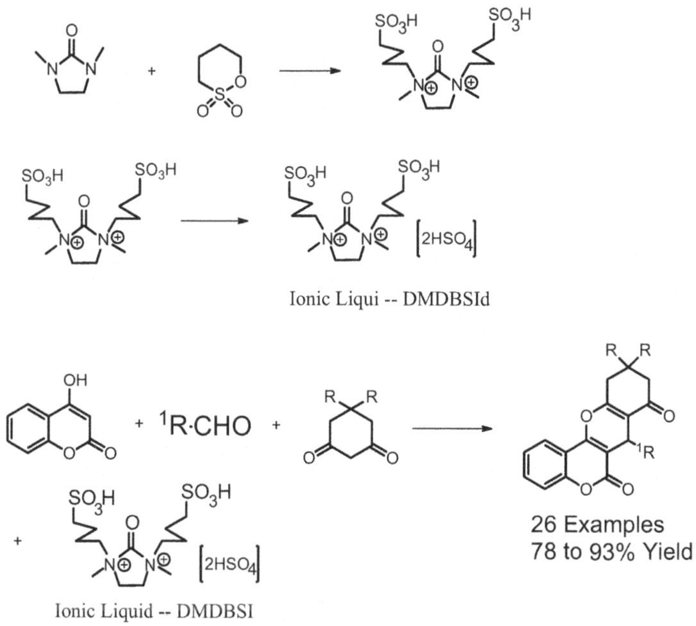

Fig. 15.75 Ionic liquid synthesis and selective formation of 10,11-dihydro-chromeno[4,3-b] chromene-6,8(7H,9H)-diones

Hamid Reza Shaterian et al. described a one-pot multicomponent reaction involving diversified aldehydes, malononitrile, 4-hydroxycoumarin, and a task-specific ionic liquid leading to the formation of 2-amino-5-oxo-4,5-dihydropyrano[3,2-c]chromene-3-carbonitrile derivatives (Fig. 15.76) (Shaterian and Honarmand 2011). The 4-hydroxyl group of coumarin directs angular cyclization leading to the products. Other researchers, Jia Zheng and Yiqun Li, also made the same skeleton of pyrano-chromene derivatives but using a different task-specific ionic liquid (Fig. 15.77) (Zheng and Li 2011). Iranian group of researchers, Shaterian and Oveisi, also reported a similar kind of transformation with another different task-specific ionic liquid under solid-state grinding conditions leading to 2-amino-5-oxo-4,5-dihydropyrano[3,2-c]chromene-3-carbonitrile derivatives (Fig. 15.78) (Shaterian and Oveisi 2011).

Synthesis of benzopyranopyrimidines and 4H-chromenes was executed by Amit Kumar Gupta et al., adopting a one-pot multicomponent reaction with the aid of ionic liquid. *o*-Hydroxybenzaldehyde, malononitrile, amine, and [Bmim]BF_4 ionic liquid were involved in a reaction at room temperature to afford the fused heterocyclic system, benzopyranopyrimidine (Fig. 15.79) (Gupta et al. 2012). The ortho-hydroxyl group of aldehyde has a role to direct the cyclization process into a linear fused manner. It was observed that other ionic liquids employed were found to be ineffective in bringing the reaction to a satisfactory completion. The amine component was replaced with phenylthiol to get totally different products—4-H-chromenes (Fig. 15.79).

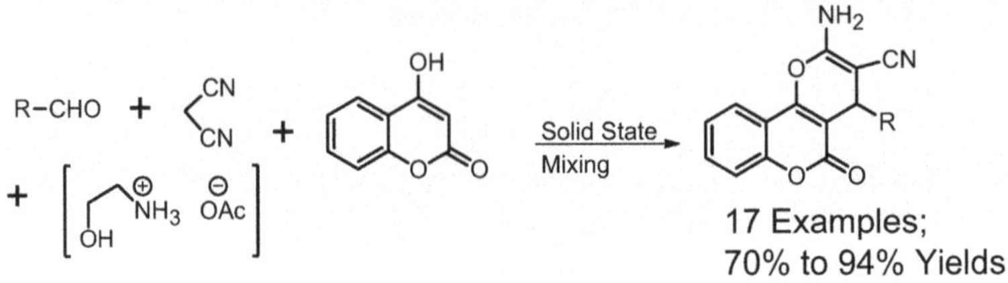

Fig. 15.76 Approach to 2-amino-5-oxo-4,5-dihydropyrano[3,2-c]chromene-3-carbonitrile derivatives

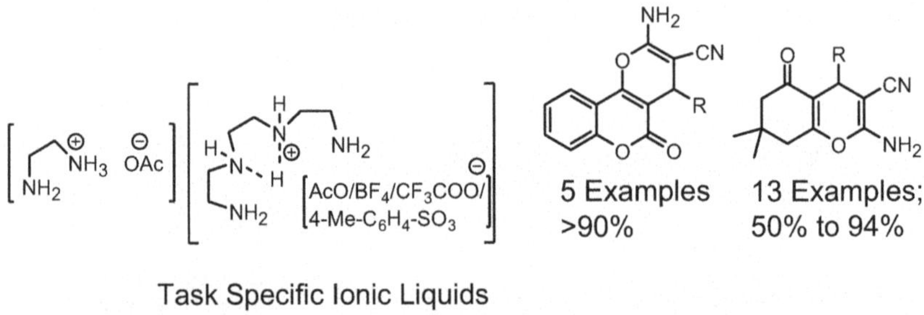

Fig. 15.77 2-Amino-5-oxo-4,5-dihydropyrano[3,2-c]chromene-3-carbonitrile derivatives

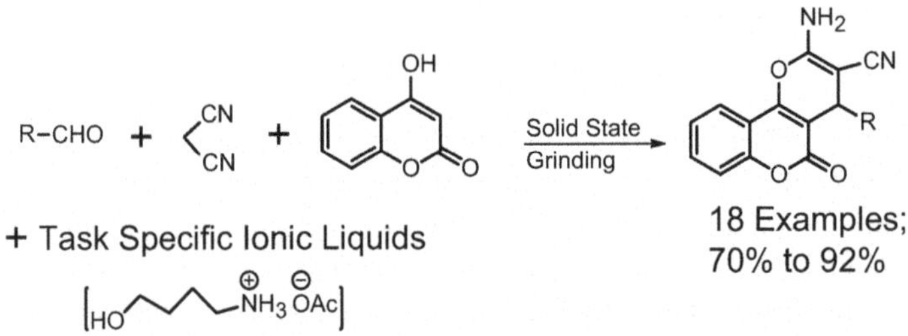

Fig. 15.78 2-Amino-5-oxo-4,5-dihydropyrano[3,2-c]chromene-3-carbonitrile derivatives

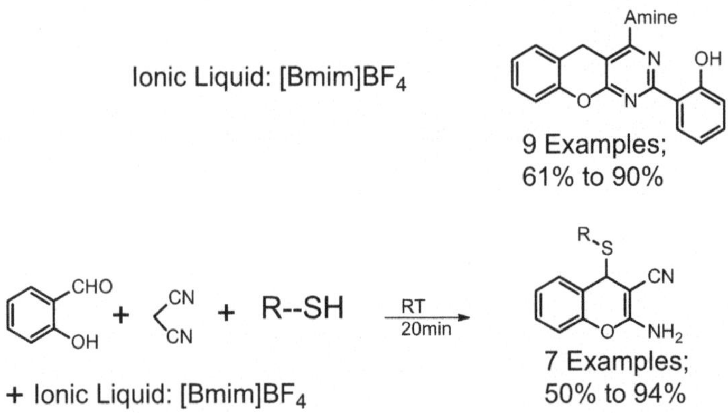

Fig. 15.79 Benzopyranopyrimidines and chromenes

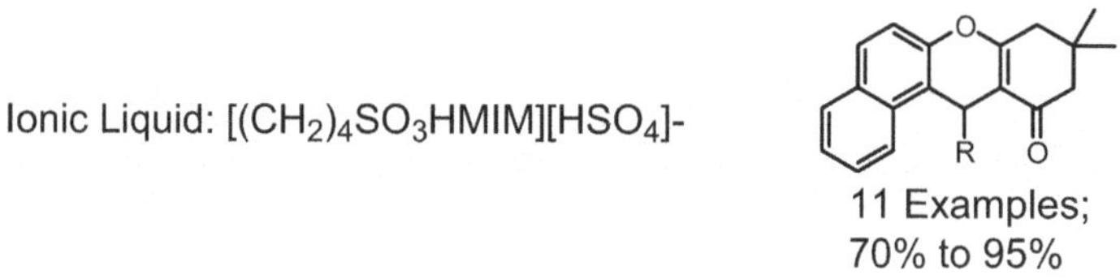

Fig. 15.80 Tetrahydrobenzo[*a*]xanthene-11-one derivatives

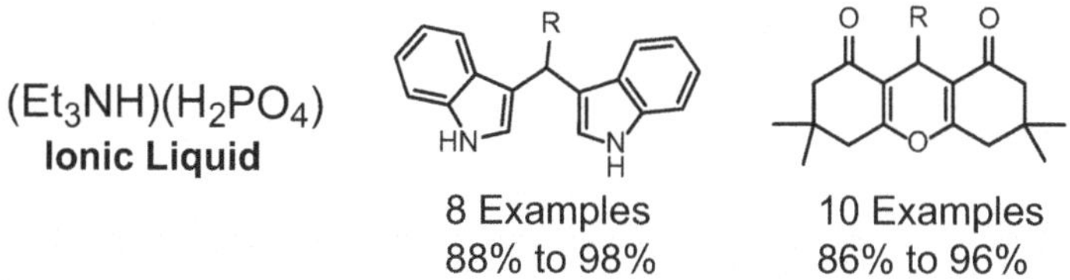

Fig. 15.81 Bis(indolyl)methanes and 1,8-dioxo-octahydroxanthenes

Majid Heravi and co-workers achieved an efficient synthesis of tetrahydrobenzo[*a*]xanthene-11-one derivatives by adopting a one-pot three-component approach involving dimedone, β-naphthol, and diversely substituted aldehydes in ionic liquid. Among the three known ionic liquids prepared, two ionic liquids with sulfonic acid residues worked well compared to the other one. Variation in the aldehyde component provided a wide range of tetrahydrobenzo[*a*]xanthene-11-one derivatives (Fig. 15.80) (Zakeri et al. 2011).

A simple, cost-effective, and mildly acidic ionic liquid was prepared from triethylamine and phosphoric acid. The same was utilized to synthesize bis(indolyl)methanes and 1,8-dioxo-octahydroxanthenes (Fig. 15.81) (Kalantari 2012). Ionic liquid utilized in these investigations was compared with the other reported ionic liquids in terms of reaction time and yields and found to be superior.

15.3 Conclusions

This decade has witnessed considerable growth in chemical research leading to the wider applications of ionic liquids, particularly in the synthesis of varied heterocyclic scaffolds aiming at broader utility of the targeted compounds. Even though standard classical methods for the heterocyclic synthesis do exist, several eco-friendly approaches are being developed with a goal of reducing the pollution burden on environment. Further research involving ionic liquids for achieving sustainable reaction protocols will ensure rapid growth in this active and potential area of heterocyclic chemistry for the construction of a broad spectrum of biologically important molecular entities.

Acknowledgements This review was authored purely out of academic interest to familiarize the young organic researchers to appraise them about the recent work appearing in the green chemistry arena, particularly related to ionic liquids. The examples covered in this review are chosen from different internationally reputed journals. The authors of this review are highly appreciative of the research articles published for their contributions towards the ever-expanding area of the ionic liquids significantly in the heterocyclic synthesis. This review is only representative in nature and not intended to be exhaustive. The authors of this review further acknowledge the original contributors and publishers of the articles cited here for their potential scientific work, with a larger interest in academic excellency. VJR thank Dr. B. Parthasaradhi Reddy, Chairman, Hetero Drugs, and Dr. K. Ratnakar Reddy, Director HRF, Hyderabad, for encouragement. VJR thank CSIR-New Delhi for Emeritus Scientist Honor.

References

Abbasi M (2015) Design, preparation and characterization of a new ionic liquid, 1,3-disulfonic acid benzimidazolium chloride, as an efficient and recyclable catalyst for the synthesis of tetrahydropyridine under solvent-free conditions. RSC Adv 5:67405–67411. https://doi.org/10.1039/C5RA10699K

Aliabadi RS, Mahmoodi NO (2016) Green and efficient synthesis of pyranopyrazoles using [bmim][OH$^-$] as an ionic liquid catalyst in water under microwave irradiation and investigation of their antioxidant activity. RSC Adv 6:85877–85884. https://doi.org/10.1039/C6RA17594E

Arya K, Prabhakar B (2013) Ionic liquid confined zeolite system: an approach towards water-mediated room temperature synthesis of spiro[pyrazolo[3,4-*e*]benzothiazepines]. Green Chem 15:2885–2894. https://doi.org/10.1039/C3GC40553B

Arya K, Rawat DS, Sasai H (2012) Zeolite supported Brønsted-acid ionic liquids: an eco approach for synthesis of spiro[indole-pyrido[3,2-*e*]thiazine] in water under ultrasonication. Green Chem 145:1956–1963. https://doi.org/10.1039/C2GC35168D

Bagdi AK, Hajra A (2014) Brønsted acidic ionic liquid catalyzed tandem reaction of 4-hydroxy-1-methyl-2-quinolone with chalcone: regioselective synthesis of pyrano[3,2-*c*]quinolin-2-ones. RSC Adv 4:23287–23291. https://doi.org/10.1039/C4RA03221G

Balalaie S, Azizian J, Shameli A, Bijanzadeh HR (2012) CuI–ionic liquids as efficient reaction media for the synthesis of pyran skeleton via domino Knoevenagel–hetero–Diels–Alder reaction with unactivated alkynes. Synth Commun 43:1787–1795. https://doi.org/10.1080/00397911.2012.669874

Balaskar RS, Gavade SN, Mane MS, Shingate BB, Shingare MS, Mane DV (2010) Greener approach towards the facile synthesis of 1,4-dihydropyrano[2,3-c]pyrazol-5-yl cyanide derivatives at room temperature. Chin Chem Lett 21:1175–1179. https://doi.org/10.1016/j.cclet.2010.06.013

Banothu J, Bavanthula R (2012) Brønsted acidic ionic liquid catalyzed highly efficient synthesis ofchromenopyrimidinone derivatives and their antimicrobial activity. Chinese Chem Lett 23:1015–1018. https://doi.org/10.1016/j.cclet.2012.06.041

Beigbaghlou SS, Marjani K, Habibi A, Atghia SV (2016) Introduction of a new ionic liquid solid acid based on clay as an efficient, recyclable and thermally stable catalyst for organic transformations. RSC Adv 6:20306–20316. https://doi.org/10.1039/C5RA25297K

Cannalire R, Tiecco M, Cecchetti V, Germani R, Manfroni G (2018) Advantageous use of ionic liquids for the synthesis of pharmaceutically relevant quinolones. Eur J Org Chem:2977–2983. https://doi.org/10.1002/ejoc.201800415

Chakraborty B, Chettri E (2018) Synthesis of some novel class of regioselective spiro isoxazolidine derivatives *via* 1,3-dipolar cycloaddition reaction of *N*-benzyl-*C*-fluorosubstituted phenyl nitrones in ionic liquid. J Hetrocyclic Chem 55:1157–1165. https://doi.org/10.1002/jhet.3148

Chanda K, Maiti B, Tseng C-C, Sun C-M (2012) Microwave-assisted linear approach toward highly substituted benzo[d]oxazol-5-yl-1H-benzo[d]imidazole on ionic liquid support. ACS Comb Sci 14:115–123. https://doi.org/10.1021/co200188g

Chavan SS, Degani MS (2012) Ionic liquid mediated one-pot synthesis of 6-aminouracils. Green Chem 14:296–299. https://doi.org/10.1039/C1GC15940B

Chen Z, Zhu Q, Weike S (2011) A novel sulfonic acid functionalized ionic liquid catalyzed multicomponent synthesis of 10,11-dihydrochromeno[4,3-b]chromene-6,8(7H,9H)-dione derivatives in water. Tetrahedron Lett 52:2601–2604. https://doi.org/10.1016/j.tetlet.2011.03.059

Chen L-H, Kao C-H, Dhole S, Barve IJ, Li-ChingShen W-SC, Sun C-M (2016) Regioselective synthesis of imidazo[1,5-*a*]quinoxalines and methyl *N*-phenylbenzimidats on an ionic liquid support. RSC Adv 6:76123–76127. https://doi.org/10.1039/C6RA11861E

Chen D-S, Liu S-J, Lu W-Q, Wang X-S (2017) Green synthesis of 6-Aryl-5,6-dihydrobenzo[4,5]imidazo[1,2-c]quinazoline derivatives in ionic liquid under catalyst-free conditions. J Heterocyclic Chem 55:166–172. https://doi.org/10.1002/jhet.3020

Choudhary S, Muthyala MK, Kumar A (2014) Ionic liquid phase synthesis (IoLiPS) of 2-aminothiazoles and imidazo[1,2-a]pyridines. RSC Adv 4:47368–47372. https://doi.org/10.1039/C4RA08009Bx.2

Cui LH, Yan Fan X (2012) A sustainable synthesis of 2-Benzoxazyl and 2-Benzothiazyl Ketones from alkynyl bromides and 2-amino(thio)phenols promoted by a recyclable catalytic system. Chin J Chem 30:992–996. https://doi.org/10.1002/cjoc.201100472

Dadhania AN, Patel VK, Raval DK (2017) Ionic liquid promoted facile and green synthesis of 1,8-dioxo-octahydroxanthene derivatives under microwave irradiation, J. Saudi Chemical Society 21:S163–S169. https://doi.org/10.1016/j.jscs.2013.12.003

Das PJ, Das J (2015) Secondary amine based ionic liquid: an efficient catalyst for solvent free one pot synthesis of xanthenes and benzoxanthenes. RSC Adv 5:11745–11752. https://doi.org/10.1039/C4RA12298D

Davarpanah J, Kiasa AR (2015) Nano Brönsted solid acid containing double-charged diazoniabicyclo[2.2.2]octane chloride supported on nano rice husk silica: an efficient catalyst for the one-pot synthesis of phthalazine compounds. RSC Adv 5:7986–7993. https://doi.org/10.1039/C4RA13263G

Deshmukh AR, Dhumal ST, Nawale LU, Khedkar VM, Sarkar D, Mane RA (2019) Dicationic liquid mediated synthesis of tetrazoloquinolinylmethoxy phenyl 4-thiazolidinones and their antibacterial and antitubercular evaluation. Synthe. Commun 49:587–601. https://doi.org/10.1080/00397911.2018.1564928

Ding M, Guo H (2016) Ionic liquid catalyzed one-pot synthesis of 2H-pyridazino[1,2-a]indazole-1,6,9(11H)-triones via three-component reaction under solvent-free conditions. J Heterocyclic Chem 53:2061–2065. https://doi.org/10.1002/jhet.1736

Ebrahimi J, Mohammadib A, Pakjoo V, Bahramzade E, Habibi A (2012) Highly efficient solvent-free synthesis of pyranopyrazoles by a Brønsted-acidic ionic liquid as a green and reusable catalyst. J Chem Sci 124:1013–1017

Guo H-Y, Li X-J, Yu Y (2012) Ionic liquid-catalyzed one-pot synthesis of 1H-pyrano[2,3-d]pyrimidin-2(8aH)-one derivatives via three-component reaction under solvent-free conditions. Synthe. Commun 42:3011–3020. https://doi.org/10.1080/00397911.2011.574246

Gupta AK, Kumari K, Singh N, Raghuvanshi DS, Singh KN (2012) An eco-safe approach to benzopyranopyrimidines and 4H-chromenes in ionic liquid at room temperature. Tetrahedron Lett 53:650–653. https://doi.org/10.1016/j.tetlet.2011.11.116

Hajipour AR, Ghayeb Y, Sheikhan N, Ruoho AE (2011) Brønsted acidic ionic liquid as an efficient and reusable catalyst for one-pot, three-component synthesis of pyrimidinone derivatives via Biginelli-type reaction under solvent-free conditions. Synthe Commun 41:2226–2233. https://doi.org/10.1080/00397911.2010.501474

Hajjami M, Tahmasbi B (2015) Synthesis and characterization of glucosulfonic acid supported on Fe_3O_4 nanoparticles as a novel and magnetically recoverable nanocatalyst and its application in the synthesis of polyhydroquinoline and 2,3-dihydroquinazolin-4(1H)-one derivatives. RSC Adv 5:59194–59203. https://doi.org/10.1039/C5RA08952B

Heravi MM, Fazeli A, Oskooie HA, Beheshtiha YS, Valizadeh H (2012) Click synthesis of 5-substituted 1H-Tetrazoles from aldehydes, hydroxylamine, and [bmim]N3 via one-pot, three-component reaction. Synlett 23:2927–2930. https://doi.org/10.1055/s-0032-1317671

Jia L, Wang Y, Qiao Y, Qi Y, Hou X (2014) Efficient one-pot synthesis of deoxyfructosazine and fructosazine from D-glucosamine hydrochloride using a basic ionic liquid as a dual solvent-catalyst. RSC Adv 4:44253–44260. https://doi.org/10.1039/C4RA06832G

Jin S-S, Wang H, Guo H-Y (2013) Ionic liquid catalyzed one-pot synthesis of novel spiro-2-amino-3-phenylsulfonyl-4H-pyran derivatives. Tetrahedron Lett 54:2353–2356. https://doi.org/10.1016/j.tetlet.2013.02.073

Kalantari M (2012) Synthesis of 1,8-dioxo-octahydroxanthenes and bis(indolyl)methanes catalyzed by [Et3NH][H2PO4]as a cheap and mild acidic ionic liquid. Arab J Chem 5:319–323. https://doi.org/10.1016/j.arabjc.2010.09

Kameshwara Rao V, Chhikara BS, Tiwarib R, Shirazi AN, Parang K, Kumar A (2012) One-pot regioselective synthesis of tetrahydroindazolones and evaluation of their antiproliferative and Src kinase inhibitory activities. Bioorg Med Chem Lett 22:410–414. https://doi.org/10.1016/j.bmcl.2011.10.124

Kanakaraju S, Prasanna B, Srinivas B, Chandramouli GVP (2012) Ionic liquid catalyzed one-pot multi-component synthesis, characterization and antibacterial activity of novel chromeno[2,3-d]pyrimidin-8-amine derivatives. J Mol Struct 1017:60–64. https://doi.org/10.1016/j.molstruc.2012.02.044

Khaligh NG (2017) Synthesis of Tetrahydrotriazoloacridines: a synergistic effect of microwave irradiation and Brönsted acidic ionic liquids. J Heterocyclic Chem 54:3350–3357. https://doi.org/10.1002/jhet.2955

Khanna G, Aggarwal K, Khurana JM (2015) An efficient and confluent approach for the synthesis of novel 3,4-dihydro-2H-naphtho[2,3-e][1,3]oxazine-5,10-dione derivatives by a three component reaction in ionic liquid. RSC Adv 5:46448–46454. https://doi.org/10.1039/C5RA06169E

Khurana JM, Nand B, Kumar S (2011) Rapid synthesis of polyfunctionalized pyrano[2,3-c]PYRAZOLES via multicomponent condensation in room-temperature ionic liquids. Synth Commun 41:405–410. https://doi.org/10.1080/00397910903576669

Khurana JM, Magoo D, Chaudhary A (2012) Efficient and green approaches for the synthesis of 4h-benzo[g]chromenes in water, under neat conditions, and using task-specific ionic liquid. Synth Commun 42:3211–3219. https://doi.org/10.1080/00397911.2011.580069

Khurana JM, Lumb A, Chaudhary A, Nand B (2013) Efficient and green syntheses of 12-aryl-2,3,4,12-tetrahydrobenzo[b] xanthene-1,6,11-triones in water and task-specific ionic liquid. Synth Commun 43:2147–2154. https://doi.org/10.1080/00397911.2012.688230

Kumar A, Kumar P, Tripathi VD, Srivastava S (2012) A novel access to indole-3-substituted dihydrocoumarins in artificial sweetener saccharin based functional ionic liquids. RSC Adv 2:11641–11644. https://doi.org/10.1039/C2RA21284F

Kumar R, Andhare NH, Shard A, Richa, Sinha AK (2014) Multicomponent diversity-oriented synthesis of symmetrical and unsymmetrical 1,4-dihydropyridines in recyclable glycine nitrate ($GlyNO_3$) ionic liquid: a mechanistic insight using Q-TOF, ESI-MS/MS. RSC Adv 4:19111–19121. https://doi.org/10.1039/C4RA02169J

Kumbhar A, Jadhav S, Shejwal R, Rashinkar G, Salunkhe R (2016) Application of novel multicationic ionic liquids in microwave assisted 2-amino-4H-chromene synthesis. RSC Adv 6:19612–19619. https://doi.org/10.1039/C6RA01062H

Lanjewar K, Rahatgaonkar A, Chorghade M, Saraf B (2011) Facile synthesis of pyrimidine–isoxazoline hybrids in a [bmim][PF6]–water biphasic system. Synthesis 43:2644–2650. https://doi.org/10.1055/s-0030-1260099

Li T-J, Yao C-S, Yu C-X, Wang X-S, Shu-Jiang T (2012) Ionic liquid-mediated one-pot synthesis of 5-(trifluoromethyl)-4,7-dihydrotetrazolo-[1,5-a]pyrimidine derivatives. Synth Commun 42:2728–2738. https://doi.org/10.1080/00397911.2011.566460

Li BL, Zhong AG, Ying AG (2014) Novel SO3H-functionalized ionic liquids-catalyzed facile and efficient synthesis of Polyhydroquinoline derivatives via Hantzsch condensation under ultrasound irradiation. J Heterocyclic Chem 52:445–449. https://doi.org/10.1002/jhet.2070

Li C-B, Li Y-W, Xu D-Z (2018) An efficient and ecofriendly three-component reaction for the rapid synthesis of 2-amino-4*H*-chromenes catalyzed by aDABCO-based ionic liquid. Synthesis 50:3708–3714. https://doi.org/10.1055/s-0037-1610123

Liang Y, Wang J, Cheng C, Jing H (2014) Lewis acidic ionic liquids of crown ether complex cations: preparation and applications in organic reactions. RSC Adv 4:93546–93550. https://doi.org/10.1039/C6RA21947K

Mahato S, Santra S, Chatterjee R, Zyryanov GV, Hajra A, Majee A (2017) Brønsted acidic ionic liquid-catalyzed tandem reaction: an efficient approach towards regioselective synthesis of pyrano[3,2-*c*]coumarins under solvent-free conditions bearing lower E-factors. Green Chem 19:3282–3295. https://doi.org/10.1039/C7GC01158J

Moghaddam M, Bazgir M, Mehdi AM, Ramin AG (2012) Alum (KAl(SO4)2•12H2O) catalyzed multicomponent transformation: simple, efficient, and green route to synthesis of functionalized spiro[chromeno[2,3-*d*]pyrimidine-5,3′-indoline]-tetraones in ionic liquid media. Chin J Chem 30:709–714. https://doi.org/10.1002/cjoc.201280014

Mohammadi K, Shirini F, Yahyazadeh A (2015) 1,3-Disulfonic acid imidazolium hydrogen sulfate: a reusable and efficient ionic liquid for the one-pot multi-component synthesis of pyrimido[4,5-*b*]quinoline derivatives. RSC Adv 5:23586–23590. https://doi.org/10.1039/C5RA02198G

Naidu S, Reddy SR (2016) Copper-catalyzed tandem reaction in ionic liquid: an efficient reusable catalyst and solvent media for the synthesis of fused poly-heterocyclic compounds. RSC Adv 6:62742–62746. https://doi.org/10.1039/C6RA13712A

Nguyen QT, Hang A-HT, Nguyen T-LH, Chau D-KN, Tran PH (2018a) Phosphonium acidic ionic liquid: an efficient and recyclable homogeneous catalyst for the synthesis of 2-arylbenzoxazoles, 2-arylbenzimidazoles, and 2-arylbenzothiazoles. RSC Adv 8:11834–11842. https://doi.org/10.1039/C8RA01709C

Nguyen HT, Le N-PT, Chau D-KN, Tran PH (2018b) New nano-Fe_3O_4-supported Lewis acidic ionic liquid as a highly effective and recyclable catalyst for the preparation of benzoxanthenes and pyrroles under solvent-free sonication. RSC Adv 18:35681–35688. https://doi.org/10.1039/C8RA04893B

Nikpassand M, Fekri LZ, Farokhian P (2015) Synthesis of novel derivatives of benzoxazole in bis-ionic liquid [BDBDIm]Br. Synth Commun 45:2303–2310. https://doi.org/10.1080/00397911.2015.1077256

Panja SK, Saha S (2013) Recyclable, magnetic ionic liquid bmim[$FeCl_4$]-catalyzed, multicomponent, solvent-free, green synthesis of quinazolines. RSC Adv 3:14495–14450. https://doi.org/10.1039/C3RA42039F

Parmar NJ, Barad HA, Pansuriya BR, Talpada NP (2013) A highly efficient, rapid one-pot synthesis of some new heteroarylpyrano[2,3-*c*]pyrazoles in ionic liquid under microwave-irradiation. RSC Adv 3:8064–8070. https://doi.org/10.1039/C3RA00068K

Patil JD, Pore DM (2014) [C_{16}MPy]$AlCl_3$Br: an efficient novel ionic liquid for synthesis of novel 1,2,4-triazolidine-3-thiones in water. RSC Adv 4:14314–14319. https://doi.org/10.1039/C3RA46916F

Patil SG, Bhadke VV, Bagul RR (2012) Synthesis of 2-Aroylbenzofuran-3-ols using basic ionic liquid [bmIm]OH. J Chem Pharm Res 4:2832–2835

Qiang W-W, Zhang M-M, Wang X-S (2016) Catalyst-free synthesis of 5-arylimidazo[1,2-c]quinazoline derivatives in ionic liquids. J Heterocyclic Chem 54:509–516. https://doi.org/10.1002/jhet.2613

Renzhong F, Yang Y, Lai W, Ma Y, Chen Z, Zhou J, Chai W, Wang Q, Yuan R (2014) Efficient and green microwave-assisted multicomponent Biginelli reaction for the synthesis of dihydropyrimidinones catalyzed by heteropolyanion-based ionic liquids under solvent-free conditions. Synth Commun 45:467–477. https://doi.org/10.1080/00397911.2014.976346

Safaei S, Mohammadpoor-Baltork I, Khosropour AR, Moghadam M, Tangestaninejad S, Mirkhani V, Kia R (2012) Application of a multi-SO_3H Brønsted acidic ionic liquid in water: a highly efficient and reusable catalyst for the regioselective and scaled-up synthesis of pyrazoles under mild conditions. RSC Adv 2:5610–5616. https://doi.org/10.1039/C2RA20624B

Safaei S, Mohammadpoor-Baltork I, Khosropour AR, Moghadam M, Tangestaninejad S, Mirkhani V (2014) A novel one-pot three component synthesis of *N*-(phenylimino)indazole-1-carbothioamides. RSC Adv 4:2251–2256. https://doi.org/10.1039/C3RA44064H

Santosh Kumar G, Pushpa Ragini S, Sanjeeva Kumar A, Meshram HM (2015) A copper-catalyzed multi-component reaction accessing fused imidazo-heterocycles *via* C–H functionalization. RSC Adv 5:51576–51580. https://doi.org/10.1039/C5RA09025C

Sharma H, Srivastava S (2018) Anion–cation co-operative catalysis by artificial sweetener saccharine-based ionic liquid for sustainable synthesis of 3,4-dihydropyrano[*c*]chromenes, 4,5-dihydropyrano[4,3-*b*]pyran and tetrahydrobenzo[*b*]pyrans in aqueous medium. RSC Adv 8:38974–38979. https://doi.org/10.1039/C8RA06889E

Shaterian HR, Honarmand M (2011) Task-specific ionic liquid as the recyclable catalyst for the rapid and green synthesis of dihydropyrano[3,2-c]chromene derivatives. Synth Commun 41:3573–3581. https://doi.org/10.1080/00397911.2010.519594

Shaterian HR, Oveisi AR (2011) A simple green approach to the synthesis of 2-Amino-5-oxo-4,5-dihydropyrano[3,2-c]chromene-3-carbonitrile derivatives catalyzed by 3-Hydroxypropanaminium acetate (HPAA) as a new ionic liquid. J Iran Chem Soc 8:545–552. https://doi.org/10.1007/BF03249089

Shiran JA, Yahyazadeh A, Yamin BM, Mamaghani M, Albadi J (2015) Basic ionic liquid as catalyst and reaction Media for the one-pot Three-component Regioselective Synthesis of various Thiazol-2-imine derivatives. J Heterocyclic Chem 53:1009–1016. https://doi.org/10.1002/jhet.2406

Simijonović D, Petrović ZD, Milovanović VM, Petrović VP, Bogdanović GA (2018) A new efficient domino approach for the synthesis of pyrazolyl-phthalazine-diones. Antiradical activity of novel phenolic products. RSC Adv 8:16663–16673. https://doi.org/10.1039/C8RA02702A

Soleimani E, Khodaei MM, Koshvandi ATK (2012) Three-component, one-pot synthesis of Benzo[*b*][1,4]oxazines in ionic liquid 1-Butyl-3-methylimidazolium bromide. Synth Commun 42:1367–1371. https://doi.org/10.1080/00397911.2010.540056

Sutariya TR, Labana BM, Parmar BD, Parmar NJ, Kant R, Gupta VK (2015) A domino synthetic approach for new, angular pyrazol- and isoxazol-heterocycles using [DBU][ac] as an effective reaction medium. RSC Adv 5:23519–23529. https://doi.org/10.1039/C5RA00493D

Veisi H, Manesh AA, Khankhani N, Ghorbani-Vaghei R (2014) Protic ionic liquid [TMG][ac] as an efficient, homogeneous and recyclable catalyst for one-pot four-component synthesis of 2H-indazolo[2,1-b]phthalazine-triones and dihydro-1H-pyrano[2,3-c]pyrazol-6-ones. RSC Adv 4:25057–25062. https://doi.org/10.1039/C4RA03514C

Wang M-Y, Song Q-W, Ma R, Xie J-N, He L-N (2016) Efficient conversion of carbon dioxide at atmospheric pressure to 2-oxazolidinones promoted by bifunctional Cu(II)-substituted polyoxometalate-based ionic liquids. RSC Adv 18:282–287. https://doi.org/10.1039/C5GC02311D

Wang Y, Liu J-Q, Wang X-S (2017) An efficient synthesis of Pyrrolo[1,2-a] or Pyrido[1,2-a]benzo[4,5]imidazo[1,2-c]quinazoline derivatives in ionic liquids catalyzed by iodine. J Heterocyclic Chem 54:3440–3446. https://doi.org/10.1002/jhet.2967

Yassaghi G, Davoodnia A, Allameh S, Zare-Bidaki A, Tavakoli-Hoseini N (2012) Preparation, characterization and first application of Aerosil silica supported acidic ionic liquid as a reusable heterogeneous catalyst for the synthesis of 2,3-dihydroquinazolin-4(1H)-ones. Bull Kor Chem Soc 33:2724–2730. https://doi.org/10.5012/bkcs.2012.33.8.2724

Yu W, Yuan R, Zhang F-R, Pang L-L, Ma R, Cai-HuiYue WL, Yin W, Bo R-C, Wu H (2011) One-pot synthesis of n2-substituted 2-amino-4-aryl-5,6,7,8-tetrahydroquinoline-3-carbonitrile in basic ionic liquid [bmim]OH. Synth Commun 41:2997–3015. https://doi.org/10.1080/00397911.2010.516459

Zakeri MH, Saeedi MM, Karimi M, Oskooie N, Tavakoli-Hoseini HA, Niloofar (2011) One-pot green procedure for synthesis of tetrahydrobenzo[*a*]-xanthene-11-one catalyzed by Brønsted ionic liquids under solvent-free conditions. Chin J Chem 129:1441–1445

Zheng J, Li Y (2011) Basic ionic liquid-catalyzed multicomponent synthesis of tetrahydrobenzo[*b*]pyrans and pyrano[*c*]chromenes. Mendeleev Commun 21:280–281. https://doi.org/10.1016/j.mencom.2011.09.017

Chapter 16
Role of Ionic Liquids in Food and Bioproduct Industries

Kasibhatta Siva Kumar

Contents

K. S. Kumar (✉)
Department of Chemistry, S.V. Arts College, TTD's, Tirupati, India

Inamuddin, A. M. Asiri (eds.), *Nanotechnology-Based Industrial Applications of Ionic Liquids*, Nanotechnology in the Life Sciences,
https://doi.org/10.1007/978-3-030-44995-7_16

16.1 Ionic Liquids, Versatile Solvent: An Introduction

One of the most significant directions of the world analysis is the hunt for new chemical compounds with special properties. Ionic liquids square measure such compounds. Their application brings new potentialities for contemporary chemical technology. The ionic liquids match well within the assumptions of inexperienced chemistry. In distinction to the previous approach, the inexperienced chemistry needs style and promotes better reaction and separation of latest methods and chemicals which permit the reduction or elimination of use and production of venturesome materials (Burczyk 2014).

Ionic liquids (ILs) square measure outlined as chemical compounds of particle structure that have the freezing point below 100 °C (Rogers and Seddon 2002; Wasserscheid and Welton 2008). This property is typically a result of an oversized size distinction between Associate in Nursing organic ion of expanded uneven structure and a tiny low organic or inorganic ion. This hinders the formation of the same Bravais lattice, so greatly reducing the natural action purpose of the compound (Huddleston et al. 2001; Gorke et al. 2010).

Figure 16.1 presents a general model of an irregular system of ions within the IL structure.

The increased interest in ILs within the last 20 years was caused by many factors. Ionic liquids quickly began to be thought of as "green" different for volatile, ignitable and sometimes hepatotoxic in style organic solvents. The ILs are recognised as additional environmentally friendly solvents and catalysts, and at a similar time their distinctive advantage, i.e. design ability, has begun to be appreciated. ILs will be designed by choosing a correct ion and ion to get a compound with desired properties. During this approach it is potential to synthesise ILs for a selected

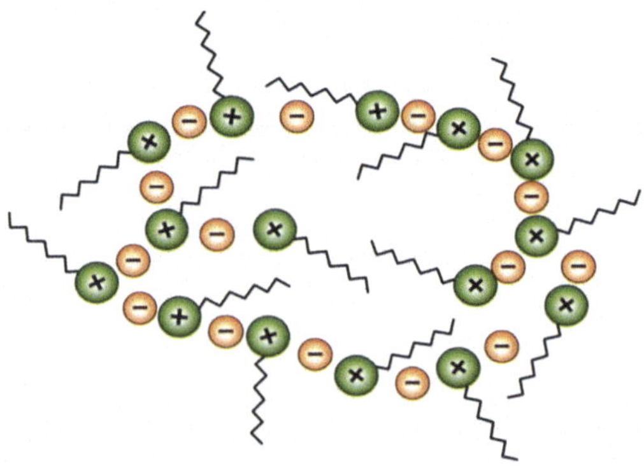

Fig. 16.1 Simplified model of an IL

application (Freemantle 1998). The term was coined within the literature for such kind of compounds—task-specific ionic liquids (TSILs).

The name ionic liquids suggests liquids containing ions (anion and cation).

In sensible terms, it is fascinating for Associate in Nursing ILs to present in the liquid form at temperatures below 100 °C and, preferably, at close temperature. However ILs are expected for a minimum of 100 years and their application in several chemical methods has solely been examined elaborately over the last three decades (Welton 1999). Analysis of ILs has ranged from their application as solvents in chemistry (MacFarlane et al. 2007), transformations of inorganic and organic (Peter and Wilhelm 2009), biotechnology processes (Dupont et al. 2002), photovoltaic cell (Roosen et al. 2008), and chemical science used as supported phases (Wang et al. 2009; Riisager et al. 2006; Gruttadauria et al. 2006). Since then, the application of ILs has increased to incorporate the event of task-specific ILs, whereby the constituent ions of the ionic liquid square measure designed to satisfy the necessities of a selected reaction (Wasserscheid and Keim 2000; Fei et al. 2006) or as an alternative biotransformation (Lee 2006; De Maria 2008). In a reaction, if IL is used once the required product is obtained, the IL is once again recycled (Van Rantwijk and Sheldon 2007; Miao and Chan 2005). However, ILs for a long time thought that they are inexperienced substitute to many organic liquids, though intensive discussion and a few differences have enclosed their "so-called" inexperienced properties. Furthermore to those issues, the union and life cycle of an IL additionally must be determined so as to totally verify its "greenness" matched to traditional organic liquids. A recent study on the "greenness" of assorted ILs found that there is plenty space for modification in terms of E-factor, energy consumption and purification (Fraga-Dubreuil and Bazureau 2001). Hence, in forthcoming industrial applications the ionic liquid will be simply and neatly synthesised, to possess entomb alia the requisite inexperienced properties of less pressure, low flammability or toxicity, and contain sensible operating methods like consistence and high recyclability, by possessing advantage of the large combinatorial and structural properties containing anion or cation composition (Deetlefs and Seddon 2010; Harjani et al. 2009a).

16.2 Classification of Ionic Liquids

It is a difficult task to classify the antecedent delineated ionic liquids, as there is no single criterion permitting it. Therefore, these ILs are classified based on the potential applications.

ILs take issue greatly in terms of their chemical science and biological properties, which results in even bigger variety of potential applications. For generations, in order to line-up the evolution history and studies of ILs contaning a cation, in which the positive charge is found on an element like nitrogen, phosphorus, sulphur or oxygen atom, hence they are divided into ammonium ion, phosphonium, sulphonium and oxonium ionic liquids. So far, ammonium ion liquids are well identified,

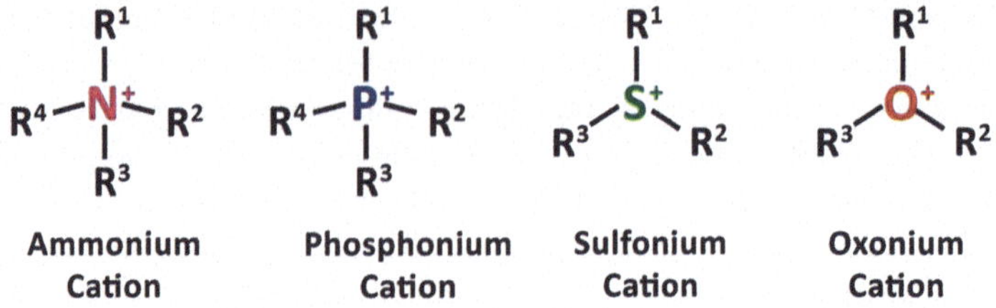

Fig. 16.2 Types of cations

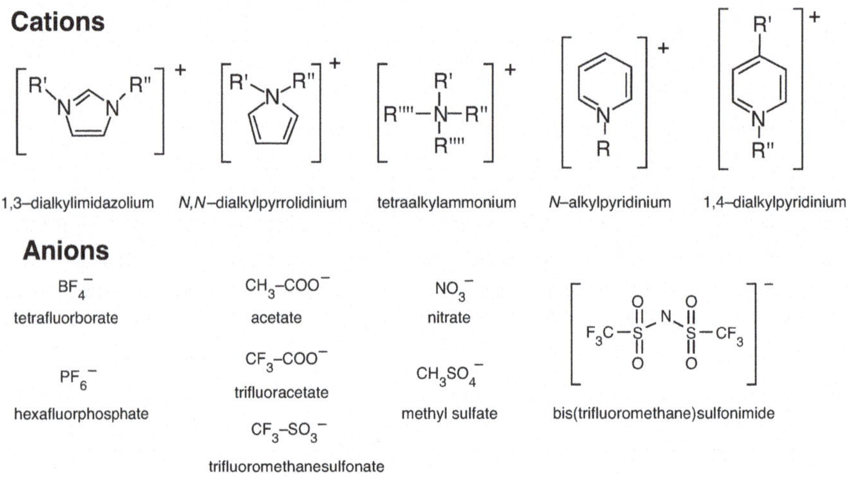

Fig. 16.3 Typical cations and anions in ionic liquids

whereas the smallest known are oxonium ones. The principle of the general classification based on the type of atom with a positive charge is presented in Fig. 16.2.

The two basic elements of ionic liquids are cations and anions which are represented in Fig. 16.3.

A general and helpful criterion was established in 2007 in designing the ILs that were divided into three generations (Fig. 16.4) depending on the types of properties and provided by the structure of an ion and ion (Harjani et al. 2009b).

The first generation includes ILs within which the ion and ion structures were chosen to get a product of specific physical properties. The result of the ion or ion structures on conduction is of special importance as it interprets into their specific applications as electrolytes (Hough et al. 2007). Ions of the first-generation ILs were additionally hand-picked to attain specific values of temperature, density, viscosity, thermal stability, hydrophilicity and refractive index.

The quality of the second-generation ionic liquids could be results of each appropriate physical parameter and chemical property. The latter includes their reactivity,

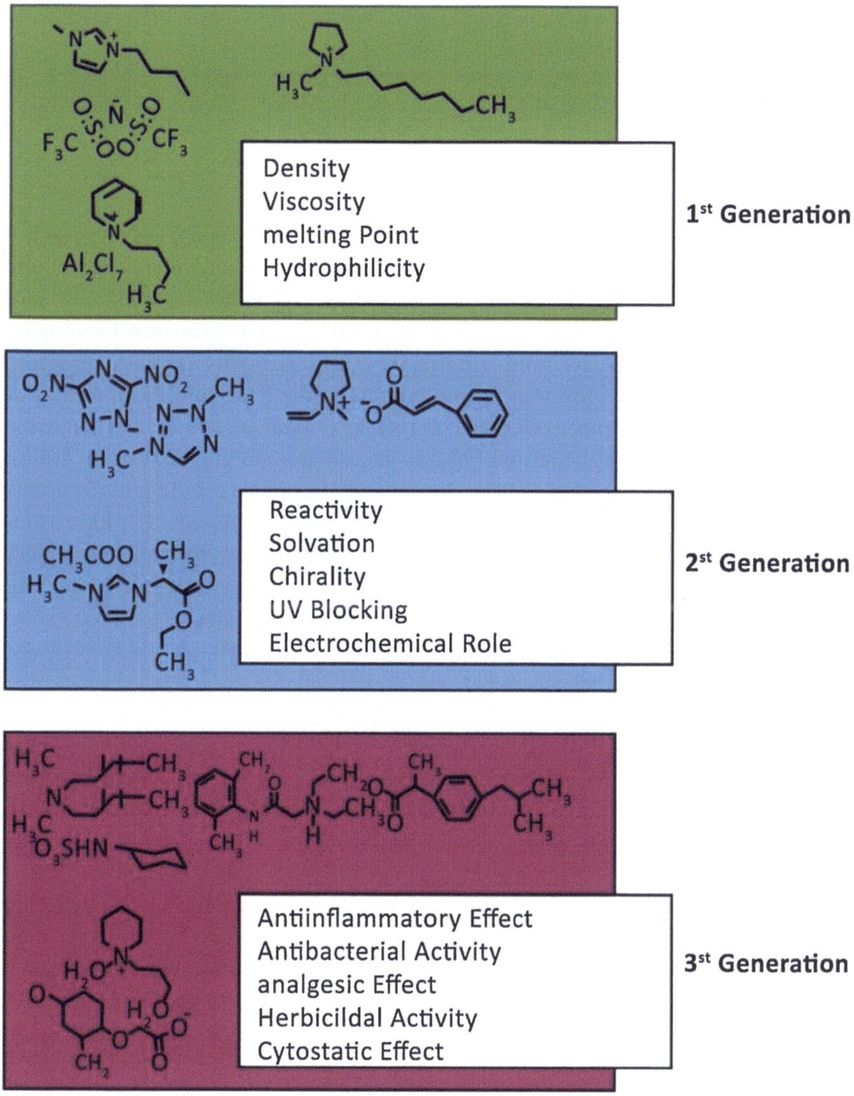

Fig. 16.4 Different generations of ionic liquids

chirality, solvating ability and talent to extract varied chemical substances. This cluster includes task-specific ionic liquids, energetic ILs and chemical change catalysts (Roszak et al. 2011).

Dynamically extensive knowledge of ILs resulted in the creation of the third generation of ILs in the early twenty-first century. It includes compounds that apart from selected physicochemical properties also exhibit a specific biological activity. One of the first third-generation ILs were salts of a strong bactericidal effect. The ions of pharmaceuticals used to serve both as cations and anions of the third-generation ILs, acting against pain and as anti-inflammatory (Lewandowski and Świderska-Mocek 2009; Erfurt et al. 2014).

16.3 Properties of Ionic Liquids

There is a variety of distinctive properties associated with the ionic structure of ILs that, together with the liquid state during a big selection of temperatures, determine their technological quality. Their benefits embody nearly non-measurable pressure underneath moderate conditions (Pernak et al. 2003). This includes a positive impact on their safety as solvents, and eliminates atmospherically vapour emission. Moreover, non-volatility, together with moderately low flammability of ionic liquids, greatly reduces fireplace or explosion risk, as critical ancient organic solvents. The bulk of ILs exhibit high thermal stability and inflammability (Earle et al. 2006). However, caution is suggested as their decomposition product could ignite and this property has been employed in the event of energetic ionic liquids. ILs have stunning useful properties like to preserve soft tissue, as feeding deterrents or rubber accelerators, that motivates their additional analysis (Ludwig 2001; Smiglak et al. 2006; Katritzky et al. 2005; Pernak et al. 2005; Pernak et al. 2013).

16.4 General Applications of Ionic Liquids

Ionic liquids (ILs) are extensively employed in several analysis and business fields, together with chemical and pharmaceutical applications (Fig. 16.5). However, ionic liquids can contribute by enhancing the reaction yield and selectivity, and can stabilise the action of a catalyst, in the separation of the catalyst during the recycling process and causing incentive in economic value.

Further ionic liquids are applied in a variety of processes consisting of homogenous catalyst in hydrogenation reactions, biotransformation, booster in the use of new feedstocks, capture and/or transformation of CO_2, removal of heavy metals in environmental samples, extraction of protein and separation of paraffin and alkenes.

Due to unique properties of ionic liquids, they are very good solvents for organic, inorganic and polymer substances, namely cellulose, proteins, lignin and DNA/RNA. During the extraction and processing of natural products, it is observed that solubility of natural products is very much higher in ionic liquids when compared to other solvents. Using a small amount of ionic liquids, large quantity of natural sources can be extracted.

The important organic reactions involving ILs as solvent/catalyst are:

- Hydroformylations
- Friedel-Crafts
- Diels-Alder

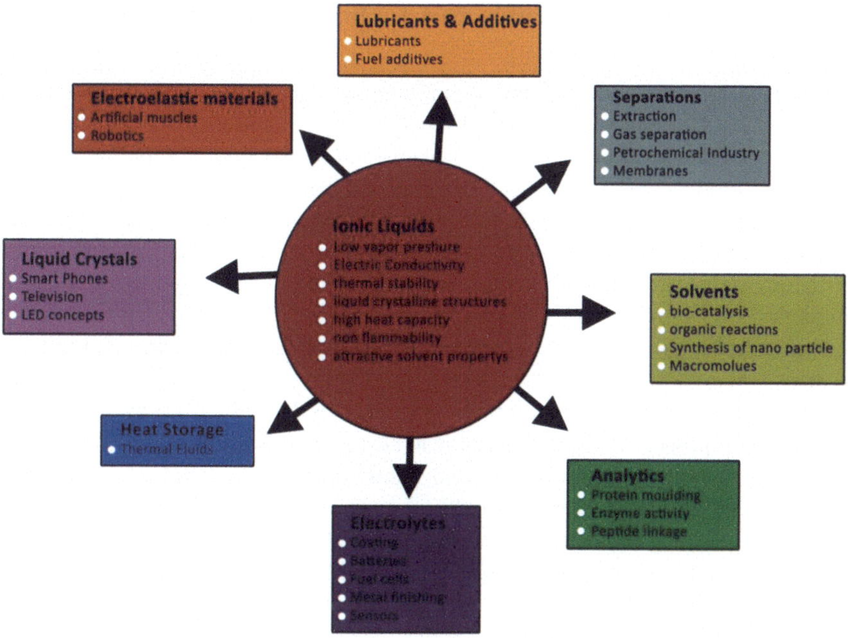

Fig. 16.5 General applications of ionic liquids

- Heck reaction
- Suzuki coupling

16.5 General Characteristics of ILs

Ionic liquids (ILs) have emerged as a novel class of molten salts possessing a unique combination of particular properties like non-volatility; non-flammability; high ionic conductivity; low vapor pressure; high thermal, chemical and electrochemical stability; and a high solvating capacity in aqueous, thereby resulting in the high potential of these liquids to be exploited as "green solvents". Moreover, ILs are considered as possible alternative for baleful conventional molecular solvents in the many chemical industrial applications due to their recyclability and eco-friendly nature. The stability and activity of a bimolecule basically depend on the solvent and solute nature to be caused by hydrophilic and hydrophobic interactions, change in structural conformation and change in temperature (Pernak et al. 2010; Yang and Dionysiou 2004). In general, the anions present in the given IL are polyatomic inorganic species, like PF_6^- and BF_4^-. The most appropriate cations are pyridinium and an imidazolium ring species with one or more alkyl groups linked to the nitrogen or carbon atoms.

16.6 Importance of Science and Technology in Food Industry

It is proved that both science and technology are competing with each other in establishing new discoveries in the product of huge quantity of food stuffs. This leads to designing of a more variety of plants and increasing the input by making agriculture more economic, productive and profitable.

The science of genetics plays a dominant role in the production of high-nutrient-content food grains. This method also improves the tolerance quality in plants to drought and diseases increased resistance and survival in the production of huge amount of food substances.

Further, the field of genetics plays a key role in agriculture science and in food security.

The huge quantity of agriculture production may be influenced by parameters like transport, power, food storage godown, best agricultural methods by providing proper facility support, information and communication systems, sanitation and marketing that will increase the agriculture production.

By strengthening the new methods and research on food systems, and support from public help and investment, agro-scientist advisory committee will not only increase agricultural production but also develop confidence in small-holder farmers for participating in agricultural activities.

16.7 Part of Food and Beverage Companies in Improving the World Population Health

A broader view of Pepsi company is to increase the quality of population health globally due to the prolong diseases in below poverty and middle income countries are becoming serious threat to mankind as well as to the economy of the country (Huddleston et al. 2001; Tenenbaum 2008; Dogliotti et al. 2014; Thi et al. 2015; De et al. 2017; Kumar et al. 2003; Ali et al. 2016; Han et al. 2015; World Health Organization 2005; Adeyi et al. 2007; Yach et al. 2004). In order to eradicate the diseases calls for prevention have increased. Many suggestions, proposals and many specific actions are laid for food industries under the broad private sector. Some of the suggestions are as follows:

Low amount of saturated fat acids, carbohydrates and salt content in the available products.

Continue to manufacture and supply affordable, good and high-nutrient-content products to consumers.

It is proper to establish the responsible marketing agencies which promote high quality and pure content of substances particularly to young children.

In general, the food waste is a main component from municipal solid waste or that is thrown out into the open area which leads to drastic damage to human health and environmental problems. Burning of waste food which contains high amount of moisture will release several toxins into air that may further deplete the natural resources. Bioenergy is produced from waste food through anaerobic process and this is clearly explained in this work; the economic growth of any country is decided by the number of food industries. Moreover, food industry is one of the effective and attractive sectors for foreign investments. All over the world, a good number of food industry units are constructed which produce their goods for households and export purposes. Catering and hotel management centres are providing highly educated and qualified persons for fast-growing private companies in the food industry sector. The availability of large quantity of agricultural products, highly qualified labour and manpower and high demand for food products are contributing to the increase of food industry share in the total industrial output.

Increasing need and necessity for the production of edible food are influencing the world climatic change. Further, the agriculture production all over the globe may affect due to the natural calamities including floods, storms for longer time period of drought, new diseases and pets could results in the food insecurity. Moreover, many challenges are influencing the food production. There are increase in the water scarcity leading to the soil infertility due to leaching of fine soil layer by wind and water erosion, immunity and resistance of pest and diseases, usage of large number of agro chemical resulting in deterioration of valuable bio diversity.

16.8 Introduction to Food

Food is defined as "consumable products obtained from soil/land or marine is (including inland waters) is utilised by human being and also to cattle. Food is nothing but bio mass used as input for survival life support, and procurebio-fuel and bio-based products. Food is linked to many aspects interms historical, environmental, social and economic dimensions. The global food varieties are affected by major societal and interrelated challenges such as climate change, migration, a growing world population, urbanisation, and resource scarcity, the triple burden of malnutrition, ageing and food poverty. Food systems produce and provide sufficient, affordable, safe, convenient, tasty and nourishing food for healthy and physical, mental and suitable diets for all citizens. Food stuffs are also need to be environmentally sustainable implying a suitable and efficient use of natural resources, and limiting negative environmental impacts. In this context, building blocks of food systems should encompass the entire value chain in its wide range and their interactions from ecosystem services. The various stages involve in the food are harvesting, cultivation, storaging, processing, protecting, packaging, distribution and retailing. The food services available at restaurants catering and various functions leads to huge waste this causes not only environmental problem but also creates diet related diseases. Hence it is necessary to introduce the concept of waste stream management, re-cycling, food and feed safety, all the way to consumers, nutrition for citizens" health and well-being, and diet related disease (World Health Organization 2005; Adeyi et al. 2007; Yach et al. 2004).

16.9 Classification of Varieties of Food Consumed by Human Beings

16.9.1 Plant Origin

Class: Cereal grainsand legumes/pulses, vegetables, fruits, nuts and seeds, spices and herbs, tea, mineral waters.

Type: Umbelliferous vegetables, solanaceous, cereal grains, mushrooms, miscellaneous vegetables, legumes/pulses, potatoes, sugar sources, cruciferous vegetables, composite vegetables, liliaceous vegetables, assorted tropical and subtropical fruits, pome fruits, stone fruits, vegetables, cucurbitaceous vegetables, legume vegetables, citrus fruits, berries, grape, oil seeds, spices, herbs.

Group: Cereal grains, peas, rice, soybeans, wheat, barley, corn, potato, sweet potato, sugarcane, cabbage, cauliflower, broccoli, onion, garlic, carrot, celery, cucumber, pumpkin, watermelon, mushroom, spinach, ginger, lemon, orange, lime, peach, cranberry, raspberry, blackberry, blueberry, strawberry, grape banana, kiwifruit, papaya, avocado, pineapple, sunflower seeds, guava mango, sesame seeds, cotton seeds.

16.9.2 Animal Origin

Class: Terrestrial mammals, poultry, aquatic animals, bee products.

Type: Terrestrial mammals (muscle, fat, liver, kidney, edible, milk), poultry (muscle, fat, liver, kidney, edible, eggs), fish.

Group: Edible offal (pig, cattle, sheep, horse, deer, goat,) chicken, duck, turkey, salmoniformes, anguilliformes, other aquatic animals.

16.10 Food Technology: Historical Approach

Wastage of food arises from carbohydrate content, namely rice and vegetables, whereas high amount of proteins and lipids may be available in meat and eggs.

16.10.1 Definition of Food Security

Ambient, safe and highly nutrient-containing food should be provided all over the globe, which is a major concern in the twenty-first century. According to Food and Agricultural Organization (2016a), four parameters influence the food security, namely availability of food, consumption/utilisation of food stability and access to food, and these factors are influenced by the overall population of the world. As defined by Food and Agricultural Organization "food security available if all the people at all times have physical , social and economic access to ample , safe and nutrient food which provided their dietary needs and for an active and healthy life" (FAO 2016).

The food and their types underwent drastic change over the past years like extending the food business, globalisation, invention of new technologies, food processing and supply areas continuously changing prices of essential commodities. Moreover, food production is influenced by rapid deforestation, urbanisation, procuring and biofuel industries by affecting the agricultural lands and this reduces the production of food.

The production of biogas from various sources like municipal, industrial and agricultural wastes may create a valuable sustainable energy. The overall energy which is obtained to get biogas is very high when compared to changes caused by other technologic processes. The method for production of large amount of methane from biowaste may result in conserving fossil fuels mainly by utilising various technologies that are attractable, adaptable methods and these have been applied by many countries globally by collecting a large amount of organic waste. Further, methane production is not confined to animal manures but it can be collected from huge amounts of biomass and waste materials which reach into the environment. Many methods have been developed to get high yield of methane from various waste materials. It is observed that methane is the ultimate product during anaerobic metabolism.

The conversion of industrial, agricultural, municipal and industrial debris into biofuel mainly methane and CO_2 is through anaerobic technique. Production of large amount of methane from biowaste may result in conserving fossil fuels, mainly coal, petroleum and natural gas. The energy procured from different wastes in the form of biogas is more when compared to other transmutation methods. The technique for methane production from wastes is well acceptable all over the world.

Once upon a time it was thought that production of biogas is only combined to animal manures. But at present majority of the biogas is collected from worthless materials and biomass.

The food system is large and highly complex and has many actors. It provides food and nutrition security; hence it is a major player in health promotion. It also provides a significant number of jobs and opportunities for growth through innovation.

Food and its related products play a key role in the survival of all types of living organisms. Further, the production of food is useful to procure biofuel or bio-based products. Different types of foods are available all over the globe and are influenced by cultural, social, economic, historical and climate changes in agriculture.

The production all over the globe is influenced by environmental changes, increase in population, urbanisation, industrialisation and non-availability of resources when needed during cultivation. Also, facts like malnutrition ageing and food poverty influence the quality of food.

In recent years, a sudden change in food production has been observed like partial changes in human diets, due to health grounds, globalisation, increase in food business, new technical discoveries, continuous production of food and processing chains and rapid ups and down of food products. The decrease in food production yield is also influenced by deforestation rate all over the globe and methods for collection of biofuel which decrease the available land food production.

16.11 Food Waste

Waste food and its deposition on land are becoming a serious issue all over the world due to huge increase in population. The increase in food waste is causing serious problems dealing with ecological changes, health and non-availability of sufficient dumping lands. It is the right time to take correct steps and decisions to control food waste by selecting the standard methods and practices. At present, anaerobic digestion method is more suitable and one of the most green methods and suitable ways to control food wastes in various components like production of energy, nutrients and product yield and this contributes to increase in world's energy production; the following figure nearly explains how the food waste is converting into world's energy requirement (Fig. 16.6).

The waste generated by different sources, particularly from food waste, is becoming more in many countries all over the globe. Various types of residues in the form of waste are divided into industrial, domestic, agricultural and urban solid waste. These wastes not only damage the environment but also severely affect the natural resources. Therefore, the society needs proper methods to enhance efficiency in the removal or reduction of wastes, leading to the production of valu-added products.

Further, the society has got a change in lifestyle and mentality thinking waste as valuable resources and not as issue. This can be achieved by the cooperation between people and governments, environmental protection agencies and voluntary organisations.

Therefore, the concept of food supply chain waste is introduced and through this it is possible to get various valuable products and high-calorific and -efficient fuels.

Applying proper methods and recycling of large quantities of waste food are necessary to decrease its environmental loads and minimise damage to human health.

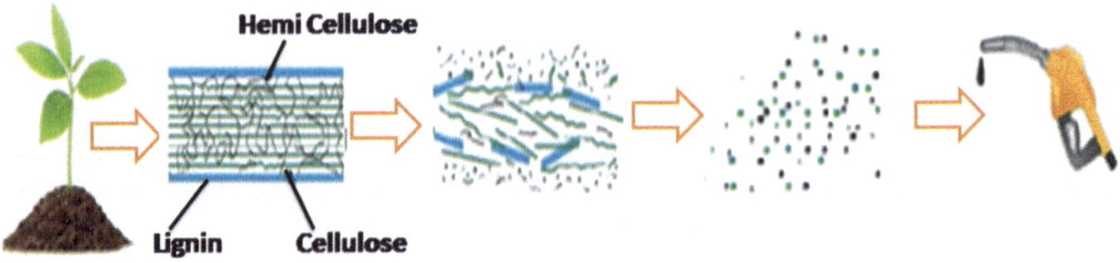

Fig. 16.6 Biodegradable wastes into energy production

According to food administration organisation it was observed that nearly 1.5 billion tonnes of food was wasted or lost in the form of human consumption (or) throughout supply chain, from production to consumption.

The food waste in general comes from vegetable, fruit, discorded items from kitchen, hotels, food processing industries, residues from sugar refining industries namely molasses, bagasse, bones, flesh and blood arising from slaughter houses and aquatic animals. Further, milk industry, distilleries, wineries, soft drink industry and breweries are also responsible for food waste.

Currently, majority of the waste food is recycled for the animal feed, and fertilisers and remaining quantities are dumped into earth and incineration. This causes severe environmental pollution by releasing of methane, and it is stronger than (23 times) carbon dioxide as a greenhouse gas that is important in leading to drastic change in global conditions.

An increase in population resulted in the quality and quantity of natural resources, depletes natural resources like fossil fuels leading to scarcity in availability of fuels. Further, day by day the need and energy utilisation go on increasing and it is inevitable to procure from other alternate sources. Hence, during the past years, conversion of food waste into energy is becoming an eco-friendly and economically attractive method.

The nature of food waste which is biodegradable comes out from many sources such as food industry, domestic and hospitality. The composition of food is starch, proteins, sugar, carbohydrate and small amounts of inorganic substances. The amount of food waste consists of vegetables and carbohydrates in abundant quantity; moreover meat and egg products will have huge amounts of lipids and proteins.

About nearly 33% of food materials are being thrown out as waste into various parts of the environment. The waste is categorised as primary and secondary wastes. Food and Agricultural Organization of the United Nations found that nearly 1.5 billion tonnes of wastes arising from food was thrown out in a year and the same may be dumped outside around 2.5 tonnes by the year 2025. This waste must be avoided and the same is distributed to hungry people in different parts of the globe. It was observed that about 20% of food waste of the total world share mainly comes from Asia. The different ways of food waste come from the following:

1. Lack of good plan for isolation of food waste from the starting point
2. Inability of appropriate management and accumulation techniques
3. No proper implementation of rules and regulations

Generally, it was demonstrated that the waste food results in the increase of solid waste material in and around of municipality areas nad or out skirts of townships. The proper administration and management will decrease the loss arising from large quantity of invested resources and this decreases the growth in socio-economic-environmental impacts.

Generally, food waste indicates an important portion of solid waste materials collected by the municipalities. However, the environmental hazards and spoil of human health issues were reduced by the implementation of suitable methodology and reuse of large volumes of food waste in the municipal and corporate towns. According to FAO (2011), around 1.3 billion tonnes of food produced in the world for human need per year are wasted through food supply chain from production to consumption. The food wastes mainly contain fruits or vegetable residues, bones, flesh and blood from meat or fish cleaning process, bagasse and molasses from sugar refining industries, discarded items and dairy wastes. In the present scenario, waste food is recycled, mainly as animal feed and compost, and the remaining quantities are burnt and disposed of in landfills, causing serious emissions of methane (CH_4), which is 23 times more harmful than carbon dioxide (CO_2) as a greenhouse gas and significantly contributes to climate change. Ethanol production from food waste involves a different approach for waste-to-energy conversion when compared to biogas. Enzymatic hydrolysis is the most common pretreatment method in ethanol production from food waste. Rapidly rising costs involved with energy supply and disposal of waste increase public concerns. In view of good-quality environment, the conversion of food wastes into energy is becoming an environmentally safe and economically attractive practice. Usually, the food waste, which is a part of municipal solid garbage, is thrown in open area which may affect health and cause environmental issues. Burning of waste food, which possesses high water content, may cause releasing of various toxins which may further lead to deteriorating of the quality of environment. The amount of waste food, which is deposited all over the globe, may be converted into different types of bioenergy through anaerobic digestion, which is explained in this work by depicting in Fig. 16.7.

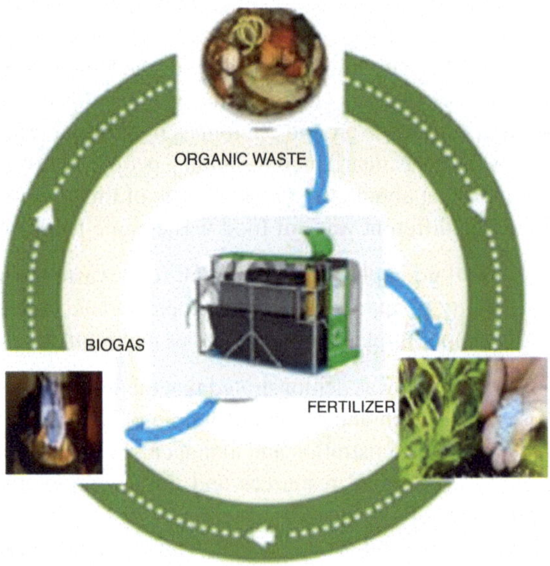

Fig. 16.7 Bioenergy production from food wastes and biomass through anaerobic digestion

16.12 The Scope and Economic Value-Added Compounds from Food Supply Chain Waste

Food supply chain waste is a vast organic matter content released into environment after consumed by man, and is discarded and thrown into open, which is observed at hotels, municipal waste, manufacturing and retail prices. A survey conducted by FAO reveals that the high wastage of food in the production process was observed before it to human beings (Partt et al. 2010).

It was observed that more than 1.5 billion tonnes per annum all over the globe was released into the environment after consumed by human beings. This leads to a major imbalance in natural resources causing economic and social problems (Gustavsson et al. 2011). Figure 16.8 gives sufficient information about the residues and by-products arising from food waste.

Food waste is seriously influenced by the surrounding environment and it is a major issue all over the world. The waste thrown on land is costly and has a major impact on environment. It is observed that greenhouse gases like methane and carbon dioxide are abundantly reaching the environment.

It is estimated that if one tonne of food waste is reaching the environment it releases nearly 4.5 tonnes of carbon dioxide. Conversion of waste food into energy by incineration is not possible not only due to huge quantity of energy loss but also because the wastes contain high moisture content. The methods are applied to processing of waste food into high-economic-value products with novel and safe industrial application (Litchifeld 1987). The designing of simple and easily operatable

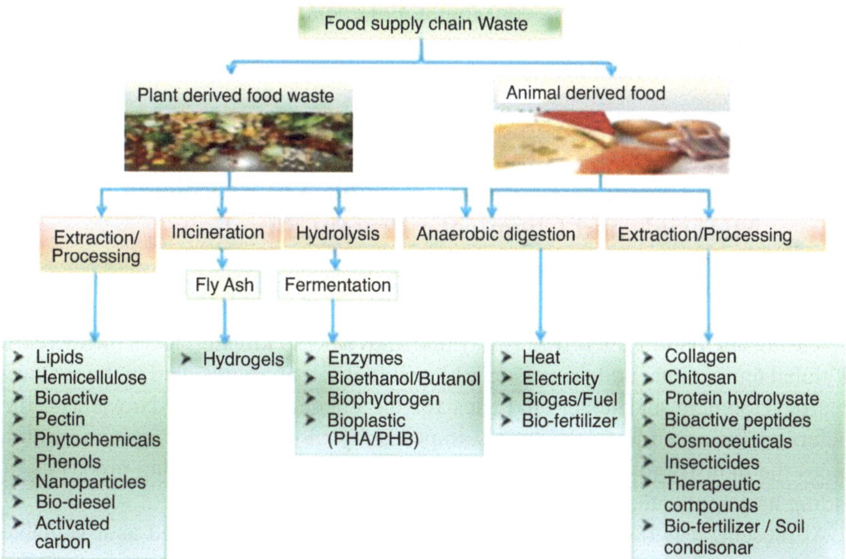

Fig. 16.8 Methods of conversion of food waste into biorefinery and valuable products

Table 16.1 The statistical data of the total food waste occurring at commercial and industrial source

Type of food wastes	Total percentage	Percentage of type	Amount in million tons
Commercial wastes	31%		23–30
Hotels and catering		16%	4–5
Wholesale (inc. food and drink)		39%	9–12
Other commercial		45%	10–15
Industrial waste	69%		48–69
Food, drink and tobacco		16%	8–11
Other industries		84%	41–58
Pine college			
Total industrial waste and commercial wastes			70–100

methods could be used to convert food waste into biofuels by collecting chemicals via biological process.

The following table (Burczyk 2014) illustrates the total waste of food generation at various stages of food processing industries.

The total food waste occurring at commercial and industrial source is as follows (Table 16.1).

The application of waste food in biorefinery is becoming an important and an urgent process. By developing proper methods, the outputs of the product could be increased by producing more valuable products including chemicals and fuels may be collected from various food waste. This not only minimises production costs but also saves time and causes less consumption of chemicals. The following provides various sources of food waste consumed by man and released into the environment (Fig. 16.9).

16.13 Waste Food as Economic Resource in Producing Fuel, Materials and Chemicals: Present Situation and Global Scenario

Increasing demand for fuels day by day due to overpopulation threatens the increase of global temperature and decreasing of fossil fuel resources, resulted in the sustainable development and novel strategies for the chemical industry. Nowadays, the society is facing serious constraints imposed by global resource system, which forces the industry to increase its overall yield by modifying existing processes or finding new applications for waste disposal.

At present, it is observed that food supply chain waste emerged as a valuable resource with a significant potential which is to be employed as a raw material for the production of fuels and chemicals which are procured in high volumes all over the globe, containing many valuable applications (as shown in Table 16.2).

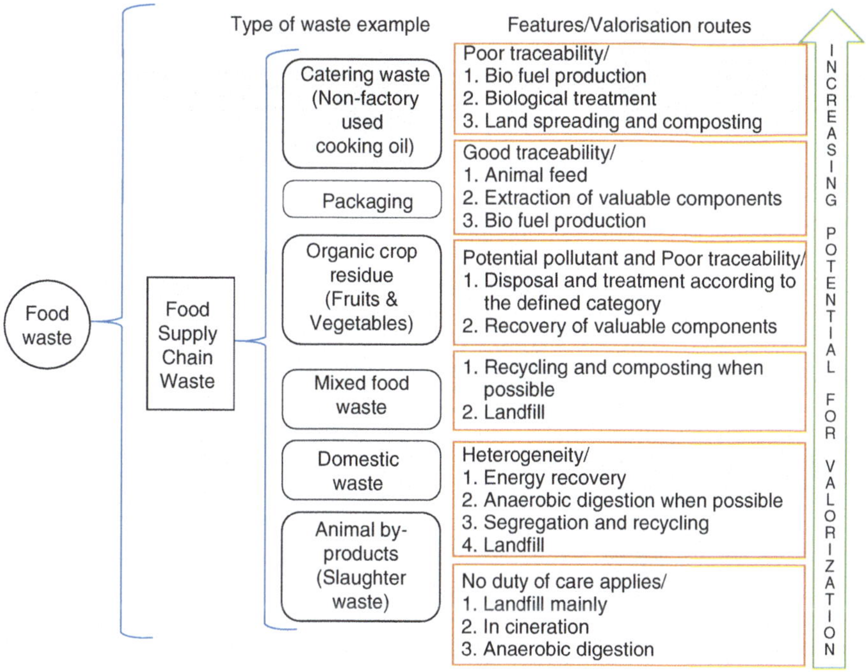

Fig. 16.9 Classification of different types of food wastes

Table 16.2 Various fuels and products that are procured from food wastes

Treatment	Valuable products developed in recycling method
Composting	Phosphorous Nitrogen Humus
Anaerobic digestion	Methane
Pre-processing Industrial fermentation and biorefinery	Animal feed Ethanol Hydrogen Biodiesel Lactic acid Organic acid cocktail Polyhydroxyalkanoate

Prof. Rethye, director of the Tucson garbage project, suggested that the waste generated by various sources, including land filling, incineration and converting them into various articles/chemicals can be reused and there by lowering the amount of bio wastes to environment. Since the depletion of conventional fossil fuels the concept of biorefinery has become an ensuing research area. The work on biofuel has become a new and innovative thought among scientists of various fields like biology, biochemistry, environmental sciences, biochemistry engineering, economics and commerce.

Figure 16.10 shows how different wastes give various forms of energy.

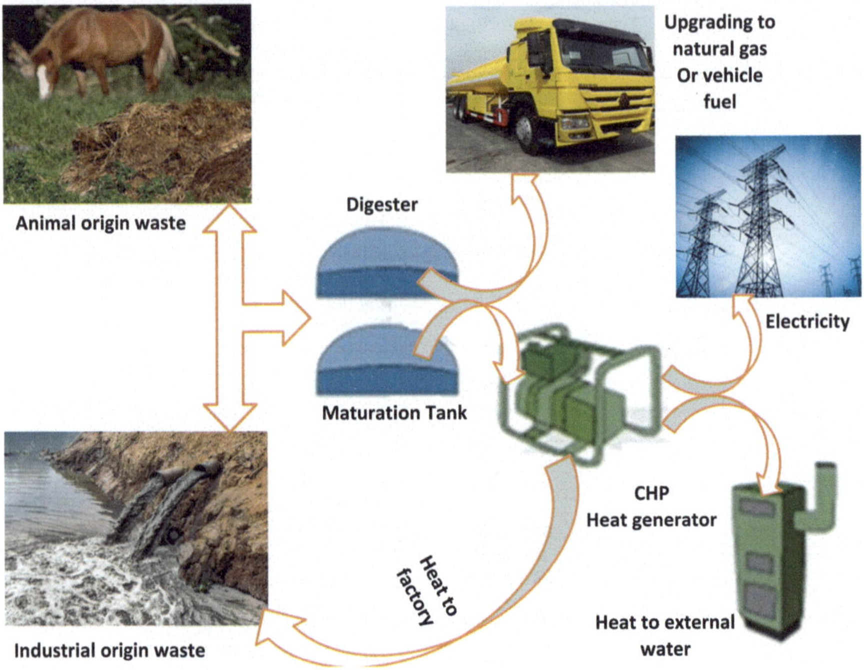

Fig. 16.10 Conversion of animal origin waste and industrial waste into various energy forms

16.14 The Origin of Food Waste

The definition for food waste is given as the ultimate products which are generated at the sources of food processing industry which are not reused or recycled for various purposes. They are not useful products because their procurement and recovery from waste are very difficult and the collection cost is more when compared to obtaining value-added economic compounds (United Nations Industrial Development Organization, Food Wastes 2002).

Every year in the European countries (B. I. Service, Preparatory Study on Food Waste across E.U.-27 for the European Commission 2010) around 89 million tonnes of food waste is produced; of 80% of the total figure 38% is contributed by the manufacturing sector and 42% by the household sector.

16.15 Methods of Conversion of Food Wastes into Useful and Economic Products

Food industry generates voluminous and serious amounts of food wastes and by-products as a result of rapid urbanisation, population growth and industrialisation. Conversion of these food wastes and by-products into value-added products is very

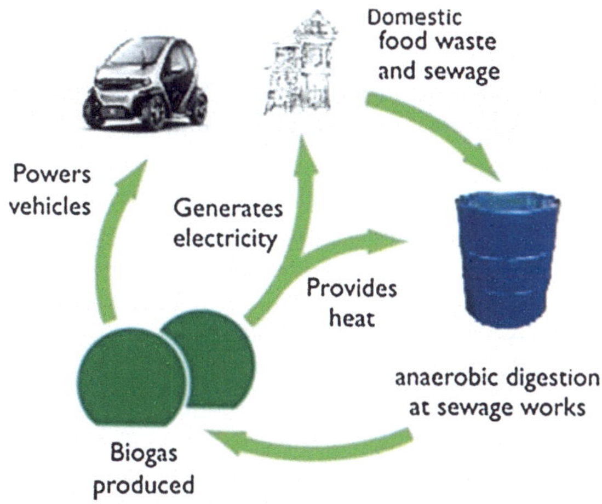

Fig. 16.11 Reduction of food waste into energy by recycling process

important for not only economic aspect but also social and environmental aspects. Food and beverage production is an industry with the main raw materials generally obtained from plant and animal sources and the industry has always aimed to convert these raw materials to value-added products. Memon (2010) and Sakai et al. (2011) proposed the 3Rs concept, namely reduce, reuse and recycle, to overcome food waste issue throughout the world (Fig. 16.11).

Moreover Riemer and Kristoffersen (1999) suggested four ways to achieve perfect waste minimisation in the industry as follows:

1. Waste prevention of waste using more efficient production technologies
2. Internal recycling of product waste
3. Source-oriented improvement of waste quality
4. Reuse of products: A number of by-products are also somewhat discarded in the food industry without any process although they can be used to produce valuable products.

Generally, in the food processing industry, hydrolysis and oxidation reactions are mostly used as chemical conversion methods for food wastes and by-products. Solvent extraction is also preferred to extract valuable organic acids, vitamins, antioxidants, phenolic compounds, colouring agents from food wastes and by-products. Valuable content of food wastes and by-products can also be extracted by green extraction techniques where water is mostly preferred as extraction medium instead of organic solvent extraction.

Food waste can be used as an ideal substrate for anaerobic digestion (AD) due to its high moisture and organic content. This process of anaerobic digestion involves four steps, namely hydrolysis, acidogenesis, acetogenesis and methanogenesis (De La Rubia et al. 2009).

The waste food and its by-products are released in enormous amount from various food processing and production industries.

During processing of food it is observed that nearly 40% of food waste is released. These include the whey, cured and other meat products from animals, slaughter houses, different parts of animal organs, meat trimming blood, aquatic animal waste. All these wastes from milk separation process and vegetable related products generates waste into the environment during the collection of edible oil, starch, juice and carbohydrate. These types of organic waste, biowaste, undergo decomposition by microorganisms exhibiting adverse effects on human health and changing the qualities of available resources. Hence there is a need for preventive steps to minimise the waste releasing from various processes, thereby safeguarding human health. Different types of waste products which are coming out from various industries may be utilised effectively in the form of valuable and economic products which raises the economy of the country and maintains the quality of environment.

Generally, the food waste consisting of valuable substances, namely b-glucans, is formed which increases the rate of lipid metabolism, lowers cholesterol percentage and acts as antiviral, antibacterial and anticancer agent in humans.

Through further processing of cereal brans certain phenolic compounds may be isolated from phenolic compound which are used as antioxidants in heart and cancer diseases.

In the case of vine and brewery industry, the waste contains a high amount of valuable substances, namely sugar, antioxidants, proteins, ethyl alcohol salts like malates, tartrates citric acid and fibre-content material (Aliyu and Bala 2011).

High-valuable protein-content lipids produce high-saturated fatty acids with high molecular weight in the form of food. The by-products generated from seafood contain nutraceuticals and bioactives which are useful for human health.

Further, the marine products contain carotenoids, antioxidants and omega-3 oils. Meat and their by-products contain high quantity of lipid proteins and carbohydrates. The bioactive compounds, namely peptides, possess antimicrobial, antithrombotic, anti-carcinogenic and antioxidative qualities. This influences the proper functioning of digestive, nervous, cardiovascular and immunity systems. It has been observed that the application of peptides can be effectively used in the cure of cancer, obesity diabetes and psychiatric problems (Lafarga and Teagase 2014). The waste liberated from milk and milk product industry contains highly economical products like lactoferrin and lactoperoxidase.

16.16 Conversion of Food Waste into Energy

As per the findings of Food and Agricultural Organization (FAO), nearly one-third of food produced all over the globe for human feeding is lost along the food supply chain. In majority of the countries, food wastes are landfilled or incinerated along with other combustible municipal wastes for possible recovery of energy. Due to its organic- and nutrient-rich composition, theoretically food waste can be utilised as a resource in producing biofuel through several fermentation processes. Till now, valorisation of food waste has gained increasing interest, with biogas, hydrogen, ethanol and biodiesel as final products.

Waste food is an organic content ejected from various sources, namely food processing plants, hotels, and domestic and commercial kitchens. FW is mainly composed of carbohydrate polymers, lignin, proteins, lipids, organic acids and smaller content of inorganic part. FW can be considered as the sole microbial feedstock for the growth of many kinds of value-added bioproducts, including methane, hydrogen, ethanol, enzymes, organic acid, biopolymers and bioplastics (Han and Shin 2004; Ohkouchi and Inoue 2007; Sakai and Ezaki 2006; Wang et al. 2005; Yang et al. 2006; Zhang et al. 2013; Koike et al. 2009; He et al. 2012; Pan et al. 2008; Rao and Singh 2004).

16.17 Conversion of Food Waste into Biofuels and Profitable Products

The economic growth of any country is influenced by food and fuel; both are correlated with each other; the traditional agriculture method may consume less amount of energy and other natural resources when compared to the current production and this increases day by day due to growth in population. A major quantity of energy (nearly 75%) is utilised all over the globe for production of fertilisers, designing of various machineries used in the agricultural industry, transportation and storage of food commodities. At an average of nearly 60% of calories are used in the consumption of human food and a very little amount is solely applied to procuring biofuels (5–10%). During the production of agricultural products, roughly 70% of water is consumed and this amount may increase in the near future. In the production of agricultural products, the processing and distribution methods may cause an increase in greenhouse gases leading to world climate change which affects or modifies food production. This process goes on, and ultimately this influences the economic growth of developing countries (Daniel et al. 2007).

The population growth is responsible for utilisation of large quantity of energy which in turn depletes the natural energy sources. This can be overcome by applying proper methods, techniques and recycling of food wastes into energy. By selecting appropriate methods in the agricultural industry, food waste management can reduce the demand for fossil fuels. Further, in the production of agricultural products, one has to think about short and longer technologies and this causes minimisation of excess energy consumption. Also the wastage of food is affected by climate change, rainfall pattern, geographical conditions, poverty, income, socio-economic aspects, awareness, and time and policies adopted by various governments.

16.18 Food Waste Biorefinery

The concept of biorefinery is influenced by choosing appropriate methods and proper implementation of food wastes into biofuels. The process of aerobic digestion is mainly involved in the procurement of biofuels and biofertilisers from food waste treatment. Conversion of food wastes into recycling element carbon leads to increase in the percentage of fossil fuels. Hence a variety of methods are adopted in throughout the globe treating food wastes as useful materials (Lu and Liu 2009). The concept of biorefinery provides an outline collecting different types of valuable and economic products from food wastes (Fig. 16.9). When compared to traditional food processing and food waste methods, modern ways are used to produce different types of high-income products by treating billions of food waste. The modern methods which include conversion of food into various steps are as follows:

1. Production of biodiesel and biorefining products from food wastes
2. Collection of hydrogen gas from food wastes through fermentation method; certain chemicals like ethanol, lactic acid and succinic acid; and conversion of food garbage into biopolymers (Sun and Cheng 2002; Wang and Ma 2008; Chen et al. 2009; Yaakob et al. 2013)

16.19 Biodiesel Production

The microbeswhich are rich in fatty acids present in lipids are suitable for biodiesel production. Fatty acids and biodiesel can be produced from food waste through the method of straight-away transesterification by applying catalysts which are in basic or acidic nature and moreover transesterification of oils is done by many oleaginous microorganisms (Zhang et al. 2003; Masoud et al. 2009).

16.20 Biodiesel

As exploitation of resources of petroleum has increased in the last few years, depletion of resources has drawn attraction for renewable energy throughout the world.

Biodiesel is recognised as an important and capable fuel, which is synthesised from vegetable oils or animal fats which are available in low cost and largely used in diesel engines and other utility systems (Du et al. 2007).

16.21 An Introduction on How to Procure Biofuel from Waste Food

Various research activities in different fields have noticed that food waste creates serious economic and environmental challenges all over the globe. Food waste is highly rich in nutritional substances containing peculiar microorganisms, leading to formation of highly valuable by-products in the form of chemicals or biofuels.

Further, as per the estimation by the majority of scientists the deposition of food waste may increase in the next 25 years due to population growth and this creates a serious threat to economic development of various countries.

Hence, many methods and efforts have been suggested in terms of new techniques for biofuels, namely ethyl alcohol, hydrogen, methane and biodiesel from waste food.

The increase in population in turn increases the utility of fossil fuels all over the globe and decreases the stock of non-renewable energy sources.

Hence an alternative, easily available and more economic fuel is required with great potential value. Biodiesel is a highly valuable fuel generated from waste food by transesterification in acidic/basic medium or transesterification in terms of microbial content. Moreover, biofuel can also be produced from animal fat, polanga, algae castor beans, pongamia, butter, oilseed radish, cotton, sunflower, babassu, palm, palm kernel, olive, sesame, corn, linseed, soybean, canola, sunflower, peanuts, coconut, jatropha, sea mango, fish oil and waste cooking oil (Bogdanov and Svinyarov 2013). The wastage amount of cooking oil generated in each country differs depending on the usage of vegetable oil by households and the method of cooking adopted by the people. The manufacture of biodiesel from waste cooking oil is one of the best methods to utilise it efficiently and cost effectively.

16.22 Production Ethanol from Food Waste

Ethanol has a global demand due to its wide range of applications in many chemical industries and forms an important renewable fuel.

The important feature of ethanol is to minimise the rate of pollution when compared to fossil fuels and the production cost is very small.

Bioethanol is isolated from carbohydrate-rich substances, namely rice, potato and sugar cane. Fermented starch in the presence of certain enzymes is converted into glucose and subsequently converted into ethanol. The cheap and abundant quantities of waste like lignocellulose and municipal, hotel and kitchen waste have been observed as an alternative for biofuel production. This can be schematically explained in Fig. 16.12.

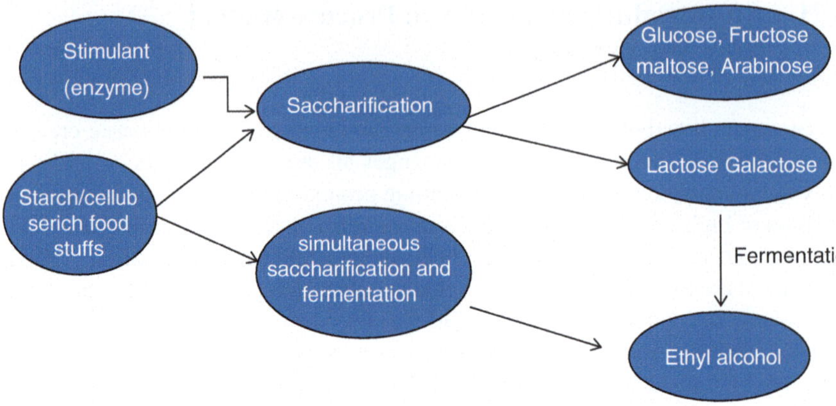

Fig. 16.12 Production of ethanol from food waste

16.23 Production of Hydrogen from Waste

Hydrogen gas is a promoting choice to fossil fuel. It is clear and renovating possessing high calorific value. The production of hydrogen is carried out by anaerobic fermentation which is an eco-friendly and energy-saving method. During the reaction process, apart from liberation of hydrogen gas, substances like organic acids and other substances are also produced. The energy produced during combustion process is about nearly 145 kg/g. In comparison this value is twice/thrice greater than that of combustion of methane and gasoline.

The essential condition for production of large amount of hydrogen gas is the presence of high amount of carbohydrate in food waste. The production of hydrogen from food waste is influenced by the amount of waste food, preliminary treatment methods and systematic arrangement of processes. It is observed scientifically that the production yield of hydrogen is 20 times more from the waste containing carbohydrate when compared to fat- and protein-content waste. Many methods have been adopted for the procurement of hydrogen from waste food which includes continuous, semi-continuous, batch-wise and one or more steps.

A process known as photo-fermentation was established to collect hydrogen from organic acids. If photo-fermentation process is coupled with dark fermentation method, the yield will become more. During the process, lactic acid is produced as a by-product, which is utilised by photo-fermentative bacteria, and it is converted into hydrogen. Figure 16.13 clearly explains generation of hydrogen from food waste.

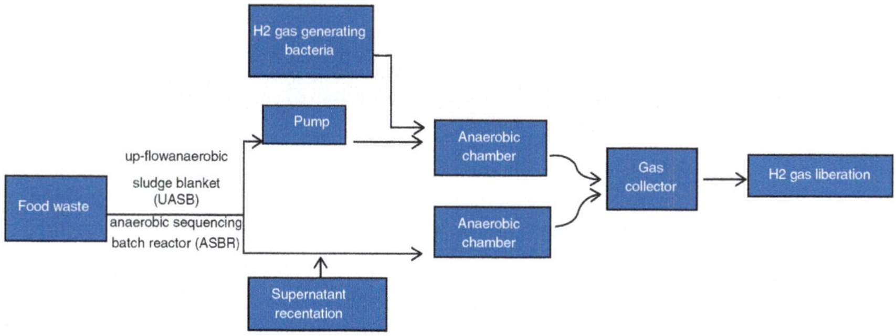

Fig. 16.13 Generation of hydrogen gas from food waste

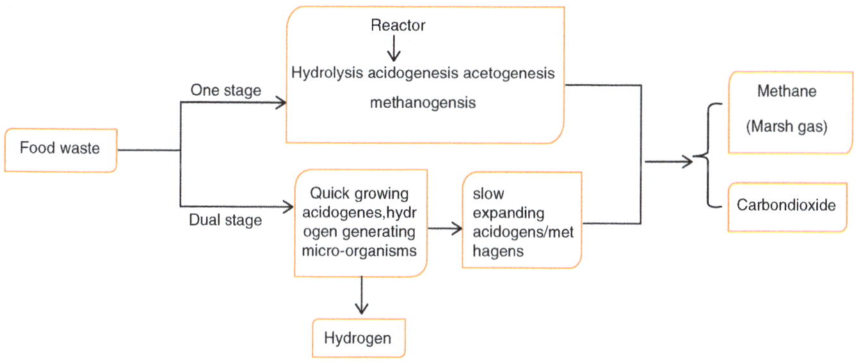

Fig. 16.14 Flow process of methane generation from food waste

16.24 Methane Production from Food Waste

The method of bio-methanation from waste food has become a novel, attractive and highly economical method. This technique has attracted attention from all over the globe due to its powerful nature; it will reduce the serious and detrimental effects on human health and environment arising from food waste. The method of bio-methanation is influencing the waste management technique, having minimum price, small amount of residual waste and also reusable energy source (Sasoki 2012). The waste generated during the production of methane is a highly nutrient-content material which can be used as fertiliser. The collection and stability method for generation of methane is influenced by characteristic properties of waste, namely amount of moisture content liable to sudden change of solid and nourishing substance and the size of waste food particle. Figure 16.14 clearly explains the generation of methane gas food waste.

16.25 Biowaste

The waste which is generated from living organisms, namely man, animal, plants and trees, is known as biowaste. This waste is biodegradable through aerobic or anaerobic state. Economic highly profitable biowaste may include forest land; residues are released from agricultural activities, animal waste, green manure drainage wastewater and commercial food. The biowaste which comes from domestic origin includes kitchen scrap garden waste, paper and pulp industry waste and natural material waste. The removal and operation of biowaste at present have become a great task for the waste coming from various industries. The best way of the biowaste removal which consists of agriculture, food factories and various food consumable activities of human beings is by anaerobic fermentation method, which is highly remarkable when compared to composting method or other processing ways of recycling. Further biowaste can be converted into biogas which is an important energy source for the production of heat and electricity. Figure 16.15 shows the conversion of biowaste into energy.

Biofuel generation from agricultural, municipal and industrial wastes is effectively converted into biogas, a mixture of methane (CH_4) and carbon dioxide (CO_2) under anaerobic condition.

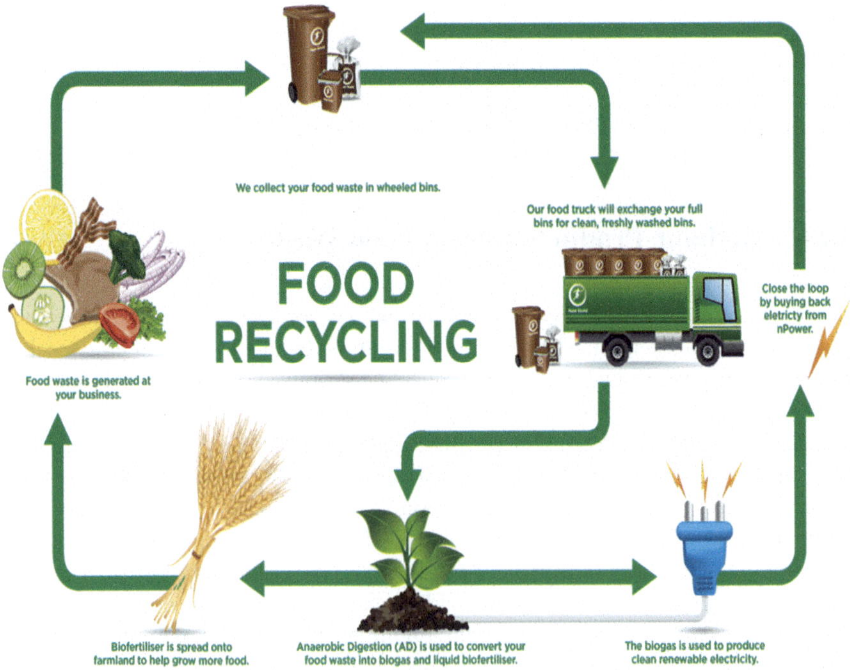

Fig. 16.15 Recycling of food waste into electricity

In general, microbial anaerobic conversion of food waste to methane is a process for both powerful waste treatment and sustainable energy production. The energy from biogas production is high when compared to other technological conversion methods.

Biogas is an easy form of renewable energy that may be used directly for producing heat and steam or converting into electricity for engines, fuel cells and gas turbines.

In many countries, biogas is considered as an environmentally safe and an economically alternative fuel to diesel and gasoline for running buses and other transport vehicles.

Biodiesel is usually prepared from either pure plant oils or waste vegetable oils through catalytic transesterification method.

16.26 Usage of Ionic Liquids in Food and Biowaste Industry

The Fig. 16.16 demonstrate the function of ionic liquids in food and particular areas consists of industrial biomass products namely, Isolation processes, analysis of food, and bioenergy from food waste sources into biofuels.

The primary role of isolation process involves getting many beneficial products from the food industry such as piperine, caffeine, essential oils, fatty acids, phenolic compounds and other food additives.

Food analysis has been extended to estimation of heavy metals in environment, antibiotics in pharma samples, dyes, herbicides in agriculture, acrylamide, vitamins, acidic food additives, phenolic compounds, preservatives, species, and folic and ascorbic acids.

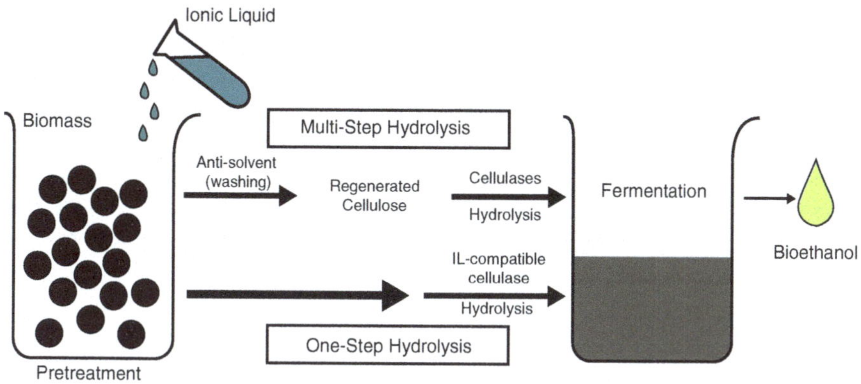

Fig. 16.16 Role of ionic liquid in the production of biofuel from biomass

It is accustomed that imidazolium-based ILs, especially based on the 1-alkyl-3-methylimidazolium cation, are widely used in the manufacture of biofuels from wheat, corncob, sugar cane bagasse, corn, vegetable oils, etc.

The important role of ionic liquids in the food processing industry is due to their low or high temperatures and higher dissolving capacity and hence they play a pivoting and dominant role when compared to traditional organic liquids.

According to modern methods which are applied in various food preparing and its related industry, highly economic value-added products are available from natural substances like polyphenols, natural dyes and essential oils.

In literature, many methods have been used for isolation of valuable chemicals from food and its wastes by using techniques like solid–liquid equilibrium (SLE), vapour–liquid extraction (VLE) or liquid–liquid extraction (LLE).

Recently, ultrasound extraction (UAE) or microwave-assisted extraction (MAE) (Ressmann et al. 2013) combined with various ionic liquids, namely 1-butyl-3-methylimidazolium acesulfamate [C4mim][Ace] (Chowdhury et al. 2010), [N111(C2O(O)C12)]-based IL (Lou et al. 2012), [N(C1)2CO2]-based IL (Bica et al. 2011), acetate (Ni et al. 2012), and amino acid- (Ribeiro et al. 2013) and cholinium-based ILs (Wang et al. 2016; Garcia et al. 2010; Zhai et al. 2009), was used for recovering aromatic oils, phenolic compounds, caffeine, piperine or preservative media.

By applying mechanical approaches, using a variety of solvents or hydrodistillation, essential oils are collected from certain plant parts, namely herbs, seeds and fruits. Microwave-assisted extraction (MAE) process (Ribeiro et al. 2013) was adopted to collect aromatic oils from dried fruits using ionic liquid, namely 1-hexyl-3-methylimidazolium hexafluorophosphate ([C6mim][PF6]).

Phenolic compounds are bioactive in nature, acquired from food and food wastes synthetically. It is suggested that polyphenols were collected from tea and mate using aqueous solutions of ionic liquids containing cholinium and imidazolium (Claudio et al. 2013).

A bioactive alkaloid, namely caffeine, is commonly found in guarana, tea, coffee and chocolate. It has been proposed in the literature (Lago et al. 2014) that yield and isolation of caffeine are increased by adding aqueous solutions of 1-butyl-3-methylimidazolium chloride ([C4mim]Cl) at different proportions from guarana seeds.

Terpenes are part of organic compounds observed in aromatic oils which possess citrus flavour. It is suggested that ionic liquids that are acetate based are more powerful and efficient to isolate terpenes when compared to other ionic liquids or other solvents (Shao et al. 2014).

16.27 Analysis of Various Food Products

Modern food analytical methods are adopted for determination and quantification of food and its related products with high efficiency, sensitivity and low cost. During the extraction methods, the ionic liquid, namely 1-alkyl-imidazolium based, is used in food analysis apart from the solutions of water or ethanol.

A method has been established in the literature (Sun et al. 2014) for estimation of antibiotics and sulphonamide in milk samples by applying 1-butyl-3-methylimidazolium tetrafluoroborate ([C4mim][BF4]). This method is applied to the treatment of animal infections, in order to avoid the risk of milk contamination.

A process has been developed for the estimation of ethyl p-hydroxybenzoate, methyl p-hydroxybenzoate, p-hydroxybenzoic, sorbic, benzoic, butyl p-hydroxybenzoate, propyl p-hydroxybenzoate and m-amino benzoic acid that are in general found as additives in cool drinks, fruit juices and components of vitamins. The addition of small amounts of these additives may arrest the development of pathogens or spoilage microorganisms (Zhu et al. 2014).

The imidazolium ionic liquid, namely 1-octyl-3- methylimidazolium chloride ([C8mim]Cl), was used to isolate various flavouring compounds, namely ethyl maltol and ethyl vanillin, from the milk powder, biscuits and chocolate(Guo et al. 2011).

16.28 Manufacture of Biofuels from Biomass and Vegetable Oils

Ionic liquids have gained an important role as environmentally safe and effective in growing more "green" and effective methods for producing biodiesel from vegetable oils and bioethanol from sugarcane. Conventionally, it is observed that biodiesel is a mixture of fatty esters which are collected from vegetable oils using the process of transesterification of triacylglycerol.

Guo et al. (Guo et al. 2012) produced biodiesel from jatropha oil and soybean oil by applying ionic liquids, namely 1-butyl-3-methylimidazolium tosylate ([C4mim][CH3SO3]) and 3-(N,N,N-triethylamino)-1-propanesulphonic hydrogen sulphate, with a yield of 93%.

In the production of bioethanol from sugar cane bagasse, ionic liquids are used as solvents, which reduce production costs (Gu et al. 2013).

In many foodstuffs, cholinium is a B-complex vitamin observed in peanuts, soybean and eggs that is known to be an important and non-toxic nutrient suggested for human diet (Blusztajn 1998). Moreover, the ionic liquids prepared from the derivatives of cholinium and amino acids are applied to preliminary operation of biomass.

In the refining of oils obtained from plants, namely soybean, palm and sun flower, fatty acids and high-molecular-weight carboxylic acids are developed.

Further, natural products like esters of fatty acids and phospholipids may also be collected from refining of vegetable oils that are in general used in the manufacture of certain important materials, namely surfactants, makeup material and certain pharma substances (Maximo et al. 2014).

Wang et al. (Wang et al. 2016) extracted flavonoids from citrus peels using cholinium- and amino acid-based ionic liquid and Garcia et al.(Zhai et al. 2009) isolated suberin, present in cell walls of cork, waste from the wine industry by applying imidazolium- and cholinium-based ionic liquids.

Qin et al. (Qin et al. 2016) reported from his work that during the processing of food waste he collected vitamin E.

16.29 Analysis of Materials Generated from Food Waste

The application of ILs in the processing of food waste is another desirable solution for minimising the environmental pollution. There is a big challenge in industry to find an effective management method to control the waste produced by food processing units. It is demonstrated that the ionic liquids could be used to isolate high-economic products from biomass.

16.30 Ionic Liquids and Their Toxic Effects

At present, an investigation about harmful effects of ionic liquid has become important due to their wide range of industrial applications. It was observed that pyridinium- and imidazolium-based ionic liquids are widely acceptable examples of ionic liquids with matched amount of toxicity (Tsarpali and Dailianis 2015; Paterno et al. 2014; Yan et al. 2012; Egorova and Ananikov 2014; Viboud et al. 2012).

But the harmful effects of ionic liquids are known abundantly by analysing their toxic estimation. It was reported in the literature (Zhao et al. 2007) that toxic effects were observed in aquatic ecosystems, cytotoxicology, animal tests, microorganisms and enzyme inhibition.

The role of ionic liquids which contain imidazolium is found in many methods involving food products, as described earlier to pursue their ecotoxicity. Some workers reported that as the chain length of ionic liquid increases it is a dominant factor in estimating their harmful effects (Kulacki and Lamberti 2008; Steudte et al. 2014; Frade and Afonso 2010).

According to the works of Peric et al. (2013) that ionic liquids containing organic acid anions are not toxic compared to traditional imidazolium molecules in terms of their biodegradable nature and ecotoxicity to the watery systems.

Genotoxicity works are applied to estimate the destruction caused by any given substance on the health of human based on gene mutation and chromosomal alteration.

In the case of human and animal cells the cytotoxic studies are providing the main techniques for assessment of risk of ionic liquids on human health, which are easy when compared to oral toxic effect on rodents and humans (Chen et al. 2014).

It is reported in the literature (Passos et al. 2014) that the ionic liquids which contain amino acids comprising pyridinium, imidazolium and cholinium cation

Table 16.3 The applications of various ionic liquids in differental food sectors

Ionic liquids	Process	Product	References
1-Butyl-3-methyl-imidazolium bromide	Solvent extractor	Acrylamide in coffee and milk	Zhang et al. (2017)
1-Butyl-3-methylimidazolium trifluoromethanesulphonate	Synthesis solvent (recycle and reuse)	Sugar fatty acid esters	Mai et al. (2014)
1-Butyl-3-methylimidazolium chloride	Synthesis co-solvent	Arylalkyl b-d-glucopyranosides	Yang et al. (2012)
1-Butyl-1-methylpyrrolidinium bis(trifluoromethylsulphonyl) imide	Solvent extractor	Vitamin D	Liang et al. (2013)
1-Butyl-3-methylimidazolium hydroxide	Catalyst and solvent	Deoxy-fructosazine and fructosazine	Jia et al. (2014)
1-Butyl-3-methylimidazolium chloride	Catalyst and solvent	Furfural	Wu et al. (2014)
1-Decatyl-3-methylimidazolium hexafluorophosphate	Synthesis co-solvent	Sucrose-6-acetate	Wei et al. (2016)
1-Butyl-3-methylimidazolium chloride	Co-solvent for Maillard reaction	Chitosan and xylan	Luo et al. (2013)
1-Vinyl-3 (10-hydroxydecyl) imidazolium bis[(trifluoromethyl) sulphonyl]imide	Microextraction	Acrylamide in coffee	Cagliero et al. (2016)
1-Butyl-3-methylimidazolium chloride	Solvent extractor	Carotenoids	Martins and de Rosso (2016)
1-Hexyl-3-methyl-imidazolium bromide	Solvent extractor	Soy food	Magiera and Sobik (2017)
Tetrabutylphosphonium chloride	Solvent extractor	Bovine serum albumin	Pereira et al. (2015)

exhibit toxic effects towards HeLa cells. The authors suggested that cholinium cation-based ionic liquid has less toxic level when compared to methyl imidazolium cation-containing ionic liquid. This suggested that cholinium and amino acid ions are naturally available causing less hazards to human health and surroundings. Moreover, the experiments on harmful effects of ionic liquids in mammals are observed to be deficient in literature (Jodynis-Liebert et al. 2010).

The lethal levels of ionic liquids depend on the estimation processes and also the size, shape and structure of anion and cation. In general, this behaviour is observed in the ionic liquids which are prepared from commonly available sources. Further still certain discovery may provide crucial knowledge on the production of non-toxic ionic liquids which in turn are applied to food products and their processing industries. Table 16.3 gives ample information about the use of ionic liquids in various food sectors.

16.31 Conclusions

Nowadays, the most prominent and promising uses of ionic liquids in food and their bioproduct industries are increasing day by day. This must be effective and applicable if the ionic liquids should behave as non-harmful additives. There is a large gap pertaining to harmful effects on human cells and mammals as advised by FDA in order to apply them in the food processing industry. It is observed that ionic liquid containing cholinium as cation is found to be appropriate and suitable for isolation and estimation of phenolic compounds, terpenes, aromatic oils and highly economic value biosubstance that may be procured from vegetable and food waste processing industries. Manufacture of biofuels from food waste is stagnant and an emerging area since much work is stressed on to investigate the traditional ionic liquids.

Although the ionic liquids which contain imidazolium are used in food processing because of their harmful nature they are not suggested for application in the food sector suggested by FDA. An extensive survey also suggests that the application of ILs in the food and biocompound development is meagre. Moreover selecting a large number of naturally available compounds that are in general used as a pair of cation and anion may improve the synthesis of novel, non-harmful and edible ionic liquids. This facilitates further application towards bioprocesses and also to preparation of food supplements and creating of products with good physical and sensorial characters.

Food wastes and by-products can be converted into valuable products through thermal, chemical and biological methods. The appropriate conversion method is selected with respect to composition of food wastes and by-products and the aim of recovery process.

Nearly one-third of the food produced globally is wasted, when more than one billion people around the world go to bed without enough food. Controlling of food waste can solve this problem to some extent. It also reduces the substantial loss of other resources like land, water, energy and labour.

Food industry by-products are a good source of proteins, minerals, fatty acids and fibre, and the important raw material is available from food and its related factories or a valuable source for fatty acids, proteins, fibre, bioactive compounds and minerals. These substances are sufficient for human beings in maintaining normal healthy and sustainable activity and this leads to nil deficiency for essential elements for growth of human beings.

The conversion of bioproduct waste coming from food factories may be converted into other useful and non-harmful products that may lower the negative cost, minimise environmental pollution and ensure sustainability of the food industry and all these methods will influence the economy of the country. At present, new innovation technique/practises are in operation to get new varieties of functional food ingredients that may be isolated from natural sources. These are important and thought-provoking area in food science and technology. This creates an innovation to make valuable food by-products and increase profitability of many industries.

The organisation pertaining to food waste has created a serious economic and environmental concern. It is feasible and practicable that bioconversion of waste food in terms of energy sources like ethanol, hydrogen, methane and biodiesel is much economic and commercial. The efficiency and production, which are based on the cost, may lead to more effective research work and optimisation studies in procuring different value-added products from industries. Still, this is not appropriate and there is a crucial need to investigate low-cost, eco-friendly and beneficial approaches for the bioconversion of waste food.

Acknowledgements

1. The author gratefully acknowledges valuable suggestions given by Dr. N Harikrishna, Assistant Professor, Annamacharya Institute of Technology and Sciences, Tirupati, and Dr. M. Ravindra, Associate Professor, IIIT, RGUKNT, R.K. Valley, Idupulapaya, A.P., for successful completion of this chapter.
2. It is a pleasure to express my deep sense of gratitude to my graduate students, Mr. U. Rambhupal Reddy and Mr. S. Iliyaz Ali, for their continuous and cheerful encouragement and enthusiastic hard work in finalising this chapter.

References

Adeyi O, Smith O, Robles S (2007) Public policy and the challenge of chronic noncommunicable diseases. World Bank, Washington, DC

Ali R, Karthikeyan OP, Hao HTN, Heimann K (2016) Hydrolysis treatments of fruit and vegetable waste for production of biofuel precursors. Bioresour Technol 217:100–103

Aliyu S, Bala M (2011) Brewr's spent grain: a review of its potentials and applications. Afr J Biotechnol 10:324–331

B. I. Service, Preparatory Study on Food Waste across E.U.-27 for the European Commission (2010). http://ec.europa.eu/environment/eussd/pdf/bio_foodwaste_report.pdf. Accessed 6 July 2012

Bica K, Gaertner P, Rogers RD (2011) Ionic liquids and fragrances - direct isolation of orange essential oil. Green Chem 13(8):1997–1999

Blusztajn JK (1998) Choline, a vital amine. Science 281(5378):794–795

Bogdanov MG, Svinyarov I (2013) Ionic liquid-supported solid–liquid extraction of bioactive alkaloids. II. Kinetics, modeling and mechanism of glaucine extraction from Glaucium flavum Cr. (Papaveraceae). Sep Purif Technol 103:279–288

Burczyk B (2014) Zielona chemia Zarys. Oficyna Wydawnicza Politechnika Wro- cławska, Wrocław

Chen Y et al (2009) Synthesis of biodiesel from waste cooking oil using immobilized lipase in fixed bed reactor. Energy Convers Manag 50(3):668–673

Chen HL, Kao HF, Wang JY, Wei GT (2014) Cytotoxicity of imidazole ionic liquids in human lung carcinoma A549 cell line. J Chin Chem Soc 61(7):763–769

Chowdhury SA, Vijayaraghavan R, MacFarlane DR (2010) Distillable ionic liquid extraction of tannins from plant materials. Green Chem 12(6):1023–1028

Claudio AFM, Ferreira AM, Freire MG, Coutinho JAP (2013) Enhanced extraction of caffeine from guarana seeds using aqueous solutions of ionic liquids. Green Chem 15(7):2002–2010

Daniel JS, Nancy DK, Ibsen N (2007) Contaminant, occurrence, identification and control in a pilot-scale corn fiber to ethanol conversion progress. Bioresour Technol 98:2942–2948

De La Rubia MA, Raposo F, Rincón B, Borja R (2009) Evaluation of the hydrolytic-acidogenic step of a two-stage mesophilic anaerobic digestion process of sunflower oil cake. Bioresour Technol 100(18):4133–4138

De Maria PD (2008) "Nonsolvent" applications of ionic liquids in biotransformations and organocatalysis. Angew Chem Int Ed 47:6960–6968
De CD, Wen Z, Gottfried O, Schmidt F, Fei F (2017) A review of global strategies promoting the conversion of food waste to bioenergy via anaerobic digestion. Renew Sust Energ Rev 79:204–221
Deetlefs M, Seddon KR (2010) Assessing the greenness of some typical laboratory ionic liquid preparations. Green Chem 12:17–30
Dogliotti S, Giller KE, VanIttersum MK (2014) Achieving global food security whilst reconciling demands on the environment: report of the first international conference on global food security. Food Secur 6(2):299–302
Du FY, Xiao XH, Li GK (2007) Application of ionic liquids in the microwave-assisted extraction of trans-resveratrol from Rhizoma Polygoni Cuspidati. J Chromatogr A 1140(1–2):56–62
Dupont J, de Souza RF, Suarez PAZ (2002) Ionic liquid (molten salt) phase organometallic catalysis. Chem Rev 102:3667–3691
Earle MJ, Esperança JMSS, Gilea MA, Canongia Lopes JN, Rebelo LPN, Seddon KR, Widegren JA (2006) Nature 439:831–834
Egorova KS, Ananikov VP (2014) Toxicity of ionic liquids: eco(cyto)activity as complicated, but unavoidable parameter for task-specific optimization. ChemSusChem 7(2):336–360
Erfurt K, Wandzik I, Walczak K, Matuszek K, Chrobok A (2014) Green Chem 16:3508–3514
FAO (Food and Agriculture Organization of the United Nations) (2016) Pulses and Climate Change: Rome
FAO (Food and Agriculture Organization of the United Nations) (2011) Global food losses and food waste - extent, causes and prevention, – The Swedish Institute for Food and Biotechnology, ISBN 978-91-7290-323-4
Fei ZF, Geldbach TJ, Zhao DB, Dyson PJ (2006) From dysfunction to bis-function: on the design and applications of functionalised ionic liquids. Chem Eur J 12:2123–2130
Frade RFM, Afonso CAM (2010) Impact of ionic liquids in environment and humans: an overview. Hum Exp Toxicol 29(12):1038–1054
Fraga-Dubreuil J, Bazureau JP (2001) Grafted ionic liquid-phase-supported synthesis of small organic molecules. Tetrahedron Lett 42:6097–6100
Freemantle M (1998) Ionic liquids may boost clean technology development. Chem Eng News 76:32–37
Garcia H, Ferreira R, Petkovic M, Ferguson JL, Leitao MC, Gunaratne HQN, Seddon KR, Rebelo LPN, Silva Pereira C (2010) Dissolution of cork biopolymers in biocompatible ionic liquids. Green Chem 12(3):367–369
Gorke J, Srienc F, Kazlauskas R (2010) Toward advanced ionic liquids. Polar, enzyme-friendly solvents for biocatalysis. Biotechnol Bioprocess Eng 15:40–53
Gruttadauria M, Riela S, Aprile C, Lo Meo P, D'Anna F, Noto R (2006) Supported ionic liquids. New recyclable materials for the l-proline-catalyzed aldol reaction. Adv Synth Catal 348:82–92
Gu T, Held MA, Faik A (2013) Supercritical CO2 and ionic liquids for the pretreatment of lignocellulosic biomass in bioethanol production. Environ Technol 34(13–14):1735–1749
Guo F, Fang Z, Tian XF, Long YD, Jiang LQ (2011) One-step production of biodiesel from Jatropha oil with high-acid value in ionic liquids. Bioresour Technol 102(11):6469–6472
Guo W, Li H, Ji G, Zhang G (2012) Ultrasound-assisted production of biodiesel from soybean oil using Brønsted acidic ionic liquid as catalyst. Bioresour Technol 125:332–334
Gustavsson J, Cederberg C, Sonesson U, van Otterdijk R, Meybeck A (2011) Global food losses and food waste. Extend, Causes and Prevention, Food and Agriculture Organization of the United Nations, Rome
Han SK, Shin HS (2004) Biohydrogen production by anaerobic fermentation of food waste. Int J Hydrog Energy 29(6):569–577
Han W, Lam WC, Melikoglu M, Wong MT, Leung HT, Ng CL, Yan P, Yeung SY, CSK L (2015) Kinetic analysis of a crude enzyme extract produced via solid-state fermentation of bakery waste. ACS Sust Chem Eng 3(9):2043–2048

Harjani JR, Singer RD, Garciac MT, Scammells PJ (2009a) Biodegradable pyridinium ionic liquids: design, synthesis and evaluation. Green Chem 11:83–90

Harjani JR, Farrell J, Garcia MT, Singer RD, Scammells PJ (2009b) Further investigation of the biodegradability of imidazolium ionic liquids. Green Chem 11:821–829

He Y et al (2012) Recent advances in membrane technologies for biorefining and bioenergy production. Biotechnol Adv 30(4):817–858

Hough WL, Smiglak M, Rodriguez H, Swatloski RP, Spear SK, Daly DT, Pernak J, Grisel JE, Carliss RD, Soutullo MD, Davis JH Jr, Rogers RD (2007) New J Chem 31:1429–1436

Huddleston JG, Visser AE, Reichert WM, Willauer HD, Broker GA, Rogers RD (2001) Characterization and comparison of hydrophilic and hydrophobic room temperature ionic liquids incorporating the imidazolium cation. Green Chem 3:156–164

Jodynis-Liebert J, Nowicki M, Murias M, Adamska T, Ewertowska M, Kujawska M, Piotrowska H, Konwerska A, Ostalska-Nowicka D, Pernak J (2010) Cytotoxicity, acute and subchronic toxicity of ionic liquid, didecyldimethylammonium saccharinate, in rats. Regul Toxicol Pharmacol 57(2–3):266–273

Katritzky AR, Singh S, Kirichenko K, Holbrey JD, Smiglak M, Reichert WM, Rogers RD (2005) Chem Commun:868–870

Koike Y et al (2009) Production of fuel ethanol and methane from garbage by high-efficiency two-stage fermentation process. J Biosci Bioeng 108(6):508–512

Kulacki KJ, Lamberti GA (2008) Toxicity of imidazolium ionic liquids to freshwater algae. Green Chem 10(1):104–110

Kumar D, Jain VK, Shanker G, Srivastava A (2003) Utilization of fruit waste for citric acid production by solid state fermentation. Process Biochem 38(12):1725–1729

Lafarga T, Teagase MH (2014) Bioactive peptides from meat muscle and by-products: generation, functionality and application as functional ingredients. Meat Sci 98:227–239

Lago S, Rodríguez H, Arce A, Soto A (2014) Improved concentration of citrus essential oil by solvent extraction with acetate ionic liquids. Fluid Phase Equilib 361:37–44

Lee SG (2006) Functionalized imidazolium salts for task-specific ionic liquids and their applications. Chem Commun:1049–1063

Lewandowski A, Świderska-Mocek A (2009) J Power Sources 194:601–609

Litchifeld JH (1987) Food Biotechnol 1:29–57

Liang R, Bao Z, Su B, Xing H, Yang Q, Yang Y (2013) Feasibility of ionic liquids as extractants for selective separation of vitamin D3 and Tachysterol3 by solvent extraction. J Agric Food Chem 61:3479–3487

Lou Z, Wang H, Zhu S, Chen S, Zhang M, Wang Z (2012) Ionic liquids based simultaneous ultrasonic and microwave assisted extraction of phenolic compounds from burdock leaves. Anal Chim Acta 716:28–33

Lu H, Liu Y (2009) Production of biodiesel from Jatropha curcas L. oil. Comput Chem Eng 33:1091–1096

Luo Y, Ling Y, Wang X, Han Y, Zeng X, Sun R (2013) Maillard reaction products from chitosanexylan ionic liquid solution. Carbohydr Polym 98(1):835–841.

Ludwig R (2001) Water Angew Chem Int Ed 40:1808–1827

Mai NL, Ahn K, Bae SW, Shin DW, Morya VK, Koo YM (2014) Ionic liquids as novel solvents for the synthesis of sugar fatty acid ester. Biom J 9(12):1565–1572.

MacFarlane DR, Forsyth M, Howlett PC, Pringle JM, Sun J, Annat G, Neil W, Izgorodina EI (2007) Ionic liquids in electrochemical devices and processes: managing interfacial electrochemistry. Acc Chem Res 40:1165–1173

Magiera S, Sobik A (2017) Ionic liquid-based ultrasound-assisted extraction coupled with liquid chromatography to determine isoflavones in soy foods. J Food Compos Anal 57:94–101

Martins PLG, De Rosso, VV (2016) Thermal and light stabilities and antioxidant activity of carotenoids from tomatoes extracted using an ultrasound assisted completely solvent-free method. Foodserv Res Int 82:156–164.

Masoud Z, Wan Mohd AWD, Mohamed K (2009) Fuel Process Technol 90:770

Maximo GJ, Costa MC, Coutinho JAP, Meirelles AJA (2014) Trends and demands in the solid-liquid equilibrium of lipidic mixtures. RSC Adv 4(60):31840–31850
Memon MA (2010) Integrated solid waste management based on the 3R approach. J Mater Cycles Waste 12(1):30–40
Miao WS, Chan TH (2005) Ionic-liquid-supported peptide synthesis demonstrated by the synthesis of leu5-enkephalin. J Org Chem 70:3251–3255
Ni X, Xing H, Yang Q, Wang J, Su B, Bao Z, Yang Y, Ren Q (2012) Selective liquid-liquid extraction of natural phenolic compounds using amino acid ionic liquids: a case of α-tocopherol and methyl linoleate separation. Ind Eng Chem Res 51(18):6480–6488
Ohkouchi Y, Inoue Y (2007) Impact of chemical components of organic wastes on L(+)-lactic acid production. Bioresour Technol 98(3):546–553
Pan J et al (2008) Effect of food to microorganism ratio on biohydrogen production from food waste via anaerobic fermentation. Int J Hydrog Energy 33(23):6968–6975
Partt J, Barthel M, Macnaughton S (2010) Philos Trans R Soc B 365:3065–3081
Passos H, Freire MG, Coutinho JAP (2014) Ionic liquid solutions as extractive solvents for value-added compounds from biomass. Green Chem 16(12):4786–4815
Paterno A, D'Anna F, Musumarra G, Noto R, Scire S (2014) A multivariate insight into ionic liquids toxicities. RSC Adv 4(46):23985–24000
Peric B, Sierra J, Martí E, Cruañas R, Garau MA, Arning J, Bottin-Weber U, Stolte S (2013) Ecotoxicity and biodegradability of selected protic and aprotic ionic liquids. J Hazard Mater 261:99–105
Pereira MM, Pedro SN, Quental MV, Lima A, Coutinho JA, Freire MG (2015) Enhanced extraction of bovine serum albumin with aqueous biphasic systems of phosphonium- and ammonium-based ionic liquids. J Biotechnol 206:17–25.
Pernak J, Sobaszkiewicz K, Mirska I (2003) Green Chem 5:52–56
Pernak A, Iwanik K, Majewski P, Grzymisławski M, Pernak J (2005) Acta Histo Chem 107(2):149–156
Pernak J, Walkiewicz F, Maciejewska M, Zaborski M (2010) Ind Eng Chem Res 49:5012–5017
Pernak J, Nawrot J, Kot M, Markiewicz B, Niemczak M (2013) RSC Adv:25019–25029
Peter W, Wilhelm K (2009) Ionic liquids - new "solutions" for transition metal catalysis. Angew Int Ed 39:3772–3789
Qin L, Zhang J, Cheng H, Chen L, Qi Z, Yuan W (2016) Selection of Imidazolium-based ionic liquids for vitamin E extraction from deodorizer distillate. ACS Sustain Chem Eng 4(2):583–590
Rao MS, Singh SP (2004) Bioenergy conversion studies of organic fraction of MSW:kinetic studies and gas yield–organic loading relationships for process optimisation. Bioresour Technol 95(2):173–185
Ressmann AK, Zirbs R, Pressler M, Gaertner P, Bica K (2013) Surface-active ionic liquids for micellar extraction of piperine from black pepper. Z. Naturforsch., B. J Chem Sci 68(10):1129–1137
Ribeiro BD, Coelho MAZ, Rebelo LPN, Marrucho IM (2013) Ionic liquids as additives for extraction of saponins and polyphenols from mate (Ilex paraguariensis) and tea (Camellia sinensis). Ind EngChem Res 52(34):12146–12153
Riemer J, Kristoffersen M (1999) Information on waste management practices. A proposed electronic framework. European Environmental Agency, Copenhagen
Riisager A, Fehrmann R, Haumann M, Wasserscheid P (2006) Supported ionic liquid phase (SILP) catalysis: an innovative concept for homogeneous catalysis in continuous fixed-bed reactors. Eur J Inorg Chem 2006:695–706
Rogers RD, Seddon KR (2002) ACS Symposium Series 818, American Chemical Society. Conference materials
Roosen C, Muller P, Greiner L (2008) Ionic liquids in biotechnology: applications and perspectives for biotransformations. Appl Microbiol Biotechnol 81:607–614
Roszak R, Trzeciak AM, Pernak J, Borucka N (2011) Appl Catal A Gen 409–410:148–155

Sakai K, Ezaki Y (2006) Open L-lactic acid fermentation of food refuse using thermophilic Bacillus coagulans and fluorescence in situ hybridization analysis of microflora. J Biosci Bioeng 101(6):457–463

Sakai SI, Yoshida H, Hirai Y, Asari M, Takigami H et al (2011) International comparative study of 3R and waste management policy developments. J Mater Cycles Waste 13(2):86–102

Sasoki MM (2012) Factor influencing the degradation of garbage is methanogenic bioreactors and impacts on biogas formation. Appl Microb Biotechnol 94:575–582

Shao M, Zhang X, Li N, Shi J, Zhang H, Wang Z, Zhang H, Yu A, Yu Y (2014) Ionic liquid-based aqueous two-phase system extraction of sulfonamides in milk. J Chromatogr B Anal Technol Biomed Life Sci 961:5–12

Smiglak M, Reichert WM, Holbrey JD, Wilkes JS, Sun L, Thrasher JS, Kirichenko K, Singh S, Katritzky AR, Rogers RD (2006) Chem Commun:2554–2556

Steudte S, Bemowsky S, Mahrova M, Bottin-Weber U, Tojo-Suarez E, Stepnowski P, Stolte S (2014) Toxicity and biodegradability of dicationic ionic liquids. RSC Adv 4(10):5198–5205

Sun Y, Cheng J (2002) Hydrolysis of lignocellulosic materials for ethanol production: a review. Bioresour Technol 83:1–11

Sun B, Qi L, Wang M (2014) Determination of preservatives in soft drinks by capillary electrophoresis with ionic liquids as the electrolyte additives. J Sep Sci 37(16):2248–2252

Tenenbaum DJ (2008) Food vs. Fuel: diversion of crops could cause more hunger. Environment Health Perspective 166(6):A254–A257

Thi NBD, Kumar G, Lin CY (2015) An overview of food waste management in developing countries: current status and future perspective. J Environ Manag 157:220–229

Tsarpali V, Dailianis S (2015) Toxicity of two imidazolium ionic liquids, [bmim][BF4] and [omim][BF4], to standard aquatic test organisms: role of acetone in the induced toxicity. Ecotoxicol Environ Saf 117:62–71

United Nations Industrial Development Organization, Food Wastes (2002). http://www.unido.org/leadmin/import/32068_35FoodWastes. Accessed 2 Feb 2012

Van Rantwijk F, Sheldon RA (2007) Biocatalysis in ionic liquids. Chem Rev 107:2757–2785

Viboud S, Papaiconomou N, Cortesi A, Chatel G, Draye M, Fontvieille D (2012) Correlating the structure and composition of ionic liquids with their toxicity on Vibrio fischeri: a systematic study. J Hazard Mater 215–216:40–48

Wang QH, Ma HZ (2008) Ethanol production from kitchen garbage using response surface methodology. J Biochem Eng 39:604–610

Wang Q et al (2005) Bioconversion of kitchen garbage to lactic acid by two wild strains of lactobacillus species. J Environ Sci Health 40(10):1951–1962

Wang ZS, Koumura N, Cui Y, Miyashita M, Mori S, Hara K (2009) Exploitation of ionic liquid electrolyte for dye-sensitized solar cells by molecular modification of organic-dye sensitizers. Chem Mater 21:2810–2816

Wang R, Chang Y, Tan Z, Li F (2016) Applications of choline amino acid ionic liquid in extraction and separation of flavonoids and pectin from ponkan peels. Sep Sci Technol 51(7):1093–1102

Wasserscheid P, Keim W (2000) Ionic liquids—new "solutions" for transition metal catalysis. Angew Chem Int Ed 39:3773–3789

Wasserscheid P, Welton T (2008) Ionic liquids in synthesis, 2nd edn. Wiley-VCH Verlag GmbH & Co. KGaA, Hoboken

Welton T (1999) Room-temperature ionic liquids. Solvents for synthesis and catalysis. Chem Rev 99:2071–2083

World Health Organization (2005) Preventing chronic diseases: a vital investment. WHO, Geneva. http://www.who.int/chp/chronic_disease_report/en/. Accessed Oct 2007

Wu C, Chen W, Zhong L, Peng X, Sun R, Fang J (2014) Conversion of xylose into furfural using lignosulfonic1 acid as catalyst in ionic liquid. J Agric Food Chem 62(30):7430–7435

Yaakob Z et al (2013) Overview of the production of biodiesel from waste cooking oil. Renew Sust Energ Rev 18:184–193

Yach D, Hawkes C, Gould CL, Hofman KJ (2004) The global burden of chronic diseases: overcoming impediments to prevention and control. JAMA 291:2616–2622

Yan F, Xia S, Wang Q, Ma P (2012) Predicting the toxicity of ionic liquids in leukemia rat cell line by the quantitative structure-activity relationship method using topological indexes. Ind Eng Chem Res 51(43):13897–13901

Yang Q, Dionysiou DD (2004) Photolytic degradation of chlorinated phenols in room temperature ionic liquids. J Photochem Photobiol A Chem 165:229–240

Yang SY et al (2006) Lactic acid fermentation of food waste for swine feed. Bioresour Technol 97(15):1858–1864

Zhai Y, Sun S, Wang Z, Cheng J, Sun Y, Wang L, Zhang Y, Zhang H, Yu A (2009) Microwave extraction of essential oils from dried fruits of Illicium verum Hook. f. and Cuminum cyminum L. using ionic liquid as the microwave absorption medium. J Sep Sci 32(20):3544–3549

Zhang Y, Dube MA, McLean DD, Kates M (2003) Bioresour Technol 89:1

Zhang C et al (2013) The anaerobic co-digestion of food waste and cattle manure. Bioresour Technol 129:170–176

Zhao D, Liao Y, Zhang ZD (2007) Toxicity of ionic liquids. Clean Soil Air Water 35(1):42–48

Zhu HB, Fan YC, Qian YL, Tang HF, Ruan Z, Liu DH, Wang H (2014) Determination of spices in food samples by ionic liquid aqueous solution extraction and ion chromatography. Chin Chem Lett 25(3):465–468

Index

Inamuddin, A. M. Asiri (eds.), *Nanotechnology-Based Industrial Applications of Ionic Liquids*, Nanotechnology in the Life Sciences,
https://doi.org/10.1007/978-3-030-44995-7

H

I

K

L

M

The manufacturer's authorised representative in the EU is Springer Nature Customer Service Centre GmbH, Europaplatz 3, 69115 Heidelberg, Germany. If you have any concerns regarding our products, please contact ProductSafety@springernature.com

Printed and bound by CPI Group (UK) Ltd, Croydon, CR0 4YY
30/09/2026
02225504-0016